T0392837

Handbook of Chemical and Biological Warfare Agents, Volume 2

Handbook of Chemical and Biological Warfare Agents, Volume 2

Nonlethal Chemical Agents and Biological Warfare Agents

Third Edition

D. Hank Ellison

CRC Press
Taylor & Francis Group
Boca Raton London New York

CRC Press is an imprint of the
Taylor & Francis Group, an **informa** business

Third edition published 2022
6000 Broken Sound Parkway NW, Suite 300, Boca Raton, FL 33487-2742

and by CRC Press
4 Park Square, Milton Park, Abingdon, Oxon, OX14 4RN

CRC Press is an imprint of Taylor & Francis Group, LLC

© 2022 Taylor & Francis Group, LLC

First edition published by CRC Press 1999
Second edition published by CRC Press 2007

ISBN: 978-1-032-13724-7 (hbk)
ISBN: 978-1-032-14894-6 (pbk)
ISBN: 978-1-003-23056-4 (ebk)

DOI: 10.4324/9781003230564

Typeset in Times
by MPS Limited, Dehradun

Contents

Incapacitating And Riot Control Agents

Biological Agents

Volume 1 contains the following classes of agents:

Preface to Volume 2

The body of information in this new edition of the handbook exceeds a size that can comfortably be handled in a single volume. Volume one of this edition contains those classes of military chemical agents that were developed for their lethal effects as well as toxic industrial agents of concern. Volume two contains chemicals that were developed as incapacitating agents or for use in riot control situations, as well as all classes of biological pathogens and mid-spectrum agents. No value or utility of the agents is implied in this grouping nor in the sequence that the classes are presented in either volume.

In this volume, two of the classes from the previous edition, bacterial and rickettsial pathogens, have been consolidated into a single class in order to provide clarity and reflect current scientific taxonomy. Two new classes of biological agents, bioregulators and non-vector entomological agents, have been added.

All materials listed in this handbook have been used on the battlefield, stockpiled as weapons, received significant interest by research programs, used or threatened to be used by terrorists, or have been assessed by qualified law enforcement and response organizations as agents of significant concern. In presenting this broad spectrum, I do not offer an evaluation of the efficacy or viability of the agent classes or any of the individual agents. I have included agents classified by the military as obsolete along with those that are still considered a potential threat on the battlefield. It is important to remember that while an agent may have been a failure on the battlefield, it could still be a very successful weapon in the hands of a terrorist. In order to assure accuracy, all data has been cross-checked over the widest variety of military, scientific, and medical sources available.

Acknowledgments

I would especially like to thank Sarah for her input, comments, and editing. She was able to clarify what I mistakenly thought was obvious.

There are numerous other individuals who have provided comments, insights, and suggestions, on both of the previous editions as well as on the manuscript for this one. I appreciate them all. I have tried to address each of them, and incorporated changes that I believe have improved this edition. Any failures or omissions are mine and not due to a lack of vigilance or effort on the part of others.

Explanatory Notes to Volume 2

In this handbook, information about the agents is divided into classes based primarily on the common military groupings of chemical and biological agents. In instances where the divisions are too broad to allow appropriate identification of the chemical or physiological properties of the individual agents, additional classes are provided. There are also classes for nontraditional agents that do not fit neatly into one of the common military groupings and for industrial materials that could be used as improvised agents. Classes are identified by a number that corresponds to the first 21 chapters in this handbook (i.e.; C01–C21). Classes contain general information about that specific group of agents. While this book covers most of the major classes of chemical and biological agents, it does not deal with anti-plant chemicals, anti-material agents, incendiary agents, or obscurants.

At the end of each class is detailed technical information about individual agents, components, or decomposition products within that class. Each of these individual materials is assigned a handbook number to allow for rapid identification and cross referencing throughout the book. The first three characters identify the agent class (e.g.; **C01**). The letter following the hyphen (e.g.; C01-**A**) indicates that the materials is primarily considered an agent (A), component or precursor of that class of agents (C), or is a significant decomposition product or impurity of that class of agents (D). The three digits that follow the letter indicate the specific agent in the order that it appears in the class (e.g.; C01-A**001**). In some cases, handbook numbers only contain the first three characters (e.g.; C01-A, C01-C, or C01-D) indicating that the material is an agent, component, or decomposition product in that class of chemical agents, but there is insufficient data published on the physical, chemical, or toxicological properties of the material to generate an individual entry.

There are four appendices that provide information on methods used to deliver chemical and biological agents, markings used to identify chemical munitions used by various nations, biological safety levels, and international and domestic lists of agents. These appendices are found at the end of this volume.

There are 10 indices to allow easy access to specific agents in this handbook. These indices are the Alphabetical Index of Names, the Chemical Abstract Service (CAS) Numbers Index, United Nations (UN) Numbers Index, European Inventory of Existing Commercial Chemical Substances (EC) Numbers Index, U.S. Food and Drug Administration Unique Ingredient Identifiers (UNII) Index, International Classification of Diseases (ICD-11) Index, European and Mediterranean Plant Protection Organization (EPPO) Codes Index, an index of viruses by their taxonomic families, a cross index of vector and the diseases that they can carry, and a cross index of agriculturally significant flora and fauna with the diseases that can impact them. These indices contain synonyms and identifying numbers for the agents in this handbook that are cross-referenced to the individual agents via the handbook number.

Information in classes for chemical agents, toxins, and bioregulators is in the following general format:

General Information
Toxicology (effects, pathways and routes of exposure, general exposure hazards, latency period)
Characteristics (physical appearance/odor, stability, persistency, environmental fate)
Additional Hazards (exposure, livestock/pets, fire, reactivity, hazardous decomposition products)
Protection (evacuation recommendations, personal protective requirements, decontamination)
Medical (Centers for Disease Control and Prevention (CDC) case definition, differential diagnosis, signs and symptoms, mass-casualty triage recommendations, casualty management, fatality management)

Information in classes for biological (i.e., pathogen) agents is in the following general format:

General Information
Response (personal protective requirements, decontamination, fatality management)

Information on the individual chemical agents is in the following general format:

Handbook Number
Name and reference numbers (CAS, RTECS, UN, ERG, EC, UNII, ICD-11)
Molecular structure and formula
Description of the agent
Additional information including mixtures with other agents, industrial uses, threat or treaty
    listing
Exposure hazards
AEGLs status and exposure values
Properties

Information on the individual toxins is in the following general format:

Handbook Number
Name and reference numbers (CAS, RTECS, EC, UNII, ICD-11)
Formula and molecular weight (if known)
Description of the toxin and source
Routes of exposure and signs and symptoms
Additional information including medicinal uses, threat or treaty listing
Exposure hazards
Because of the molecular size of many of the toxins, graphics of the structures are contained
    in a separate listing at the end of the chapter.

Information on the individual pathogens is in the following general format:

Handbook Number
Name, disease, and reference numbers (ICD-11, UNII, EPPO)
Description of the disease including where it is endemic, natural transmission, natural reservoir,
    survivability in the environment outside a host, and a biosafety level if established.
Additional information including threat or treaty listing.
The disease as it appears in people including the Centers for Disease Control and Prevention
    (CDC) case definition, communicability, normal routes of exposure, infectious dose,
    secondary hazards, incubation period, potential signs and symptoms, suggested alternatives
    for differential diagnosis, and the untreated mortality rate.
The disease as it appears in animals including agricultural target species, communicability,
    normal routes of exposure, secondary hazards, incubation period, potential signs,
    potential postmortem lesions, suggested alternatives for differential diagnosis, and the
    untreated mortality rate.
The disease as it appears in plants including agricultural target species, normal routes of
    exposure, secondary hazards, and potential signs.

Information on the individual bioregulators is in the following general format:

Handbook Number
Name and reference numbers (CAS, RTECS, EC, UNII)

Formula and molecular weight (if known)

Routes of exposure, potential signs and symptoms of exposure to a high acute dose

Because of the molecular size of many of the toxins, graphics of the structures are contained in a separate listing at the end of the chapter.

Information on the individual entomological agents is in the following general format:

Handbook Number

Common name, scientific name, and EPPO

Description of adults, eggs, larva, and pupa/nymphs

Information on the life cycle, where it is indigenous to, and key habitat issues

Agricultural impacts, means of migration and secondary transport, and signs of infestation

Decontamination information contained in the class indices is directed at minimizing the threat during an emergency and may not be sufficient to allow unrestricted access by the general public. Thorough sampling should be conducted to determine additional remediation requirements.

Levels of protection identified in this volume follow the convention of the U.S. Environmental Protection Agency, which has established four basic levels for responding to hazardous material emergencies. Each level has benefits and limitations for the responder.

Level A provides the highest level of skin, eye, and respiratory protection. It is selected when the hazardous material requires the highest level of protection, and there is a splash or vapor hazard that will be harmful to the skin or the material may be absorbed through the skin causing immediate serious injury or harm. Level A consist of a totally encapsulating suit and a supplied air respirator (i.e., SCBA or airline).

Level B provides the highest level of respiratory protection but has some exposed skin. It is selected when the hazardous material requires the highest level of protection, a totally encapsulating suit is not necessary, and some areas of unprotected skin will not adversely affect the responder. Level B consist of a supplied air respirator (i.e., SCBA or airline) and a chemical protective suit.

Level C provides some degree of skin and respiratory protection. It is selected when the hazardous material does not require the highest level of protection, a totally encapsulating suit is not necessary, and some areas of unprotected skin will not adversely affect the responder. In addition, the ambient oxygen level must be above 19.5% and, the identity as well as the concentration of the hazardous material must be known and measurable in real time. Level C consists of an air purifying respirator (i.e., on demand or powered) and a chemical protective suit.

Level D provides the least protection from potential chemical exposure. It is selected when there is no need for respiratory protection and minimal skin protection is needed. Level D may consist of coveralls or a chemical protective suit with some form of eye (e.g., glasses, goggles, face shield), hand (e.g., chemical resistant, abrasion or thermal protection), and foot protection.

Protective clothing can be made from any number of chemical resistant fabrics and must be selected based the chemicals involved in the emergency and the potential for exposure. Selection criteria include the possibility that chemical can degrade the protective fabric (i.e., physical destruction or reduction in one or more of the physical properties of the chemical protective clothing) or permeate through the fabric (i.e., process by which the hazardous material moves through the chemical protective clothing on a molecular level). Proper design and construction of the suit (e.g., splash guards, zipper covers, taped seams) is also essential to prevent penetration of the chemical through zippers, stitched seams, or imperfection in the ensemble.

Abbreviations used in identifying individual agents are listed below. For any given parameter, a dash (i.e., —) means that the value is unavailable because it has not been determined or has not been published. An "NA" designation means that the parameter is not applicable or does not apply to that specific material. If the value of a parameter is an estimate, it is indicated by (est.).

CAS: Chemical Abstracts Service registry number. It is unique for each chemical without inherent meaning that is assigned by the Chemical Abstracts Service, a division of the American Chemical Society. It allows for efficient searching of computerized databases.

EC: a unique seven-digit identifier that was assigned to substances for regulatory purposes within the European Union by the European Commission. These materials were in the European Inventory of Existing Commercial chemical Substances, an inventory of chemicals that were deemed to be on the European Community market between 1 January 1971 and 18 September 1981.

EPPO: Formerly BAYER codes. Encoded identifier maintained by the European and Mediterranean Plant Protection Organization. It is a system designed to uniquely identify plants, pests, and pathogens important to agriculture and crop protection. It allows for efficient searching of computerized databases.

ICD-11: 11th Revision of the International Statistical Classification of Diseases and Related Health Problems. It is the international standard diagnostic classification for all general epidemiological and many health management purposes.

RTECS: Registry of Toxic Effects of Chemical Substances number is a unique and unchanging identifier used to cross reference the RTECS database, which is a compendium of data extracted from the open scientific literature.

UN: United Nations identification number used in transportation of hazardous materials. If the substance is only listed under a generic class or group of materials (e.g., UN 2810: Toxic Liquids. Organic, N.O.S.), then it is not reported. The *ERG: 2020 Emergency Response Guidebook* number is reported following the UN number. As in the *Guidebook*, the letter "P" following the guide number indicates that the material has a significant risk of violent polymerization if not properly stabilized.

UNII: U.S. Food and Drug Administration Unique Ingredient Identifier numbers are unique, permanent identifiers for substances in regulated products for any material, and can from an atom to an organism.

Unless otherwise indicated, civilian exposure hazards are for a standardized individual (i.e., a male weighting 70 kg/154 lbs.) with a respiratory tidal volume of 15 liters/minute (i.e.; resting). The military, because of the expected greater physical activity level of deployed soldiers, uses a standard 20 liters/minute respiratory tidal volume for their exposure guidelines (see MEG below). Other potential respiratory rates include walking (25 liters/minute), hard labor such as digging (60 liters/minute), or jogging (75 liters/minute). If a different breathing rate is used, then it is indicated in brackets. If temperature is a factor, then the critical values are indicated. The military typically classifies moderate temperatures as 65°F to 85°F. Temperature above 85°F are classified as hot.

Exposure parameters described below that have not been determined or have not been published for a specific material are not listed in those individual entries in this handbook. Older values of individual parameters from historic manuals that have not been updated in more current literature, are identified. In some cases, published toxicological endpoints allow relative comparisons to know chemical agents (i.e., the nerve agents tabun, sarin, or VX). These are identified in individual entries when no other significant information on exposure hazards is available.

AEGL: Acute Exposure Guideline Levels describe the human health effects from once-in-a-lifetime, or rare, exposure to airborne chemicals and are intended for use by emergency responders when dealing with chemical spills or other catastrophic exposure events involving the general populace. In this handbook, only values for 1, 4, and 8 hours are denoted. Values that have not been developed or not recommended are annotated as "Not Developed". AEGLs appear in three categories:

AEGL-1 is the airborne concentration above which the general population could experience notable discomfort, irritation, or certain asymptomatic non-sensory effects. Effects are not disabling; they are transient and reversible.

AEGL-2 is the airborne concentration above which the general population could experience irreversible or other serious, long-lasting adverse health effects or an impaired ability to escape.

AEGL-3 is the airborne concentration above which the general population could experience life-threatening health effects or death.

The status indicates where the chemical is in the review process and includes Draft, Proposed, Interim, and Final. For more information, see the U.S. Environmental Protection Agency website at http://www.epa.gov/oppt/aegl/index.htm.

Ceiling: Exposure limit that specifies the concentration of vapor, dust or aerosol that should not be exceeded at any time during the workday. In some instances, a time limit for exposure to the ceiling value is established and is indicated in brackets. A [Skin] notation indicates that percutaneous absorption of the material is a potential hazard and may contribute to the overall exposure. The organization establishing the exposure limit is designated in subscript: NIOSH: National Institute for Occupational Safety and Health; ACGIH: American Conference of Governmental Industrial Hygienists; OARS: Occupational Alliance for Risk Safety. It is important to realize that these standards are intended to be applied to individuals who are adults (i.e., 16 years to 64 years old) and healthy enough to be employed full-time. These standards are not appropriate for residential scenarios or for susceptible populations (e.g., children, the elderly, the infirmed, etc.).

Conversion Factor: Ratio of parts per million to milligrams per cubic meter at 77°F. Parts per million can be also be converted to milligrams per cubic meter at other temperatures by the following formula:

$$mg/m^3 = (ppm)(gmw)/RT$$

where:

mg/m^3 = the concentration in milligrams per cubic meter

ppm = the concentration in parts per million

gmw = the gram molecular weight of the agent

RT = theoretical molar volume of agent vapor in liters. The value of RT is determined by multiplying the temperature in degrees Kelvin by 0.08205. For 77°F, this value is 24.45.

ICt_{50}: Is an expression of the dose of vapor or aerosolized agent necessary to incapacitate half of an exposed population. These values are expressed as milligram-minutes per cubic meter (mg-min/m^3). The incapacitating concentration (IC_{50}) is determined by dividing the ICt_{50} by the duration of exposure in minutes. Values are for inhalation (Inh) and percutaneous (Per) exposures; and in the case of vesicants, damage to the skin (Skin) and eyes (Eyes). These dose-response values are not universally valid over all exposure periods and can lead to dangerous misinterpretations if used incorrectly. For inhalation of agent, time parameters are generally 2to 8minutes. For percutaneous absorption of agent, as well as for exposures that cause damage to the skin and eyes, time parameters are generally 2minutes to 6 hours. Typically, an incapacitating concentration in parts per million (ppm) for a set exposure time is included in brackets following the mg-min/m^3 value.

ID_{50}: Amount of liquid or solid material required to incapacitate half of an exposed population. Values are for ingestion (Ing), percutaneous (Per) exposures, and subcutaneous injection (Sub). These values are expressed as total milligrams per individual.

IDLH: Immediately dangerous to life or health levels indicate that exposure to the listed concentrations of airborne contaminants is likely to cause death, immediate or delayed permanent adverse health effects, or prevent escape from the contaminated environment in a short period of time, typically 30 minutes or less. They indicate a maximum exposure level above which only individuals wearing a highly reliable breathing apparatus, providing maximum worker protection (e.g., SCBA), are permitted to enter.

Irritation values for eyes, skin, and respiratory system. These values are expressed as a concentration (ppm for gases, mg/m^3 for aerosols) for a 2-minute exposure. Concentrations identified in the literature as "intolerable" to exposed individuals are also noted.

Infectious Dose: Number of viable organisms (e.g., cells, viral particles, colony forming units, plaque forming units) needed to cause infection in 50% of the exposed individuals or animals.

LCt$_{50}$: Is an expression of the dose of vapor or aerosolized agent necessary to kill half of an exposed population. These values are expressed as milligram-minutes per cubic meter (mg-min/m^3). The lethal concentration (LC$_{50}$) is determined by dividing the LCt$_{50}$ by the duration of exposure in minutes. For example, the lethal concentration necessary to kill approximately 50% (LC$_{50}$) of people exposed for 2-minute to sarin vapor is approximately 18 mg/m^3 (6 ppm), while the LC$_{50}$ for an 8-minute exposure is only 4.4 mg/m^3 (1.5 ppm). Values are for inhalation (Inh) and percutaneous (Per) exposures. These dose-response values are not universally valid over all exposure periods and can lead to dangerous misinterpretations if used incorrectly. For inhalation of agent, time parameters are generally valid within a range of 2to 8minutes. For percutaneous absorption of agent, time parameters are generally valid within a range of 30 minutes to 6 hours. Typically, a lethal concentration in parts per million (ppm) for a set exposure time is identified in brackets following the mg-min/m^3 value.

LC$_{50}$: Concentration of vapor or aerosolized agent necessary to kill half of an exposed population. Used when a specific set of exposure conditions (i.e., concentration and duration of exposure) are known but a more generalized dose-response (LCt$_{50}$) is not available.

LD$_{50}$: Amount of liquid or solid material required to kill half of an exposed population. Values are for ingestion (Ing), percutaneous (Per) exposures, and subcutaneous injection (Sub). These values are expressed as total milligrams per individual.

MEG: Military exposure guidelines for deployed military personnel. Levels reported in this handbook are for a 1-hour exposure and generally consider three possible health endpoints related to a military operation:

Negligible (Neg): Continuous exposure to vapor concentrations above these levels could begin to produce mild, non-disabling, transient, reversible effects (e.g., mild irritation).

Marginal (Mar): Continuous exposure to vapor concentrations above these levels could produce some performance degradation, especially for tasks requiring extreme mental/visual acuity or physical dexterity/strength, in a portion of exposed individuals.

Critical (Crit): Continuous exposure to vapor concentrations above these levels could produce some potential life-threatening or lethal effects; particularly in personnel with underlying susceptibility factors.

A fourth health endpoint, Catastrophic (Cat), is listed only for standard military chemical warfare agents. Continuous exposure to vapor concentrations these levels is anticipated to result in deaths and/or severe incapacitating effects that will likely require medical treatment.

Miosis: Concentration in parts per million (ppm) required to induce significant constriction of the pupil of the eye following a 2-minute exposure to the agent.

PACs: U.S. Department of Energy Protective Action Criteria designed to provide essential components for planning and response to uncontrolled releases of hazardous chemicals. PACs appear in three categories:

PAC-1 is the airborne concentration above which the general population could experience mild, transient health effects.

PAC-2 is the airborne concentration above which the general population could experience irreversible or other serious health effects that could impair the ability to take protective action.

PAC-3 is the airborne concentration above which the general population could experience life-threatening health effects.

PEL: Permissible exposure limits adopted by the Occupational Safety and Health Administration (OSHA) and based on a time-weighted average exposure over an 8-hour period and 40-hour workweek. A [Skin] notation indicates that percutaneous absorption of the material is a potential hazard and may contribute to the overall exposure. It is important to realize that these standards are intended to be applied to individuals who are adults (i.e., 16 years to 64 years old) and healthy enough to be employed full-time. These standards are not appropriate for residential scenarios or for susceptible populations (e.g., children, the elderly, the infirmed, etc.).

REL: Recommended exposure limits recommended by the National Institute for Occupational Safety and Health (NIOSH) and based on a time-weighted average exposure over a 10-hour period and 40-hour workweek. A [Skin] notation indicates that percutaneous absorption of the material is a potential hazard and may contribute to the overall exposure. It is important to realize that these standards are intended to be applied to individuals who are adults (i.e., 16 years to 64 years old) and healthy enough to be employed full-time. These standards are not appropriate for residential scenarios or for susceptible populations (e.g., children, the elderly, the infirmed, etc.).

STEL: Short-term exposure limits are based on a time-weighted average exposure of 15 minutes (unless otherwise noted). A [Skin] notation indicates that percutaneous absorption of the material is a potential hazard and may contribute to the overall exposure. The organization establishing the exposure limit is designated in subscript: NIOSH: National Institute for Occupational Safety and Health; ACGIH: American Conference of Governmental Industrial Hygienists; OARS: Occupational Alliance for Risk Safety. It is important to realize that these standards are intended to be applied to individuals who are adults (i.e., 16 years to 64 years old) and healthy enough to be employed full-time. These standards are not appropriate for residential scenarios or for susceptible populations (e.g., children, the elderly, the infirmed, etc.).

TLV: Threshold limit values recommended by the American Conference of Governmental Industrial Hygienists (ACGIH) and based on a time-weighted average exposure over an 8-hour period and 40-hour workweek. A [Skin] notation indicates that percutaneous absorption of the material is a potential hazard and may contribute to the overall exposure. It is important to realize that these standards are intended to be applied to individuals who are adults (i.e., 16 years to 64 years old) and healthy enough to be employed full-time. These standards are not appropriate for residential scenarios or for susceptible populations (e.g., children, the elderly, the infirmed, etc.).

Vomiting: Inhaled concentration of vapor or aerosolized agent necessary to induce significant nausea and vomiting in half of an exposed population. These values are expressed as a concentration (ppm for gases, mg/m^3 for aerosols) for a 2-minute exposure.

WEEL: Workplace environmental exposure levels were originally developed by the American Industrial Hygiene Association (AIHA). Since 2013, management of these exposure levels has been managed by the nonprofit organization **Toxicology Excellence for Risk Assessment** and published by the Occupational Alliance for Risk Science (OARS). These exposure limits are based on a time-weighted airborne exposure averaged over an 8-hour workday and 40-hour workweek, and designed to protect most workers from adverse health effects related to occupational chemical exposures. It is important to realize that these standards are intended to be applied to individuals who are adults (i.e., 16 years to 64 years old) and healthy enough to be employed full-time. These standards are not appropriate for residential scenarios or for susceptible populations (e.g., children, the elderly, the infirmed, etc.).

In addition to the lists of reportable or controlled materials established by various treaties, governments and international organizations (i.e., the Chemical Weapons Convention schedules of chemicals, Australia's chemicals of security concern, the United States' select agents and toxins list, the U.S. Department of Homeland Security chemicals of concern list, the World Health Organization for Animals listed diseases, and the Australia Group common control lists), a number of agencies have established lists of commercially available materials that could be used by terrorists or insurgents as improvised chemical agents. These lists are identified where applicable in individual entries of the agent index. These lists are: The U.S. Federal Bureau of Investigation (FBI) list of highly toxic chemicals and pesticides judged likely to be used by terrorists, the International Taskforce-25 (ITF-25) list of agricultural and commercial industrial chemicals that could be significant in a military operation, and the U.S. Naval Research Laboratory (NRL) list of industrial chemicals for evaluation of chemical/biological defense systems. For more information on these and other sources of the agents detailed in this handbook see appendix 4.

Unless otherwise indicated, chemical and physical properties are for the pure or production quality material. The literature values reported for these properties often reflect optimal or laboratory conditions and may differ from the behavior of the agent in the environment. Also, persistence and degradation of agents released inside buildings or confined spaces will vary from agents released in the open environment where they will be exposed to sunlight and the weather. Properties of mixed, binary, thickened, or dusty agents, even those in solutions, will have physical and chemical properties that vary from the listed values. These variations will depend on the proportion of agent to other materials (e.g., solvents, thickener, etc.) and the properties of these other materials. If available, data on mixtures or modified agents (e.g.; salts) is included. For any given parameters, a dash (i.e., —) means that the value is unavailable because it has not been determined or has not been published.

BP: Boiling point of the material in degrees Fahrenheit at standard pressure (760 mmHg). If a reduced pressure is reported, it is indicated in brackets following the boiling point. A designation of "decomposes" indicates that the agent will thermally decompose before it reaches its boiling point. A designation of "sublimes" indicates that the material goes directly from a solid to a gas without melting.

D: Density of the solid or liquid material at 68°F. If the density is reported for another temperature, it is indicated in brackets following the density. Liquefied gases are also indicated.

ER: Expansion ratio of a volatile liquid or liquified gas. Indicates the ratio of the volume of a given mass of liquified material compared to the theoretical volume of the same mass of material in the gaseous state at the same temperature and pressure.

FlP: Flashpoint of the material.

H_2O: Solubility of the agent in water. Solubilities are generally given in percentages indicating the weight of agent that will dissolve in the complementary amount of water. When quantitative solubility data is not available, qualitative terms (e.g., negligible, slight) are used to provide an intuitive evaluation of agent solubility. A designation of "miscible" indicates that the agent is soluble in water in all proportions. A designation of "insoluble" indicates that no appreciable amount of the agent will dissolve in water. A designation of "reacts" indicates that the agent reacts with water and will decompose into other materials which may or may not be hazardous. If known, the rate of decomposition is indicated.

IP: Ionization potential is the amount of energy needed to remove an electron from a molecule of chemical vapor. The resultant ion is a charged particle that is detectable by various instrumentation such as photo ionization or flame ionization detectors.

LEL: Lower explosive limit in air, expressed as a percentage by volume.

MP: Melting point of the material in degrees Fahrenheit. A designation of "decomposes" indicates that the agent will thermally decompose before it reaches its melting point.

MW: Molecular or formula weight of the material.

RP: Relative persistency is a mathematical comparison of the evaporation and diffusion rates of water at 68°F to the evaporation and diffusion rates of the agent. The value represents an estimate of the ratio of the time required for a liquid or solid agent to dissipate as compared to the amount of time required for an equal mass of liquid water to dissipate. A value of "1" indicates that a puddle of the agent and a similar puddle of water will evaporate at about the same rate. The greater the value, the greater the proportional amount of time required for the agent to evaporate. For example, a value of "2" would mean that it would theoretically take approximately twice as long for the agent to evaporate as compared to a similar volume of water. Relative persistency is calculated by the formula:

$$RP = 4.34/p \ (K/M)^{1/2}$$

where:
   RP = the relative persistency of the agent
   p = the vapor pressure of the agent in mmHg at the temperature K
   K = the ambient temperature in degrees Kelvin
   M = the gram molecular weight of the agent

The relative persistency value does not account for additional factors that could impact the stability and persistence of a given agent, such as wind currents or decomposition due to reaction with other chemicals in the environment (e.g., water).

Sol: List of common organic solvents in which the material has appreciable solubility.

UEL: Upper explosive limit in air, expressed as a percentage by volume.

VD: Relative vapor density of the gaseous agent as compared to air. Unless otherwise indicated, these values are calculated based on the standard reference weight of air (i.e., 29).

Vlt: Volatility is the mass of agent in a unit of air that is saturated with the agent vapor. The volatility of an agent varies with temperature and it is often used to estimate the tendency of a chemical to vaporize or give off fumes. It can be calculated using the following formula:

$$Vlt = 16,040 \, (gmw)p/T$$

where:
   Vlt = the volatility of the agent in mg/m^3
   gmw = the gram molecular weight of the agent
   p = the vapor pressure of the agent in mmHg at the ambient temperature
   T = the ambient temperature in degrees Kelvin

The volatility of a material can also be estimated in ppm by multiplying its vapor pressure in millimeters of mercury (mmHg) by 1310.

Volatility is also sometimes used to estimate the persistency of an agent. However, it does not account for the migration (diffusion) of that vapor out of the area to allow more agent to evaporate. A better estimate of persistency is relative persistency (RP).

VP: Vapor pressure of the material in millimeters of mercury (mmHg) at 68°F. If another temperature is used, it is indicated in parentheses following the vapor pressure.

Vsc: Dynamic viscosity of the material in centistokes (cS). Obtained when the absolute viscosity, expressed as centipoise, is divided by the specific gravity of the material.

About the Author

D. Hank Ellison served in the U.S. Army as a chemical officer and has worked for the U.S. Environmental Protection Agency as both a remedial project manager and federal on-scene coordinator under the Superfund Program. He currently is president of Cerberus & Associates, Inc., a consulting firm that specializes in response to technological disasters. As a private consultant, Ellison has responded to hazardous material incidents involving highly poisonous materials, chemical fires, water reactive substances, and shock-sensitive materials throughout the state of Michigan. He has provided chemical and biological counterterrorism training to members of emergency medical service (EMS) units, hazardous materials (hazmat) teams, police special weapons and tactics (SWAT) teams, and explosive ordinance disposal (EOD) teams. During the anthrax events of 2001, he helped state and local governments as well as *Fortune* 500 companies to develop and implement response plans for biological threats. He currently advises clients on issues of hazardous materials, and related safety and security concerns.

He also served as a safety officer and then the training officer for the U.S. Department of Health and Human Services DMORT-WMD emergency response team, which has the primary mission for recovery and decontamination of fatalities contaminated with radiological, biological, or chemical materials. He deployed with the team to New Orleans in response to Hurricane Katrina in 2005. He has a master of science degree in chemistry from the University of California, Irvine. His graduate research involved methods to synthesize poisons extracted from Colombian poison-dart frogs. He has a bachelor of science in chemistry from the Georgia Institute of Technology. In addition to his works on weapons of mass destruction, he is the author of *Chemical Warfare During the Vietnam War: Riot Control Agents in Combat*, published in 2011, as well as a chapter on the hazardous properties of materials in *Managing Hazardous Materials: A Definitive Text,* published in 2002.

Incapacitating and Riot Control Agents

12 Mind-Altering Agents

GENERAL INFORMATION

This class of agents is also referred to as incapacitating agents, psychomimetic agents, or calmatives. Mind-altering agents became popular during the Cold War. Agent BZ (C12-A001) and two key components necessary to synthesize BZ, 3-quinuclidinol (C12-C032), and benzilic acid (C12-C034), are listed in Schedule 2 of the Chemical Weapons Convention. In addition to military specific agents, materials in this class encompass a wide variety of commercially available medicinal dugs that interfere with higher functions of the brain such as attention, orientation, perception, memory, motivation, conceptual thinking, planning, and judgment. They produce their effects mainly by altering or disrupting the higher regulatory activity of the central nervous system.

Mind-altering agents differ from other chemical agents in that the lethal dose is theoretically many times greater than the incapacitating dose (i.e., a large therapeutic index). Under normal battlefield conditions, they should not pose a serious danger to the life of an exposed individual and should not produce any permanent injury. Soldiers should recover without medical intervention in a relatively short time (i.e., hours to days). In reality, this proves to be much more problematic. While the therapeutic index is relatively large, the actual dose required to produce either incapacitation or death is very small; and, although relatively easy to disperse, it is difficult to effectively control the dose delivered to the target population and prevent fatalities.

Mind-altering agents are relatively easy to isolate from natural sources or to synthesize. Several are clandestinely synthesized and used as recreational drugs. For information on some of the chemicals used to manufacture military mind-altering agents, see the component section (C12-C) following information on the individual agents.

Mind-altering agents have been stockpiled by numerous countries and there have been unverified reports that they have been utilized on the battlefield. In addition, they have been employed by police and special forces as a way to end hostage situations (e.g., the September 2002 counter-terrorism raid on a Moscow theater). These operations have met with mixed success.

TOXICOLOGY

EFFECTS

Mind-altering agents produce their effects mainly by altering or disrupting the higher regulatory activity of the central nervous system. Military mind-altering agents can be separated into four fairly discrete categories: deliriants (producing confusion, hallucinations, and disorganized behavior), stimulants (essentially flooding the brain with too much information), calmatives (depressants that induce passivity or even sleep), and psychedelics (producing abnormal psychological effects resembling mental illness).

In normal usage, mind-altering agents will not cause permanent or long-lasting injury. Unlike lachrymatory agents (C13) or vomiting/sternatory agents (C14), mind-altering agents produce effects that may last up to several days post-exposure.

PATHWAYS AND ROUTES OF EXPOSURE

Mind-altering agents are primarily a hazard through inhalation. However, exposure to liquid or solid agents may be hazardous through skin absorption (especially if the agent is dissolved in an

DOI: 10.4324/9781003230564-2

appropriate solvent), ingestion, introduction through abraded skin (e.g., breaks in the skin or penetration of skin by debris), and may also produce local skin/eye effects (e.g., dilation of the pupil).

GENERAL EXPOSURE HAZARDS

Mind-altering agents do not have good warning properties. They have little or no odor, and the vapors do not irritate the eyes. Contact with liquid or solid agents neither irritates the skin and nor causes cutaneous injuries.

LATENCY PERIOD

Vapors/Aerosols (mists or dusts)
Depending on the specific agent and the concentration of vapor or aerosol, the effects begin to appear in seconds or may be delayed up to several hours.

Liquids
Typically, there is a latent period with no visible effects between the time of exposure and the sudden onset of symptoms. Effects from dermal exposure may be delayed up to several days. Some factors affecting the length of time before the onset of symptoms are the amount of agent involved, the amount of skin surface in contact with the agent, previous exposure to materials that chap or dry the skin (e.g., organic solvents such as gasoline or alcohols), and addition of additives designed to enhance the rate of percutaneous penetration by the agents.

Another key factor affecting the rate of percutaneous penetration by the agent is the part of the body that is exposed. It takes the agent longer to penetrate thicker and tougher skin. The regions of the body that allow the fastest percutaneous penetration are the groin, head, and neck. The least susceptible body regions are the hands, feet, front of the knee, and outside of the elbow.

Solids (non-aerosol)
Typically, there is a latent period with no visible effects between the time of exposure and the sudden onset of symptoms. Effects from dermal exposure may be delayed up to several days and is affected by such factors as the amount of agent involved, the amount of skin surface in contact with the agent, previous exposure to materials that chap or dry the skin, and the area of the body exposed (see above). Moist, sweaty areas of the body are more susceptible to percutaneous penetration by solid agents.

CHARACTERISTICS

PHYSICAL APPEARANCE/ODOR

Laboratory Grade
Laboratory grade agents are typically colorless liquids or solids. Depending on the specific agent, liquids may be mobile, viscous, or even waxy in nature. Many solids are salts of the free-base liquid that are colorless to white to beige crystalline materials. In either state, these materials typically have little or no odor when pure.

Munition Grade
Munition grade agents are typically white to light brown powders, waxy solids, or viscous liquids. Production impurities and decomposition products in these agents may give them an odor. Odors for all agents may become more pronounced during storage.

Modified Agents

Solvents have been added to mind-altering agents to facilitate handling, stabilize the agents or to increase the ease of percutaneous penetration by the agents. Percutaneous enhancement solvents include dimethyl sulfoxide, N,N-dimethylformamide, N,N-dimethylpalmitamide, N,N-dimethyldecanamide, and saponin. Color and other properties of these solutions may vary from the pure agent. Odors will vary depending on the characteristics of the solvent(s) used and concentration of mind-altering agent in the solution.

STABILITY

Military specific agents are stable even under tropical conditions and can be stored in glass, aluminum or steel. Stabilizers are not required. Some potentially dual-use agents are sensitive to heat and light, and may require stabilizers for long-term storage.

PERSISTENCY

Depending on the properties of the specific agent, unmodified mind-altering agents are classified as either non-persistent or persistent by the military.

Addition of solvents may alter the persistency of these agents. Salts of agents have negligible vapor pressure and will not evaporate. Depending on the size of the individual particles and on any encapsulation or coatings applied to the particles, they can be re-aerosolized by ground traffic or strong winds.

ENVIRONMENTAL FATE

Vapors of volatile mind-altering agents have a density greater than air and tend to collect in low places. Most mind-altering agents are nonvolatile and produce negligible amounts of vapor.

Most of these agents are only slightly soluble or insoluble in water. However, the solubility of any agent may be modified (either increased or decreased) by solvents, components or impurities. The specific gravities of unmodified liquid agents are slightly greater than that of water. Mind-altering agents are typically soluble in most organic solvents including gasoline, alcohols, and oils. Salts of agents are water soluble.

ADDITIONAL HAZARDS

EXPOSURE

All foodstuffs in the area of a release should be considered contaminated. Unopened items packaged in glass, metal or heavy-duty plastic and exposed only to agent vapors, aerosols or to solid agents may be used after decontamination of the container. Unopened items exposed to solid agents or solutions of agents should be decontaminated within a few hours post exposure or destroyed. Opened or unpackaged items, or those packaged only in paper or cardboard, should be destroyed.

LIVESTOCK/PETS

Animals can be decontaminated with shampoo/soap and water (see the Decontamination section below). If the animal's eyes have been exposed to agent, they should be irrigated with water or saline solution for a minimum of 30 minutes.

The topmost layer of unprotected feedstock (e.g., hay or grain) should be destroyed. The remaining material should be quarantined until tested. Leaves of forage vegetation could still retain sufficient agent to produce effects for several weeks post-release, depending on the level of contamination and the weather conditions.

FIRE

Because of their low vapor pressures, heat from a fire will destroy most mind-altering agents before generating any significant concentration of agent vapor. However, it is possible that hot gases from the fire may distill some of the agent and create a hazardous aerosol within the smoke. Actions taken to extinguish the fire can spread the agent. Salts are water soluble and runoff from firefighting efforts will pose a significant threat. Some of the decomposition products resulting from hydrolysis or combustion of mind-altering agents are water soluble and highly toxic (see hazardous decomposition products below). Other potential decomposition products include toxic and/or corrosive gases.

REACTIVITY

Some mind-altering agents decompose slowly in water. Raising the pH of an aqueous solution of these agents significantly increases the rate of decomposition.

HAZARDOUS DECOMPOSITION PRODUCTS

Hydrolysis
Varies depending on the specific agent but may include complex alkaloids and organic acids.

Combustion
Varies depending on the specific agent but volatile decomposition products may include hydrogen fluoride (HF), hydrogen chloride (HCl), phosgene ($COCl_2$), nitrogen oxides (NO_x), aromatic hydrocarbons such as benzene, as well as potentially toxic lower molecular weight hydrocarbons.

PROTECTION

EVACUATION RECOMMENDATIONS

Isolation and protective action distances listed below are taken from the *2020 Emergency Response Guidebook*. BZ is the only mind-altering agent addressed and recommendations are based on a release scenario involving direct aerosolization of the solid agents with a particle size between 2 and 10 microns. A small release involves 2 kilograms (approximately 3,900 cubic centimeters, a box approximately 16 centimeters on a side) of bulk agent and a large release involves 25 kilograms (approximately 0.05 cubic meters, a box approximately 37 centimeters on a side) of bulk agent.

BZ *C12-A001*

	Initial Isolation	Downwind Day	Downwind Night
Small device (2 kg)	60 m	0.4 km	1.7 km
Large device (25 kg)	400 m	2.2 km	8.1 km

PERSONAL PROTECTIVE REQUIREMENTS

Structural Firefighters' Gear
Structural firefighters' protective clothing is recommended for fire situations only; it is not effective in spill situations or release events. If chemical protective clothing is not available and it is

necessary to rescue casualties from a contaminated area, then structural firefighters' gear will provide very limited skin protection against agent vapors and aerosols. Contact with solid and liquid agents should be avoided.

Respiratory Protection

Self-contained breathing apparatuses (SCBAs) or air purifying respirators (APRs) should have a National Institute for Occupational Safety and Health (NIOSH) Chemical/Biological/Radiological/Nuclear (CBRN) certification. However, during emergency operations, other NIOSH-approved SCBAs or APRs that have been specifically tested by the manufacturer against chemical warfare agents may be used if deemed necessary by the incident commander. APRs should be equipped with a NIOSH-approved CBRN filter or a combination organic vapor/acid gas/particulate cartridge.

Immediately dangerous to life or health (IDLH) levels are the ceiling limit for respirators other than SCBAs. Any exposures approaching the IDLH level should be regarded with extreme caution and the use of SCBAs for respiratory protection should be considered.

Chemical Protective Clothing

Use only chemical protective clothing that has undergone material and construction performance testing against mind-altering agents. Reported permeation rates may be affected by solvents, components, or impurities in munition grade or modified agents.

DECONTAMINATION

General

Apply universal decontamination procedures using soap and water.

Vapors

Casualties/Personnel

Remove all clothing as it may contain trapped agent. To avoid further exposure of the head, neck, and face to the agent, cut off potentially contaminated clothing that must be pulled over the head. Shower using copious amounts of soap and water. Ideally, showers will be high volume with low pressure. Ensure that the hair has been washed and rinsed to remove potentially trapped agent. If there is a potential that the eyes have been exposed to the agent, irrigate with water or 0.9% saline solution for a minimum of 15 minutes.

Small areas

Ventilate to remove the vapors. If deemed necessary, wash the area with copious amounts of soap and water. Collect and place into containers lined with high-density polyethylene. Removal of porous material, including painted surfaces, may be required because agents that has been absorbed into these materials may migrate back to the surface and pose a residual hazard.

Liquids/Solutions or Liquid Aerosols

Casualties/Personnel

Cover all open wounds during the decontamination process. Remove all clothing immediately. To avoid further exposure of the head, neck, and face to the agent, cut off potentially contaminated clothing that must be pulled over the head. Use a sponge or cloth with liquid soap and copious amounts of water to wash the skin surface and hair at least three times. Ideally, showers will be high volume with low pressure. Do not delay decontamination to find warm or hot water if it is not readily available. Avoid rough scrubbing as this could abrade the skin and increase percutaneous absorption of residual agent. Rinse with copious amounts of water. If there is a potential that the eyes have been exposed to mind-altering agents, irrigate with water or 0.9% saline solution for a minimum of 15 minutes.

Small areas

Ventilate to remove the aerosol. Small puddles of liquid can be contained by covering with absorbent material such as vermiculite, diatomaceous earth, clay, sponges, or towels. Place the absorbed material into containers lined with high-density polyethylene. Larger puddles can be collected using vacuum equipment made of materials inert to the released material and equipped with a high-efficiency particulate air (HEPA) filter and appropriate vapor filters. Wash the area with copious amounts of the soap and water. Collect and containerize the rinseate. Ventilate the area to remove residual vapors.

Solids or Particulate Aerosols

Casualties/Personnel

Do not attempt to brush the agent off of the individual or their clothing as this can aerosolized the agent. If possible, dampen the agent with a water mist to help prevent aerosolization. Cover all open wounds during the decontamination process. Remove all clothing immediately. To avoid further exposure of the head, neck, and face to the agent, cut off potentially contaminated clothing that must be pulled over the head. Wash the skin surface and hair at least three times with copious amounts of soap and water. Ideally, showers will be high volume with low pressure. Do not delay decontamination to find warm or hot water if it is not readily available. Rinse with copious amounts of water. If there is a potential that the eyes have been exposed to mind-altering agents, irrigate with water or 0.9% saline solution for a minimum of 15 minutes.

Small areas

If indoors, close windows and doors in the area and turn off anything that could create air currents (e.g., fans, air conditioner, etc.). Allow aerosol to settle. Avoid actions that could aerosolize the agent such as sweeping or brushing. Collected the agent using a vacuum cleaner equipped with a high-efficiency particulate air (HEPA) filter. Do not use a standard home or industrial vacuum. Do not allow the vacuum exhaust to stir the air in the affected area. Vacuum all surfaces with extreme care in a very slow and controlled manner to minimize aerosolizing the agent. Place the collected material into containers lined with high-density polyethylene. Wash the area with copious amounts of the soap and water. Collect and containerize the rinseate in containers lined with high-density polyethylene.

MEDICAL

CDC CASE DEFINITION

1) A case in which mind-altering agents are detected in the urine. Routine toxicologic screens might not detect all of these drugs. OR 2) Detection of mind-altering agents in environmental samples. The case can be confirmed if laboratory testing is not performed because either a predominant amount of clinical and nonspecific laboratory evidence is present or an absolute certainty of the etiology of the agent is known.

DIFFERENTIAL DIAGNOSIS

The following factors have been suggested as alternatives to consider when presented with a potential case of exposure to mind-altering agents: conduct disorder, personality disorders, dysthymic disorder, attention deficit hyperactivity disorder, panic disorders, delirium, dementia, amnesia, anxiety, headache, migraine, brain abscess; encephalitis, CNS infection, acute respiratory distress syndrome; heat exhaustion/heat stroke, hypoxia, hypoglycemia, electrolyte abnormalities, myocardial infarction, myocarditis, diabetic ketoacidosis; substance abuse (alcohol, plant and mushroom poisoning, scopolamine-tainted heroin), withdrawal syndromes, delirium tremens, methemoglobinemia, exposure to organophosphate pesticides, carbamate pesticides, carbon monoxide, or cyanides.

Signs and Symptoms

Varies according to the type of mind-altering agent. Care must be taken in that many signs and symptoms associated with exposure to mind-altering agents are also associated with anxiety or physical trauma. Potential indications of exposure include apprehension, restlessness, dizziness, confusion, erratic behavior, inappropriate smiling or laughing, irrational fear, difficulty in communicating (mumbling, slurred, or nonsensical speech), euphoria, lethargy, trembling, pleading, crying, perceptual distortions, hallucinations, disrobing, stumbling or staggering, blurred vision, dilated or pinpointed pupils, flushed face and skin, elevated temperature, dry mouth and skin, foul breath, stomach cramps, vomiting, difficulty urinating, change in pulse rate (slow or elevated), change in blood pressure (lowered or elevated), changes in breathing rate, stupor, or coma.

Mass-Casualty Triage Recommendations

Priority 1

A casualty with cardiovascular collapse or severe hyperthermia. Immediate attention to ventilation, hemodynamic status, and temperature control could be life-saving.

Priority 2

A casualty with severe or worsening signs after exposure.

Priority 3

A casualty with mild peripheral or central nervous system effects. However, these patients will not be able to manage themselves and should be controlled.

Priority 4

A casualty with severe cardiorespiratory compromise when treatment or evacuation resources are unavailable.

Casualty Management

Decontaminate the casualty ensuring that all mind-altering agent has been removed. If mind-altering agents have gotten into the eyes, irrigate the eyes with water or 0.9% saline solution for at least 15 minutes. Irrigate open wounds with water or 0.9% saline solution for at least 10 minutes. However, do not delay treatment if thorough decontamination cannot be undertaken immediately.

Once the casualty has been decontaminated, including the removal of foreign matter from wounds, medical personnel do not need to wear a chemical-protective mask.

Antidotes are available for some mind altering agents. Prior to administering antidotes or other drugs, ensure that the signs and symptoms (e.g., coma, seizures, etc.) are due to chemical exposure and not the result of head trauma or other physical injury. Otherwise, general treatment consists of observation, supportive care with fluids, and possibly restraint or confinement. Casualties should be isolated in a safe area. Remove any potentially harmful material from individuals suspected of being exposed to mind-altering agents including such items as cigarettes, matches, medications, and other small items they might attempt to ingest. Observe casualties for signs of heat stroke as some mind-altering agents eliminate the ability of exposed individuals to sweat. Monitor to ensure that casualties are breathing. Full recovery from effects may take several days.

FATALITY MANAGEMENT

Remove all clothing and personal effects and decontaminate with soap and water. While it may be possible to decontaminate durable items, it may be safer and more efficient to destroy non-durable items rather than attempt to decontaminate them. Items that will be retained for further processing should be double sealed in impermeable containers, ensuring that the inner container is decontaminated before placing it in the outer one.

To remove agents on the outside of the body, wear appropriate respiratory and dermal protective clothing while washing the remains with soap and water. Pay particular attention to areas where agent may get trapped, such as hair, scalp, pubic areas, fingernails, folds of skin, and wounds. If remains are heavily contaminated with residue, wash and rinse waste should be contained for proper disposal.

Once the remains have been thoroughly decontaminated, no further protective action is necessary. Body fluids removed during the embalming process do not pose any additional risks and should be contained and handled according to established procedures. Use standard burial procedures.

C12-A AGENTS

C12-A001

3-Quinuclidinyl Benzilate (Agent BZ)
CAS: 6581-06-2
RTECS: DD4638000
UN: —
EC: —
UNII: E69DLR7470
ICD-11: —

$C_{21} H_{23} N O_3$

White to beige crystalline solid that is odorless. Various salts have been reported.

Also reported as a mixture with Agent CS (C13-A009).

Exposure Hazards

$LCt_{50\ (Inh)}$: 200,000 mg-min/m^3. This value is from an older source (circa 1960) and is not supported by modern data. No updated toxicity estimates have been proposed.
$ICt_{50\ (Inh)}$: 100 mg-min/m^3
Mild incapacitation: 90 mg-min/m^3 (some hallucinations)
Severe incapacitation: 135 mg-min/m^3 (marked hallucinations)
$MEG_{(1hr):}$ Neg: 0.020 mg/m^3; Mar: 0.037 mg/m^3; Crit: 0.69 mg/m^3; Cat: —
PAC-1: 0.0010 mg/m^3; *PAC-2:* 0.011 mg/m^3; *PAC-3:* 0.21 mg/m^3

Final AEGLs
AEGL-1: Not Developed
AEGL-2: 1 hr – 0.011 mg/m^3; Not Developed
AEGL-3: 1 hr – 0.21mg/m^3; Not Developed

Properties:

MW: 337.4	VP: Negligible	FIP: 475°F
D: 1.3 g/cm^3 (crystalline)	VD: —	FIP: 428°F (munition grade)
D: 0.51 g/cm^3 (munition grade)	Vlt: —	LEL: —
MP: 334°F	ER: —	UEL: —
BP: 822°F	H$_2$O: 0.0012% (77°F)	RP: 200,000
Vsc: —	Sol: Common organic solvents	IP: —

C12-A002

1-Methyl-4-piperidyl Cyclopentyl-1-propynylglycolate (Agent 302196 (B))

CAS: 53034-67-6

RTECS: GY2453300

UN: —

EC: —

UNII: —

ICD-11: —

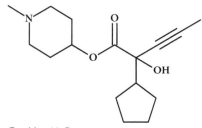

C$_{16}$ H$_{25}$ N O$_3$

Ivory colored powder. Various salts have been reported.

Exposure Hazards

Human toxicity values have not been established or have not been published. However, based on available information, this agent appears to be approximately one-fourth as potent as BZ (C12-A001) with a more rapid onset and shorter duration of effects.

Properties:

MW: 279.4	VP: 4.8 × 10^{-7} mmHg (77°F)	FIP: 376°F
D: 1.1 g/cm^3	VD: —	LEL: —
MP: 253°F	Vlt: —	UEL: —
BP: 671°F (est.)	ER: —	RP: 13,000,000
Vsc: —	H$_2$O: 2.25% (77°F)	IP: —
	Sol: Chloroform	

C12-A003

3-Quinuclidinyl Cyclopentylphenylglycolate (Agent EA 3167)

CAS: 26758-53-2

RTECS: —

UN: —

EC: —

UNII: —
ICD-11: —

$C_{20} H_{27} N O_3$

Straw-colored, extremely viscous and tacky liquid that is odorless. Various salts have been reported.

Exposure Hazards

Human toxicity values have not been established or have not been published. However, based on available information, this agent appears to be as potent as BZ (C12-A001) with a longer duration of effects.

Properties:

MW: 329.5	*VP:* 5×10^{-7} mmHg (77°F)	*FlP:* 448°F
D: 1.1 g/mL (77°F)	*VD:* —	*LEL:* —
MP: —	*Vlt:* —	*UEL:* —
BP: 613°F (est.)	*ER:* —	*RP:* 11,000,000
Vsc: 15,000 cS (131°F)	*H₂O:* <0.001%	*IP:* —
	Sol: Ethanol, chloroform	

C12-A004

N-Methyl-4-piperidyl Cyclopentylphenylglycolate (Agent EA 3443)

CAS: 37830-21-0
RTECS: —
UN: —
EC: —
UNII: —
ICD-11: —

$C_{19} H_{27} N O_3$

Waxy solid or viscous liquid that is odorless. Various salts have been reported.

Exposure Hazards

$ICt_{50\ (Inh)}$: 54.4 mg-min/m³

Properties:

MW: 317.5	*VP:* Negligible	*FIP:* 424°F
D: 1.1 g/cm³	*VD:* —	*LEL:* —
D: 0.5 g/cm³ (munition grade)	*Vlt:* —	*UEL:* —
MP: 119°F	*ER:* —	*RP:* 11,000,000
BP: 714°F	*H₂O:* 0.12% (77°F)	*IP:* —
Vsc: 86,000 cS (86°F)	*Sol:* Most organic solvents	

C12-A005

1-Methyl-4-piperidyl Cyclobutylphenylglycolate (Agent EA 3580B; hydrochloride salt is Agent EA 3580A)

CAS: 54390-94-2

RTECS: —

UN: —

EC: —

UNII: —

ICD-11: —

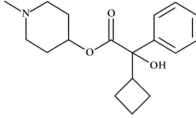

$C_{18}\ H_{25}\ N\ O_3$

White crystalline or waxy solid that is odorless. Hydrochloride salt is a white to yellow crystalline solid.

An aerosol cloud from the hydrochloride salt can be thermally generated with little decomposition of the agent.

Exposure Hazards

$ICt_{50\ (Inh)}$: 71 mg-min/m³

$ICt_{50\ (Inh)}$: 79 mg-min/m³ (Hydrochloride salt)

Properties:

MW: 303.4	*VP:* 5 × 10⁻⁶ mmHg (77°F)	*FIP:* —
D: 1.1 g/cm³ (77°F)	*VD:* 11	*LEL:* —
MP: 133°F	*Vlt:* —	*UEL:* —
BP: 710°F	*ER:* —	*RP:* 1,200,000
Vsc: 4,060 cS (77°F)	*H₂O:* 9.42% (77°F)	*IP:* —
	Sol: Most organic solvents	

Hydrochloride Salt
MW: 339.9
D: 1.2 g/cm^3 (77°F)
D: 0.5 g/cm^3 (munition grade)
MP: 387°F
H$_2$O: 39% (77°F)

C12-A006
1-Methyl-4-piperidyl Isopropylphenylglycolate (Agent EA 3834B; hydrochloride salt is Agent EA 3834A)
CAS: 75321-25-4; 137444-35-0 (Hydrochloride salt)
RTECS: —
UN: —
EC: —
UNII: —
ICD-11: —

C$_{17}$ H$_{25}$ N O$_3$

White crystals to brown oil that is odorless. Hydrochloride salt is a white crystalline solid.

An aerosol cloud from the hydrochloride salt can be thermally generated with little decomposition of the agent. Salt has greater absorption via lungs than the free base.

Also reported as a mixture with Agent CH (C13-A014).

Exposure Hazards
ICt$_{50}$ (Inh): 73.4 mg-min/m^3
ICt$_{50}$ (Inh): 82.6 mg-min/m^3 (Hydrochloride salt)

Properties:

MW: 291.4	*VP:* 1.5 × 10^{-5} mmHg (77°F)	*FlP:* 397°F
D: 1.1 g/cm^3	*VD:* 10	*LEL:* —
MP: 120°F	*Vlt:* —	*UEL:* —
BP: 639°F	*ER:* —	*RP:* 390,000
Vsc: 13,525 cS (77°F)	*H$_2$O:* <0.002%	*IP:* —
	Sol: Hexane, ether	

Hydrochloride Salt
MW: 327.9
D: 1.2 g/cm^3
MP: 399°F
H$_2$O: 43.9% (77°F)

C12-A007

Phencyclidine (Agent SN)

CAS: 77-10-1; 2981-31-9 (Hydrobromide salt)

RTECS: TN2272590

UN: —

EC: —

UNII: J1DOI7UV76

ICD-11: XM5M84

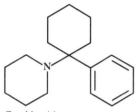

$C_{17} H_{25} N$

Colorless or white crystalline substance. Recreational material may be a powder or gummy solid with a color from tan to brown. Various salts have been reported.

Used as a veterinary anesthetic. Also used as an illegal street drug.

Exposure Hazards

$ICt_{50\ (Inh)}$: 1,000 mg-min/m^3. Concentration of 25–50 mg-min/m^3 produce anesthetic effects, and concentrations >100 mg-min/m^3 cause mental disturbances.

$ID_{50\ (Ing)}$: 0.010 g–0.020 g

$LD_{50\ (Ing)}$: 7 g

Properties:

MW: 243.4	VP: 1.9×10^{-4} mmHg (77°F) (est.)	FIP: 52°F
D: 1.0	VD: —	LEL: —
MP: 115°F	Vlt: —	UEL: —
BP: 644°F	ER: —	RP: —
Vsc: —	H$_2$O: Insoluble	IP: —
	Sol: Ethanol	

Hydrochloride Salt
MW: 279.9
MP: 451°F

Hydrobromide Salt
MW: 324.3
MP: 417°F

C12-A008

Alfentanil

CAS: 71195-58-9; 69049-06-5 (Hydrochloride salt)

RTECS: TX1452480; TX1452500 (Hydrochloride salt)

UN: —

EC: 273-846-3 (Hydrochloride salt)

UNII: 1N74HM2BS7; 11S92G0TIW (Hydrochloride salt)

ICD-11: XM4G88

$C_{21} H_{32} N_6 O_3$

Solid. Hydrochloride is a colorless solid with no odor.

Used medicinally as a short-acting general anesthetic. Effects dissipate after approximately 20 minutes.

Exposure Hazards

Human toxicity values have not been established or have not been published. However, based on available information, this agent appears to be approximately one-fourth as potent an analgesic as fentanyl (C12-A013) with only one-tenth as severe respiratory depression.

Properties:

MW: 416.5	VP: 3.6×10^{-8} mmHg (est.)	FIP: —
D: 1.2 g/cm^3	VD: —	LEL: —
MP: —	Vlt: —	UEL: —
BP: —	ER: —	RP: —
Vsc: —	H$_2$O: 0.0035%	IP: —
	Sol: —	

Hydrochloride Salt
MW: 453.0
MP: 285°F
H$_2$O: Soluble
Sol: DMSO, alcohols

C12-A009

Carfentanil
CAS: 59708-52-0; 61380-27-6 (Citrate salt); 61086-44-0 (Oxalate salt)
RTECS: TM6406000
UN: —
EC: 262-748-6 (Citrate salt)
UNII: LA9DTA2L8F; 7LG286J8GV (Citrate salt)
ICD-11: —

$C_{24} H_{30} N_2 O_3$

Solid. Various salts (solids) have been reported.

Used as a large animal tranquilizer with doses ranging from 0.005 to 0.020 mg/kg of body weight. Also used as an illegal street drug.

Exposure Hazards

Conversion Factor: 1 ppm = 16.14 mg/m^3 at 77°F

LD$_{50}$: 238 mg (route unspecified)

Lowest effective dose: 0.238 mg (route unspecified)

Human toxicity values have not been established or have not been published. However, based on available information, carfentanil is approximately 10,000 times more potent than Morphine (C12-A022) and 100 more potent than Fentanyl (C12-A013).

Properties:

MW: 394.5	VP: 2.3×10^{-10} mmHg (est.)	FIP: —
D: 1.1 g/cm^3	VD: —	LEL: —
MP: —	Vlt: —	UEL: —
BP: 947°F	ER: —	RP: —
Vsc: —	H$_2$O: 0.0004% (est.)	IP: —
	Sol: Alcohols	

Citrate Salt

MW: 586.6

MP: 306°F

Sol: DMSO

Oxalate Salt

MW: 484.5

MP: 373°F

C12-A010

Cocaine

CAS: 50-36-2; 53-21-4 (Hydrochloride salt); 5913-62-2 (Nitrate salt)

RTECS: YM2800000

UN: —

EC: 200-032-7; 200-167-1 (Hydrochloride salt); 227-635-8 (Nitrate salt)

UNII: I5Y540LHVR; XH8T8T6WZH (Hydrochloride salt)

ICD-11: XM7UN8

$C_{17} H_{21} N O_4$

Colorless to white crystals or powder with no odor. Various salts have been reported.

Used medicinally and by veterinarians as a local anesthetic. Also used as an illegal street drug.

Exposure Hazards

$LD_{50\ (Ing)}$: 1–1.2 g (est.)

Properties:

MW: 303.4	VP: 1.91×10^{-7} mmHg (77°F)	FIP: —
D: —	VD: —	LEL: —
MP: 208°F	Vlt: —	UEL: —
BP: 369°F (1 mmHg)	ER: —	RP: —
Vsc: —	H_2O: 0.17%	IP: —
	Sol: Most organic solvents	

Hydrochloride Salt
MW: 339.9
MP: 383°F
H_2O: 250%

Nitrate Salt
MW: 366.4
MP: 136°F
H_2O: Soluble

C12-A011

Dexamphetamine

CAS: 51-64-9; 51-63-8 (Sulfate salt); 1462-73-3 (Hydrochloride salt)
RTECS: SH9100000
UN: —
EC: 200-112-1; 200-111-6 (Sulfate salt)
UNII: TZ47U051FI
ICD-11: XM6LD5

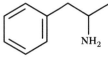

$C_9 H_{13} N$

Colorless oily liquid. The salts are white, odorless crystalline solids.

Used by veterinarians as a stimulant of the central nervous system. Also used as an illegal street drug.

Exposure Hazards

Conversion Factor: 1 ppm = 5.53 mg/m^3 at 77°F

$LD_{50\ (Ing)}$: 1.4 – 1.8 g (est.)

Properties:

MW: 135.2	*VP:* 0.2 mmHg (est.)	*FlP:* —
D: 0.95 g/mL (59°F)	*VD:* 4.7	*LEL:* —
MP: 64°F (est.)	*Vlt:* —	*UEL:* —
BP: 397°F	*ER:* —	*RP:* —
Vsc: —	*H$_2$O:* Slight	*IP:* —
	Sol: Alcohol, ether	

Sulfate Salt   **Hydrochloride Salt**

MW: 233.3   *MW:* 171.7

D: 1.2 g/cm^3 *MP:* 304°F

MP: 572°F

H$_2$O: 10%

C12-A012

Ecstasy

CAS: 42542-10-9; 64057-70-1 (Hydrochloride salt)
RTECS: DF4914490; SH5700000 (Hydrochloride salt)
UN: —
EC: —
UNII: KE1SEN21RM
ICD-11: XM07Y4

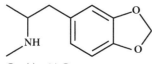

C$_{11}$ H$_{15}$ N O$_2$

Oily liquid. The hydrochloride salt is a white crystalline solid.

There is no approved medical use in the United States, but it is used as an illegal street drug.

Exposure Hazards

Conversion Factor: 1 ppm = 7.90 mg/m^3 at 77°F

LD$_{50 (Ing)}$: 1.4 g – 1.8 g (est.)

This material is approximately 3 times more potent than mescaline (C12-A019). Males are more sensitive to the acute toxicity of this agent than females.

Properties:

MW: 193.2	*VP:* 0.0021 mmHg (77°F) (est.)	*FlP:* —
D: 1.1 g/mL	*VD:* 6.7	*LEL:* —
MP: —	*Vlt:* —	*UEL:* —
BP: 221°F (0.4 mmHg)	*ER:* —	*RP:* —
Vsc: —	*H$_2$O:* 0.7% (est.)	*IP:* —
	Sol: —	

Hydrochloride Salt

MW: 229.7

MP: 297°F

H₂O: Soluble

C12-A013

Fentanyl

CAS: 437-38-7; 990-73-8 (Citrate salt)

RTECS: UE5550000

UN: —

EC: 207-113-6; 213-588-0 (Citrate salt)

UNII: UF599785JZ; MUN5LYG46H (Citrate salt)

ICD-11: XM76M8

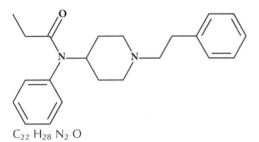

$C_{22} H_{28} N_2 O$

Crystalline solid that is odorless. Various salts have been reported.

Used medicinally as an anesthetic, by veterinarians as a tranquilizer. Also used as an illegal street drug and as an illegal stimulant for racehorses.

Exposure Hazards

Human toxicity values have not been established or have not been published. However, this material is between 50 and 100 times more potent than morphine (C12-A022).

Properties:

MW: 336.5	*VP:* 4.4×10^{-9} mmHg (77°F) (est.)	*FlP:* —
D: 1.1 g/cm³	*VD:* —	*LEL:* —
MP: 181°F	*Vlt:* —	*UEL:* —
BP: 871°F	*ER:* —	*RP:* —
Vsc: —	*H₂O:* 0.02%	*IP:* —
	Sol: Alcohols	

Citrate Salt

MW: 528.6

MP: 307°F

H₂O: 2.5%

Sol: Alcohols

C12-A014

Haloperidol (302034)

CAS: 52-86-8; 1511-16-6 (Hydrochloride salt)

RTECS: EU1575000

UN: —

EC: 200-155-6

UNII: J6292F8L3D; UM06W2ADRY (Hydrochloride salt)

ICD-11: XM9580

$C_{21} H_{23} Cl F N O_2$

White to faintly yellowish crystalline solid that is odorless. Various salts (solids) have been reported.

Exposure Hazards

Conversion Factor: 1 ppm = 15.37 mg/m^3 at 77°F

Human toxicity values have not been established or have not been published. The main features of severe overdosage are extrapyramidal reactions (e.g., Parkinson-like symptoms, akathisia, and dystonia), hypotension, respiratory difficulty and impairment of consciousness.

Properties:

MW: 375.9	*VP:* 5×10^{-11} mmHg (77°F) (est.)	*FlP:* —
D: 1.3 g/cm^3	*VD:* 13	*LEL:* —
MP: 306°F	*Vlt:* —	*UEL:* —
BP: —	*ER:* —	*RP:* —
Vsc: —	*H$_2$O:* 0.0014% (77°F)	*IP:* —
	Sol: Chloroform, acetone, benzene, methanol	

Hydrochloride Salt

MW: 412.4

MP: 439°F

H$_2$O: 0.3%

C12-A015

Halothane

CAS: 151-67-7

RTECS: KH6550000

UN: —

EC: 205-796-5

UNII: UQT9G45D1P

ICD-11: XM2FH0

C_2 H Br Cl F_3

Clear, colorless liquid with a sweetish, pleasant odor. This material is hazardous through inhalation.

Used medicinally as an inhalation anesthetic.

Exposure Hazards

Conversion Factor: 1 ppm = 8.07 mg/m^3 at 77°F

MEG$_{(1hr)}$: Neg: —; Mar: 12 ppm; Crit: 50 ppm; Cat: —

TLV: 50 ppm

Ceiling$_{(NIOSH)}$: 2.0 ppm

Properties:

MW: 197.4	*VP:* 243 mmHg	*FIP:* NA
D: 1.9 g/mL	*VD:* 6.8	*LEL:* NA
MP: −180°F	*Vlt:* 330,000 ppm	*UEL:* NA
BP: 122°F	*ER:* 230	*RP:* 0.029
Vsc: —	*H$_2$O:* 0.345%	*IP:* 11.0 eV
	Sol: Hydrocarbon solvents	

C12-A016

Ketamine

CAS: 6740-88-1; 1867-66-9 (Hydrochloride salt)

RTECS: GW1300000; GW1400000 (Hydrochloride salt)

UN: —

EC: 229-804-1; 217-484-6 (Hydrochloride salt)

UNII: 690G0D6V8H; O18YUO0I83 (Hydrochloride salt)

ICD-11: XM7C11

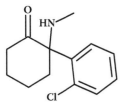

C_{13} H_{16} Cl N O

Solid. Various salts (solids) have been reported.

Also reported as a mixture with cocaine (C12-A010), ecstasy (C12-A012), dexamphetamine (C12-A011), and methamphetamine (C12-A020).

Exposure Hazards

Conversion Factor: 1 ppm = 9.72 mg/m^3 at 77°F

Human toxicity values have not been established or have not been published. It produces hypnotic, analgesic, and hallucinogenic effects. Deaths have been reported in adults following doses of 900 to 1,000 mg.

Properties:

MW: 237.7	*VP:* 5.2 × 10^{-5} mmHg (77°F) (est.)	*FlP:* —
D: —	*VD:* —	*LEL:* —
MP: 199°F	*Vlt:* —	*UEL:* —
BP: —	*ER:* —	*RP:* —
Vsc: —	*H$_2$O:* Soluble	*IP:* —
	Sol: —	

Hydrochloride Salt

MW: 274.2
MP: 504°F
H$_2$O: 20%
Sol: Alcohols

C12-A017

Lofentanyl

CAS: 61380-40-3; 61380-41-4 (Oxalate salt)
RTECS: —
UN: —
EC: 262-750-7 (Oxalate salt)
UNII: 7H7YQ564XV
ICD-11: —

C$_{25}$ H$_{32}$ N$_2$ O$_3$

Solid. Oxalate is a colorless solid with no odor.

Used medicinally as an anesthetic.

Exposure Hazards

Human toxicity values have not been established or have not been published. However, one of the most potent opioid analgesics known and based on available information, lofentanyl is slightly more potent than carfentanil (C12-A009) and with a longer duration of action.

Properties:

MW: 408.5	VP: 4 × 10⁻¹⁰ mmHg (77°F) (est.)	FlP: —
D: 1.1 g/cm³	VD: —	LEL: —
MP: —	Vlt: —	UEL: —
BP: —	ER: —	RP: —
Vsc: —	H₂O: 0.00007%	IP: —
	Sol: —	

Oxalate Salt

MW: 498.6

MP: 323°F (est.)

H₂O: 0.002% (est.)

C12-A018

LSD

CAS: 50-37-3; 15232-63-0 (Tartrate salt)

RTECS: KE4100000; KE4378000 (Tartrate salt)

UN: —

EC: 200-033-2

UNII: 8NA5SWF92O; 7WNP51KA7M (Tartrate salt)

ICD-11: XM04H7

$C_{20} H_{25} N_3 O$

Colorless crystalline solid that is odorless. Various salts have been reported.

Used in biochemical research as an antagonist to serotonin. Also used as an illegal street drug.

Exposure Hazards

$LD_{50\ (Ing)}$: 0.021 g–3.5 g estimates. There are no known human cases of overdose.

Properties:

MW: 323.4	VP: 2 × 10⁻⁸ mmHg (77°F) (est.)	FlP: —
D: 1.2 g/cm³	VD: —	LEL: —
MP: 176°F–185°F	Vlt: —	UEL: —
BP: 830°F (est.)	ER: —	RP: —
Vsc: —	H₂O: 0.0002%	IP: 7.25 eV
	Sol: Chloroform, methanol, benzene	

Tartrate Salt

MW: 861.0

MP: 388°F

H₂O: 0.03% (est.)

C12-A019

Mescaline

CAS: 54-04-6; 832-92-8 (Hydrochloride salt); 1152-76-7 (Sulfate salt)

RTECS: SI2625000; SI2800000 (Hydrochloride salt)

UN: —

EC: 200-190-7; 212-626-3 (Hydrochloride salt)

UNII: RHO99102VC; XQG91WTZ12 (Hydrochloride salt); ADE45928XQ (Sulfate salt)

ICD-11: XM0T53

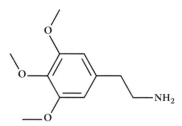

$C_{11} H_{17} N O_3$

Crystalline solid. Reacts with carbon dioxide in the air to form the carbonate. Various salts have been reported.

Used in religious ceremonies by the North American Church of Native Americans. Also used as an illegal street drug.

Exposure Hazards

Conversion Factor: 1 ppm = 8.64 mg/m^3 at 77°F

Human toxicity values have not been established or have not been published.

Properties:

MW: 211.3	VP: 3 × 10^{-4} mmHg (77°F) (est.)	FIP: —
D: 1.1 g/cm^3	VD: 7.3	LEL: —
MP: 97°F	Vlt: —	UEL: —
BP: 356°F (12 mmHg)	ER: —	RP: —
Vsc: —	H$_2$O: Moderate	IP: 8.18 eV
	Sol: Ethanol, chloroform, benzene	

Hydrochloride Salt
MW: 247.7
MP: 358°F
BP: 594°F

Sulfate Salt
MW: 309.3
MP: 361°F

C12-A020

Methamphetamine

CAS: 537-46-2; 51-57-0 (Hydrochloride salt)

RTECS: SH4910000

UN: —

EC: 208-668-7; 200-106-9 (Hydrochloride salt)

UNII: 44RAL3456C; 997F43Z9CV (Hydrochloride salt)

ICD-11: XM5B49

$C_{10} H_{15} N$

Clear, colorless liquid with a characteristic odor resembling geranium leaves. The hydrochloride salt is a crystalline solid that is odorless.

Used medicinally as an anesthetic. Also used as an illegal street drug.

Exposure Hazards

Conversion Factor: 1 ppm = 6.10 mg/m^3 at 77°F

LD$_{50\ (Ing)}$: 1.4 g – 1.8 g (est.)

Properties:

MW: 149.2	*VP:* 0.163 mmHg (77°F)	*FIP:* —
D: 0.9 g/mL	*VD:* 5.1	*LEL:* —
MP: 38°F (est.)	*Vlt:* 210 ppm (77°F)	*UEL:* —
BP: 414°F	*ER:* —	*RP:* 50
Vsc: —	*H$_2$O:* 1.33%	*IP:* —
	Sol: Ethanol, ether, chloroform	

Hydrochloride Salt

MW: 185.7

MP: 338°F

H$_2$O: 50%

C12-A021

Methoxyflurane

CAS: 76-38-0

RTECS: KN7820000

UN: —

EC: 200-956-0

UNII: 30905R8O7B

ICD-11: XM3P63

$C_3 H_4 Cl_2 F_2 O$

Clear colorless liquid with a sweet, fruity odor.

Used medicinally as an inhalation anesthetic.

Exposure Hazards

Conversion Factor: 1 ppm = 6.75 mg/m^3 at 77°F

Ceiling$_{(NIOSH)}$: 2.0 ppm

Properties:

MW: 165.0	*VP:* 23 mmHg	*FlP:* 145°F
D: 1.4 g/mL	*VD:* 5.7	*LEL:* 7%
MP: –31°F	*Vlt:* 65,000 ppm	*UEL:* —
BP: 221°F	*ER:* —	*RP:* 0.16
Vsc: —	*H₂O:* 2.8% (100°F)	*IP:* 11.0 eV
	Sol: —	

C12-A022

Morphine

CAS: 57-27-2; 64-31-3 (Sulfate salt)

RTECS: QC7875000

UN: —

EC: 200-320-2; 200-582-8 (Sulfate salt)

UNII: 76I7G6D29C; X3P646A2J0 (Sulfate salt)

ICD-11: XM69R4

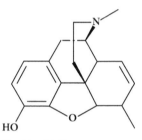

$C_{17} H_{19} N O_3$

White crystalline solid that is odorless. Acetate has a slightly acetic odor. Various salts have been reported.

Used medicinally as an anesthetic. Also used as an illegal street drug.

Exposure Hazards

$LD_{50\ (IV)}$: 0.03 g

Properties:

MW: 285.3	*VP:* 2×10^{-10} mmHg (77°F) (est.)	*FlP:* —
D: 1.3 g/cm³	*VD:* —	*LEL:* —
MP: 374°F	*Vlt:* —	*UEL:* —
BP: Sublimes	*ER:* —	*RP:* —
Vsc: —	*H₂O:* 0.015%	*IP:* 8.3 eV
	Sol: Methanol	

Sulfate Salt

MW: 668.8

MP: 491°F

BP: 889°F

C12-A023

Fluphenazine

CAS: 69-23-8; 146-56-5 (Hydrochloride salt); 5002-47-1 (Decanoate)

RTECS: TL9730000; TL9800000 (Hydrochloride salt); HE0525000 (Decanoate)

UN: —

EC: 200-702-9; 205-674-1 (Hydrochloride salt); 225-672-4 (Decanoate)

UNII: S79426A41Z; ZOU145W1XL (Hydrochloride salt); FMU62K1L3C (Decanoate)

ICD-11: XM6Z10

$C_{22} H_{26} F_3 N_3 O S$

Light yellow to pale brown solid. The hydrochloride salt is a white crystalline solid, while the decanoate ester is pale yellow to yellowish orange viscous liquid or oily solid.

Used medicinally as an antipsychotic medication and tranquilizer.

Exposure Hazards

Human toxicity values have not been established or have not been published. The main features of severe overdosage are extrapyramidal reactions (e.g., Parkinson-like symptoms, akathisia, and dystonia), hypotension, and sedation.

Properties:

MW: 437.5	VP: 1×10^{-10} mmHg (77°F) (est.)	FIP: —
D: 1.3 g/cm^3	VD: —	LEL: —
MP: 340°F (est.)	Vlt: —	UEL: —
BP: 482°F (0.3 mmHg)	ER: —	RP: —
Vsc: —	H$_2$O: 0.003% (99°F)	IP: 8.64 eV
	Sol: DMSO, DMF	

Hydrochloride Salt	**Decanoate**
MW: 510.4	MW: 591.8
MP: 436°F	MP: 86°F
H$_2$O: Soluble	H$_2$O: Practically insoluble

C12-A024

Psilocybin

CAS: 520-52-5

RTECS: NM3150000

UN: —

EC: 208-294-4

UNII: 2RV7212BP0

ICD-11: XM7642

$C_{12} H_{17} N_2 O_4 P$

White to off-white crystalline solid with a slight ammonia odor.

Exposure Hazards

Conversion Factor: 1 ppm = 11.63 mg/m^3 at 77°F

LD_{50} is estimated to be 6 g for an adult, with an effective dose of 6 mg.

Properties:

MW: 284.3	VP: 1.5×19^{-10} mmHg (est.)	FlP: —
D: 1.5 g/cm^3 (est.)	VD: —	LEL: —
MP: 435°F	Vlt: —	UEL: —
BP: 760°F (est.)	ER: —	RP: —
Vsc: —	H_2O: 1.3% (est.)	IP: —
	Sol: Methanol	

C12-A025

Scopolamine

CAS: 51-34-3; 55-16-3 (Hydrochloride salt); 114-49-8 (Hydrobromide salt)

RTECS: VR3675000; CY1634501(Hydrochloride salt); YM4550000 (Hydrobromide salt)

UN: —

EC: 200-090-3; 200-225-6 (Hydrochloride salt); 204-050-6 (Hydrobromide salt)

UNII: DL48G20X8X; Q2P66EIP1F (Hydrochloride salt); 451IFR0GXB (Hydrobromide salt)

ICD-11: XM1MW1

$C_{17} H_{21} N O_4$

Thick viscous liquid or white solid that is odorless. Various salts (solids) have been reported. Salts are colorless or white crystalline powders

Used to prevent the nausea and vomiting associated with motion sickness.

Exposure Hazards

Conversion Factor: 1 ppm = 12.41 mg/m^3 at 77°F

LD$_{50}$: 280 mg (pathway unspecified)

ID$_{50}$: 1.4 mg (pathway unspecified)

Properties:

MW: 303.4	*VP:* 7 × 10^{-9} mmHg (77°F) (est.)	*FIP:* —
D: 1.3 g/cm^3	*VD:* —	*LEL:* —
MP: 138°F	*Vlt:* —	*UEL:* —
BP: 860°F	*ER:* —	*RP:* —
Vsc: —	*H$_2$O:* 10%	*IP:* —
	Sol: Alcohol, ether, chloroform, acetone	

Hydrochloride Salt

MW: 339.8

MP: 392°F

H$_2$O: Soluble

Hydrobromide Salt

MW: 384.3

MP: 387°F (decomposes)

H$_2$O: Soluble

C12-A026

Sufentanil

CAS: 56030-54-7; 60561-17-3 (Citrate salt)

RTECS: TX1474020; UE5295000 (Citrate salt)

UN: —

EC: 262-295-4 (Citrate salt)

UNII: AFE2YW0IIZ; S9ZFX8403R (Citrate salt)

ICD-11: XM1EF3

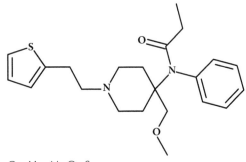

C$_{22}$ H$_{30}$ N$_2$ O$_2$ S

Off white crystalline solid that is odorless. Various salts have been reported.

Used medicinally as an anesthetic.

Exposure Hazards

Human toxicity values have not been established or have not been published. However, this material is approximately 700 times more potent than morphine (C12-A022).

Properties:

MW: 386.6	*VP:* 7 × 10^{-9} mmHg (77°F) (est.)	*FIP:* —
D: 1.2 g/cm^3	*VD:* —	*LEL:* —
MP: 206°F	*Vlt:* —	*UEL:* —

BP: —   ER: —   RP: —
Vsc: — H_2O: 0.0076%   IP: —
                                Sol: DMSO, alcohols

Citrate Salt
MW: 578.7
MP: 277°F
H_2O: Soluble

C12-A027

W-18
CAS: 93101-02-1
RTECS: —
UN: —
EC: —
UNII: 04WOYJF7QH
ICD-11: —

$C_{19} H_{20} N_3 O_4 Cl S$

Crystalline yellow solid.

Used as an illegal street drug.

Exposure Hazards
Conversion Factor: 1 ppm = 17.26 mg/m^3 at 77°F

Human toxicity values have not been established or have not been published. However, this material is reported to be approximately 100 times more potent than fentanyl (C12-A013).

Properties:

MW: 421.9	VP: 7×10^{-9} mmHg (77°F) (est.)	FIP: —
D: 1.4 g/cm^3 (est.)	VD: —	LEL: —
MP: 315°F	Vlt: —	UEL: —
BP: —	ER: —	RP: —
Vsc: —	H_2O: 0.001% (est.)	IP: —
	Sol: —	

C12-A028

1-Methyl-4-piperidyl Phenyl-(3-methylbut-1-yn-3-enyl)-glycolate (302282)

CAS: 62869-67-4

RTECS: —

UN: —

EC: —

UNII: —

ICD-11: —

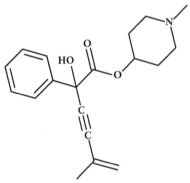

$C_{19} H_{23} N O_3$

Specific information is not available for this material.

Exposure Hazards

Conversion Factor: 1 ppm = 12.82 mg/m^3 at 77°F

LD_{50}: 42 mg (route unspecified)

ID_{50}: 0.2 mg (route unspecified)

Properties:

MW: 313.4	*VP:* —	*FIP:* —
D: 1.2 g/cm^3	*VD:* —	*LEL:* —
MP: —	*Vlt:* —	*UEL:* —
BP: 856°F	*ER:* —	*RP:* —
Vsc: —	*H$_2$O:* —	*IP:* —
	Sol: —	

C12-A029

2-alpha-Tropanyl Benzilate (CS 27349)

CAS: 64471-12-1; 64520-33-8 (Hydrochloride salt)

RTECS: YM4405000

UN: —

EC: —

UNII: RD1A447UC0

ICD-11: —

$C_{22} H_{25} N O_3$

Specific information is not available for this material. Various salts (solids) have been reported.

Exposure Hazards

Conversion Factor: 1 ppm = 14.37 mg/m^3 at 77°F

LD$_{50}$: 24 mg/man (route unspecified)

ID$_{50}$: 0.2 mg (route unspecified)

Properties:

MW: 351.4	*VP:* 1×10^{-9} mmHg (est.)	*FIP:* —
D: 1.2 g/cm^3	*VD:* 12	*LEL:* —
MP: 280°F (est.)	*Vlt:* —	*UEL:* —
BP: 833°F	*ER:* —	*RP:* —
Vsc: —	*H$_2$O:* Insoluble	*IP:* —
	Sol: —	

Hydrochloride Salt

MW: 387.9

MP: 388°F (est.)

C12-A030

3-Methylfentanyl

CAS: 42045-86-3

RTECS: TX1478600

UN: —

EC: —

UNII: QVU94XE61A; 6C6W9LJZ84 (Hydrochloride salt)

ICD-11: —

$C_{23} H_{30} N_2 O$

Crystalline solid. Various salts (solids) have been reported.

There are two stereoisomers of this agent based on the relative relationship of the methyl group to the amide group on the cyclohexyl ring. Of these two isomers, the cis isomer (both groups on the same face of the ring) is the most bioactivity potent.

Exposure Hazards

Conversion Factor: 1 ppm = 14.34 mg/m^3 at 77°F

Human toxicity values have not been established or have not been published. However, based on available information, this agent appears to be up to 6,000 times as potent as morphine (C12-A022); and up to 15 times as potent as fentanyl (C12-A013).

Properties:

MW: 350.5	*VP:* 2.9 × 10^{-9} mmHg (77°F) (est.)	*FIP:* —
D: 1.1 g/cm^3	*VD:* 12	*LEL:* —
MP: 523°F	*Vlt:* —	*UEL:* —
(hydrochloride salt)		
BP: 883.4°F	*ER:* —	*RP:* —
Vsc: —	*H_2O:* Insoluble	*IP:* —
	Sol: Alcohol	

C12-A031

4-(1-Methyl-1,2,3,6-tetrahydropyridyl)-methyl-isopropylphenylglycolate (302668)

CAS: 93101-83-8 (Hydrochloride salt)

RTECS: —

UN: —

EC: —

UNII: —

ICD-11:

$C_{18} H_{25} N O_3$

Specific information is not available for this material. Various salts (solids) have been reported.

Exposure Hazards

Conversion Factor: 1 ppm = 12.41 mg/m^3 at 77°F

LD_{50}: 35 mg (route unspecified)

ID_{50}: 0.6 mg (route unspecified)

Properties:

MW: 303.4	*VP:* —	*FIP:* —
D: —	*VD:* —	*LEL:* —
MP: —	*Vlt:* —	*UEL:* —

BP: —	*ER:* —	*RP:* —
Vsc: —	*H₂O:* —	*IP:* —
	Sol: —	

Hydrochloride:

MW: 339.9

MP: 340°F (est.)

VP: 7×10^{-8} mmHg (est.)

H₂O: 0.02% (est.)

C12-C   COMPONENTS AND PRECURSORS

C12-C032

3-Quinuclidinol

CAS: 1619-34-7

RTECS: VD6191700

UN: —

EC: 216-578-4

UNII: 974MVZ0WOK

ICD-11: —

$C_7 H_{13} N O$

White to beige crystalline powder that is odorless. This material produces local skin/eye impacts.

Used industrially as an intermediate in the synthesis of pharmaceuticals.

This material is on Schedule 2 of the CWC, the Australia Group Export Control list and DHS Chemicals of Interest list.

This material is also a component for several organophosphorus nerve agents (C01)

Exposure Hazards

Conversion Factor: 1 ppm = 5.20 mg/m³ at 77°F

Human toxicity values have not been established or have not been published.

Properties:

MW: 127.2	*VP:* 3×10^{-4} mmHg (77°F) (est.)	*FIP:* —
D: —	*VD:* —	*LEL:* —
MP: 424°F	*Vlt:* —	*UEL:* —

BP: Sublimes	*ER:* —	*RP:* —
Vsc: —	*H$_2$O:* Soluble	*IP:* ≤8.1 eV
	Sol: —	

C12-C033

3-Quinuclidinone

CAS: 3731-38-2; 1193-65-3 (Hydrochloride salt)

RTECS: —

UN: —

EC: 223-087-9; 214-776-5 (Hydrochloride salt)

UNII: —

ICD-11: —

C$_7$ H$_{11}$ N O

Solid. Various salts (solids) have been reported. This material produces local skin/eye impacts. This material is on the Australia Group Export Control list.

Exposure Hazards

Conversion Factor: 1 ppm = 5.12 mg/m^3 at 77°F

Human toxicity values have not been established or have not been published.

Properties:

MW: 125.2	*VP:* 0.2 mmHg (est.)	*FlP:* —
D: 1.1 g/cm^3	*VD:* 4.3	*LEL:* —
MP: 266°F	*Vlt:* —	*UEL:* —
BP: 400°F	*ER:* —	*RP:* —
Vsc: —	*H$_2$O:* 30% (est.)	*IP:* 8.20 eV
	Sol: —	

C12-C034

Benzilic Acid

CAS: 76-93-7

RTECS: DD2064000

UN: —

EC: 200-993-2

UNII: 8F6J993XXR

ICD-11: —

$C_{14} H_{12} O_3$

White to cream powder. This material produces local skin/eye impacts.

Used industrially as an intermediate in the synthesis of pharmaceuticals.

This material is on the Australia Group Export Control list and Schedule 2 of the CWC.

Exposure Hazards

Conversion Factor: 1 ppm = 9.33 mg/m^3 at 77°F

Human toxicity values have not been established or have not been published.

Properties:

MW: 228.2	*VP:* 2×10^{-7} mmHg (est.)	*FIP:* —
D: 1.1 g/cm^3	*VD:* —	*LEL:* —
MP: 300°F	*Vlt:* —	*UEL:* —
BP: 356°F (13 mmHg)	*ER:* —	*RP:* —
Vsc: —	*H$_2$O:* 0.14%	*IP:* —
	Sol: Ethanol, ethyl ether	

C12-C035

3-Hydroxy-1-methylpiperidine
CAS: 3554-74-3
RTECS: —
UN: —
EC: 222-609-2
UNII: —
ICD-11: —

$C_6 H_{13} N O$

Clear yellow liquid. This material produces local skin/eye impacts.

This material is on the Australia Group Export Control list.

Exposure Hazards

Conversion Factor: 1 ppm = 4.71 mg/m^3 at 77°F

Human toxicity values have not been established or have not been published.

Properties:

MW: 115.2	*VP:* 0.06 mmHg (est.)	*FIP:* 158°F
D: 1.0 g/mL (77°F)	*VD:* 4.0	*LEL:* —
MP: 103°F (est.)	*VIt:* —	*UEL:* —
BP: 171°F (11 mmHg)	*ER:* —	*RP:* —
Vsc: —	*H$_2$O:* Miscible	*IP:* —
	Sol: —	

C12-C036

Methyl Benzilate

CAS: 76-89-1

RTECS: —

UN: —

EC: 200-991-1

UNII: 692D4TH3BM

ICD-11: —

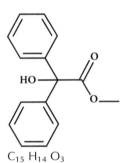

C$_{15}$ H$_{14}$ O$_3$

White to pale yellow crystalline solid. This material is hazardous through ingestion.

Used industrially as an intermediate in the synthesis of pharmaceuticals.

This material is on the Australia Group Export Control list.

Exposure Hazards

Conversion Factor: 1 ppm = 9.91 mg/m^3 at 77°F

Human toxicity values have not been established or have not been published.

Properties:

MW: 242.3	*VP:* 1.2×10^{-4} (est.)	*FIP:* —
D: 1.2 g/cm^3 (est.)	*VD:* —	*LEL:* —
MP: 163°F	*VIt:* —	*UEL:* —
BP: 369°F (13 mmHg)	*ER:* —	*RP:* —
Vsc: —	*H$_2$O:* Insoluble	*IP:* —
	Sol: Methanol	

BIBLIOGRAPHY

American Conference of Governmental Industrial Hygienists. *2020 TLVs and BEIs Based on the Documentation of the Threshold Limit Values for Chemical Substances and Physical Agents & Biological Exposure Indices*. Cincinnati, Ohio: ACGIH Signature Publications, 2020.

Bernstein, Jack and Kathryn A. Losee. "Dihydrodibenzocycloheptene Derivatives," United States Patent 3052721, September 04, 1962.

Centers for Disease Control and Prevention. "Biological and Chemical Terrorism: Strategic Plan for Preparedness and Response. Recommendations of the CDC Strategic Planning Workgroup." *Morbidity and Mortality Weekly Report* 49 (RR-4) (2000): 1–14.

Centers for Disease Control and Prevention "Case Definition: 3-Quinuclidinyl Benzilate (BZ)." March 11, 2005.

Centers for Disease Control and Prevention "Case Definition: Opioids (Fentanyl, Etorphine, or Others)" March 11, 2005.

Compton, James A.F. *Military Chemical and Biological Agents: Chemical and Toxicological Properties*. Caldwell, New Jersey: The Telford Press, 1987.

Gupta, Ramesh C., ed. *Handbook of Toxicology of Chemical Warfare Agents. 2nd Edition*. London, United Kingdom: Elsevier, 2015.

Harkins, Deanna, Rose Overturf, Veronique Hauschild and Scott Goodison. *Safety and Health Guidance for Mortuary Affairs Operations: Infectious Materials and CBRN Handling, Technical Guide 195*. Washington, D.C.: Government Printing Office, May 2009.

Judd, C. I., Helen A. Leiser, J. LaFrentz, and W. Hoya. "Chemical Study-Synthesis of Incapacitating Agents." Lakeside Laboratories, October 18, 1964.

Ketchum, James S. *Chemical Warfare: Secrets Almost Forgotten*. Santa Rosa, California: ChemBooks Inc., 2006.

Lakoski, Joan M., W. Bosseau Murray, and John M. Kenny. *The Advantages and Limitations of Calmatives for Use as Non-Lethal Technique*. College of Medicine, Applied Research Laboratory, The Pennsylvania State University, October 3, 2000.

Lenhart, Martha K., ed. *Medical Aspects of Chemical Warfare, Textbooks of Military Medicine Series*. Washington, D.C.: Office of the Surgeon General, Department of the Army, 2007.

Marrs, Timothy C., Robert L. Maynard, and Frederick R. Sidell, ed. *Chemical Warfare Agents: Toxicology and Treatment. 2nd Edition*. West Sussex, United Kingdom: John Wiley & Sons, Ltd., 2007.

National Institute for Occupational Safety and Health. *NIOSH Pocket Guide to Chemical Hazards*. Washington, D.C.: Government Printing Office, September 2007.

National Research Council. *Anticholinesterases and Anticholinergics*. Volume 1 of Possible Long-Term Health Effects of Short-Term Exposure to Chemical Agents. Washington, D.C.: National Academy Press, 1982.

National Research Council. *Cholinesterase Reactivators, Psychochemicals, and Irritants and Vesicants*. Volume 2 of Possible Long-Term Health Effects of Short-Term Exposure to Chemical Agents. Washington, D.C.: National Academy Press, 1984.

Olson, Kent R., ed. *Poisoning & Drug Overdose*. 4th Edition. New York, New York: Lange Medical Books/McGraw-Hill, 2004.

Perrine, Daniel M. *The Chemistry of Mind-Altering Drugs: History, Pharmacology, and Cultural Context*. Washington, D.C.: American Chemical Society, 1996.

Sidell, Fredrick R., Ernest T. Takafuji, and David R. Franz, ed. *Medical Aspects of Chemical and Biological Warfare, Textbook of Military Medicine Series, Part 1, Warfare, Weaponry, and the Casualty*. Washington, D.C.: Office of the Surgeon General, Department of the Army, 1997.

Sifton, David W. ed. *PDR Guide to Biological and Chemical Warfare Response*. Montvale, New Jersey: Thompson/Physicians Desk Reference, 2002.

Sommer, Harold Z., and Jacob I. Miller. "Quaternary Quinuclidinones," United States Patent 3919240, November 11, 1975.

Transport Canada, United States Department of Transportation, Secretariat of Transport and Communications of Mexico. *2020 Emergency Response Guidebook*. Neenah, Wisconsin: J.J. Keller and Associates, 2020.

True, Bey-Lorraine and Robert H. Dreisbach. *Dreisbach's Handbook of Poisoning: Prevention, Diagnosis and Treatment*. 13th Edition. London, United Kingdom: The Parthenon Publishing Group, 2002.

United States Army Headquarters. *Chemical Agent Data Sheets Volume I, Edgewood Arsenal Special Report No. EO-SR-74001*. Washington, D.C.: Government Printing Office, December 1974.

United States Army Headquarters. *Chemical Agent Data Sheets Volume II, Edgewood Arsenal Special Report No. EO-SR-74002*. Washington, D.C.: Government Printing Office, December 1974.

United States Army Headquarters. *Potential Military Chemical/Biological Agents and Compounds, Field Manual No. 3-11.9*. Washington, D.C.: Government Printing Office, January 10, 2005.

United States Army Medical Research Institute of Chemical Defense. *Medical Management of Chemical Casualties Handbook*. 3rd Edition. Aberdeen Proving Ground, Maryland: United States Army Medical Research Institute of Chemical Defense, July 2000.

United States Military Joint Chiefs of Staff. *Joint Tactics, Techniques, and Procedures for Mortuary Affairs in Joint Operations, Joint Publication No. 4-06*. Washington, D.C.: Government Printing Office, August 28, 1996.

United States National Institute of Health, National Library of Medicine. PubChem. 2020 [https://pubchem.ncbi.nlm.nih.gov/]. December 31, 2020.

Williams, Kenneth E. *Detailed Facts About Psychedelic Agent 3-Quinuclidinyl Benzilate (BZ)*. Aberdeen Proving Ground, Maryland: United States Army Center for Health Promotion and Preventive Medicine, 1996.

World Health Organization. International. *Health Aspects of Chemical and Biological Weapons: Report of A WHO Group of Consultants*. Geneva, Switzerland: World Health Organization, 1970.

World Health Organization. *Public Health Response to Biological And Chemical Weapons: WHO Guidance*. Geneva, Switzerland: World Health Organization, 2004.

13 Irritating and Lachrymatory Agents

GENERAL INFORMATION

The majority of these materials are alkylating agents that react with the moisture in the eyes to cause irritation. Under normal battlefield conditions, they do not pose a serious danger to the life of an exposed individual and do not produce any permanent injury. They were the first agents deployed in World War I. Many of these agents would not pass the current criteria to be classified as riot control agents under the Chemical Weapons Convention due to their toxicity. Since the end of World War I, numerous new agents have been developed, typically with greater irritating power and less toxicity. Under the general-purpose criterion of the Chemical Weapons Convention the use of irritating and lachrymatory agents is banned during a war. However, they may still be used by the military during operations other than war such as when responding to incidents of civil unrest. Lachrymatory agents are also used by police forces throughout the world to control rioters and disband unruly crowds. In some countries, agents can be purchased by individuals for personal protection.

TOXICOLOGY

EFFECTS

Irritating and lachrymatory agents cause intense eye pain and tears. They may also irritate the respiratory tract, causing the sensation that the casualty has difficulty breathing (dyspnea). In high concentrations, they can cause nausea, vomiting, and skin irritation with a burning or itching sensation. Very high concentrations can blister the skin. Direct contact with bulk agent can cause chemical burns. In an enclosed or confined space, very high aerosol concentrations can be lethal.

PATHWAYS AND ROUTES OF EXPOSURE

Irritating and lachrymatory agents are primarily an eye-contact and inhalation hazard. Aerosols and vapors are irritating to the eyes and skin at low concentrations but are otherwise relatively nontoxic via these routes. However, exposure to bulk liquid or solid agents may be hazardous through skin absorption, ingestion, and introduction through abraded skin (e.g., breaks in the skin or penetration of skin by debris).

GENERAL EXPOSURE HAZARDS

Irritating and lachrymatory agents have excellent warning properties. In general, they produce eye, respiratory, and/or skin irritation at concentrations well below lethal levels.

This class of agents docs not scriously endanger life except at exposures greatly exceeding an effective dose, usually only achieved in a confined or enclosed space.

LATENCY PERIOD

Exposure to irritating and lachrymatory agents produce immediate effects.

DOI: 10.4324/9781003230564-3

CHARACTERISTICS

Physical Appearance/Odor

Laboratory Grade

Laboratory grade agents are typically colorless to yellow liquids or solids. They typically have little or no odor when pure. If present, odors range from sweetish to floral to pepper-like. Most simply cause a burning sensation in the nose and nasal passages.

Munition Grade

Munition grade agents are typically off-white to yellow or even brown. As the agent ages and decomposes it continues to discolor. Production impurities and decomposition products in these agents may give them additional odors.

Modified Agents

Solvents have been added to these materials to increase the efficacy of the agent, facilitate handling, stabilize the agents or to aid in dispersing the agents. Typical solvents include propylene glycol, benzene, carbon tetrachloride, chloroform, and/or trioctylphosphite. Solvents may pose toxic hazards themselves (e.g., chloroform, carbon tetrachloride, and benzene). Color and other properties of these solutions may vary from the pure agent. Odors will vary depending on the characteristics of the solvent(s) used and concentration of agent in the solution.

Agents have also been micro pulverized, encapsulated, or treated with flowing agents such as silica aerogel to facilitate their dispersal and increase their persistency. Color and other physical properties of the agent may be affected by these additives.

Stability

Modern irritating and lachrymatory agents are stable and can be stored even under tropical conditions. Older agents, typically simple halogenated acyl or aryl compounds, tend to be sensitive to air, moisture and/or light. Some older agents are also prone to polymerization during storage. Stabilizers may be added to enhance stability and increase shelf life. Stabilizers include butyl-phenol, butylhydroquinone, amyl nitrate, and calcium carbonate. Modern agents can typically be stored in aluminum, glass, or steel containers. Older agents typically require glass, enamel-lined, or lead-lined containers. Many halogenated agents tend to deteriorate in the presence of metals such as iron and aluminum.

Persistency

As typically deployed, unmodified irritating and lachrymatory agents are classified as non-persistent by the military. However, bulk solid agents deployed for the purpose of area denial may persist for weeks or months. Depending on the size of the individual particles and on any encapsulation or coatings applied to the particles, they can be re-aerosolized by ground traffic or strong winds.

Environmental Fate

Many irritating and lachrymatory agents are nonvolatile and produce negligible amounts of vapor. Vapors of volatile agent have a density greater than air and tend to collect in low places.

Most agents are insoluble in water and have specific gravities that range from near water to greater than water. Lack of solubility inhibits reaction of these agents with water. Further, solvents used to disperse irritating and lachrymatory agents are generally insoluble in water and will help prevent interaction of the agent with water. Solvents may have densities less than or greater than water and may cause agents to either float or sink in a water column. Most of these agents are soluble in organic solvents including gasoline, alcohols and ketones.

Agents may be absorbed into porous material, including painted surfaces, and these materials may be difficult to decontaminate.

ADDITIONAL HAZARDS

EXPOSURE

All foodstuffs in the area of a release should be considered contaminated. Unopened items packaged in glass, metal, or heavy-duty plastic and exposed only to agent vapors or aerosols may be used after decontamination of the container. Unopened items exposed to solid or liquid agents, or solutions of agents, should be decontaminated within a few hours post-exposure or destroyed. Opened or unpackaged items, or those packaged only in paper or cardboard, should be destroyed.

Plants, fruits, and vegetables should be washed thoroughly with soap and water. Skins should be removed prior to use.

LIVESTOCK/PETS

Animals can be decontaminated with shampoo/soap and water (see the decontamination section below). If the animal's eyes have been exposed to agent, they should be irrigated with water or saline solution for a minimum of 30 minutes.

The topmost layer of unprotected feedstock (e.g., hay or grain) should be destroyed. The remaining material should be quarantined until tested.

FIRE

Heat from a fire will increase the amount of agent vapor in the area. A significant amount of the agent could be volatilized and escape into the surrounding environment before the agent is consumed by the fire. Actions taken to extinguish the fire can also spread the agent. Although many irritating and lachrymatory agents are only slightly soluble or insoluble in water, runoff from firefighting efforts will still pose a potential contact threat. Many decompose to produce toxic and/or corrosive gases such as hydrogen chloride (HCl), hydrogen cyanide (HCN), and phosgene ($COCl_2$). Some of the decomposition products resulting from hydrolysis or combustion of these agents are water soluble and highly toxic (see hazardous decomposition products below). In addition, solvents used in many formulations are highly flammable.

REACTIVITY

Irritating and lachrymatory agents either do not react with water or are very slowly decomposed by it. Some agents may be corrosive and react with metal. In some cases, these reactions may be violent. Most of these agents are incompatible with strong oxidizers, including household bleach, and may produce toxic decomposition products. Solvents used to disperse agents may be incompatible with strong oxidizers and may also decompose to form toxic and/or corrosive decomposition products.

HAZARDOUS DECOMPOSITION PRODUCTS

Hydrolysis

Irritating and lachrymatory agents are generally stable in or very slowly decomposed by water. Further, solvents used to disperse these agents are generally insoluble in water and will help prevent interaction of the agent with water. However, should hydrolysis occur, decomposition products may include hydrogen chloride (HCl), hydrogen cyanide (HCN), hydrogen bromide (HBr), and/or aromatic hydrocarbons, as well as other complex products.

Combustion

Volatile decomposition products may include hydrogen chloride (HCl), hydrogen bromide (HBr), hydrogen cyanide (HCN), phosgene ($COCl_2$), nitrogen oxides (NO_x), aromatic hydrocarbons such as benzene, and/or halogenated aromatic compounds.

PROTECTION

EVACUATION RECOMMENDATIONS

Isolation and protective action distances listed below are taken from the *2020 Emergency Response Guidebook*. For irritating and lachrymator agents, these recommendations are based on a release scenario involving direct aerosolization of the solid agents with a particle size between two and ten microns. For agents CA and CN, a small release involves 10 kilograms (approximately 0.02 cubic meters, a box approximately 29 centimeters on a side) of bulk agent and a large release involves 500 kilograms (approximately 1.7 cubic meters) of bulk agent. For CS, a small release involves ten kilograms (approximately 0.04 cubic meters, a box approximately 35 centimeters on a side) of bulk agent and a large release involves 100 kilograms (approximately 0.42 cubic meters, a box approximately 75 centimeters on a side) of bulk agent.

CA (Bromobenzyl Cyanide) *C13-A004*

	Initial Isolation	Downwind Day	Downwind Night
Small device (10 kg)	30 m	0.1 km	0.4 km
Large device (500 kg)	100 m	0.5 km	2.6 km

CN (Chloroacetophenone) *C13-A008*

	Initial Isolation	Downwind Day	Downwind Night
Small device (10 kg)	30 m	0.1 km	0.2 km
Large device (500 kg)	60 m	0.3 km	1.2 km

CS (o-Chlorobenzylmalononitrile) *C13-A009*

	Initial Isolation	Downwind Day	Downwind Night
Small device (10 kg)	30 m	0.1 km	0.6 km
Large device (100 kg)	100 m	0.4 km	1.9 km

PERSONAL PROTECTIVE REQUIREMENTS

Structural Firefighters' Gear

Structural firefighters' protective clothing is recommended for fire situations only; it is not effective in spill situations or release events. If chemical protective clothing is not available and it is necessary to rescue casualties from a contaminated area, then structural firefighters' gear will provide very limited skin protection against agent vapors and aerosols. Contact with solid and liquid agents should be avoided.

Respiratory Protection

Self-contained breathing apparatuses (SCBAs) or air purifying respirators (APRs) should have a National Institute for Occupational Safety and Health (NIOSH) Chemical/Biological/Radiological/ Nuclear (CBRN) certification. However, during emergency operations, other NIOSH-approved SCBAs or APRs that have been specifically tested by the manufacturer against chemical warfare agents may be used if deemed necessary by the incident commander. APRs should be equipped with a NIOSH-approved CBRN filter or a combination organic vapor/acid gas/particulate cartridge.

Immediately dangerous to life or health (IDLH) levels are the ceiling limit for respirators other than SCBAs. Any exposures approaching the IDLH level should be regarded with extreme caution and the use of SCBAs for respiratory protection should be considered.

Chemical Protective Clothing

Irritating and lachrymatory agents are primarily an eye and respiratory hazard; however, at elevated vapor/aerosol concentrations or in contact with bulk material, agents may also pose a dermal hazard. In addition, solvents used in agent formulations may pose respiratory or contact hazards.

Use only chemical protective clothing that has undergone material and construction performance testing against the specific agent that has been released. Since chemical protective clothing is tested against relatively pure agents, reported permeation rates may be affected by solvents, components, or impurities in munition grade agents.

DECONTAMINATION

General

Apply universal decontamination procedures using soap and water. If available, an alkaline soap/ detergent works best. Do not use bleach or detergents containing bleach as they may interact with agents to produce toxic decomposition products. Alternatively, an aqueous solution of sodium bicarbonate (i.e., baking soda) may be used.

Casualties should be warned that there may be a mild reaction between water and the agent and that they could experience a burning sensation during the decontamination process.

Vapors

Casualties/Personnel

Aeration and ventilation. If decontamination is deemed necessary, remove all clothing as it may contain trapped agent. Flush skin with cool water followed by showering with copious amounts of soap and warm water. If available, an alkaline soap/detergent works best. Do not use hot water as it will increase the burning sensation on the skin. Ensure that the hair has been washed and rinsed to remove potentially trapped vapor. For severe eye irritation, irrigate with water or 0.9% saline solution for a minimum of 15 minutes. Do not allow casualties to rub their eyes or skin as this may exacerbate agent effects.

Small areas

Ventilate to remove the vapors. If deemed necessary, wash the area with copious amounts of alkaline soap/detergent and water. Collect and place into containers lined with high-density polyethylene. Removal of porous material, including painted surfaces, may be required because these materials may be difficult to decontaminate.

Liquids, Solutions, or Liquid Aerosols

Casualties/Personnel

Remove all clothing immediately. To avoid further exposure of the head, neck, and face to the agent, cut off potentially contaminated clothing that must be pulled over the head. Flush skin with cool water. Ideally, showers will be high volume with low pressure. After flushing, use a sponge or cloth with liquid soap and copious amounts of water to wash the skin surface and hair at least three times. If available, an alkaline soap/detergent works best. Do not use hot water as it will increase the burning sensation on the skin. Avoid rough scrubbing as this could abrade the skin and increase discomfort. Rinse with copious amounts of water. For severe eye irritation, irrigate with water or 0.9% saline solution for a minimum of 15 minutes. Do not allow casualties to rub their eyes or skin as this may exacerbate agent effects.

Small areas

Ventilate to remove the aerosol. Small puddles of liquid can be contained by covering with absorbent material such as vermiculite, diatomaceous earth, clay, sponges, or towels. Place the absorbed material into containers lined with high-density polyethylene. Larger puddles can be collected using vacuum equipment made of materials inert to the agent and equipped with appropriate vapor filters. Wash the area with copious amounts of an alkaline soap/detergent and water. Collect and containerize the rinseate. Removal of porous material, including painted surfaces, may be required because these materials may be difficult to decontaminate. Ventilate the area to remove residual vapors.

Solids or Particulate Aerosols

Casualties/Personnel

Do not attempt to brush the agent off of the individual or their clothing as this can aerosolized the agent. If possible, dampen the agent with a water mist to help prevent aerosolization. Remove all clothing immediately. To avoid further exposure of the head, neck, and face to the agent, cut off potentially contaminated clothing that must be pulled over the head. Flush skin with cool water. Ideally, showers will be high volume with low pressure. After flushing, use a sponge or cloth with liquid soap and copious amounts of water to wash the skin surface and hair at least three times. If available, an alkaline soap/detergent works best. Do not use hot water as it will increase the burning sensation on the skin. Avoid rough scrubbing as this could abrade the skin and increase discomfort. Rinse with copious amounts of water. For severe eye irritation, irrigate with water or 0.9% saline solution for a minimum of 15 minutes. Do not allow casualties to rub their eyes or skin as this may exacerbate agent effects.

Small areas

If indoors, close windows and doors in the area and turn off anything that could create air currents (e.g., fans, air conditioner, etc.). Allow aerosol to settle. Avoid actions that could aerosolize the agent such as sweeping or brushing. Collected the agent using a vacuum cleaner equipped with a high-efficiency particulate air (HEPA) filter. Do not use a standard home or industrial vacuum. Do not allow the vacuum exhaust to stir the air in the affected area. Vacuum all surfaces with extreme care in a very slow and controlled manner to minimize aerosolizing the agent. Place the collected material into containers lined with high-density polyethylene. Wash the area with copious amounts of an alkaline soap/detergent and water. Collect and containerize the rinseate in containers lined with high-density polyethylene. Removal of porous material, including painted surfaces, may be required because these materials may be difficult to decontaminate.

MEDICAL

CDC CASE DEFINITION

No specific biologic marker/test is available for irritating and lachrymatory agents as a class; however, the case can be confirmed if they are detected in environmental samples. The case can be confirmed if laboratory testing is not performed by either a predominant amount of clinical and nonspecific laboratory evidence is present or an absolute certainty of the etiology of the agent is known.

DIFFERENTIAL DIAGNOSIS

The following factors have been suggested as alternatives to consider when presented with a potential case of exposure to irritating agents: anxiety, anaphylaxis, conjunctivitis, pneumonia, ultraviolet keratitis; thermal or chemical burns; inhalation of smoke, hydrocarbons, ammonia, hydrogen sulfide, phosgene, halogens (e.g., chlorine), sulfuric acid, hydrogen chloride, or nickel carbonyl; sodium azide; acute respiratory distress syndrome, chronic obstructive pulmonary disease, and emphysema; congestive heart failure; and pulmonary edema.

SIGNS AND SYMPTOMS

Vapors/Aerosols

Irritating and lachrymatory agents produce intense eye pain and tearing. They may also produce burning or stinging sensations of exposed mucous membranes (e.g., nose and mouth) and skin. Symptoms may also include rhinorrhea (runny nose), sneezing, coughing, respiratory discomfort (e.g., tightness of the chest or inability to breath), nausea, and/or vomiting. Increases in ambient temperature and/or humidity exacerbate agent effects. Effects from solvents will be minimal in comparison to the impacts caused by the actual agents themselves.

Solids/Solutions

General signs and symptoms may include intense pain of the eyes and mucous membranes, tearing, as well as localized irritation and burning of the skin.

MASS-CASUALTY TRIAGE RECOMMENDATIONS

Typically not required. Casualties will usually recover unassisted shortly after removal from the contaminated atmosphere. Consult the base station physician or regional poison control center for advice on specific situations.

CASUALTY MANAGEMENT

Decontaminate the casualty ensuring that all the irritating and lachrymatory agent has been removed. For severe eye irritation, irrigate with water or 0.9% saline solution for a minimum of 15 minutes. Do not allow casualties to rub their eyes or skin as this may exacerbate agent effects. Irrigate open wounds with water or 0.9% saline solution for at least 10 minutes.

Once the casualty has been decontaminated, including the removal of foreign matter from wounds, medical personnel do not need to wear a chemical-protective mask.

Casualties will usually recover unassisted from exposure to irritating agents within 15 minutes after removal from the contaminated atmosphere. Most patients can be discharged safely. Rarely a patient with significant respiratory findings may merit admission.

FATALITY MANAGEMENT

Remove all clothing and personal effects and decontaminate with soap and water. Do not use bleach or detergents containing bleach as they may interact with agents to produce toxic decomposition products.

If heavily contaminated, personnel may need to wear respiratory protection, gloves, and an apron during the decontamination process. Wash the remains with soap and water. Pay particular attention to areas where agent may get trapped, such as hair, scalp, pubic areas, fingernails, folds of skin, and wounds. If remains are heavily contaminated with residue, wash and rinse waste should be contained for proper disposal.

Once the remains have been thoroughly decontaminated, no further protective action is necessary. Body fluids removed during the embalming process do not pose any additional risks and should be contained and handled according to established procedures. Use standard burial procedures.

C13-A   AGENTS

C13-A001

Acrolein (Papite)

CAS: 107-02-8

RTECS: AS1050000

UN: 1092; 131P

EC: 203–453-4

UNII: 7864XYD3JJ

ICD-11: XM4HP5; XM2MJ1 (Gas)

$C_3 H_4 O$

Colorless to greenish-yellow liquid with a pungent, piercing, disagreeable odor detectable at ≥0.3 ppm. Unstable and prone to polymerization; often stabilized with amyl nitrate or hydroquinone. May form shock-sensitive peroxides during storage.

Used industrially as a pesticide, warning agent in refrigerants, and in the manufacturing of glycerol, polyurethane, polyester resins, and pharmaceuticals.

Exposure Hazards

Conversion Factor: 1 ppm = 2.29 mg/m^3 at 77°F

LCt$_{50}$ (Inh): 3,500 mg-min/m^3 (760 ppm for a 2-minute exposure)

Eye Irritation: 0.06 ppm; exposure duration unspecified

Intolerable Irritation: 22 ppm; exposure duration unspecified

These values are from older sources (circa 1937). No updated toxicity estimates have been proposed.

MEG$_{(1hr)}$: Neg: 0.03 ppm; Mar: 0.10 ppm; Crit: 1.4 ppm; Cat: —

PAC-1: 0.03 ppm; *PAC-2:* 0.10 ppm; *PAC-3:* 1.4 ppm

PEL: 0.1 ppm

STEL$_{(NIOSH)}$: 0.3 ppm

Ceiling$_{(ACGIH)}$: 0.1 ppm [Skin]
IDLH: 2.0 ppm

Final AEGLs
AEGL-1: 1 hr – 0.030 ppm; 4 hr – 0.030 ppm; 8 hr – 0.030 ppm
AEGL-2: 1 hr – 0.10 ppm; 4 hr – 0.10 ppm; 8 hr – 0.10 ppm
AEGL-3: 1 hr – 1.4 ppm; 4 hr – 0.48 ppm; 8 hr – 0.27 ppm

Properties:

MW: 56.1	*VP:* 210 mmHg	*FlP:* –15°F
D: 0.84 g/mL (59°F)	*VD:* 1.9	*LEL:* 2.8%
MP: –126°F	*Vlt:* 280,000 ppm	*UEL:* 31%
BP: 127°F	*ER:* 350	*RP:* 0.063
Vsc: 0.35 cS	*H$_2$O:* 40%	*IP:* 10.13 eV
	Sol: Alcohol, ether, acetone	

C13-A002

Benzyl Bromide (Cyclite)
CAS: 100-39-0
RTECS: XS7965000
UN: 1737; 156
EC: 202–847-3
UNII: XR75BS721D
ICD-11: —

C$_7$ H$_7$ Br

Colorless to yellow to brownish liquid with a pleasant and aromatic odor, resembling water cress.

Used industrially as a chemical intermediate.

Exposure Hazards
Conversion Factor: 1 ppm = 6.99 mg/m^3 at 77°F
LCt$_{50\ (Inh)}$: 45,000 mg-min/m^3 (3,220 ppm for a 2-minute exposure)
Eye Irritation: 0.6 ppm; exposure duration unspecified
Intolerable Irritation: 4.3 ppm; exposure duration unspecified

These values are from older sources (circa 1937). No updated toxicity estimates have been proposed.

MEG$_{(1hr)}$: Neg: 0.21 ppm; Mar: 1.9 ppm; Crit: 8.5 ppm; Cat: —

Properties:

MW: 171.0	*VP:* 0.450 mmHg	*FlP:* 174°F
D: 1.4 g/mL	*VD:* 5.9	*LEL:* —
MP: 27°F	*Vlt:* 600 ppm	*UEL:* —

BP: 394°F	*ER:* —	*RP:* 17
Vsc: —	*H₂O:* 0.0385% (slowly decomposes)	*IP:* 8.99 eV
	Sol: Alcohol, ether, benzene	

C13-A003

Bromoacetone (Agent BA)

CAS: 598-31-2
RTECS: UC0525000
UN: 1569; 131
EC: 209–928-2
UNII: 3O8L0EWR5Q
ICD-11: —

$C_3 H_5 Br O$

Colorless to violet liquid with a pungent odor. It is unstable and decomposed by heat and light.

Used industrially as a chemical intermediate.

Exposure Hazards

Conversion Factor: 1 ppm = 5.60 mg/m³ at 77°F

LCt₅₀ (Inh): 32,000 mg-min/m³ (2,860 ppm for a 2-minute exposure)

Eye Irritation: 0.27 ppm; exposure duration unspecified

Intolerable Irritation: 1.8 ppm; exposure duration unspecified

Liquid agent produces blisters on exposed skin.

These values are from older sources (circa 1937). No updated toxicity estimates have been proposed.

MEG₍₁ₕᵣ₎: Neg: 0.01 ppm; Mar: 0.32 ppm; Crit: 0.98 ppm; Cat: —

PAC-1: 0.011 ppm; *PAC-2:* 0.32 ppm; *PAC-3:* 0.98 ppm

Properties:

MW: 137.0	*VP:* 9 mmHg	*FIP:* 124°F
D: 1.6 g/mL (73°F)	*VD:* 4.7	*LEL:* —
MP: −34°F	*Vlt:* 13,000 ppm	*UEL:* —
BP: 279°F	*ER:* —	*RP:* 0.9
BP: 146°F (50 mmHg)	*H₂O:* Slight	*IP:* 9.73 eV
Vsc: —	*Sol:* Alcohol, acetone, ether, benzene	

C13-A004

Bromobenzyl Cyanide (Agent CA)

CAS: 5798-79-8
RTECS: AL8090000
UN: 1694; 159
EC: 227–348-8

UNII: 7JP1R2F6C6
ICD-11: XM9FN5

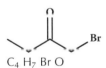

$C_8 H_6 Br N$

Yellow solid or yellow to brown liquid with an odor like soured or rotting fruit detectable at ≥0.01 ppm. Undergoes considerable decomposition when a large explosive charge is used to disseminate the agent.

Exposure Hazards

Conversion Factor: 1 ppm = 8.02 mg/m^3 at 77°F

LCt$_{50 (Inh)}$: 8,000–11,000 mg-min/m^3 (500–690 ppm for a 2-minute exposure)

ICt$_{50 (Eyes)}$: 30 mg-min/m^3 (1.9 ppm for a 2-minute exposure)

Eye Irritation: 0.04 ppm; exposure duration unspecified

Properties:

MW: 196.0	*VP:* 0.012 mmHg	*FIP:* NA
D: 1.5 g/cm^3	*VD:* 6.8	*LEL:* NA
MP: 84°F	*Vlt:* 34 ppm (86°F)	*UEL:* NA
BP: 468°F (decomposes)	*ER:* —	*RP:* 640
Vsc: —	*H$_2$O:* Slightly	*IP:* <10 eV
	Sol: Common organic solvents	

C13-A005

Bromomethylethyl Ketone (Bn-Stoff)

CAS: 816-40-0
RTECS: EL7000000
UN: —
EC: 212–431-3
UNII: —
ICD-11: —

$C_4 H_7 Br O$

Colorless to light yellow liquid.

Exposure Hazards

Conversion Factor: 1 ppm = 6.18 mg/m^3 at 77°F

LCt$_{50 (Inh)}$: 20,000 mg-min/m^3 (1,600 ppm for a 2-minute exposure)

Eye Irritation: 2 ppm; exposure duration unspecified

Intolerable Irritation: 2.6 ppm; exposure duration unspecified

These values are from older sources (circa 1937). No updated toxicity estimates have been proposed.

Properties:

MW: 151.0	*VP:* 15 mmHg (57°F)	*FIP:* 154°F
D: 1.5 g/mL	*VD:* 5.2	*LEL:* —
MP: —	*Vlt:* 20,000 ppm	*UEL:* —
BP: 293°F (decomposes)	*ER:* —	*RP:* 0.53
Vsc: —	*H₂O:* Insoluble	*IP:* —
	Sol: —	

C13-A006

Capsaicin (Agent OC)

CAS: 404-86-4

RTECS: RA8530000

UN: —

EC: 206–969-8

UNII: S07O44R1ZM

ICD-11: XM9TT3; XE203 (Pepper spray)

$C_{18} H_{27} N O3$

White crystalline solid.

Used medicinally and as a food flavoring.

Exposure Hazards

Conversion Factor: 1 ppm = 12.49 mg/m^3 at 77°F

$LD_{50 (Ing)}$: 350 g (estimate)

Properties:

MW: 305.4	*VP:* 1 × 10^{-8} mmHg (77°F) (est.)	*FIP:* 235°F
D: —	*VD:* —	*LEL:* —
MP: 149°F	*Vlt:* —	*UEL:* —
BP: 410°F (0.01 mmHg)	*ER:* —	*RP:* 38,000,000
Vsc: —	*H₂O:* Practically insoluble	*IP:* —
	Sol: Alcohol, ether, benzene, chloroform	

C13-A007

Chloroacetone (Tonite)

CAS: 78–95-5

RTECS: UC0700000

UN: 1695; 131

EC: 201-161-1

UNII: 60ZTR74268

ICD-11: XM2N89

C₃ H₅ Cl O

Colorless to light yellow liquid with a pungent odor similar to hydrochloric acid. Readily breaks down during storage; often stabilized with water or calcium carbonate. Turns dark and forms resins on prolonged exposure to light.

Used industrially as a chemical intermediate in the manufacture of couplers for color photography, as a photo polymerization agent for vinyl compounds, as a solvent, and as an enzyme inactivator in biological research.

This material is on the ITF-25 Medium Threat list.

Exposure Hazards

Conversion Factor: 1 ppm = 3.78 mg/m^3 at 77°F

LCt$_{50}$ $_{(Inh)}$: 23,000 mg-min/m^3 (3,040 ppm for a 2-minute exposure)

Eye Irritation: 4.8 ppm; exposure duration unspecified

Intolerable Irritation: 26 ppm; exposure duration unspecified

These values are from older sources (circa 1937). No updated toxicity estimates have been proposed.

MEG$_{(1hr)}$: Neg: 0.53 ppm; Mar: 4.5 ppm; Crit: 13 ppm; Cat: —

PAC-1: 0.40 ppm; *PAC-2:* 4.5 ppm; *PAC-3:* 13 ppm

Ceiling$_{(ACGIH)}$: 1 ppm [Skin]

Final AEGLs

AEGL-1: Not Developed

AEGL-2: 1 hr – 4.4 ppm; 4 hr – 1.1 ppm; 8 hr – 0.53 ppm

AEGL-3: 1 hr – 13 ppm; 4 hr – 3.3 ppm; 8 hr – 1.6 ppm

Properties:

MW: 92.5	VP: 12.0 mmHg (77°F)	FlP: 104°F
D: 1.2 g/mL (77°F)	VD: 3.7	LEL: 3.4%
MP: −48°F	Vlt: 15,800 ppm (77°F)	UEL: —
BP: 248°F	ER: —	RP: 0.88
Vsc: 0.96 cS (77°F)	H₂O: 12.4%	IP: 9.92 eV
	Sol: Alcohols, ether, chloroform	

C13-A008

Chloroacetophenone (Agent CN)

CAS: 532-27-4

RTECS: AM6300000

UN: 1697; 153

EC: 208–531-1

UNII: 88B5039IQG

ICD-11: XM7J14

$C_8 H_7 Cl O$

Colorless to white or gray crystalline solid with a sharp, fragrant odor like apple blossoms that is detectable at ≥0.02 ppm.

Used industrially as a chemical intermediate for pharmaceuticals and as a denaturant for alcohol. Also reported as a mixture with various solvents including CNB (mixture of 10% chloroaceto-phenone, 45% benzene and 45% carbon tetrachloride), CNC (mixture of 30% chloroacetophenone and 70% chloroform), CND (mixture of chloroacetophenone and ethylene dichloride), and CNS (C13-A015).

This material is on the NRL Critical Threat list.

Exposure Hazards
Conversion Factor: 1 ppm = 6.32 mg/m^3 at 77°F
LCt$_{50}$ (Inh): 7,000 mg-min/m^3 when dispersed as a solution of the agent dissolved in a solvent.
LCt$_{50}$ (Inh): 14,000 mg-min/m^3 when dispersed as an aerosol from a thermal device.
ICt$_{50}$ (Eyes): 80 mg-min/m^3
Eye Irritation: 0.13 – 0.38 mg/m^3; exposure duration unspecified

Exposure to high concentrations of aerosolized agent can cause severe skin irritation and even produce blistering similar to vesicants (C03 – C04).

PEL: 0.3 mg/m^3
REL: 0.3 mg/m^3
TLV: 0.05 ppm
IDLH: 15 mg/m^3

Properties:

MW: 154.6	*VP:* 0.0041 mmHg	*FlP:* 244°F
D: 1.3 g/cm^3	*VD:* 5.3	*LEL:* —
MP: 129°F	*Vlt:* 5.4 ppm	*UEL:* —
BP: 470°F	*ER:* —	*RP:* 2,000
Vsc: —	*H$_2$O:* Insoluble	*IP:* 9.44 eV
	Sol: Most organic solvents	

C13-A009

o-Chlorobenzylmalononitrile (Agent CS)
CAS: 2698-41-1
RTECS: OO3675000
UN: —
EC: 220–278-9
UNII: D8317IAV7Q
ICD-11: XM1N99

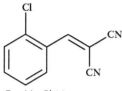

$C_{10} H_5 Cl N_2$

Colorless solid with a pungent, pepper-like odor.

Also reported as a 1% mixture in trioctylphosphite (Agent CSX).

Exposure Hazards

Conversion Factor: 1 ppm = 7.71 mg/m^3 at 77°F

LCt$_{50\ (Inh)}$: 52,000 mg-min/m^3–61,000 mg-min/m^3. This is a provisional update from an older value that has not been formally adopted.

Intolerable Irritation: 7 mg/m^3; exposure duration unspecified

MEG$_{(1hr)}$: Neg: 0.050 mg/m^3; Mar: 0.50 mg/m^3; Crit: 11 mg/m^3; Cat: —

PAC-1: 0.0050 mg/m^3; *PAC-2:* 0.083 mg/m^3; *PAC-3:* 11 mg/m^3

PEL: 0.4 mg/m^3

Ceiling$_{(ACGIH)}$: 0.4 mg/m^3 [Skin]

Ceiling$_{(NIOSH)}$: 0.4 mg/m^3 [Skin]

IDLH: 2.0 mg/m^3 [Skin]

Final AEGLs

AEGL-1: Not Developed

AEGL-2: 1 hr – 0.083 mg/m^3; 4 hr – 0.083 mg/m^3; 8 hr – 0.083 mg/m^3

AEGL-3: 1 hr – 11 mg/m^3; 4 hr – 1.5 mg/m^3; 8 hr – 1.5 mg/m^3

Properties:

MW: 188.6	*VP:* 3.4 × 10^{-5} mmHg	*FIP:* 387°F
D: 1.0 g/cm^3	*VD:* 6.5	*MEC:* 25 g/m^3
D: 0.24 g/cm^3 (munition grade)	*Vlt:* —	*UEL:* —
MP: 203°F	*ER:* —	*RP:* 21,000
BP: 590°F (decomposes)	*H$_2$O:* 0.008% (77°F)	*IP:* <10 eV
Vsc: —	*Sol:* Acetone, hexane, benzene, methylene chloride	

C13-A010

Chloromethyl Chloroformate (Palite)

CAS: 22128-62-7

RTECS: —

UN: 2745; 157

EC: 244–793-3

UNII: —

ICD-11:

$C_2 H_2 Cl_2 O_2$

Colorless to pale yellow liquid with a pungent, irritating, acrid odor.

Used industrially as a chemical intermediate.

Also reported as a mixture with dichloromethyl chloroformate (C13-A018) and also stannic chloride.

Exposure Hazards

Conversion Factor: 1 ppm = 5.27 mg/m^3 at 77°F

LCt$_{50}$ (Inh): 10,000 mg-min/m^3 (950 ppm for a 2-minute exposure)

ICt$_{50}$ (Eyes): 50 mg-min/m^3 (4.7 ppm for a 2-minute exposure)

These values are from older sources (circa 1937) and are not supported by modern data. No updated toxicity estimates have been proposed.

Eye Irritation: 0.38 ppm; exposure duration unspecified

Properties:

MW: 128.9	*VP:* 26.5 mmHg (77°F)	*FIP:* 203°F
D: 1.4 g/mL	*VD:* 4.4	*LEL:* —
MP: −74°F	*Vlt:* 7,600 ppm	*UEL:* —
BP: 225°F	*ER:* —	*RP:* 0.97
Vsc: —	*H$_2$O:* Reacts	*IP:* —
	Sol: Hydrocarbon solvents	

C13-A011

Dibenz-(b,f)-1,4-oxazepine (Agent CR)

CAS: 257-07-8

RTECS: HQ3950000

UN: —

EC: —

UNII: C1Q77A87V1

ICD-11: —

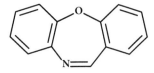

C$_{13}$ H$_9$ N O

Yellow needles or brown powder with a peppery odor.

Exposure Hazards

Conversion Factor: 1 ppm = 7.98 mg/m^3 at 77°F

ICt$_{50}$ (Eyes): 0.15 mg-min/m^3

Eye Irritation: 0.002 mg/m^3–0.004 mg/m^3 for a 1-minute exposure

Properties:

MW: 195.2	*VP:* 5.9 x 10^{-5} mmHg	*FIP:* 370°F
D: 1.6 g/cm^3	*VD:* 6.7	*LEL:* —
MP: 160°F	*Vlt:* —	*UEL:* —

BP: 634°F   ER: —   RP: 120,000
Vsc: —   H₂O: 0.008%   IP: <9 eV
                Sol: Ethanol, ether, benzene, chloroform

C13-A012

Ethyl Iodoacetate (Agent SK)
CAS: 623-48-3
RTECS: AI3575000
UN: —
EC: 210–796-3
UNII: —
ICD-11: XM41V6

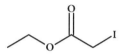

$C_4 H_7 I O_2$

Colorless oily liquid that is light and moisture sensitive. Becomes brown during storage due to liberation of iodine.

Also reported as a mixture with chloropicrin (C10-A006), and as a solution dissolved in ethyl acetate and ethanol (agent KSK)

Exposure Hazards
Conversion Factor: 1 ppm = 8.75 mg/m³ at 77°F
LCt$_{50}$ $_{(Inh)}$: 15,000 mg-min/m³ (860 ppm for a 2-minute exposure)
Intolerable Irritation: 1.7 ppm; exposure duration unspecified

These values are from older sources (circa 1939) and are not supported by modern data. No updated toxicity estimates have been proposed.

Properties:
MW: 214.0   VP: 0.54 mmHg   FIP: 169°F
D: 1.8 g/mL   VD: 7.4   LEL: —
MP: –7°F   Vlt: 350 ppm   UEL: —
BP: 355°F   ER: —   RP: 13
Vsc: —   H₂O: Reacts   IP: —
                 Sol: —

C13-A013

Iodoacetone (Bretonite)
CAS: 3019-04-3
RTECS: —
UN: —
EC: 221-161-5
UNII: —
ICD-11: —

C₃ H₅ I O

Faintly yellow liquid. Becomes brown during storage due to liberation of iodine. Decomposes on standing to 1,3-Diiodopropanone.

Also reported as a mixture with stannic chloride.

Exposure Hazards

Conversion Factor: 1 ppm = 7.53 mg/m^3 at 77°F

LCt$_{50}$ (Inh): 19,000 mg-min/m^3 (1,300 ppm for a 2-minute exposure)

ICt$_{50}$ (Eyes): 100 mg-min/m^3 (6.6 ppm for a 2-minute exposure)

Intolerable Irritation: 13 ppm; exposure duration unspecified

These values are from older sources (circa 1937) and are not supported by modern data. No updated toxicity estimates have been proposed.

Properties:

MW: 184.0	*VP:* 7 mmHg (est.)	*FIP:* 126°F
D: 1.8 g/mL	*VD:* 6.3	*LEL:* —
MP: −19°F (est.)	*Vlt:* 400 ppm	*UEL:* —
BP: 216°F	*ER:* —	*RP:* —
Vsc: —	*H₂O:* —	*IP:* 9.3 eV
	Sol: —	

C13-A014

1-Methoxy-1,3,5-cycloheptatriene (Agent CH)

CAS: 1714-38-1

RTECS: —

UN: —

EC: —

UNII: —

ICD-11: —

C₈ H₁₀ O

Colorless to brown liquid with a sweetish odor. Various salts have been reported.

Also reported as a mixture with Agent EA 3834 (C12-A006).

Exposure Hazards

Conversion Factor: 1 ppm = 5.00 mg/m^3 at 77°F

Human toxicity values have not been established or have not been published.

Properties:

MW: 122.2	*VP:* 1.3 mmHg (77°F)	*FIP:* 133°F
D: 0.97 g/mL (77°F)	*VD:* 4.2	*LEL:* —
MP: −153°F	*Vlt:* 1,700 ppm (77°F)	*UEL:* —
BP: 345°F	*ER:* —	*RP:* 7
Vsc: 1.50 cS (77°F)	H_2O: 0.072%	*IP:* 7.23 eV
	Sol: Aromatic organic solvents	

C13-A015

CNS

CAS: 675600-78-9
RTECS: —
UN: —
EC: —
UNII: —
ICD-11: —

Liquid mixture of 38.4% chloropicrin (C10-A006), 23% chloroacetophenone (C13-A008), and 38.4% chloroform. Historically, the odor of this mixture has been compared to flypaper.

Exposure Hazards

$LCt_{50\ (Inh)}$: 11,400 mg-min/m^3

$ICt_{50\ (Eyes)}$: 60 mg-min/m^3

In addition to lacrimation, the chloropicrin component may cause pulmonary edema, vomiting, nausea, and diarrhea.

Properties:

MW: Mixture	*VP:* 78 mmHg	*FIP:* NA
D: 1.5 g/mL	*VD:* 5	*LEL:* NA
MP: 36°F (precipitation occurs)	*Vlt:* —	*UEL.* NA
BP: 140°F	*ER:* —	*RP:* —
Vsc: —	H_2O: Insoluble	*IP.* —
	Sol: —	

C13-A016

o-Nitrobenzyl Chloride (Cedenite)
CAS: 612-23-7
RTECS: XS9092000
UN: —
EC: 210–300-5

UNII: HB8U484NPM

ICD-11: —

$C_7\ H_6\ Cl\ N\ O_2$

Pale yellow crystalline solid that is odorless.

Used industrially for photographic imaging and in medicinal research.

Exposure Hazards

Conversion Factor: 1 ppm = 7.02 mg/m^3 at 77°F

ICt$_{50\ (Eyes)}$: 15 mg-min/m^3

Eye Irritation: 1.8 mg/m^3; exposure duration unspecified

These values are from older sources (circa 1939) and are not supported by modern data. No updated toxicity estimates have been proposed.

Properties:

MW: 171.6	*VP:* 0.03 mmHg (est.)	*FlP:* 234°F
D: 1.6 g/cm^3	*VD:* 5.9	*LEL:* —
MP: 118°F	*Vlt:* —	*UEL:* —
BP: 261°F (10 mmHg)	*ER:* —	*RP:* —
Vsc: —	*H$_2$O:* Insoluble	*IP:* <10 eV
	Sol: Alcohol, ether, benzene	

C13-A017

Thiophosgene (Lacrymite)

CAS: 463-71-8

RTECS: XN2450000

UN: 2474; 157

EC: 207–341-6

UNII: 067FQP576P

ICD-11: —

$C\ Cl_2\ S$

Reddish-yellow liquid with a sharp, choking, foul odor. Air and moisture sensitive; decomposes above 392°F to carbon disulfide and carbon tetrachloride.

Used industrially as a chemical intermediate.

Exposure Hazards

Conversion Factor: 1 ppm = 4.70 mg/m^3 at 77°F

Human toxicity values have not been established or have not been published.

Properties:

MW: 115.0	*VP:* 127 mmHg (77°F)	*FlP:* 144°F
D: 1.5 g/mL	*VD:* 4.0	*LEL:* —
MP: –20°F (est.)	*Vlt:* 170,000 ppm (77°F)	*UEL:* —
BP: 163°F	*ER:* 320	*RP:* 0.07
Vsc: —	*H$_2$O:* Insoluble (decomposes)	*IP:* 9.68 eV
	Sol: Ether	

C13-A018

Dichloromethyl Chloroformate

CAS: 22128-63-8

RTECS: —

UN: —

EC: —

UNII: —

ICD-11: —

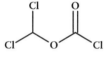

C$_2$ H Cl$_3$ O$_2$

Colorless liquid.

Also reported as a mixture with chloromethyl chloroformate (C13-A010)

Exposure Hazards

Conversion Factor: 1 ppm = 6.68 mg/m^3 at 77°F

Intolerable Irritation: 11 ppm; exposure duration unspecified

This value is from older sources (circa 1939) and is not supported by modern data. No updated toxicity estimates have been proposed.

Propertles:

MW: 163.4	*VP:* 5 mmHg	*FlP:* —
D: 1.8 g/mL (59°F)	*VD:* 5.6	*LEL:* —
MP: —	*Vlt:* 6,700 ppm	*UEL:* —
BP: 230°F	*ER:* —	*RP:* 1.6
Vsc: —	*H$_2$O:* —	*IP:* —
	Sol: —	

C13-A019

Benzyl Chloride

CAS: 100-44-7

RTECS: XS8925000

UN: 1738; 156
EC: 202–853-6
UNII: 83H19HW7K6
ICD-11: —

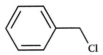

$C_7 H_7 Cl$

Colorless to slightly yellow liquid with a pungent, aromatic odor.

Used industrially as a chemical intermediate in the manufacture of pharmaceuticals, perfumes, dyes, synthetic tannins, and artificial resins.

Exposure Hazards

Conversion Factor: 1 ppm = 5.18 mg/m^3 at 77°F

Intolerable Irritation: 16 ppm; exposure duration unspecified

This value is from older sources (circa 1939) and is not supported by modern data. No updated toxicity estimates have been proposed.

MEG$_{(1hr)}$: Neg: 1.0 ppm; Mar: 10 ppm; Crit: 50 ppm; Cat: —

PAC-1: 1.0 ppm; *PAC-2:* 10 ppm; *PAC-3:* 50 ppm

PEL: 1 ppm

TLV: 1 ppm

Ceiling$_{(NIOSH)}$: 1 ppm

IDLH: 10 ppm

Properties:

MW: 126.6	*VP:* 1.23 mmHg (77°F)	*FIP:* 153°F
D: 1.1 g/mL	*VD:* 4.4	*LEL:* 1.1%
MP: −49°F	*Vlt:* 1,200 ppm	*UEL:* 7.1%
BP: 354°F	*ER:* —	*RP:* 7.3
Vsc: 1.15 cS	*H$_2$O:* 0.0525%	*IP:* 9.14 eV
	Sol: Most organic solvents	

C13-A020

Benzyl Iodide
CAS: 620-05-3
RTECS: —
UN: 2653; 156
EC: 210–623-1
UNII: —
ICD-11: —

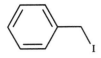

$C_7 H_7 I$

White crystalline solid or colorless liquid.

Exposure Hazards

Conversion Factor: 1 ppm = 8.92 mg/m^3 at 77°F

LCt$_{50\ (Inh)}$: 30,000 mg-min/m^3 (1,700 ppm for a 2-minute exposure)

Eye Irritation: 0.2 ppm; exposure duration unspecified

Intolerable Irritation: 3.4 ppm for a 1-minute exposure

These values are from older sources (circa 1939) and are not supported by modern data. No updated toxicity estimates have been proposed.

Properties:

MW: 218.0	*VP:* 0.125 mmHg	*FIP:* 187°F
D: 1.7 g/cm^3	*VD:* 7.5	*LEL:* —
MP: 75°F	*Vlt:* 130 ppm	*UEL:* —
BP: 438°F (decomposes)	*ER:* —	*RP:* 64
Vsc: —	*H$_2$O:* Insoluble	*IP:* 8.73 eV
	Sol: Alcohol, ether, benzene	

C13-A021

Ethyl Bromoacetate

CAS: 105-36-2

RTECS: AF6000000

UN: 1603; 155

EC: 203–290-9

UNII: D20KFB313W

ICD-11: —

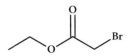

C$_4$ H$_7$ Br O$_2$

Clear colorless liquid with a pungent odor.

Used industrially as a chemical intermediate.

Exposure Hazards

Conversion Factor: 1 ppm = 6.83 mg/m^3 at 77°F

LCt$_{50\ (Inh)}$: 23,000 mg-min/m^3 (1,700 ppm for a 2-minute exposure)

Eye Irritation: 0.44 ppm; exposure duration unspecified

Intolerable Irritation: 5.9 ppm; exposure duration unspecified

These values are from older sources (circa 1937) and are not supported by modern data. No updated toxicity estimates have been proposed.

MEG$_{(1hr)}$: Neg: 0.0030 ppm; Mar: 0.019 ppm; Crit: 0.088 ppm; Cat: —

PAC-1: 0.18 ppm; *PAC-2:* 1.9 ppm; *PAC-3:* 12 ppm

Properties:

MW: 167.0	*VP:* 2.6 mmHg (77°F)	*FIP:* 118°F
D: 1.5 g/mL	*VD:* 5.8	*LEL:* —

MP: –36°F	*Vlt:* 3,100 ppm	*UEL:* —
BP: 318°F	*ER:* —	*RP:* 2.3
Vsc: 1.13 cS	*H₂O:* 0.7%	*IP:* —
	Sol: Acetone, benzene, ethanol, ether	

C13-A022

N-Ethylcarbazole
CAS: 86-28-2
RTECS: FE6225700
UN: —
EC: 201–660-4
UNII: 6AK165L0RO
ICD-11: —

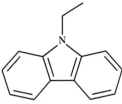

C₁₄ H₁₃ N

White to brown flaky solid.

Exposure Hazards
Conversion Factor: 1 ppm = 7.99 mg/m³ at 77°F

Human toxicity values have not been established or have not been published.

Properties:

MW: 195.3	*VP:* 0.02 mmHg (167°F)	*FlP:* 367°F
D: 1.2 g/cm³	*VD:* 6.7	*LEL:* —
MP: 154°F	*Vlt:* —	*UEL:* —
BP: 374°F	*ER:* —	*RP:* 420
Vsc: —	*H₂O:* Insoluble	*IP:* —
	Sol: Alcohol, ether	

C13-A023

Xylyl Bromide (T-Stoff)
CAS: 89–92-9 (Ortho); 620-13-3 (Meta); 104-81-4 (Para)
RTECS: —
UN: 1701; 152
EC: 201–951-6 (Ortho); 210–625-2 (Meta); 203–240-6 (Para)
UNII: 3NDQ4L309U (Ortho); U49150ER3L (Meta); V2DIK2AP5I (Para)
ICD-11: —

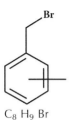

$C_8 H_9 Br$

Mixture of the ortho, meta, and para isomers. Colorless to light yellow, slightly viscous liquid with a pungent and aromatic odor, resembling lilacs or elder blossoms. Weapons grade material is a black liquid.

Used industrially as a chemical intermediate.

Exposure Hazards

Conversion Factor: 1 ppm = 7.57 mg/m^3 at 77°F

LCt$_{50\ (Inh)}$: 56,000 mg-min/m^3 (3,700 ppm for a 2-minute exposure)

Eye Irritation: 0.24 ppm; exposure duration unspecified

Intolerable Irritation: 2.0 ppm; exposure duration unspecified

These values are from older sources (circa 1937) and are not supported by modern data. No updated toxicity estimates have been proposed.

Properties:

MW: 185.1	*VP:* —	*FIP:* —
D: 1.4 g/mL	*VD:* 6.4	*LEL:* —
MP: —	*Vlt:* 80 ppm	*UEL:* —
BP: 410°–428°F	*ER:* —	*RP:* —
Vsc: —	*H$_2$O:* Insoluble (decomposes)	*IP:* —
	Sol: —	

C13-A024

Xylylene Dibromide

CAS: 91-13-4 (Ortho); 626-15-3 (Meta); 623-24-5 (Para)

RTECS: —

UN: —

EC: 202-042-7 (Ortho); 210–931-6 (Meta); 210–781-1 (Para)

UNII: 9FRJ55E5UL (Ortho)

ICD-11: —

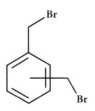

$C_8 H_8 Br_2$

Mixture of the ortho, meta and para isomers. White to beige crystalline solid.

Exposure Hazards
Human toxicity values have not been established or have not been published.

Properties:

MW: 264.0	*VP:* Negligible	*FlP:* —
D: 2 g/cm³	*VD:* —	*LEL:* —
MP: —	*Vlt:* —	*UEL:* —
BP: —	*ER:* —	*RP:* —
Vsc: —	*H₂O:* Insoluble	*IP:* —
	Sol: —	

BIBLIOGRAPHY

Agency for Toxic Substances and Disease Registry. *Toxicological Profile for Acrolein.* Washington, D.C.: Government Printing Office, August 2007.

American Conference of Governmental Industrial Hygienists. *2020 TLVs and BEIs Based on the Documentation of the Threshold Limit Values for Chemical Substances and Physical Agents & Biological Exposure Indices.* Cincinnati, Ohio: ACGIH Signature Publications, 2020.

Brophy, Leo P., Wyndham D. Miles and Rexmond C. Cohrane. *The Chemical Warfare Service: From Laboratory to Field.* Washington, D.C.: Government Printing Office, 1968.

Compton, James A.F. *Military Chemical and Biological Agents: Chemical and Toxicological Properties.* Caldwell, New Jersey: The Telford Press, 1987.

Edgewood Research Development, and Engineering Center, Department of the Army. *Material Safety Data Sheet (MSDS) for Agent CS.* Aberdeen Proving Ground, Maryland: Chemical Biological Defense Command, Revised 30 June 1995.

Edgewood Research Development, and Engineering Center, Department of the Army. *Material Safety Data Sheet (MSDS) for Riot Control Agent CR.* Aberdeen Proving Ground, Maryland: Chemical Biological Defense Command, Revised 30 June 1995.

Fries, Amos A. and Clarence J. West. *Chemical Warfare.* New York, New York: McGraw-Hill Book Company, Inc., 1921.

Grant, George A. "Safe Sensory Irritant," United States Patent 4598096, July 1, 1986.

International Labour Organization. *International Chemical Safety Cards (ICSCs)* [https://www.ilo.org/safework/info/publications/WCMS_113134/lang--en/index.htm]. December 31, 2020.

Jackson, Kirby E., and Margaret A. Jackson. "Lachrymators." *Chemical Reviews* 16 (1935): 195–242.

Langford, Gordon E. "Scale-up and Synthesis of 1-Methoxycyglohepta-1,3,5-triene," United States Patent 4978806, December 18, 1990.

Lenhart, Martha K., ed. *Medical Aspects of Chemical Warfare, Textbooks of Military Medicine Series.* Washington, D.C.: Office of the Surgeon General, Department of the Army, 2007.

Marrs, Timothy C., Robert L. Maynard, and Frederick R. Sidell. *Chemical Warfare Agents: Toxicology and Treatment.* Chichester, England: John Wiley & Sons, 1997.

National Institute for Occupational Safety and Health. *NIOSH Pocket Guide to Chemical Hazards.* Washington, D.C.: Government Printing Office, September 2005.

Olson, Kent R., ed. *Poisoning & Drug Overdose.* 4th Edition. New York, New York: Lange Medical Books/McGraw-Hill, 2004.

Prentiss, Augustin M. *Chemicals in War: A Treatise on Chemical Warfare.* New York, New York: McGraw-Hill Book Company, Inc., 1937.

Sartori, Mario F. *The War Gases: Chemistry and Analysis.* Translated byL. W. Marrison. London, United Kingdom: J. & A. Churchill, Ltd., 1939.

Sidell, Frederick R. *Medical Management of Chemical Warfare Agent Casualties: A Handbook for Emergency Medical Services.* Bel Air, Maryland: HB Publishing, 1995.

Sidell, Fredrick R., Ernest T. Takafuji, and David R. Franz, ed. *Medical Aspects of Chemical and Biological Warfare, Textbook of Military Medicine Series, Part 1, Warfare, Weaponry, and the Casualty.* Washington, D.C.: Office of the Surgeon General, Department of the Army, 1997.

Sifton, David W. ed. *PDR Guide to Biological and Chemical Warfare Response*. Montvale, New Jersey: Thompson/Physicians' Desk Reference, 2002.

Somani, Satu M., ed. *Chemical Warfare Agents*. New York, New York: Academic Press, 1992.

Somani, Satu M. and James A. Romano, Jr., ed. *Chemical Warfare Agents: Toxicity at Low Levels*. Boca Raton, Florida: CRC Press, 2001.

Swearengen, Thomas F. *Tear Gas Munitions: An Analysis of Commercial Riot Gas Guns, Tear Gas Projectiles, Grenades, Small Arms Ammunition, and Related Tear Gas Devices*. Springfield, Illinois: Charles C Thomas, Publisher, 1966.

Transport Canada, United States Department of Transportation, Secretariat of Transport and Communications of Mexico. *2020 Emergency Response Guidebook*. Neenah, Wisconsin: J.J. Keller and Associates, 2020.

True, Bey-Lorraine and Robert H. Dreisbach. *Dreisbach's Handbook of Poisoning: Prevention, Diagnosis and Treatment*. 13th Edition. London, United Kingdom: The Parthenon Publishing Group, 2002.

Tuorinsky, Shirley D., ed. *Medical Aspects of Chemical Warfare*. Washington, D.C.: Office of the Surgeon General, Department of the Army, 2008.

United States Army Headquarters. *Chemical Agent Data Sheets Volume II, Edgewood Arsenal Special Report No. EO-SR-74002*. Washington, D.C.: Government Printing Office, December 1974.

United States Army Headquarters. *Potential Military Chemical/Biological Agents and Compounds, Field Manual No. 3-11.9*. Washington, D.C.: Government Printing Office, January 10, 2005.

United States Army Medical Research Institute of Chemical Defense. *Medical Management of Chemical Casualties Handbook*. 3rd Edition. Aberdeen Proving Ground, Maryland: United States Army Medical Research Institute of Chemical Defense, July 2000.

Wachtel, Curt. *Chemical Warfare*. Brooklyn, New York: Chemical Publishing Co., Inc., 1941.

Waitt, Alden H. Gas Warfare: *The Chemical Weapon, Its Use, and Protection Against It*. Revised Edition. New York, New York: Duell, Sloan and Pearce, 1944.

Williams, Kenneth E. *Detailed Facts About Tear Agent 2-Chloroacetophenone (CN)*. Aberdeen Proving Ground, Maryland: United States Army Center for Health Promotion and Preventive Medicine, 1996.

Williams, Kenneth E. *Detailed Facts About Tear Agent Bromobenzylcyanide (CA)*. Aberdeen Proving Ground, Maryland: United States Army Center for Health Promotion and Preventive Medicine, 1996.

Williams, Kenneth E. *Detailed Facts About Tear Agent Chloracetophenone in Benzene and Carbon Tetrachloride (CNB)*. Aberdeen Proving Ground, Maryland: United States Army Center for Health Promotion and Preventive Medicine, 1996.

Williams, Kenneth E. *Detailed Facts About Tear Agent Chloroacetophenone and Chloropicrin in Chloroform (CNS)*. Aberdeen Proving Ground, Maryland: United States Army Center for Health Promotion and Preventive Medicine, 1996.

Williams, Kenneth E. *Detailed Facts About Tear Agent o-Chlorobenzylidene Malononitrile (CS)*. Aberdeen Proving Ground, Maryland: United States Army Center for Health Promotion and Preventive Medicine, 1996.

World Health Organization. *Health Aspects of Chemical and Biological Weapons: Report of A WHO Group of Consultants*. Geneva, Switzerland: World Health Organization, 1970.

World Health Organization. *Public Health Response to Biological and Chemical Weapons: WHO Guidance*. Geneva, Switzerland: World Health Organization, 2004.

14 Vomiting/Sternatory Agents

GENERAL INFORMATION

The majority of these materials are halo or cyano organoarsines. Under normal battlefield conditions, they do not pose a serious danger to the life of an exposed individual and do not produce any permanent injury. Although the only agent in this class that is specifically banned under the Chemical Weapons Convention is lewisite 2 (C14-A004), which is listed in Schedule 1 because it can be readily converted into lewisite 1 (C04-A002, Volume 1), the use of all the other vomiting/sternatory agents is banned during a war under the general-purpose criterion of the convention. They may still be used by law enforcement and the military during operations other than war such as when responding to incidents of civil unrest or to disband unruly crowds. However, because of their toxicity, this class of agents has been abandoned for other riot control agents (see C13).

Vomiting/sternatory agents were employed during World War I in an attempt to defeat the existing masks filters. Toward the end of the war, adamsite (C14-A003) was developed, produced, and weaponized but never used. It was first used when the British dropped thermal generators containing a mixture of adamsite and diphenylchloroarsine (C14-A001) from aircraft during the Russian civil war. This was the first reported deployment of air delivered chemical munitions.

These agents are moderately difficult to synthesize. They are relatively easy to disperse as thermally generated aerosols or as aerosolized solutions.

TOXICOLOGY

EFFECTS

Vomiting/sternatory agents are primarily respiratory irritants. Effects from exposure are usually delayed for several minutes and include violent, uncontrolled sneezing and coughing; pain in the nose and throat; nausea; vomiting; chills; abdominal cramps; nasal discharge; and/or tears. Severe headaches and depression often follow exposure to vomiting/sternatory agents. Effects may persist for several hours post-exposure. They may produce dermatitis on exposed skin. When released in an enclosed or confined space, they can cause serious illness or death. Most vomiting/sternatory agents contain arsenic as a constituent and decomposition products may pose a serious health hazard.

PATHWAYS AND ROUTES OF EXPOSURE

Vomiting/sternatory agents are primarily a hazard through inhalation. Aerosols are very irritating to the skin and eyes at low concentrations but relatively nontoxic via these routes. However, exposure to bulk liquid or solid agents may be hazardous through skin and eye exposure, ingestion, or introduction through abraded skin (e.g., breaks in the skin or penetration of skin by debris). Ingestion of some decomposition products may pose a significant hazard.

GENERAL EXPOSURE HAZARDS

Most bulk agents have very little odor in pure form although impurities may give some agents an odor of garlic or bitter almonds. Aerosols of vomiting/sternatory agents have excellent

DOI: 10.4324/9781003230564-4

warning properties, producing eye, respiratory, and/or skin irritation at concentrations well below lethal levels.

Lethal concentrations (LC_{50}s) for inhalation of these agents are as low as 5,000 mg/m^3 for a 2-minute exposure.

Incapacitating concentrations (ICt_{50}) due to sneezing and regurgitation for inhalation of these agents are as low as 6 mg/m^3 for a 2-minute exposure.

Eye irritation from exposure to agent aerosols occurs at concentrations as low as 0.3 mg/m^3.

LATENCY PERIOD

Depending on dose, the effects from exposure may be delayed from 30 seconds to several minutes, and may last up to several hours. Mild effects may persist for several days.

CHARACTERISTICS

PHYSICAL APPEARANCE/ODOR

Laboratory Grade

Laboratory grade agents are typically colorless to yellow or green liquids or solids. Pure materials are typically odorless.

Munition Grade

Munition grade agents are typically colorless to yellow, green, or brown. As the agent ages colors may become more pronounced. Production impurities and decomposition products in these agents may give them an odor. Odors for some agents have been described as similar to garlic or bitter almonds. Odors may become more pronounced during storage.

Modified Agents

Solvents have been added to these materials to facilitate handling. Color and other properties of these solutions may vary from the pure agent. Odors will vary depending on the characteristics of the solvent(s) used and concentration of agent in the solution.

Agents have also been micro pulverized, encapsulated, or treated with flowing agents to facilitate their dispersal and increase their persistency. Color and other physical properties of may be affected by these additives.

Mixtures with Other Agents

Vomiting/sternatory agents have been mixed with other agents such as phosgene (C10-A003, Volume 1), diphosgene (C10-A004, Volume 1), phenyl dichloroarsine (C04-A004, Volume 1), N-ethylcarbazole (C13-A022), arsenic trichloride (C04-C006, Volume 1), and triphenylarsine.

STABILITY

Most vomiting/sternatory agents are stable at typical temperatures. Apomorphine (C14-A005) is stable as a salt but is air and light sensitive. It is unstable in solution.

Agents can be stored in glass, steel or Teflon-coated containers when pure. If moisture is present, they may rapidly corrode aluminum, iron, bronze, and brass.

PERSISTENCY

When vomiting/sternatory agents are employed as aerosols they are not classified as persistent by the military. Normally, there is minimal secondary risk once the initial aerosol has settled.

However, depending on the size of the individual particles and on any encapsulation or coatings applied to the particles, they may be re-aerosolized by ground traffic or strong winds. Bulk liquid or solid agents can persist in the environment for extended periods. Decomposition products from the breakdown of these agents can pose a persistent hazard.

ENVIRONMENTAL FATE

Most vomiting/sternatory agents are nonvolatile and produce negligible amounts of vapor. They are deployed as dust aerosols. Once the aerosols settle, there is minimal extended hazard from the agents unless the dusts are re-suspended. Decomposition products can be persistent hazards.

Most of these agents are insoluble in water. However, the solubility of any agent may be increased by solvents, components or impurities. The specific gravities of unmodified agents are greater than that of water. With the exception of adamsite (C14-A003), vomiting/sternatory agents are soluble in most organic solvents. Adamsite has only limited solubility in common organic solvents except acetone.

ADDITIONAL HAZARDS

EXPOSURE

All foodstuffs in the area of a release should be considered contaminated. Unopened items packaged in glass, metal or heavy-duty plastic and exposed only to agent aerosols may be used after decontamination of the container. Unopened items exposed to solid or liquid agents, or solutions of agents, should be decontaminated within a few hours post-exposure or destroyed. Opened or unpackaged items, or those packaged only in paper or cardboard, should be destroyed.

Plants, fruits, vegetables, and grains should be washed thoroughly with soap and water.

LIVESTOCK/PETS

Animals can be decontaminated with shampoo/soap and water. If the animal's eyes have been exposed to agent, they should be irrigated with water or saline solution for a minimum of 30 minutes.

The topmost layer of unprotected feedstock (e.g., hay or grain) should be destroyed. The remaining material should be quarantined until tested.

FIRE

Although not normally volatile, heat from a fire will increase the amount of vomiting/sternatory agent vapor in the area. Actions taken to extinguish the fire can also spread the agent and steam generated in combating the fire could create agent aerosols. Combustion of vomiting/sternatory agents will produce volatile toxic metal (i.e., arsenic, antimony, lead) decomposition products. In addition, combustion of these agents may produce toxic and/or corrosive gases such as hydrogen chloride (HCl) and hydrogen cyanide (HCN).

REACTIVITY

Vomiting/sternatory agent decompose slowly in water. Some agents are self-protecting and form an oxide coating that delays further hydrolysis. Agents may be corrosive to some metals.

HAZARDOUS DECOMPOSITION PRODUCTS

Hydrolysis

Varies depending on the specific agent but may include hydrogen chloride (HCl) and hydrogen cyanide (HCN). Several produce diphenylarsenious oxide. Other organic oxides of arsenic, antimony, or lead may also be present.

Combustion

Volatile decomposition products may include hydrogen chloride (HCl), hydrogen cyanide (HCN), nitrogen oxides (NO_x), benzene, and oxides of arsenic, antimony, or lead.

PROTECTION

EVACUATION RECOMMENDATIONS

Isolation and protective action distances listed below are taken from the *2020 Emergency Response Guidebook*. For vomiting/sternatory agents, these recommendations are based on a release scenario involving direct aerosolization of the solid agents with a particle size between two and ten microns. For DA, a small release involves 10 kilograms (approximately 7 liters) of bulk agent and a large release involves 500 kilograms (approximately 360 liters) of bulk agent. For DC, a small release involves ten kilograms (approximately 20,000 cubic centimeters, a box approximately 27 centimeters on a side) of bulk agent and a large release involves 100 kilograms (approximately 0.2 cubic meters, a box approximately 58 centimeters on a side) of bulk agent. For DM, a small release involves five kilograms (approximately 9,000 cubic centimeters, a box 21 centimeters on a side) of bulk agent and a large release involves 100 kilograms (approximately 0.2 cubic meters, a box approximately 57 centimeters on a side) of bulk agent.

DA (Diphenylchloroarsine) *C14-A001*

	Initial Isolation	Downwind Day	Downwind Night
Small device (10 kg)	30 m	0.2 km	0.8 km
Large device (500 kg)	300 m	1.9 km	7.5 km

DC (Diphenylcyanoarsine) *C14-A002*

	Initial Isolation	Downwind Day	Downwind Night
Small device (10 kg)	30 m	0.1 km	0.6 km
Large device (100 kg)	60 m	0.4 km	1.8 km

DM (Adamsite) *C14-A003*

	Initial Isolation	Downwind Day	Downwind Night
Small device (5 kg)	30 m	0.1 km	0.3 km
Large device (100 kg)	60 m	0.2 km	1.4 km

Personal Protective Requirements

Structural Firefighters' Gear

Structural firefighters' protective clothing is recommended for fire situations only; it is not effective in spill situations or release events. If chemical protective clothing is not available and it is necessary to rescue casualties from a contaminated area, then structural firefighters' gear will provide very limited skin protection against agent vapors and aerosols. Contact with solid and liquid agents should be avoided.

Respiratory Protection

Self-contained breathing apparatuses (SCBAs) or air purifying respirators (APRs) should have a National Institute for Occupational Safety and Health (NIOSH) Chemical/Biological/Radiological/Nuclear (CBRN) certification. However, during emergency operations, other NIOSH-approved SCBAs or APRs that have been specifically tested by the manufacturer against chemical warfare agents may be used if deemed necessary by the incident commander. APRs should be equipped with a NIOSH-approved CBRN filter or a combination organic vapor/acid gas/particulate cartridge.

Immediately dangerous to life or health (IDLH) levels are the ceiling limit for respirators other than SCBAs. However, IDLH levels have not been established for vomiting/sternatory agents. Therefore, any potential exposure to elevated concentrations of these agents should be regarded with extreme caution and the use of SCBAs for respiratory protection should be considered.

Chemical Protective Clothing

Vomiting/sternatory agents are primarily an eye and respiratory hazard; however, at elevated aerosol concentrations or in contact with bulk material, agents may also pose a dermal hazard. Currently, there is no information on performance testing of chemical protective clothing against vomiting/sternatory agent.

Decontamination

General

Apply universal decontamination procedures using soap and water.

Solutions or Liquid Aerosols

Casualties/Personnel

Remove all clothing immediately. To avoid further exposure of the head, neck, and face to the agent, cut off potentially contaminated clothing that must be pulled over the head. Use a sponge or cloth with liquid soap and copious amounts of water to wash the skin surface and hair at least three times. Do not use hot water as it may increase skin irritation. Avoid rough scrubbing as this could abrade the skin and increase discomfort. Rinse with copious amounts of water. For severe eye irritation, irrigate with water or 0.9% saline solution for a minimum of 15 minutes. Do not allow casualties to rub their eyes or skin as this may exacerbate agent effects.

Small areas

Ventilate to remove the aerosol. Small puddles of liquid can be contained by covering with absorbent material such as vermiculite, diatomaceous earth, clay, sponges, or towels. Place the absorbed material into containers lined with high-density polyethylene. Larger puddles can be collected using vacuum equipment made of materials inert to the released material and equipped with appropriate vapor filters. Wash the area with copious amounts of an alkaline soap/detergent and water. Collect and containerize the rinseate. Removal of porous material, including painted surfaces, may be required because these materials may be difficult to decontaminate. Arsenic or antimony metal and/or oxides, due to decomposition of the agents, may be present and require additional decontamination.

Solids or Particulate Aerosols

Casualties/Personnel

Do not attempt to brush the agent off of the individual or their clothing as this can aerosolized the agent. If possible, dampen the agent with a water mist to help prevent aerosolization. Remove all clothing immediately. To avoid further exposure of the head, neck and face to the agent, cut off potentially contaminated clothing that must be pulled over the head. Use a sponge or cloth with liquid soap and copious amounts of water to wash the skin surface and hair at least three times. Do not use hot water as it may increase skin irritation. Avoid rough scrubbing as this could abrade the skin and increase discomfort. Rinse with copious amounts of water. For severe eye irritation, irrigate with water or 0.9% saline solution for a minimum of 15 minutes. Do not allow casualties to rub their eyes or skin as this may exacerbate agent effects.

Small areas

If indoors, close windows and doors in the area and turn off anything that could create air currents (e.g., fans, air conditioner, etc.). Allow aerosol to settle. Avoid actions that could aerosolize the agent such as sweeping or brushing. Collected the agent with a vacuum cleaner equipped with a high-efficiency particulate air (HEPA) filter. Do not use a standard home or industrial vacuum. Do not allow the vacuum exhaust to stir the air in the affected area. Vacuum all surfaces with extreme care in a very slow and controlled manner to minimize aerosolizing the agent. Place the collected material into containers lined with high-density polyethylene. Wash the area with copious amounts of an alkaline soap/detergent and water. Collect and containerize the rinseate in containers lined with high-density polyethylene. Removal of porous material, including painted surfaces, may be required because these materials may be difficult to decontaminate. Arsenic or antimony metal and/or oxides, due to decomposition of the agents, may be present and require additional decontamination.

MEDICAL

CDC CASE DEFINITION

No specific biologic marker/test is available for vomiting/sternatory agents as a class; however, the case can be confirmed if they are detected in environmental samples. The case can be confirmed if laboratory testing is not performed by either a predominant amount of clinical and nonspecific laboratory evidence is present or an absolute certainty of the etiology of the agent is known.

DIFFERENTIAL DIAGNOSIS

The following factors have been suggested as alternatives to consider when presented with a potential case of exposure to irritating agents: anxiety, anaphylaxis, conjunctivitis, pneumonia, ultraviolet keratitis; inhalation of smoke, hydrocarbons, ammonia, hydrogen sulfide, phosgene, halogens such as chlorine, sulfuric acid, hydrogen chloride, or nickel carbonyl; sodium azide, street drugs; acute respiratory distress syndrome, chronic obstructive pulmonary disease, and emphysema; congestive heart failure; pulmonary edema; and anthrax.

SIGNS AND SYMPTOMS

Progression of symptoms is generally irritation of the eyes and mucous membranes, viscous discharge from the nose similar to that caused by a cold, violent uncontrollable sneezing and coughing, severe headache, acute pain and difficulty breathing (tightness of the chest), nausea, and vomiting. Mental depression may occur. Severe effects last from 30 minutes to several hours. Minor effects may persist for over 24 hours.

MASS-CASUALTY TRIAGE RECOMMENDATIONS

Typically not required. Casualties will usually recover unassisted after removal from the contaminated atmosphere. Consult the base station physician or regional poison control center for advice on specific situations.

CASUALTY MANAGEMENT

Decontaminate the casualty ensuring that all the agent has been removed. For severe eye irritation, irrigate with water or 0.9% saline solution for a minimum of 15 minutes. Do not allow casualties to rub their eyes or skin as this may exacerbate agent effects. Irrigate open wounds with water or 0.9% saline solution for at least 10 minutes.

Once the casualty has been decontaminated, including the removal of foreign matter from wounds, medical personnel do not need to wear a chemical-protective mask.

Casualties will usually recover unassisted from exposure to vomiting/sternatory agent although it may take several hours after removal from the contaminated atmosphere. Vigorous exercise may lessen and shorten symptoms. Most patients can be discharged safely. Rarely a patient with significant respiratory findings may merit admission.

FATALITY MANAGEMENT

Remove all clothing and personal effects. Because of the potential for hazardous residual metal content (i.e., arsenic, lead, antimony), it may be appropriate to ship non-durable items to a hazardous waste disposal facility. Otherwise, decontaminate with soap and water.

If heavily contaminated, personnel may need to wear respiratory protection, gloves, and an apron during the decontamination process. Wash the remains with soap and water. Pay particular attention to areas where the agent may get trapped, such as hair, scalp, pubic areas, fingernails, folds of skin, and wounds. If remains are heavily contaminated with residue, wash and rinse waste should be contained for proper disposal.

Once the remains have been thoroughly decontaminated, no further protective action is necessary. Body fluids removed during the embalming process do not pose any additional risks and should be contained and handled according to established procedures. Use standard burial procedures.

C14-A AGENTS

C14-A001

Diphenylchloroarsine (Agent DA)
CAS: 712-48-1
RTECS: CG9900000
UN: 1769; 156
EC: 211–921-4
UNII: 1H39V3559B
ICD-11: XM9YA9

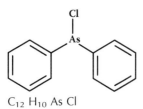

$C_{12} H_{10} As Cl$

Colorless odorless crystals. Weapons grade material is a dark brown, thick, viscous, semisolid resembling shoe polish. It has in the past been used industrially as a wood preservative, pesticide, and herbicide for cacti.

Also reported as a mixture with adamsite (C14-A003); phosgene (C10-A003); diphosgene (C10-A004); phenyldichloroarsine (C04-A004), N-Ethylcarbazole (C13-A022); arsenic trichloride (C04-C006); and triphenylarsine

Exposure Hazards

Conversion Factor: 1 ppm = 10.82 mg/m^3 at 77°F

LCt$_{50 (Inh)}$: 15,000 mg-min/m^3. This value is from older sources (circa 1942) and is not supported by modern data. No updated toxicity estimate has been proposed.

ICt$_{50 (Inh)}$: 12 mg-min/m^3

Eye Irritation: 0.3 mg/m^3; exposure duration unspecified

The hydrolysis product, diphenylarsenious oxide, is very poisonous if ingested.

MEG$_{(1hr)}$: Neg: 0.039 mg/m^3; Mar: 0.39 mg/m^3; Crit: 1.2 mg/m^3; Cat: —
PAC-1: 0.035 mg/m^3; *PAC-2:* 0.39 mg/m^3; *PAC-3:* 1.2 mg/m^3

Interim AEGLs
AEGL-1: Not Developed
AEGL-2: 1 hr – 0.39 mg/m^3; 4 hr – 0.098 mg/m^3; 8 hr – 0.049 mg/m^3
AEGL-3: 1 hr – 1.2 mg/m^3; 4 hr – 0.30 mg/m^3; 8 hr – 0.15 mg/m^3

Properties:

MW: 264.5	*VP:* 4.5 × 10^{-4} mmHg	*FlP:* 662°F
D: 1.4 g/mL (122°F)	*VD:* 9.1	*LEL:* —
MP: 113°F	*Vlt:* 0.65 ppm	*UEL:* —
MP: 100°F (weapons grade)	*ER:* —	*RP:* 13,000
BP: 631°F (decomposes)	*H$_2$O:* Insoluble	*IP:* <9 eV
Vsc: —	*Sol:* Acetone, ethanol, ether, chloroform	

C14-A002

Diphenylcyanoarsine (Agent DC)
CAS: 23525-22-6
RTECS: —
UN: —
EC: 245–716-6
UNII: SF5K94TPOF
ICD-11: —

$C_{13} H_{10} As N$

Colorless crystalline solid or clear liquid with an odor like garlic or bitter almonds detectable at ≥0.0005 ppm. Undergoes considerable decomposition when explosively disseminated.
Also reported as a mixture with phenyldichloroarsine (C04-A004).

Exposure Hazards

Conversion Factor: 1 ppm = 10.44 mg/m^3 at 77°F

$LCt_{50\ (Inh)}$: 10,000 mg-min/m^3 (480 ppm for a 2-minute exposure)

$IC_{50\ (Inh)}$: 60 mg/m^3 (5.7 ppm) for a 30-second exposure

$IC_{50\ (Inh)}$: 4 mg/m^3 (0.38 ppm) for a 5-minute exposure

These values are from older sources (circa 1942) and are not supported by modern data. No updated toxicity estimates have been proposed.

Properties:

MW: 255.2	VP: 0.0002 mmHg	FIP: —
D: 1.5 g/cm^3	VD: 8.8	LEL: —
MP: 89°F	Vlt: —	UEL: —
BP: 662°F (decomposes)	ER: —	RP: 31,000
Vsc: —	H$_2$O: Insoluble	IP: <9 eV
	Sol: Most organic solvents	

C14-A003

Adamsite (Agent DM)
CAS: 578-94-9
RTECS: SG0680000
UN: 1698; 154
EC: 209–433-1
UNII: QI287G628L
ICD-11: —

$C_{12} H_9 As Cl N$

Bright canary-yellow crystals that are odorless but produce irritation. Weapons grade material is a dark green to brown solid. When dispersed as a particulate cloud from a thermal munition, it has a characteristic smoky odor.

Also reported as a mixture with diphenylchloroarsine (C14-A001).

Exposure Hazards

Conversion Factor: 1 ppm = 11.35 mg/m^3 at 77°F

LCt$_{50\ (Inh)}$: 11,000 mg-min/m^3

ICt$_{50\ (Inh)}$: 22–150 mg-min/m^3

Eye & Respiratory Irritation: 0.38 mg/m^3 for a 2-minute exposure

These are provisional updates from older values that have not been formally.

MEG$_{(1hr)}$: Neg: 0.016 mg/m^3; Mar: 2.6 mg/m^3; Crit: 6.4 mg/m^3; Cat: —

PAC-1: 0.016 mg/m^3; *PAC-2:* 2.6 mg/m^3; *PAC-3:* 6.4 mg/m^3

The hydrolysis product, diphenylaminoarsenious oxide, is very poisonous if ingested.

Interim AEGLs

AEGL-1: 1 hr – 0.016 mg/m^3; 4 hr – 0.0022 mg/m^3; 8 hr – 0.00083 mg/m^3

AEGL-2: 1 hr – 2.6 mg/m^3; 4 hr – 0.36 mg/m^3; 8 hr – 0.14 mg/m^3

AEGL-3: 1 hr – 6.4 mg/m^3; 4 hr – 0.91 mg/m^3; 8 hr – 0.34 mg/m^3

Properties:

MW: 277.6	*VP:* 2 × 10^{-13} mmHg	*FIP:* NA
D: 1.6 g/cm^3	*VD:* —	*LEL:* NA
MP: 383°F	*Vlt:* —	*UEL:* NA
BP: 770°F (decomposes)	*ER:* —	*RP:* —
Vsc: —	*H$_2$O:* 0.006%	*IP:* —
	Sol: Acetone	

C14-A004

Lewisite 2 (Agent L2)

CAS: 40334-69-8

RTECS: —

UN: —

EC: —

UNII: 14O3QZ9K1F

ICD-11: —

C$_4$ H$_4$ As Cl$_3$

Clear yellow to yellowish-brown liquid. In conjunction with lewisite 3 (C04-C007), Lewisite 2 is a major synthetic by-product in the production of lewisite (C04-A002). It is readily converted into lewisite.

This material is on Schedule 1 of the CWC

Exposure Hazards

Conversion Factor: 1 ppm = 9.55 mg/m^3 at 77°F

Although lewisite 2 acts primarily as a vomiting/sternatory agent, it does have limited vesicant power similar to lewisite (C04-A002).

MEG$_{(1hr)}$: Neg: —; Mar: 0.12 mg/m^3; Crit: 0.74 mg/m^3; Cat: —

PAC-1: 0.0024 ppm; PAC-2: 0.026 ppm; PAC-3: 0.077 ppm

AEGLs have been proposed only for mixtures of this compound with lewisite (C04-A002).

Properties:

MW: 233.4	VP: —	FIP: —
D: 1.7 g/mL	VD: 8.1	LEL: —
MP: 50°F (est.)	Vlt: —	UEL: —
BP: Decomposes	ER: —	RP: —
Vsc: —	H$_2$O: Insoluble (slowly decomposes)	IP: —
	Sol: Most organic solvents	

C14-A005

Apomorphine

CAS: 58-00-4; 314-19-2 (Hydrochloride salt)

RTECS: —

UN: —

EC: 200–360-0; 206–243-0 (Hydrochloride salt)

UNII: N21FAR7B4S; F39049Y068 (Hydrochloride salt)

ICD-11: XM5PE4

C$_{17}$ H$_{17}$ N O$_2$

White to grayish crystalline solid that is odorless. It is air and light sensitive; rapidly becomes green on standing. Solutions are unstable. Various salts have been reported.

Used in the treatment of Parkinson's disease or erectile dysfunction, and as an emetic in veterinary medicine.

Exposure Hazards

Conversion Factor: 1 ppm = 10.93 mg/m^3 at 77°F

Human toxicity values have not been established or have not been published.

Properties:

MW: 267.3	VP: 1 × 10^{-9} mmHg (77°F) (est.)	FIP: —
D: 1.3 g/cm^3 (est.)	VD: —	LEL: —
MP: 383°F (decomposes)	Vlt: —	UEL: —
BP: —	ER: —	RP: —
Vsc: —	H$_2$O: 1.66% (61°F)	IP: —
	Sol: Alcohol, acetone, ether, chloroform	

BIBLIOGRAPHY

Banks, Daniel E., ed. *Medical Management of Chemical Warfare Casualties Handbook*, 5th Edition. Washington, D.C.: Office of the Surgeon General, Department of the Army, 2014.

Brophy, Leo P., Wyndham D. Miles and Rexmond C. Cohrane. *The Chemical Warfare Service: From Laboratory to Field*. Washington, D.C.: Government Printing Office, 1968.

Compton, James A.F. *Military Chemical and Biological Agents: Chemical and Toxicological Properties*. Caldwell, New Jersey: The Telford Press, 1987.

Fries, Amos A. and Clarence J. West. *Chemical Warfare*. New York, New York: McGraw-Hill Book Company, Inc., 1921.

Green, Stanley Joseph and Thomas Slater Price. "LVI. The Chlorovinylchloroarsines." *Journal of the Chemical Society* 119 (1921): 452.

International Labour Organization. *International Chemical Safety Cards (ICSCs)* [https://www.ilo.org/safework/info/publications/WCMS_113134/lang--en/index.htm]. December 31, 2020.

Jackson, Kirby E. "Sternutators." *Chemical Reviews* 17 (1935): 251–292.

Jackson, Kirby E., and Margaret A. Jackson. "The Chlorovinylarsines." *Chemical Reviews* 16 (1935): 439–452.

Lenhart, Martha K., ed. *Medical Aspects of Chemical Warfare, Textbooks of Military Medicine Series*. Washington, D.C.: Office of the Surgeon General, Department of the Army, 2007.

Lewis, W. Lee, and G. A. Perkins. "The Beta-Chlorovinyl Chloroarsines." *Industrial & Engineering Chemistry* 15 (March 1923): 290–295.

Munro, Nancy B., Sylvia S. Talmage, Guy D. Griffin, Larry C. Waters, Annetta P. Watson, Joseph F. King, and Veronique Hauschild. "The Sources, Fate and Toxicity of Chemical Warfare Agent Degradation Products." *Environmental Health Perspectives* 107 (1999): 933–974.

Olson, Kent R., ed. *Poisoning & Drug Overdose*. 4th Edition. New York, New York: Lange Medical Books/McGraw-Hill, 2004.

Prentiss, Augustin M. *Chemicals in War: A Treatise on Chemical Warfare*. New York, New York: McGraw-Hill Book Company, Inc., 1937.

Sartori, Mario F. *The War Gases: Chemistry and Analysis*. Translated byL. W. Marrison. London, United Kingdom: J. & A. Churchill, Ltd., 1939.

Sidell, Frederick R. *Medical Management of Chemical Warfare Agent Casualties: A Handbook For Emergency Medical Services*. Bel Air, Maryland: HB Publishing, 1995.

Sidell, Fredrick R., Ernest T. Takafuji and David R. Franz, ed. *Medical Aspects of Chemical and Biological Warfare, Textbook of Military Medicine Series, Part 1, Warfare, Weaponry, and the Casualty*. Washington, D.C.: Office of the Surgeon General, Department of the Army, 1997.

Sifton, David W. ed. *PDR Guide to Biological and Chemical Warfare Response*. Montvale, New Jersey: Thompson/Physicians Desk Reference, 2002.

Somani, Satu M., ed. *Chemical Warfare Agents*. New York, New York: Academic Press, 1992.

Somani, Satu M. and James A. Romano, Jr., ed. *Chemical Warfare Agents: Toxicity at Low Levels*. Boca Raton, Florida: CRC Press, 2001.

Swearengen, Thomas F. *Tear Gas Munitions: An Analysis of Commercial Riot Gas Guns, Tear Gas Projectiles, Grenades, Small Arms Ammunition, and Related Tear Gas Devices*. Springfield, Illinois: Charles C Thomas, Publisher, 1966.

Transport Canada, United States Department of Transportation, Secretariat of Transport and Communications of Mexico. *2020 Emergency Response Guidebook*. Neenah, Wisconsin: J.J. Keller and Associates, 2020.

True, Bey-Lorraine and Robert H. Dreisbach. *Dreisbach's Handbook of Poisoning: Prevention, Diagnosis and Treatment*. 13th Edition. London, United Kingdom: The Parthenon Publishing Group, 2002.

United States Army Headquarters. *Chemical Agent Data Sheets Volume I, Edgewood Arsenal Special Report No. EO-SR-74001*. Washington, D.C.: Government Printing Office, December 1974.

United States Army Headquarters. *Potential Military Chemical/Biological Agents and Compounds, Field Manual No. 3-11.9*. Washington, D.C.: Government Printing Office, January 10, 2005.

United States Army Medical Research Institute of Chemical Defense. *Medical Management of Chemical Casualties Handbook*. 3rd Edition. Aberdeen Proving Ground, Maryland: United States Army Medical Research Institute of Chemical Defense, July 2000.

United States National Institute of Health, National Library of Medicine. PubChem. 2020 [https://pubchem.ncbi.nlm.nih.gov/]. December 31, 2020.

Wachtel, Curt, *Chemical Warfare*. Brooklyn, New York: Chemical Publishing Co., Inc., 1941.

Waitt, Alden H. *Gas Warfare: The Chemical Weapon, Its Use, and Protection Against It*. Revised Edition. New York, New York: Duell, Sloan and Pearce, 1944.

Williams, Kenneth E. *Detailed Facts About Vomiting Agent Adamsite (DM)*. Aberdeen Proving Ground, Maryland: United States Army Center for Health Promotion and Preventive Medicine, 1996.

World Health Organization. *Health Aspects of Chemical and Biological Weapons: Report of A WHO Group of Consultants*. Geneva, Switzerland: World Health Organization, 1970.

World Health Organization. *Public Health Response to Biological And Chemical Weapons: WHO Guidance*. Geneva, Switzerland: World Health Organization, 2004.

15 Malodorants

GENERAL INFORMATION

The majority of these materials are organic sulfur compounds that may also contain an odor intensifier. These chemicals are generally volatile liquids at room temperature with odors that are detectable at very low levels. Under normal battlefield conditions, these materials do not pose a serious danger to the life of an exposed individual and do not produce any permanent injury. Since approximately 0.2% of the population is unable to detect various odors (anosmic), compositions may contain multiple malodorant components.

Obnoxious smelling agents have been used throughout history to provide camouflage and assist in breaching hardened defensive positions. Research into development of new and more effective malodorants began after World War I. During World War II, an agent known as "Who Me?", which smelled of rotting food and carcasses, was developed by the Allies and tested by resistance fighters on German and Japanese soldiers in an effort to humiliate and embarrass them. However, the agent was volatile and difficult to deliver and did not always produce the desired response. As a result of this failure, malodorants were largely abandoned in the United States until the 1980s. Modern malodorants are use in riot control, to clear facilities, to deny an area, or as a taggant. Stench weapons in development include concentrates of natural odors such as the those produced by skunks, rotting meat, excrement, and body odor. Since odors can provoke varying reactions in people based on their social and cultural conditioning, malodorants offer the possibility of ethnic or cultural targeting.

Natural malodorants derived from a biological entity are prohibited under the Biological Weapons Convention. Malodorants comprised strictly of synthetic chemicals would be defined as a riot control agent and the Chemical Weapons Convention bans the use of such agents during a war. However, they may still be used by the military during operations other than war such as when responding to incidents of civil unrest. The military also has used many of these materials as simulants for lethal chemical agents. Malodorants have also been developed for use by police to control rioters and disband crowds.

TOXICOLOGY

EFFECTS

Malodorants induce strong, repulsive reflexes including nausea, gagging, and/or vomiting. Unpleasant odors impede cognitive performance, increase feelings of discomfort, and heighten a perception of illness. If an odor is perceived to be harmful or hazardous, they can stimulate a feeling of panic and fear, and cause exposed personnel to want to flee.

Reactions vary in different people based on the concentration of the odor and on social and cultural conditioning. Extended exposure may desensitize individuals. High concentrations may cause severe physiological trauma. In an enclosed or confined space, very high concentration can be lethal.

PATHWAYS AND ROUTES OF EXPOSURE

Malodorants are primarily an inhalation hazard. Aerosols and vapors are extremely foul smelling at low concentrations but are otherwise relatively nontoxic. However, exposure to bulk liquid or solid

DOI: 10.4324/9781003230564-5

agents may be hazardous through skin absorption, ingestion, and introduction through abraded skin (e.g., breaks in the skin or penetration of skin by debris).

GENERAL EXPOSURE HAZARDS

Malodorants do not seriously endanger life except at exposures greatly exceeding an effective dose, usually only achieved in a confined or enclosed space.

LATENCY PERIOD

Exposure to malodorants produces immediate effects.

CHARACTERISTICS

PHYSICAL APPEARANCE/ODOR

Laboratory Grade
Laboratory grade agents are typically colorless to yellow liquids or solids.

Munition Grade
Munition grade agents typically consist of at least one malodorant agent (10 to 90%) and an odor intensifier (0.5 to 5%) dissolved in a liquid carrier. Solvents include volatile hydrocarbons, plant/vegetable oils, and water. Solvents typically pose minimal toxic hazards themselves. Compositions are typically colorless to yellow liquids. As the agent ages and decomposes it may discolor and become brown.

Modified Agents
Agents have been microencapsulated to facilitate their dispersal and increase their persistency.

STABILITY

Malodorants compositions are stable during storage but some ingredients are sensitive to light and air.

PERSISTENCY

As typically deployed, malodorants are classified as non-persistent by the military. However, when used as taggants, malodorants are intended to last for days. Contact with heavy aerosols, sprays, or bulk material may cause contamination that may persist for weeks.

ENVIRONMENTAL FATE

Malodorant vapors have a density greater than air and tend to collect in low places. The majority of these agents are insoluble in water and have a wide range of specific gravities that cause them to either float or sink in water. Further, solvents used in the formulation are often insoluble in water and can change the specific gravities of the active agents. Most of these agents are soluble in common organic solvents including oils, alcohols, and ketones.

Agents may be absorbed into porous material, including painted surfaces, and these materials may be difficult to decontaminate.

ADDITIONAL HAZARDS

EXPOSURE

All foodstuffs in the area of a release should be considered contaminated. Unopened items packaged in glass, metal, or heavy-duty plastic may be used after decontamination of the container. Opened or unpackaged items, or those packaged only in paper or cardboard, should be destroyed.

Plants, fruits, vegetables, and grains should be washed thoroughly with soap and water and aerated to remove residual odor.

LIVESTOCK/PETS

Animals can be decontaminated with shampoo/soap and water or an aqueous solution of mild oxidants (see the decontamination section below).

The topmost layer of unprotected feedstock (e.g., hay or grain) should be destroyed. The remaining material should be aerated to remove residual odor.

FIRE

Heat from a fire will increase the amount of agent vapor in the area. A significant amount of the agent could be volatilized and escape into the surrounding environment before the agent is consumed by the fire. Actions taken to extinguish the fire can also spread the agent. Although many malodorants are only slightly soluble or insoluble in water, runoff from firefighting efforts will still pose a potential contact threat.

REACTIVITY

Malodorants generally do not react with water or are very slowly decomposed by water. Most of these agents are incompatible with strong oxidizers. Solvents used to disperse agents may also be incompatible with strong oxidizers.

HAZARDOUS DECOMPOSITION PRODUCTS

Hydrolysis

Malodorants are generally stable or very slowly decomposed by water.

Combustion

Volatile decomposition products may include sulfur oxides (SO_x), hydrogen sulfide (H_2S), and nitrogen oxides (NO_x).

PROTECTION

EVACUATION RECOMMENDATIONS

There are no published recommendations for isolation or protective action distances for malodorants released in mass casualty situations. However, traditional isolation and protective action distances for many of these materials can be found in the Department of Transportation *2020 Emergency Response Guidebook*. These recommendations are based upon an accidental release during transportation of the material and involving a small spill (i.e., a commercial gas cylinder or 208 liters or less of liquid material), or a large spill (i.e., more than one gas cylinder, a large gas container such as a railcar, or more than 208 liters of liquid material). In the event of terrorism, sabotage, or a catastrophic release of these materials, the *Guidebook* recommends doubling these distances.

PERSONAL PROTECTIVE REQUIREMENTS

Structural Firefighters' Gear

Structural firefighters' protective clothing is recommended for fire situations only; it is not effective in spill situations or release events. If chemical protective clothing is not available and it is necessary to rescue casualties from a contaminated area, then structural firefighters' gear will provide very limited skin protection against agent vapors and aerosols. Contact with solid and liquid agents should be avoided.

Respiratory Protection

Self-contained breathing apparatuses (SCBAs) or air purifying respirators (APRs) should have a National Institute for Occupational Safety and Health (NIOSH) Chemical/Biological/Radiological/Nuclear (CBRN) certification. However, during emergency operations, other NIOSH-approved SCBAs or APRs that have been specifically tested by the manufacturer against chemical warfare agents may be used if deemed necessary by the incident commander. APRs should be equipped with a NIOSH-approved CBRN filter or a combination organic vapor/acid gas/particulate cartridge.

Immediately dangerous to life or health (IDLH) levels are the ceiling limit for respirators other than SCBAs. Any exposures approaching the IDLH level should be regarded with extreme caution and the use of SCBAs for respiratory protection should be considered.

Chemical Protective Clothing

Malodorants are primarily an inhalation hazard; however, at elevated vapor/aerosol concentrations or in contact with bulk material, agents may also pose a dermal hazard.

Use only chemical protective clothing that has undergone material and construction performance testing against the specific agent that has been released. Reported permeation rates may be affected by solvents, components or impurities in munition grade agents.

DECONTAMINATION

General

Apply universal decontamination procedures using soap and water.

Some malodorants are insoluble in water and difficult to remove with soap and water. In such situations, an aqueous solution of mild oxidants may be effective in destroying the odorous ingredients of the agent. Published examples include:

One liter of 3% household hydrogen peroxide with 140 grams (60 milliliters) of baking soda and 5 milliliters of liquid dish soap.

OR

259 milliliters of a sodium perborate bleach solution and 45 milliliters of liquid dish soap in 3.8 liters of water.

The mixture is applied with a sponge and allowed to remain in contact with the agent for several minutes. It is then rinsed off with water. Either of these decontamination mixtures could cause discoloration of hair or fabrics.

For Skunk (C15-A027), a specialty soap is available from the vendor to neutralize the malodorant characteristics of the agent.

Vapors

Casualties/Personnel

Aeration and ventilation. If decontamination is deemed necessary, remove all clothing as it may continue to emit trapped agent vapor after contact with the vapor cloud has ceased. Shower using copious amounts of soap and water. Ideally, showers will be high volume with low pressure. Ensure that the hair has been washed and rinsed to remove potentially trapped vapor. If necessary, an aqueous solution containing a mild oxidant can be used (see general section above). Avoid any contact with sensitive areas such as the eyes. Rinse with copious amounts of water. If eye irritation occurs, irrigate with water or 0.9% saline solution for a minimum of 15 minutes.

Small areas

Ventilate to remove the vapors. If deemed necessary, wash the area with copious amounts of soap and water. Collect and place into containers lined with high-density polyethylene. If necessary, an aqueous solution containing a mild oxidant can be used (see general section above). Removal of porous material, including painted surfaces, may be required because agents that have been absorbed into these materials may migrate back to the surface and pose a contact hazard.

Liquids, Solutions or Liquid Aerosols

Casualties/Personnel

Remove all clothing immediately. Even clothing that has not come into direct contact with the agent may contain trapped vapor. To avoid further exposure of the head, neck, and face to the agent, cut off potentially contaminated clothing that must be pulled over the head. Use a sponge or cloth with liquid soap and copious amounts of water to wash the skin surface and hair at least three times. Ideally, showers will be high volume with low pressure. Avoid rough scrubbing. Rinse with copious amounts of water. If necessary, an aqueous solution containing a mild oxidant can be used (see general section above). Avoid any contact with sensitive areas such as the eyes. Rinse with copious amounts of water. If eye irritation occurs, irrigate with water or 0.9% saline solution for a minimum of 15 minutes.

Small areas

Ventilate to remove the aerosol and vapors. Small puddles of liquid can be contained by covering with absorbent material such as vermiculite, diatomaceous earth, clay, sponges, or towels. Place the absorbed material into containers lined with high-density polyethylene. Larger puddles can be collected using vacuum equipment made of materials inert to the released material and equipped with appropriate vapor filters. Wash the area with copious amounts of soap and water. If necessary, an aqueous solution containing a mild oxidant can be used (see general section above). Collect and containerize the rinseate. Removal of porous material, including painted surfaces, may be required because these materials may be difficult to decontaminate. Ventilate the area after decontamination is complete to remove residual vapors.

MEDICAL

CDC CASE DEFINITION

The CDC has not published a specific case definition for intoxication by malodorants.

DIFFERENTIAL DIAGNOSIS

Anyone exposed to malodorants will be immediately identifiable.

SIGNS AND SYMPTOMS

Vapors

Nausea, gagging, vomiting, dizziness, loss of coordination, disorientation, difficulty breathing (dyspnea), and headache.

Liquids

Irritation and burning of the skin, eyes, mucous membrane, and respiratory system.

MASS-CASUALTY TRIAGE RECOMMENDATIONS

Typically not required. Casualties will usually recover unassisted after removal from the contaminated atmosphere and decontamination. Consult the base station physician or regional poison control center for advice on specific situations.

CASUALTY MANAGEMENT

Decontaminate the casualty ensuring that all the agent has been removed. For severe eye irritation, irrigate with water or 0.9% saline solution for a minimum of 15 minutes. Irrigate open wounds with water or 0.9% saline solution for at least 10 minutes.

Once the casualty has been decontaminated, including the removal of foreign matter from wounds, medical personnel do not need to wear a chemical-protective mask.

Casualties will usually recover from exposure to malodorants shortly after removal from the contaminated atmosphere. Most patients can be discharged safely. Rarely a patient with significant respiratory findings may merit admission.

FATALITY MANAGEMENT

Remove all clothing and personal effects and decontaminate with soap and water. In cases of heavy contamination, treatment with a mild oxidizing agent may be required. While it may be possible to decontaminate porous items, it may not be realistic given the low odor threshold of these agents.

Wash the remains with soap and water. If necessary, an aqueous solution containing a mild oxidant can be used (see general section above). Pay particular attention to areas where agent may get trapped, such as hair, scalp, pubic areas, fingernails, folds of skin, and wounds. If remains are heavily contaminated with residue, wash and rinse waste should be contained for proper disposal.

Once the remains have been thoroughly decontaminated, no further protective action is necessary. Use standard burial procedures.

C15-A   AGENTS

C15-A001

Mercaptoethanol
CAS: 60-24-2
RTECS: KI5600000
UN: 2966; 153
EC: 200–464-6
UNII: 14R9K67URN
ICD-11: —

$C_2 H_6 O S$

Clear, colorless mobile liquid with a strong disagreeable odor (stench) detectable at ≥0.64 ppm.

Use in industry as a chemical intermediate for dyestuffs, pharmaceuticals, rubber chemicals, flotation agents, insecticides, and plasticizers; used as a water-soluble reducing agent and reagent in biochemical research.

Exposure Hazards

Conversion Factor: 1 ppm = 3.20 mg/m^3 at 77°F

MEG$_{(1hr)}$: Neg: 1.9 ppm; Mar: 13 ppm; Crit: 190 ppm; Cat: —

PAC-1: 0.59 ppm; *PAC-2:* 3.4 ppm; *PAC-3:* 29 ppm

WEEL: 0.2 ppm [Skin]

Properties:

MW: 78.1	*VP:* 1.0 mmHg	*FlP:* 165°F
D: 1.1 g/mL	*VD:* 2.7	*LEL:* 2.3%
MP: −148°F	*Vlt:* 2,300 ppm	*UEL:* 18%
BP: Decomposes	*ER:* —	*RP:* 6.5
Vsc: 3.1 cS	*H$_2$O:* Miscible	*IP:* 9.10 eV
	Sol: Most organic solvents	

C15-A002

Butyl Mercaptan

CAS: 109-79-5

RTECS: EK6300000

UN: 2347; 130

EC: 203–705-3

UNII: 77OY909F30

ICD-11: —

$C_4 H_{10} S$

Colorless mobile liquid with a strong, obnoxious odor like garlic, cabbage, or a skunk depending on the concentration. Various salts (solids) have been reported. This material has been used by the military as a simulant of a non-persistent agent to evaluate chemical equipment. It is also used to activate detector kits and alarms. It has been used during field exercises to simulate the threat posed by toxic agent vapor and assist in training soldiers on the proper use of protective equipment.

Used industrially as a solvent and chemical intermediate for insecticides, acaricides, herbicides, and defoliants. It is a flavoring agent and also used in agriculture as a deer repellant.

Exposure Hazards

Conversion Factor: 1 ppm = 3.69 mg/m^3 at 77°F

MEG$_{(1hr)}$: Neg: 5.4 ppm; Mar: 41 ppm; Crit: 410 ppm; Cat: —

PAC-1: 4.3 ppm; *PAC-2:* 49 ppm; *PAC-3:* 2,500 ppm

PEL: 10 ppm

TLV: 0.5 ppm

Ceiling(NIOSH): 0.5 ppm

IDLH: 500 ppm

Properties:

MW: 90.2	*VP:* 35 mmHg	*FlP:* 35°F
D: 0.84 g/mL (77°F)	*VD:* 3.1	*LEL:* 1.4%
MP: −176°F	*Vlt:* 60,000 ppm (77°F)	*UEL:* 10.2%
BP: 209°F	*ER:* —	*RP:* 0.23
Vsc: 0.67 cS	*H₂O:* 0.06%	*IP:* 9.15 eV
	Sol: Alcohol, ether	

C15-A003

s-Butyl Mercaptan

CAS: 513-53-1

RTECS: —

UN: 2347; 130

EC: 208-165-2

UNII: Y2D731QBYN

ICD-11: —

$C_4 H_{10} S$

Colorless, mobile liquid with a heavy skunk-like odor.

Used industrially as a chemical intermediate for cadusafos and as an odorant for natural gas.

Exposure Hazards

Conversion Factor: 1 ppm = 3.69 mg/m³ at 77°F

MEG(1hr): Neg: 5.4 ppm; Mar: 41 ppm; Crit: 203 ppm; Cat: —

PAC-1: 4.3 ppm; *PAC-2:* 46 ppm; *PAC-3:* 270 ppm

Properties:

MW: 90.2	*VP:* 80.7 mmHg (77°F)	*FlP:* −10°F
D: 0.83 g/mL (63°F)	*VD:* 3.1	*LEL:* 1.7%
MP: −265°F	*Vlt:* 110,000 ppm	*UEL:* 9.6%
BP: 185°F	*ER:* 220	*RP:* 0.13
Vsc: 0.554 cS	*H₂O:* 0.13%	*IP:* 9.10 eV
	Sol: Alcohol, ether, benzene	

C15-A004

Isobutyl Mercaptan

CAS: 513-44-0

RTECS: TZ7630000

UN: 2347; 130

EC: 208-162-6

UNII: 9H070UFP2X

ICD-11: —

C$_4$ H$_{10}$ S

Clear, colorless to pale yellow liquid with a heavy skunk-like odor.

It is a flavoring agent.

Exposure Hazards

Conversion Factor: 1 ppm = 3.69 mg/m^3 at 77°F

Human toxicity values have not been established or have not been published.

Properties:

MW: 90.2	*VP:* 69.79 mmHg (77°F)	*FIP:* 14°F
D: 0.83 g/mL (77°F)	*VD:* 3.1	*LEL:* —
MP: –110°F	*Vlt:* 92,000 ppm	*UEL:* —
BP: 191°F	*ER:* 230	*RP:* 0.15
Vsc: —	*H$_2$O:* 0.17%	*IP:* 9.12 eV
	Sol: Alcohol, ether	

C15-A005

t-Butyl Mercaptan

CAS: 75-66-1

RTECS: TZ7660000

UN: 2347; 130

EC: 200–890-2

UNII: 489PW92WIV

ICD-11: —

C$_4$ H$_{10}$ S

Colorless mobile liquid with a strong, offensive, skunk-like odor.

Used industrially as an odorant for natural gas, chemical intermediate, and bacterial nutrient

Exposure Hazards

Conversion Factor: 1 ppm = 3.69 mg/m^3 at 77°F

Human toxicity values have not been established or have not been published.

Properties:

MW: 90.2	*VP:* 143 mmHg	*FIP:* -15°F
D: 0.79 g/mL (77°F)	*VD:* 3.1	*LEL:* —
MP: 31°F	*Vlt:* 240,000 ppm	*UEL:* —

BP: 147°F   *ER:* 210   *RP:* 0.059
Vsc: 0.81 cS   *H₂O:* 0.2% (77°F)   *IP:* 9.03 eV
                                  Sol: Alcohol, acetone, ether

---

C15-A006

Mercaptoethyl Sulfide
CAS: 3570-55-6
RTECS: —
UN: —
EC: 222–671-0
UNII: OEU4AZC07S
ICD-11: —

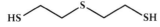

$C_4 H_{10} S_3$

Colorless to yellow liquid with a strong, offensive, skunk-like odor.

Exposure Hazards
Conversion Factor: 1 ppm = 6.31 mg/m³ at 77°F

Human toxicity values have not been established or have not been published.

Properties:

MW: 154.3	*VP:* 0.05 mmHg (est.)	*FlP:* 194°F
D: 1.2 g/mL (77°F)	*VD:* 5.3	*LEL:* —
MP: 12°F	*Vlt:* —	*UEL:* —
BP: 276°F (10 mmHg)	*ER:* —	*RP:* —
Vsc: —	*H₂O:* Insoluble	*IP:* —
	Sol: Ether, alcohols, chloroform, toluene	

C15-A007

Butyl Sulfide
CAS: 544-40-1
RTECS: ER6417000
UN: —
EC: 208–870-5
UNII: 3E3H471GA3
ICD-11: —

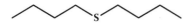

$C_8 H_{18} S$

Clear, colorless to slightly yellow liquid with a skunk-like stench. This material is hazardous through inhalation and produces local skin/eye impacts.

Exposure Hazards

Conversion Factor: 1 ppm = 5.98 mg/m^3 at 77°F

Human toxicity values have not been established or have not been published.

Properties:

MW: 146.3	*VP:* 1.2 mmHg (77°F)	*FlP:* 169°F
D: 0.84 g/mL	*VD:* 5.0	*LEL:* —
MP: −112°F	*Vlt:* 1,600 ppm (77°F)	*UEL:* —
BP: 372°F	*ER:* —	*RP:* 7.0
Vsc: —	*H$_2$O:* 0.004% (77°F)	*IP:* 8.40 eV
	Sol: Alcohol, ether	

C15-A008

Amyl Mercaptan

CAS: 110-66-7
RTECS: SA3150000
UN: 1111; 130
EC: 203–789-1
UNII: 9A3YK965F3
ICD-11: —

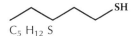

C$_5$ H$_{12}$ S

Colorless to yellow liquid with a strong, disagreeable odor like garlic detectable at ≥0.070 ppm.

Used industrially as a chemical intermediate, synthetic flavoring ingredient, and odorant to locate gas leaks.

Exposure Hazards

Conversion Factor: 1 ppm = 4.26 mg/m^3 at 77°F

Ceiling$_{(NIOSH)}$: 0.5 ppm

Properties:

MW: 104.2	*VP:* 13.8 mmHg (77°F)	*FlP:* 65°F
D: 0.84 g/mL	*VD:* 3.6	*LEL:* —
MP: −104°F	*Vlt:* 18,000 ppm	*UEL:* —
BP: 260°F	*ER:* —	*RP:* 0.72
Vsc: —	*H$_2$O:* 0.016%	*IP:* —
	Sol: Alcohol, ether	

C15-A009

s-Amyl Mercaptan

CAS: 2084-19-7
RTECS: —
UN: 1111; 130
EC: 218–224-4
UNII: 529XY6M4QX
ICD-11: —

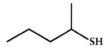

C$_5$ H$_{12}$ S

Colorless liquid with a disagreeable odor (stench).

Exposure Hazards
Conversion Factor: 1 ppm = 4.26 mg/m^3 at 77°F

Human toxicity values have not been established or have not been published.

Properties:

MW: 104.2	*VP:* 23.5 mmHg	*FIP:* 59°F
D: 0.83 g/mL	*VD:* 3.6	*LEL:* —
MP: −272°F	*Vlt:* 31,000 ppm	*UEL:* —
BP: 235°F	*ER:* −	*RP:* 0.41
Vsc: —	*H$_2$O:* Slight	*IP:* —
	Sol: Alcohol	

C15-A010

Isoamyl Mercaptan
CAS: 541-31-1
RTECS: —
UN: 1111; 130
EC: 208–774-3
UNII: MMK4SUN45E
ICD-11: —

C$_5$ H$_{12}$ S

Clear, colorless to slightly yellow liquid with a strong disagreeable odor (stench).

Exposure Hazards
Conversion Factor: 1 ppm = 4.26 mg/m^3 at 77°F

Human toxicity values have not been established or have not been published.

Properties:

MW: 104.2	*VP:* 41.4 mmHg (100°F)	*FIP:* 64°F
D: 0.84 g/mL	*VD:* 3.6	*LEL:* —
MP: —	*Vlt:* 52,000 ppm	*UEL:* —
BP: 244°F	*ER:* −	*RP:* 2.5
Vsc: —	*H$_2$O:* Insoluble	*IP:* —
	Sol: Alcohol	

C15-A011

t-Amyl Mercaptan
CAS: 1679-09-0
RTECS: EK6570500
UN: 1111; 130
EC: 216–843-4
UNII: 26IU670827
ICD-11: —

$C_5 H_{12} S$

Colorless to pale yellow hygroscopic oil with a strong disagreeable odor (stench).

Used industrially as a chemical intermediate, odorant, and bacterial nutrient.

Exposure Hazards
Conversion Factor: 1 ppm = 4.26 mg/m^3 at 77°F

Human toxicity values have not been established or have not been published.

Properties:

MW: 104.2	*VP:* 87.2 mmHg (100°F)	*FlP:* 30°F
D: 0.81 g/mL	*VD:* 3.6	*LEL:* —
MP: –155°F	*Vlt:* 110,000 ppm (100°F)	*UEL:* —
BP: 210°F	*ER:* 190	*RP:* 0.12
Vsc: —	*H₂O:* —	*IP:* —
	Sol: Chloroform	

C15-A012

Allyl Sulfide
CAS: 592-88-1
RTECS: BC4900000
UN: —
EC: 209–775-1
UNII: 60G7CF7CWZ
ICD-11: —

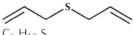

$C_6 H_{10} S$

Colorless liquid with a strong odor like garlic or horseradish.

Used industrially for the manufacture of food flavors.

Exposure Hazards
Conversion Factor: 1 ppm = 4.67 mg/m^3 at 77°F

Human toxicity values have not been established or have not been published.

Properties:

MW: 114.2

D: 0.89 g/mL (81°F)

MP: −117°F

BP: 282°F

Vsc: —

VP: 6.75 mmHg

VD: 3.9

Vlt: 12,000 ppm

ER: —

H$_2$O: Practically insoluble

Sol: Alcohol, chloroform, ether

FIP: 115°F

LEL: —

UEL: —

RP: 1.0

IP: 8.52 eV

C15-A013

1-Hexanethiol

CAS: 111-31-9

RTECS: MO4550000

UN: —

EC: 203–857-0

UNII: 1KH2129A0Q

ICD-11: —

C$_6$H$_{14}$S

Clear colorless liquid with a strong disagreeable odor (stench). Reacts with air on long storage.

Exposure Hazards

Conversion Factor: 1 ppm = 4.83 mg/m^3 at 77°F

MEG$_{(1hr)}$: Neg: 0.072 ppm; Mar: 0.52 ppm; Crit: 100 ppm; Cat: —

PAC-1: 0.046 ppm; *PAC-2:* 0.50 ppm; *PAC-3:* 48 ppm

Ceiling$_{(NIOSH)}$: 0.5 ppm

Properties:

MW: 118.2

D: 0.84 g/mL

MP: −113°F

BP: 304°F

Vsc: —

VP: 4.2 mmHg

VD: 4.1

Vlt: 5,600 ppm

ER: —

H$_2$O: Insoluble

Sol: Alcohol

FIP: 68°F

LEL: 0.7%

UEL: 5.5%

RP: 2.2

IP: —

C15-A014

1-Dodecanethiol

CAS: 112-55-0

RTECS: JR3155000

UN: —

EC: 203–984-1

UNII: S8ZJB6X253

ICD-11: —

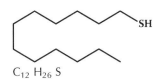

C$_{12}$ H$_{26}$ S

Colorless or pale-yellow oily liquid with a mild, skunk-like odor detectable at ≥0.5 ppm.

Used industrially as a chemical intermediate in the manufacture of synthetic rubber, plastics, pharmaceuticals, insecticides, fungicides, and nonionic detergents; used in industry as a flotation reagent for the removal of metals from wastes.

This material has also been used as a simulant in government tests.

Exposure Hazards
Conversion Factor: 1 ppm = 8.28 mg/m^3 at 77°F
MEG$_{(1hr)}$: Neg: 0.10 ppm; Mar: 0.48 ppm; Crit: 2.4 ppm; Cat: —
PAC-1: 0.30 ppm; *PAC-2:* 0.50 ppm; *PAC-3:* 3.0 ppm
TLV: 0.1 ppm
Ceiling$_{(NIOSH)}$: 0.5 ppm

Properties:

MW: 202.4	*VP:* 0.00853 mmHg	*FlP:* 262°F
D: 0.84 g/mL	*VD:* 7.0	*LEL:* —
MP: 19°F	*Vlt:* —	*UEL:* —
BP: 500°F–514°F	*ER:* —	*RP:* 820
Vsc: 3.55 cS	*H$_2$O:* 2 × 10^{-5}% (77°F)	*IP:* —
	Sol: Methanol, acetone ethyl acetate	

C15-A015

1-Octadecanethiol
CAS: 2885-00-9
RTECS: —
UN: —
EC: 220–744-1
UNII: —
ICD-11: —

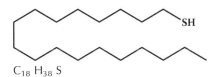

C$_{18}$ H$_{38}$ S

White solid with a strong disagreeable odor (stench).

This material has also been used as a simulant in government tests.

Exposure Hazards
Conversion Factor: 1 ppm = 11.72 mg/m^3 at 77°F
Ceiling$_{(NIOSH)}$: 5.9 mg/m^3

Properties:

MW: 286.6	*VP:* 4.0×10^{-7} mmHg	*FlP:* 365°F
D: 0.85 g/cm^3	*VD:* 9.9	*LEL:* —
MP: 86°F	*Vlt:* —	*UEL:* —
BP: 680°F	*ER:* —	*RP:* —
Vsc: —	*H$_2$O:* Insoluble	*IP:* —
	Sol: —	

C15-A016

Cadaverine

CAS: 462-94-2; 1476-39-7 (Hydrochloride salt)

RTECS: SA0200000

UN: —

EC: 207–329-0; 216-022-0 (Hydrochloride salt)

UNII: L90BEN6OLL; 84700R84PB (Hydrochloride salt)

ICD-11: —

$C_5 H_{14} N_2$

Clear colorless to slightly yellow, syrupy, hygroscopic liquid with the odor of rotting flesh. Various salts have been reported.

Exposure Hazards

Conversion Factor: 1 ppm = 4.18 mg/m^3 at 77°F

Human toxicity values have not been established or have not been published.

Properties:

MW: 102.2	*VP:* 0.666 mmHg	*FlP:* 144°F
D: 0.87 g/mL	*VD:* 3.5	*LEL:* —
MP: 53°F	*Vlt:* 890 ppm	*UEL:* —
BP: 354°F	*ER:* —	*RP:* 15
Vsc: —	*H$_2$O:* Miscible	*IP:* —
	Sol: Ethanol	

C15-A017

Putrescine

CAS: 110-60-1; 333-93-7 (Hydrochloride salt)

RTECS: EJ6800000

UN: —

EC: 203–782-3; 206–375-9 (Hydrochloride salt)

UNII: V10TVZ52E4; X45SUR7RHY {Hydrochloride Salt}

ICD-11: —

$C_4 H_{12} N_2$

White solid or colorless oily liquid with a putrid odor like rotting flesh. Various salts have been reported.

Exposure Hazards

Conversion Factor: 1 ppm = 3.61 mg/m^3 at 77°F

Human toxicity values have not been established or have not been published.

Properties:

MW: 88.2	*VP:* 2.5 mmHg	*FlP:* 145°F
D: 0.88 g/mL	*VD:* 3.0	*LEL:* 0.7%
MP: 81°F	*Vlt:* 5,400 ppm	*UEL:* 11.2%
BP: 316°F	*ER:* —	*RP:* 2.6
Vsc: —	*H$_2$O:* Miscible	*IP:* —
	Sol: Ethanol, ether	

C15-A018

Butyric Acid
CAS: 107-92-6
RTECS: ES5425000
UN: 2820; 153
EC: 203–532-3
UNII: 40UIR9Q29H
ICD-11: —

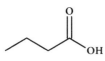

C$_4$ H$_8$ O$_2$

Colorless oily liquid with a pungent putrid odor like rancid butter or vomit detectable at ≥2.5 ppm.

This material has been used by the military as a simulant of a non-persistent agent to evaluate chemical equipment. It has been used during field exercises to simulate the threat posed by toxic agents and assist in training soldiers on the proper use of protective equipment.

A mixture of butyric acid (2%), magnesium silicate (5%), and water (93%) was once fielded as a simulant for mustard gas (C03-A001), with the military designation AS.

Used industrially as a chemical intermediate for pharmaceuticals, emulsifiers, disinfectants, perfumes; artificial flavorings; in the manufacture of cellulose derivatives in lacquers and plastics; used as a leather tanning agent, as a sweetening agent for gasoline; in the food industry to add body to butter, cheese, butterscotch, caramel, fruit, and nut flavors; and in some countries it is used as an antifungal agent in the preservation of high moisture grains.

Exposure Hazards

Conversion Factor: 1 ppm = 3.60 mg/m^3 at 77°F

MEG$_{(1hr)}$: Neg: 42 ppm; Mar: —; Crit: 210 ppm; Cat: —

PAC-1: 1.4 ppm; *PAC-2:* 16 ppm; *PAC-3:* 110 ppm

Properties:

MW: 88.1	*VP:* 0.43 mmHg	*FlP:* 161°F
D: 0.96 g/mL	*VD:* 3.0	*LEL:* 2.0%
MP: 18°F	*Vlt:* 580 ppm	*UEL:* 10%
BP: 329°F	*ER:* —	*RP:* 25
Vsc: 1.7 cS	*H₂O:* Miscible	*IP:* 10.17 eV
	Sol: Alcohol, ether	

C15-A019

Isobutyric Acid

CAS: 79-31-2
RTECS: NQ4375000
UN: 2529; 132
EC: 201-195-7
UNII: 8LL21OO1U0
ICD-11: —

C₄ H₈ O₂

Colorless liquid with a light odor like rancid butter.

Used industrially as a chemical intermediate for the synthesis of esters, manufacture of varnish, as a tanning agent, grain preservative, and food additive (flavor).

Exposure Hazards

Conversion Factor: 1 ppm = 3.60 mg/m³ at 77°F
MEG₍₁ₕᵣ₎: Neg: 0.97 ppm; Mar: 6.9 ppm; Crit: 36 ppm; Cat: —
PAC-1: 0.23 ppm; *PAC-2:* 2.6 ppm; *PAC-3:* 15 ppm

Properties:

MW: 88.1	*VP:* 1.81 mmHg (77°F)	*FlP:* 131°F
D: 0.95 g/mL	*VD:* 3.0	*LEL:* 2.0%
MP: −53°F	*Vlt:* 2,400 ppm	*UEL:* 9.2%
BP: 306°F (decomposes)	*ER:* —	*RP:* 5.9
Vsc: 1.5 cS	*H₂O:* 16.7%	*IP:* 10.24 eV
	Sol: Alcohol, chloroform, ether	

C15-A020

2-Methylbutyric Acid
CAS: 116-53-0
RTECS: —
UN: —
EC: 204-145-2

UNII: PX7ZNN5GXK

ICD-11: —

$C_5 H_{10} O_2$

Colorless to slightly yellow liquid with a strong disagreeable odor (stench).

Exposure Hazards

Conversion Factor: 1 ppm = 4.18 mg/m^3 at 77°F

Human toxicity values have not been established or have not been published.

Properties:

MW: 102.1	*VP:* 0.375 mmHg	*FIP:* 171°F
D: 0.94 g/mL	*VD:* 3.5	*LEL:* 1.2%
MP: −94°F	*Vlt:* 650 ppm	*UEL:* 5.7%
BP: 352°F	*ER:* —	*RP:* 20
Vsc: 2.2 cS	*H$_2$O:* 4.5%	*IP:* 10.27 eV
	Sol: Alcohol, propylene glycol	

C15-A021

3-Methylindole

CAS: 83-34-1

RTECS: NM0350000

UN: —

EC: 201–471–7

UNII: 9W945B5H7R

ICD-11: —

$C_9 H_9 N$

White crystalline solid that darkens and turns brown with exposure to air. At very low levels, it has a pleasant, sweet, warm odor similar to over-ripe fruit. Otherwise, a putrid odor like fecal matter.

This material acts as an odor intensifier.

Used industrially as a fixative for perfumes, artificial civet, food additive (flavoring), and medication.

Exposure Hazards

Conversion Factor: 1 ppm = 5.37 mg/m^3 at 77°F

Human toxicity values have not been established or have not been published.

Properties:

MW: 131.2	VP: 0.00555 mmHg	FIP: 270°F
D: 0.2 g/cm³	VD: 4.5	LEL: —
MP: 203°F	Vlt: —	UEL: —
BP: 509°F	ER: —	RP: 1,600
Vsc: —	H₂O: 0.05% (77°F)	IP: 7.54 eV
	Sol: Acetone, alcohol, benzene, ether	

MW: 131.2

D: 0.2 g/cm³

MP: 203°F

BP: 509°F

Vsc: —

VP: 0.00555 mmHg

VD: 4.5

Vlt: —

ER: —

H₂O: 0.05% (77°F)

Sol: Acetone, alcohol, benzene, ether

FlP: 270°F

LEL: —

UEL: —

RP: 1,600

IP: 7.54 eV

C15-A022

4-Methyl Indole

CAS: 16096-32-5

RTECS: —

UN: —

EC: 240–262-5

UNII: 3338387XEA

ICD-11: —

C₉ H₉ N

Clear, yellow to brown, light-sensitive liquid with a disagreeable odor.

This material acts as an odor intensifier.

Exposure Hazards

Conversion Factor: 1 ppm = 5.37 mg/m³ at 77°F

Human toxicity values have not been established or have not been published.

Properties:

MW: 131.2

D: 1.1 g/mL

MP: 41°F

BP: 513°F

Vsc: —

VP: 0.0087 mmHg (est.)

VD: 4.5

Vlt: —

ER: —

H₂O: 0.1% (est.)

Sol: —

FlP: >234°F

LEL: —

UEL: —

RP: —

IP: 7.6 eV

C15-A023

6-Methyl Indole

CAS: 3420-02-8

RTECS: —

UN: —

EC: —

UNII: QTM6OJZ56S

ICD-11: —

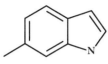

$C_9 H_9 N$

Pale yellow to light-brown, light-sensitive solid with a disagreeable odor.

This material acts as an odor intensifier.

Exposure Hazards

Conversion Factor: 1 ppm = 5.37 mg/m^3 at 77°F

Human toxicity values have not been established or have not been published.

Properties:

MW: 131.2	*VP:* 0.01 mmHg (est.)	*FlP:* 230°F
D: 1.1 g/cm^3	*VD:* 4.5	*LEL:* —
MP: 84°F	*Vlt:* —	*UEL:* —
BP: 234°F (5 mmHg)	*ER:* —	*RP:* —
Vsc: —	*H$_2$O:* Insoluble	*IP:* 7.54 eV
	Sol: —	

C15-A024

Who Me

CAS: —
RTECS: —
UN: —
EC: —
UNII: —
ICD-11: —

Mixture of sulfur and volatile agents with a strong smell of fecal matter.

Exposure Hazards

Human toxicity values have not been established or have not been published.

C15-A025

Standard Bathroom Malodor

CAS: —
RTECS: —
UN: —
EC: —
UNII: —
ICD-11: —

Mixture of 62.82% dipropylene glycol, 21.18% mercaptoacetic acid, 6% n-hexanoic acid, 6% N-methyl morpholine, 2.18% p-cresyl isovalerate, 0.91% 2-naphthalenethiol, and 0.91% skatole.

Exposure Hazards
Human toxicity values have not been established or have not been published.

C15-A026

Dippel's Oil
CAS: 8001-85-2
RTECS: —
UN: —
EC: 232-294-3
UNII: 7A7G0PQI12
ICD-11: —

Dark-colored, highly viscous liquid with a potent smell, described as somewhere between wood-creosote and rancid meat. Made by the destructive distillation of bones and contains pyrrole, aniline, stearonitrile, valeronitrile, pyridine, methylamine, and other nitrogenous compounds.

Use as an animal deterrent, insect attractant, to denature alcohol, and for production of pyrrole

It is combustible, has a density of 0.97 g/mL, and is insoluble in water.

Exposure Hazards
Human toxicity values have not been established or have not been published.

C15-A027

Skunk
CAS: —
RTECS: —
UN: —
EC: —
UNII: —
ICD-11: —

A proprietary liquid agent containing a blend of yeast and sodium bicarbonate developed by the Israel Defense Force. It has an odor described as a cross between rotting meat and human excrement. While nontoxic and biodegradable, unless washed with a special decontaminating soap, the odor persists for an extended period.

Exposure Hazards
Human toxicity values have not been established or have not been published.

BIBLIOGRAPHY

Dalton, Pamela and Gary K. Beauchamp. *Establishment of Odor Response Profiles: Ethnic, Racial and Cultural Influences*. Philadelphia, Pennsylvania: Monell Chemical Senses Center, February 4, 1999.

International Labour Organization. *International Chemical Safety Cards (ICSCs)* [https://www.ilo.org/safework/info/publications/WCMS_113134/lang--en/index.htm]. December 31, 2020.

National Institute for Occupational Safety and Health. *NIOSH Pocket Guide to Chemical Hazards*. Washington, D.C.: Government Printing Office, September 2007.

Pinney, Virginia Ruth. "Malodorant Compositions," United State Patent 6242489, June 5, 2001.

Pinney, Virginia Ruth. "Non-lethal Weapon Systems," United State Patent 6386113, May 14, 2002.

Prentice, John A. "Defensive and Offensive Projector Composition," United States Patent 1643954, October 4, 1927.

Transport Canada, United States Department of Transportation, Secretariat of Transport and Communications of Mexico. *2020 Emergency Response Guidebook.* Neenah, Wisconsin: J.J. Keller and Associates, 2020.

United States Air Force. *Development of Candidate Chemical Simulant List: The Evaluation of Candidate Chemical Simulants Which May Be Used in Chemically Hazardous Operations, Technical Report AFAMRL-TR-82-87.* Washington, D.C.: Government Printing Office, 1982.

United States National Institute of Health, National Library of Medicine. PubChem. 2020 [https://pubchem.ncbi.nlm.nih.gov/]. December 31, 2020.

Yaws, Carl L. *Matheson Gas Data Book.* 7th Edition. Parsippany, New Jersey: Matheson Tri-Gas, 2001.

Biological Agents

16 Toxins

GENERAL INFORMATION

Toxins are any poisonous substance that is produced by an animal, plant or microbe. Because of their complexity, most toxins are difficult to synthesize in large quantities by traditional chemical means. However, they can be harvested from cultured sources or produced by genetically engineered species. Toxins are odorless, tasteless, and nonvolatile.

Toxins are often referred to as mid-spectrum agents; meaning they fall between classical manmade chemical agents (e.g., tabun (C01-A001, Volume 1)) and living biological agents (e.g., *Bacillus anthracis* (C17-A001)). They are covered under the general-purpose criteria of both the Chemical Weapons Convention and the Biological Weapons Convention. Ricin (C16-A031) and saxitoxin (C16-A015) are the only toxins specifically listed in the Chemical Weapons Convention (Schedule 1). Unlike the Chemical Weapons Convention, the Biological Weapons Convention does not attempt to identify or list individual pathogens or toxins.

Several countries have stockpiled a limited number of toxins. Their use on the battlefield has been alleged (e.g., Laos, Kampuchea, and Afghanistan) but not documented to the extent that it is universally accepted. Toxins have been used for political assassinations (e.g., 1978 murder of Georgi Markov with ricin) and terrorists have threatened the use of toxins, usually through contamination of food or water supplies.

TOXICOLOGY

EFFECTS

Toxins present a variety of both incapacitating and lethal effect. Most toxins of military significance can be broadly classified in one of two ways. Neurotoxins disrupt the nervous system and interfere with nerve impulse transmission similar to nerve agents (C01–C02, Volume 1). However, all neurotoxins do not operate through the same mechanism of action nor do they produce the same symptoms. Cytotoxins are poisons that destroy cells or impair cellular activities. Symptoms may resemble those of vesicants (C03–C05, Volume 1) or they may resemble food poisoning or other diseases. Toxins may also produce effects that are a combination of these general categories. The consequences of intoxication from any individual toxin can vary widely with route of exposure and dose. In addition, some toxins act as biomediators and cause the body to release excessive, and therefore harmful, amounts of chemicals that are normally produced by the body (see bioregulators, C20).

Although classified as biological weapons, toxins are chemicals. They are not alive and do not replicate themselves like pathogens (C17–C19). They are not communicable; to be affected an individual must come into direct contact with the toxin.

PATHWAYS AND ROUTES OF EXPOSURE

Toxins are primarily hazardous through inhalation, ingestion, and broken, abraded, or lacerated skin (e.g., penetration of skin by debris). A small number of toxins, such as the mycotoxins, are dermally hazardous and produce skin lesions as well as systemic illness through skin and eye exposure. Some toxins of appropriate molecular size and structure may also be dissolved in select solvents and delivered as solutions that pose a percutaneous hazard.

DOI: 10.4324/9781003230564-7

General Exposure Hazards

In general, toxins do not have good warning properties. They are nonvolatile and do not have an odor. Although some toxins irritate the skin and eyes, in most cases they do not. Many neurotoxins will produce severe pain in contact with any abrasion or laceration.

Individual toxins may be effective through multiple pathways and the route of exposure may significantly change the signs and symptoms associated with any given toxin. In most cases, effects are most severe when the toxin is inhaled.

Latency Period

Effects from exposure to toxins can appear within minutes or be delayed for days. The impacts from some toxins, especially cytotoxins, may occur within minutes but symptoms may not appear for hours. The route of exposure to the toxin can significantly change the latency period.

CHARACTERISTICS

Physical Appearance/Odor

Laboratory Grade

Pure toxins are typically colorless, white, tan, or yellow solids. Venoms – crude mixtures of toxins and other natural chemical produced by animals such as snakes, spiders and scorpions – are colorless to yellow or brown liquids.

Modified Agents

Toxins have been dissolved in solvents to facilitate handling, stabilize them, or to create a per-cutaneous hazard. Percutaneous enhancement solvents include dimethyl sulfoxide, N,N-dimethylformamide, N,N-dimethylpalmitamide, N,N-dimethyldecanamide, and saponin. Color and other properties of these solutions may vary from the pure agent. Odors will vary depending on the characteristics of the solvent(s) used.

Toxins have also been micropulverized and microencapsulated to facilitate their dispersal and increase their persistency. Color and other physical properties of the toxin may be affected by these modifications.

Stability

Very dependent on the specific toxin. Many are sensitive to heat and/or light. Freeze drying (lyophilization) or isolation as salts increases their stability and shelf life.

Persistency

For military purposes, toxins are generally non-persistent. In cases where toxins have been mi-cropulverized, microencapsulated, or otherwise modified to facilitate their dispersal, re-aerosolization by ground traffic or strong winds may be a concern.

Environmental Fate

All toxins are nonvolatile. Once the initial aerosol has settled, there is minimal inhalation hazard unless the toxin is released as an aerosolized powder that has been modified to increase the po-tential of re-aerosolization. Solubility in water depends on the specific toxin, presence of solvents, and isolation as salts.

ADDITIONAL HAZARDS

EXPOSURE

All foodstuffs in the area of a release should be considered contaminated. Unopened items packaged in glass, metal, or heavy plastic and exposed only to aerosols may be used after decontamination of the container. All unopened items exposed to bulk agents should be decontaminated within a few hours post-exposure or destroyed. Opened or unpackaged items, or those packaged only in paper or cardboard, should be destroyed.

Meat, milk and animal products from animals affected or killed by toxins should be destroyed.

LIVESTOCK/PETS

Animals can be decontaminated with shampoo/soap and water, or a 0.5% household bleach solution (see the decontamination section below). If the animal's eyes have been exposed to agent, they should be irrigated with water or saline solution for a minimum of 30 minutes.

Unprotected feedstock (e.g., hay or grain) should be destroyed. Depending on the specific toxin released, the level of contamination and the weather conditions, leaves of forage vegetation could still retain sufficient agent to produce effects for several days post-release.

FIRE

Toxins are not volatile and the heat from a fire will destroy these agents. However, actions taken to extinguish the fire may spread the agent before it is destroyed. Runoff from firefighting efforts may pose a potential contact threat through exposure of broken, abraded, or lacerated skin, or though accidental ingestion. Smoke from a fire may contain acrid, irritating, and/or toxic decomposition products.

REACTIVITY

Varies depending on the specific toxin. Some decompose rapidly after they become wet. Many react with strong acids, bases, or oxidizing agents.

PROTECTION

EVACUATION RECOMMENDATIONS

There are no published recommendations for isolation or protective action distances for toxins released in mass casualty situations.

PERSONAL PROTECTIVE REQUIREMENTS

Structural Firefighters' Gear

Structural firefighters' protective clothing is recommended for fire situations only; it is not effective in spill situations or release events. However, toxins have negligible vapor pressure and do not pose a vapor hazard. The primary risk of exposure is through contact with aerosolized agents, bulk agents (e.g., spilled liquids or solids) or solutions of agents. If chemical protective clothing is not available and it is necessary to rescue casualties from a contaminated area, then structural firefighters' gear will provide some skin protection against most toxin aerosols. Contact with bulk material and solutions should be avoided. However, any responder with preexisting areas of cut, abraded, or lacerated skin should not make entry because this places the individual at extreme risk of subcutaneous exposure.

Structural firefighters' protective clothing should never be used as the primary chemical protective garment to enter an area contaminated with dermally hazardous toxins.

There is also a significant hazard posed by injection of toxins through contact with contaminated debris. Appropriate protection to avoid any potential laceration or puncture of the skin is essential.

Respiratory Protection

Self-contained breathing apparatuses (SCBAs) or air purifying respirators (APRs) should have a NIOSH CBRN certification. However, during emergency operations, other NIOSH-approved SCBAs or APRs that have been specifically tested by the manufacturer against chemical warfare agents may be used if deemed necessary by the incident commander. APRs should be equipped with a NIOSH-approved Chemical/Biological/Radiological/Nuclear (CBRN) filter or a combination organic vapor/acid gas/particulate cartridge.

Immediately dangerous to life or health (IDLH) levels are the ceiling limit for respirators other than SCBAs. However, IDLH levels have not been established for toxins. Therefore, any potential exposure to aerosols of these agents should be regarded with extreme caution and the use of SCBAs for respiratory protection should be considered.

Chemical Protective Clothing

Use only chemical protective clothing that has undergone material and construction performance testing against toxins. Reported permeation rates may be affected by solvents or impurities in these agents.

In the event that dermally hazardous toxins have been released, responders should wear a Level A protective ensemble. Also, because of the extreme hazard posed by toxin aerosols to any area of cut or lacerated skin, responders should wear a Level A protective ensemble whenever there is any potential for exposure to an airborne agent.

Because there is a significant hazard posed by injection of toxins through contact with debris, appropriate protection to avoid any potential abrasion, laceration, or puncture of the skin is essential.

DECONTAMINATION

General

Apply universal decontamination procedures using soap and water.

Most toxins are readily destroyed by high pH (i.e., basic solutions), especially when used in combination with a strong oxidizing agent. For this reason, undiluted household bleach is an excellent agent for decontamination of these agents. Ensure that the bleach solution remains in contact with the toxin for a minimum of 10 minutes.

Liquids, Solutions, or Liquid Aerosols

Casualties/Personnel

Cover all open wounds during the decontamination process. Remove all clothing immediately. To avoid further exposure of the head, neck and face to the agent, cut off potentially contaminated clothing that must be pulled over the head. Use a sponge or cloth with liquid soap and copious amounts of water to wash the skin surface and hair at least three times. Ideally, showers will be high volume with low pressure. Do not delay decontamination to find warm or hot water if it is not readily available. Avoid rough scrubbing as this could abrade the skin and increase the potential for movement of any residual toxin through the skin barrier. Rinse with copious amounts of water. If there is a potential that the eyes have been exposed to the agent, irrigate with water or 0.9% saline solution for a minimum of 15 minutes.

Although the CDC does not recommend the use of hypochlorite bleach directly on skin for decontamination, the military still recognizes household bleach as an effective alternative to soap and water. The bleach solution should be no more than 0.5% sodium hypochlorite (i.e., no more than one-part household bleach in nine-parts water) to avoid damaging the skin. Avoid any contact with sensitive areas such as the eyes or open wounds. Ensure that the bleach solution remains in contact with the agent for a minimum of 10 minutes. Rinse with copious amounts of water.

Small areas

Ventilate to remove the aerosol. Puddles of liquid can be absorbed by covering with absorbent material such as vermiculite, diatomaceous earth, clay, sponges, or towels. Place the absorbed material into containers lined with high-density polyethylene. Wash the area with copious amounts of soap and water or undiluted household bleach (see general section above). If bleach is used, then rewash the area with soap and water. Collect and containerize the rinseate in containers lined with high-density polyethylene.

Solids or Particulate Aerosols

Casualties/Personnel

Do not attempt to brush the agent off of the individual or their clothing as this can aerosolized the agent. Cover all open wounds during the decontamination process. If possible, dampen the agent with a water mist to help prevent aerosolization. Remove all clothing immediately. To avoid further exposure of the head, neck, and face to the agent, cut off potentially contaminated clothing that must be pulled over the head. Wash the skin surface and hair at least three times with copious amounts of soap and water. Ideally, showers will be high volume with low pressure. Do not delay decontamination to find warm or hot water if it is not readily available. Rinse with copious amounts of water. If there is a potential that the eyes have been exposed to toxins, irrigate with water or 0.9% saline solution for a minimum of 15 minutes.

Although the CDC does not recommend the use of hypochlorite bleach directly on skin for decontamination, the military still recognizes household bleach as an effective alternative to soap and water. The bleach solution should be no more than 0.5% sodium hypochlorite (i.e., no more than one-part household bleach in nine-parts water) to avoid damaging the skin. Avoid any contact with sensitive areas such as the eyes or open wounds. Ensure that the bleach solution remains in contact with the agent for a minimum of 10minutes. Rinse with copious amounts of water.

Small areas

Extreme care must be exercised when dealing with dry or powdered agents as toxins may adhere to the skin or clothing then be spread to other areas. Because of the minute quantities need to produce a response in an exposed individual, cross contamination can pose a significant inhalation or puncture hazard later.

If indoors, close windows and doors in the area and turn off anything that could create air currents (e.g., fans, air conditioner, etc.). Allow aerosol to settle. Avoid actions that could aerosolize the agent such as sweeping or brushing. Collected the agent with a vacuum cleaner equipped with a high-efficiency particulate air (HEPA) filter. Do not use a standard home or industrial vacuum. Do not allow the vacuum exhaust to stir the air in the affected area. Vacuum all surfaces with extreme care in a very slow and controlled manner to minimize aerosolizing the agent. Place the collected material into containers lined with high-density polyethylene. Wash the area with copious amounts of soap and water or undiluted household bleach (see general section above). If bleach is used, then rewash the area with soap and water. Collect and containerize the rinseate in containers lined with high-density polyethylene.

MEDICAL

CDC Case Definition

1) A case in which the toxin or appropriate metabolite is detected in clinical specimens (e.g., serum, plasma, urine). OR 2) Detection of the specific toxin in environmental samples unless there could be a local source of the toxin (e.g., molds that produce mycotoxins have been found in some residential and industrial settings, and the toxins have been implicated in some cases of sick building syndrome).

A case should not be considered due to toxin poisoning if another confirmed diagnosis exists to explain the signs and symptoms. However, the case can be confirmed if either a predominant amount of clinical and nonspecific laboratory evidence is present or an absolute certainty of the etiology of the agent is known.

Differential Diagnosis

Varies greatly by individual toxin.

Signs and Symptoms

Highly variable depending on the specific toxin, route of exposure and dose. Even symptoms presented by toxins with the same general classification (i.e., neurotoxin or cytotoxin) may vary depending on the specific mechanism of action within the body.

Mass-Casualty Triage Recommendations

Because of the wide variety of potential toxins and their symptoms, there are no universal recommendations for triaging casualties exposed to toxins as a class. However, in general, anyone who has been exposed should be transported to a medical facility for evaluation. Individuals who are asymptomatic and have not been directly exposed to the agent can be discharged after their names, addresses, and telephone numbers have been recorded. They should be told to seek medical care immediately if symptoms develop.

Casualty Management

Save clinical and environmental samples for diagnosis. The best early diagnostic sample for most toxins is a swab of the nasal mucosa.

Decontaminate the casualty ensuring that all the toxin has been removed. Extreme care must be exercised when dealing with dry or powdered agents as toxins may adhere to the skin or clothing and present an inhalation hazard. If any agent has gotten into the eyes, irrigate the eyes with water or 0.9% saline solution for at least 15 minutes. Irrigate open wounds with water or 0.9% saline solution for at least 10 minutes.

Although these agents do not produce any significant vapor, aerosolization of residual dusts on casualties could cause impacts to medical responders. Once the casualty has been decontaminated, including the removal of foreign matter from wounds, medical personnel do not need to wear a chemical-protective mask.

Treatment primarily consists of supportive care. Ventilate patient if there is difficulty breathing and administer oxygen. Be prepared to treat for shock. Monitor and support cardiac and respiratory functions as necessary. If the identity of the toxin is known, administer antidote if available. Unlike chemical agents, toxins can cause an immune response. Vaccines are available for some toxins but generally require more than four weeks for the body to produce antibodies. Passive immunotherapy is effective for some neurotoxins but must be instituted shortly after exposure. The utility of antibody therapy drops sharply at or shortly after the onset of the first signs of disease.

FATALITY MANAGEMENT

Remove all clothing and personal effects segregating them as either durable or non-durable items. While it may be possible to decontaminate durable items, it may be safer and more efficient to destroy non-durable items rather than attempt to decontaminate them. Items that will be retained for further processing should be double sealed in impermeable containers, ensuring that the inner container is decontaminated before placing it in the outer one.

Extreme care must be exercised when dealing with dry or powdered agents as toxins may adhere to the skin or clothing and present an inhalation hazard.

To remove agents on the outside of the body, wear appropriate respiratory and dermal protective clothing while washing the remains with a two percent sodium hypochlorite bleach solution (i.e., two liters of water for every liter of household bleach), ensuring the solution is introduced into the ears, nostrils, mouth and any wounds. This concentration of bleach will not affect remains but will destroy all toxins on the skin surface, greatly reducing the risk of secondary exposure. Higher concentrations of bleach can harm remains. Pay particular attention to areas where agent may get trapped, such as hair, scalp, pubic areas, fingernails, folds of skin, and wounds. The bleach solution should remain on the cadaver for a minimum of 15 minutes. Wash with soap and water. Ensure that all the bleach solution is removed prior to embalming as it will react with embalming fluid. All wash and rinse waste must be contained for proper disposal.

Once the remains have been thoroughly decontaminated, no further protective action is necessary. Use standard burial procedures.

C16-A   NEUROTOXINS

C16-A001

Aconitine

CAS: 302-27-2; 6034-57-7 (Hydrobromide salt); 6055-69-2 (Hydrochloride salt)

RTECS: AR5960000

EC: 206-121-7; 227–910-2 (Hydrobromide salt); 227–978-3 (Hydrochloride salt)

UNII: X8YN71D5WC; US306163EJ (Hydrobromide salt); DFU5RIN74Y (Hydrochloride salt)

ICD-11: XM2WR7

$C_{34} H_{47} N O_{11}$
Molecular Weight: 645.7
Melting Point: 399°F

Rapid acting channel-activating neurotoxin. It is obtained from the leaves and roots of various plants including wolfbane (*Aconitum lycoctonum*) and monkshood (*Aconitum napellus*). It is an off-white powder that is insoluble in water (0.03%) but soluble in diethyl ether, chloroform, benzene, and alcohol. Various salts have been reported.

This material is hazardous through inhalation, penetration through broken skin, and ingestion. Once in the body, the toxin can become stored in body fat and may have a cumulative effect. Symptoms include nausea, vomiting, diarrhea, slow heart rate (bradycardia), restlessness, incoordination (ataxia), vertigo, difficulty breathing (dyspnea), low body temperature (hypothermia), convulsions, headache, and pallor. Ingestion causes a burning or tingling sensation with subsequent numbness on the lips, tongue, mouth, and throat. It produces similar effects on other mucous membranes. Death results from respiratory or cardiac failure.

Used in biomedical research to study heart arrhythmia.

Toxicology

$LD_{50\ (Ing)}$: 2–60 mg

C16-A002

Anatoxin A (VFDF)

CAS: 64285-06-9; 64314-16-5 (Hydrochloride salt)

RTECS: KM5528500

EC: —

UNII: 80023A73NK

ICD-11: XM9TZ4

$C_{10} H_{15} N O$

Molecular Weight: 165.2

Very rapid-acting paralytic neurotoxin (post-synaptic depolarizing neuromuscular blocker) that binds to the same receptor as acetylcholine (nicotinic cholinergic receptor agonist) producing neural and muscular stimulation. However, acetylcholinesterase does not hydrolyze the toxin. It is an oil obtained from blue-green algae (*Anabaena* spp.). It is water soluble, but decomposes after several days as an aqueous solution. Chlorination, at concentrations used for water purification, is ineffective for inactivating the toxin. It is destroyed by heat, light, and high pH. Various salts (solids) have been reported.

This material is hazardous through inhalation, penetration through broken skin, and ingestion. Symptoms mimic nerve agent exposure and include twitching, incoordination, tremors, paralysis, and respiratory arrest. Symptoms typically begin within 5 minutes. Death from ingestion can be delayed up to 3 hours, depending on the dose. There is no specific treatment.

Toxicology

Human toxicity values have not been established or have not been published. However, based on available information, this material appears to be approximately one-fifth as toxic as the nerve agent VX (C01-A017, Volume 1).

C16-A003

Anatoxin-A(S)

CAS: 103170-78-1

RTECS: —

EC: —

UNII: 4258P76E76

ICD-11: XM9TZ4

$C_7 H_{17} N_4 O_4 P$

Molecular Weight: 252.2

Neurotoxin obtained from freshwater blue-green cyanobacteria (*Anabaena*. spp., *Aphanizomenon* spp., and *Oscillatoria* spp.). It is the only known naturally occurring organophosphate toxin.

This material is hazardous through inhalation, penetration through broken skin, and ingestion. It is an irreversible inhibitor of acetylcholinesterase and produces symptoms similar to an organophosphate pesticide or military nerve agent, including difficulty breathing with a feeling of shortness of breath or tightness of the chest, sweating, nausea, vomiting, involuntary urination/defecation, and a feeling of weakness.

Toxicology

Human toxicity values have not been established or have not been published. However, based on available information, this material appears to be approximately twice as toxic as the nerve agent VX (C01-A017, Volume 1).

C16-A004

Batrachotoxin
CAS: 23509-16-2
RTECS: CR3990000
EC: —
UNII: TSG6XHX09R
ICD-11: XM4BD8

$C_{31} H_{42} N_2 O_6$
Molecular Weight: 538.7

Very rapid-acting paralytic neurotoxin that binds to sodium channels of nerve and muscle cells depolarizing neurons by increasing the sodium channel permeability. It is obtained from South American poison-dart frogs (*Phyllobates* spp.), as well as some beetles (*Choresine* spp.) and several species of birds indigenous to New Guinea. It is insoluble in water but soluble in hydrocarbons and other non-polar solvents. The dried toxin can remain active for at least a year. However, it is relatively non-persistent in the environment.

This material is hazardous through inhalation and penetration through broken skin. Once in the body, the toxin can become stored in body fat and may have a cumulative effect. Symptoms include loss of balance and coordination, profound weakness, irregular heart rhythms, convulsions, and cyanosis. These symptoms occur quickly and are produced in rapid succession. Batrachotoxin has no effect on the skin but produces a long-lasting painful stinging sensation in contact with the smallest scratch. If ingested, it is only toxic if a lesion exists in the gastrointestinal tract.

Used in biomedical research to study sodium channels in the nervous system.

Toxicology

Human toxicity values have not been established or have not been published. However, based on available information, this material appears to be approximately 1,000 times more toxic than the nerve agent VX (C01-A017, Volume 1).

C16-A005

Botulinum Toxin A (Agent X)
CAS: 93384-43-1
RTECS: ED9300000
EC: 297-253-4
UNII: E211KPY694
ICD-11: —

Large proteins
Molecular Weight: 150,000

Delayed-action paralytic neurotoxin that blocks the release of acetylcholine causing a symmetric, descending flaccid paralysis of motor and autonomic nerves. Paralysis always begins with the cranial nerves. Obtained from anaerobic bacteria (*Clostridium* spp.). It is a white powder or crystalline solid that is readily soluble in water. It is stable for up to 7 days as an aqueous solution. It is destroyed by heat and decomposes when exposed to air for more than 12 hours.

This material is hazardous through inhalation, penetration through broken skin, and ingestion. Botulinum toxins are unusual in that they are more toxic when ingested than when inhaled. They are most toxic when injected. Symptoms include dizziness, difficulty swallowing and speaking, blurred or double vision, sensitivity to light, and muscular weakness progressing from the head downward. In some cases, may produce nausea and profuse vomiting. Symptoms from ingestion usually begin within 12 to 72 hours, but can be delayed for up to 8 days. Symptoms from inhalation have a more rapid onset, usually 3 to 6 hours. Onset of symptoms from wound botulism is usually 3 days or longer. Death results from respiratory failure.

Used medicinally as a muscle relaxant.

They are on the Australia Group Core list and Tier-1 agents on the U.S. Select Agents and Toxins list.

Toxicology

$LD_{50\ (Inh)}$: 0.0009 mg (estimate)

$LD_{50\ (Ing)}$: 0.07 mg (estimate)

$MEG_{(1hr)}$: Neg: 4.0×10^{-9} mg/m^3; Mar: 3.0×10^{-8} mg/m^3; Crit: 3.0×10^{-5} mg/m^3; Cat: —

PAC-1: 3.3×10^{-9} mg/m^3; PAC-2: 3.6×10^{-8} mg/m^3; PAC-3: 2.2×10^{-7} mg/m^3

C16-A006

Brevetoxins

ICD-11: —

Paralytic neurotoxins that binds to sodium channels of nerve and muscle cells causing muscle contractions. They are obtained from the dinoflagellate that causes "red tide" (*Gymnodinium breve*). Toxins are typically light tan crystalline solids. They are insoluble in water and very unstable.

These materials are hazardous through inhalation and ingestion, and causes eye irritation. Once in the body, the toxin can become stored in body fat and may have a cumulative effect. Symptoms from ingestion include burning, prickling, or tingling sensations in the mouth as well as reversal of temperature sensations progressing to paralysis of the lips and extremities, with dizziness, incoordination (ataxia), muscle pain (myalgia), nausea, and diarrhea. Inhalation causes constriction or spasms of the airway (bronchospasm) and difficulty breathing (dyspnea).

Toxicology
Human toxicity values have not been established or have not been published.

Brevetoxin PbTx-1 $C_{49} H_{70} O_{13}$
CAS: 98112-41-5
RTECS: —
EC: —
UNII: —
Molecular Weight: 867.1

Brevetoxin PbTx-2 $C_{50} H_{70} O_{14}$
CAS: 79580-28-2
RTECS: XW5886000
EC: —
UNII: —
Molecular Weight: 895.1

Brevetoxin PbTx-3 $C_{50} H_{72} O_{14}$
CAS: 85079-48-7
RTECS: —
EC: —
UNII: —
Molecular Weight: 897.1

C16-A007

Bungarotoxins
ICD-11: XM2RD0

Mixture of neurotoxins that block the acetylcholine receptors. Alpha-bungarotoxin is a postsynaptic neural toxin irreversibly binding nicotinic acetylcholine receptors, beta- and gamma-bungarotoxins act presynaptically causing acetylcholine release and depletion, and κ-bungarotoxin is specific to the neuronal receptors in ganglions. They are obtained from the venom of some krait snakes (*Bungarus* spp.). Various salts have been reported.

This material is hazardous through inhalation and penetration through broken skin. Symptoms include paralysis and respiratory failure.

Used as a research tool in neurophysiology.

Toxicology
$LD_{50 (Sub)}$: 7.6 mg

α-Bungarotoxin $C_{50} H_{70} O_{14}$
CAS. 11032-79-4
RTECS: EI6201400
EC: 234–266-6
UNII: —
Molecular Weight: 895.1

β-Bungarotoxin $C_{50} H_{70} O_{14}$
CAS: 12778-32-4
RTECS: EI6201000
EC: 235–821-5

UNII: —
Molecular Weight: 895.1

C16-A008

Ciguatoxins

ICD-11: XM7V29

Lipid soluble polycyclic ethers with 13 or 14 rings fused into rigid ladder-like structures. There are over 42 known congeners in this family of toxins.

Paralytic neurotoxins that binds to sodium channels of nerve and muscle cells causing muscle contractions. They are heat stable (not destroyed by cooking or freezing) solids obtained from dinoflagellates (*Gambierdiscus* spp.).

These materials are hazardous through ingestion. Symptoms from ingestion include abdominal cramps, nausea, vomiting, diarrhea, severe itching (pruritus), burning, prickling, or tingling skin sensation (paresthesia), headache, muscle pain (myalgia), joint pain (arthralgia), and paradoxical reversal of temperature sensations, paralysis of the lips and extremities. Severe cases may progress to low blood pressure (hypotension), slow heart rate (bradycardia), coma, and respiratory paralysis.

Ciguatoxin-1 $C_{60} H_{86} O_{19}$
CAS: 11050-21-8
RTECS: —
EC: —
UNII: 2UKQ3B7696
Molecular Weight: 1111

Toxicology

Most toxic; 0.007 mg causes mild intoxication in people.

Ciguatoxin-2 $C_{60} H_{86} O_{18}$
CAS: 142185-85-1
RTECS: —
EC: —
UNII: 9FEM1MON3P
Molecular Weight: 1095

Ciguatoxin-3 $C_{60} H_{86} O_{18}$
CAS: 139341-09-6
RTECS: —
EC: —
UNII: 19A86ZW520
Molecular Weight: 1095

C16-A009

Cobrotoxin

CAS: 12584-83-7
RTECS: *GG4220000*
EC: —

UNII: M9RL1JF99O

ICD-11: XM1R49

$C_{277} H_{443} N_{97} O_{98} S_8$

Molecular Weight: 6957

Neurotoxin obtained from cobra venom (*Naja* spp.). It is a relatively heat stable, water soluble, crystalline solid.

This material is hazardous by penetration through broken skin; can cause localized pain, tearing, and blurred vision in contact with the eyes. Symptoms include drooping upper eyelid (ptosis), paralysis of the motor nerves of the eye (ophthalmoplegia), difficulty in swallowing (dysphagia), difficulty speaking (dysphasia), profuse salivation, nausea, vomiting, abdominal pain, muscular weakness followed by flaccid paralysis, chest pain or tightness, shortness of breath, and respiratory failure.

Toxicology

$LD_{50\ (Sub)}$: 2.9 mg

C16-A010

Conotoxins

CAS: —

RTECS: —

EC: —

UNII: —

ICD-11: XM48L3

Oligopeptides, typically 10–30 amino acids long. Obtained from sea snails (*Conus* spp.), over 2,000 individual toxins have been identified. Toxins typically fall into five major families: alpha (bind to and inhibit the nicotinic acetylcholine receptor); delta (block voltage-gated sodium channels in muscles); kappa (block voltage-gated potassium channels in muscles); mu (block voltage-gated sodium channels in muscles); and omega (block voltage-gated calcium channels and also block conduction at the neuromuscular junctions of skeletal muscles. Venoms are white, gray, yellow, or black viscous liquids. Toxins are water soluble freeze-dried solids that are highly stable.

These materials are hazardous through inhalation, penetration through broken skin, and ingestion. Symptoms vary widely based on the toxin, but may include vague feeling of bodily discomfort (malaise), incoordination (ataxia), weakness, headache, nausea, vomiting, difficulty in swallowing (dysphagia), inability to speak (aphonia), numbness, absence of reflexes (areflexia), double vision (diplopia), severe itching (pruritus) or other burning, prickling, tingling skin sensations (paresthesia), rapid heart rate (tachycardia), paralysis, temporary cessation of breathing (apnea), respiratory failure, cerebral swelling (edema), coma, and cardiac failure.

Used as a research tool in molecular biology. Many are being assessed as potential pharmaceuticals for a wide variety of diseases. Omega conotoxins have potential as analgesics. A synthetic version of ω-conotoxin MVIIA has been approved as the analgesic drug ziconotide, and is 1,000 as potent as morphine.

They are on the Australia Group Core list and the U.S. Select Agents and Toxins list.

Toxicology

$LD_{50\ (Inj)}$: 0.07–0.21 mg (*Conus geographus* venom)

C16-A011

Gonyautoxins

ICD-11: —

Natural sulfate homologues of Saxitoxin (C16-A015) obtained from dinoflagellates (*Gonyaulax* spp., *Alexandrium* spp., *Protogonyaulax* spp.) and cyanobacteria (*Anabaena* spp.).

Rapid-acting paralytic neurotoxins that blocks transient sodium channels and inhibits depolarization of nerve cells. They are some of the causative agents of paralytic shellfish poisoning (PSP).

These materials are hazardous through inhalation, penetration through broken skin, and ingestion. Symptoms include numbness or burning, prickling, itching, or tingling sensations (paresthesia) of the lips, tongue, and gums spreading rapidly to the extremities; headache, dizziness, incoordination (ataxia), drowsiness, difficulty swallowing (dysphagia), incoherent speech, flaccid paralysis, and possibly cardiovascular collapse.

Toxicology

Human toxicity values have not been established or have not been published. However, toxicity equivalent factors, comparing the individual toxins to saxitoxin (C16-A015), for the most commonly found toxins have been determined: GTX1 = 100%; GTX2 = 40%; GTX3 = 60%; GTX4 = 70%; GTX5 = 10% and GTX6 = 60%.

Gonyautoxin-1 ($C_{10} H_{17} N_7 O_9 S$)

CAS: 60748-39-2

RTECS: —

EC: —

UNII: —

Molecular Weight: 411.4

Gonyautoxin-2 ($C_{10} H_{17} N_7 O_8 S$)

CAS: 60508-89-6

RTECS: —

EC: —

UNII: YU77Z3NY6Z

Molecular Weight: 395.4

Gonyautoxin-3 ($C_{10} H_{17} N_7 O_8 S$)

CAS: 60537-65-7

RTECS: —

EC: —

UNII: 1FFI43DL2K

Molecular Weight: 395.4

Gonyautoxin-4 ($C_{10} H_{17} N_7 O_9 S$)

CAS: 64296-26-0

RTECS: —

EC: —

UNII: 7Z2LPD8YGH
Molecular Weight: 411.4

Gonyautoxin-5 ($C_{10} H_{17} N_7 O_7 S$)
CAS: 64296-25-9
RTECS: —
EC: —
UNII: 3833H3TURF
Molecular Weight: 379.4

Gonyautoxin-6 ($C_{10} H_{17} N_7 O_8 S$)
CAS: 82810-44-4
RTECS: —
EC: —
UNII: —
Molecular Weight: 395.4

Gonyautoxin-8 ($C_{10} H_{17} N_7 O_{11} S_2$)
CAS: 80226-62-6
RTECS: —
EC: —
UNII: —
Molecular Weight: 475.4

C16-A012

α-Latrotoxin
CAS: 65988-34-3
RTECS: OE9020000
EC: —
UNII: —
ICD-11: XM7JS2

Protein
Molecular Weight: 130,000

Produces a massive release of transmitters from cholinergic and adrenergic nerve endings resulting in continuous stimulation of muscles. It also induces formation of an ion channel allowing the inward flow of calcium ions into the nerve cell. It is a water soluble, white, freeze-dried powder obtained from the venom of the widow spiders (*Latrodectus* spp.).

This material is hazardous by penetration through broken skin. Symptoms include excessive sweating (diaphoresis), salivation, drooping upper eyelid (ptosis), headache, dizziness, nausea, vomiting, contractions, cramps, abdominal pain, severe itching (pruritus), respiration is shallow and rapid, high blood pressure (hypertension), rapid heart rate (tachycardia), and cardiac arrhythmia. May be mistaken for an allergic reaction, acute abdomen, or heart attack. Produces intense pain that becomes excruciating over time.

Used as a research tool in molecular biology.

Toxicology
Human toxicity values have not been established or have not been published.

C16-A013

Neosaxitoxin
CAS: 64296-20-4
RTECS: —
EC: —
UNII: 6YRL8BWD9H
ICD-11: —

$C_{10} H_{17} N_7 O_5$
Molecular Weight: 315.3
Natural analogue of saxitoxin (C16-A015).

Rapid-acting paralytic neurotoxin that reversibly blocks the voltage-gated sodium channels at neuronal level thus stopping the propagation of the nerve impulse. One of the causative agents of paralytic shellfish poisoning (PSP). It is a white solid that is water soluble, heat stable (not destroyed by cooking or freezing) and obtained from dinoflagellates (*Gonyaulax* spp., *Alexandrium* spp., *Protogonyaulax* spp.) and cyanobacteria (*Anabaena* spp.).

This material is hazardous through inhalation, penetration through broken skin, and ingestion. Symptoms include numbness or burning, prickling, itching, or tingling sensations (paresthesia) of the lips, tongue, and gums spreading rapidly to the extremities; headache, dizziness, incoordination (ataxia), drowsiness, difficulty swallowing (dysphagia), incoherent speech, flaccid paralysis, and possibly cardiovascular collapse.

Used as a research tool in molecular biology. It is being tested for medicinal use as a local anesthetic.

Toxicology
Human toxicity values have not been established or have not been published. However, a toxicity equivalent factor comparing it to saxitoxin (C16-A015) has been developed that specifies it is approximately twice as toxic.

C16-A014

Palytoxin
CAS: 11077-03-5
RTECS: RT6475000
EC: —
UNII: OQ17NC0MOV
ICD-11: XM0CW4

$C_{129} H_{223} N_3 O_{54}$
Molecular Weight: 2,680

Rapid-acting neurotoxin that causes irreversible depolarization of neural and muscular tissue by altering their normal ion equilibrium (homeostasis). This causes tissue to breakdown and leak their intracellular contents into the blood (rhabdomyolysis). It is also a potent vasoconstrictor of renal and coronary arteries. It is a non-crystalline solid obtained from soft corals (*Palythoa* spp.) or animals that bioaccumulate the poison. It is soluble in water and alcohol, and stable to heat (not destroyed by cooking or freezing).

This material is hazardous through inhalation, skin absorption (high doses), penetration through broken skin, and ingestion. Symptoms may vary with the route of exposure. Symptoms include fever, cough, sore throat, nasal discharge (rhinorrhea), difficulty breathing (dyspnea), burning, prickling, itching, or tingling sensations (paresthesia) or numbness, bitter metallic taste, rapid breathing (tachypnea), low oxygen levels in the blood (hypoxia), blue lips (cyanosis), low blood pressure (hypotension), rapid heart rate (tachycardia), abdominal cramps, nausea, vomiting, diarrhea, discharge of black urine, muscle spasms, muscle pain (myalgia), coma, and cardiac arrest.

Used as a research tool in molecular biology.

Toxicology

Human toxicity values have not been established or have not been published. However, based on available information, this material appears to be several hundred times as toxic as the nerve agent VX (C01-A017, Volume 1).

C16-A015

Saxitoxin (Agent SS)
CAS: 35523-89-8; 35554-08-6 (Dihydrochloride salt); 220355-66-8 (Diacetate salt)
RTECS: UY8708500
EC: —
UNII: Q0638E899B; VKS19V6FQN (Dihydrochloride salt)
ICD-11: —

$C_{10} H_{17} N_7 O_4$
Molecular Weight: 299.3

Rapid acting channel-activating neurotoxin that blocks transient sodium channels and inhibits depolarization of nerve cells. It is one of the causative agents of paralytic shellfish poisoning (PSP). It is obtained from dinoflagellates (*Gonyaulax* spp., *Alexandrium* spp., *Protogonyaulax* spp.) and cyanobacteria (*Anabaena* spp.). It is soluble and stable in water. It is resistant to chlorine at 10 ppm and iodine at 16 ppm has no effect. Various salts (solids) have been reported. The dihydrochloride salt is a white hygroscopic powder that is water soluble.

This material is hazardous through inhalation, penetration through broken skin, and ingestion. Symptoms include numbness or burning, prickling, itching, or tingling sensations (paresthesia) of the lips, tongue, and gums spreading rapidly to the extremities; headache, dizziness, incoordination (ataxia), drowsiness, difficulty swallowing (dysphagia), incoherent speech, flaccid paralysis, and possibly cardiovascular collapse.

Used as a research tool in molecular biology.

It is on Schedule 1 of the CWC, the Australia Group Core list, and the U.S. Select Agents and Toxins list.

Toxicology

$LC_{50\ (Inh)}$: 2.5 mg/m^3 for a 2-minute exposure
$LD_{50\ (Inh)}$: 0.49 mg
$LD_{50\ (Sub)}$: 0.04 mg
$LD_{50\ (Ing)}$: 0.4 mg
$MEG_{(1hr)}$: Neg: 0.0001 mg/m^3; Mar: 0.0006 mg/m^3; Crit: 0.0035 mg/m^3; Cat: —
PAC-1: 2.1×10^{-5} mg/m^3; PAC-2: 2.3×10^{-4} mg/m^3; PAC-3: 1.4×10^{-3} mg/m^3

C16-A016

Taipoxin

CAS: 52019-39-3

RTECS: WW2200000

EC: —

UNII: —

ICD-11: XM4SM0

Protein

Molecular Weight: 45,600

Obtained from the venom of taipan snakes (*Oxyuranus* spp.) that blocks the release of acetylcholine and cause degeneration of nerve terminal and intramuscular axons leading to severe muscle necrosis (myotoxicity).

This material is hazardous by penetration through broken skin. Symptoms include dizziness, nausea, vomiting, sweating, headache, drowsiness, weakness, excessive salivation, low blood pressure (hypotension), difficulty breathing (dyspnea), slurred speech, blurred vision, flaccid paralysis, and convulsions. Death is due to asphyxia following paralysis of the respiratory muscles.

Toxicology

$LD_{50\ (Sub)}$: 0.0074 g

C16-A017

Tetanus Toxin

CAS: 676570-37-9

RTECS: XW5807000

EC: —

UNII: RS7296A9LB

ICD-11: —

Protein

Molecular Weight: 150,000

Delayed-action neurotoxin that blocks the release of acetylcholine. Once the toxin becomes fixed to neurons, it cannot be neutralized with antitoxin. Recovery of nerve function from tetanus toxins requires sprouting of new nerve terminals and formation of new synapses. It is a crystalline solid obtained from bacteria (*Clostridium tetani*). Dried material is stable for years when stored between 39°F and 45°F; otherwise it is relatively unstable and very sensitive to heat.

This material is hazardous through inhalation, penetration through broken skin, and ingestion. Symptoms include muscle spasms (frequently of the jaw muscle) progressing to rigid paralysis. Generalized spasms can be induced by sensory stimulation. Even minor stimulation may trigger these spasms. Spasms may be so severe as to cause bone fractures. Spasms affecting the larynx, diaphragm, and intercostal muscles lead to respiratory failure. Involvement of the autonomic nervous system results in an irregular heartbeat (cardiac arrhythmias), rapid heart rate (tachycardia), and high blood pressure (hypertension).

Toxicology

$LD_{50\ (Ing)}$: 0.0002 mg (estimate)

PAC-1: 6.6×10^{-9} mg/m^3; *PAC-2:* 7.2×10^{-8} mg/m^3; *PAC-3:* 4.3×10^{-7} mg/m^3

C16-A018

Tetrodotoxin (TTX)
CAS: 4368-28-9; 18660-81-6 (Citrate salt)
RTECS: IO1450000
EC: 224–458-8
UNII: 3KUM2721U9
ICD-11: XM7FH1

$C_{11} H_{17} N_3 O_8$
Molecular Weight: 319.3

Rapid-acting neurotoxin that inhibits sodium-ion channels in neural and muscular tissue. It does not affect the neuromuscular junction. It is colorless crystals or a white powder that is obtained from puffer fish (*Arothron* spp.), the blue-ringed octopus (*Hapalochlaena* spp.), some frogs, newts, dinoflagellates (*Takifugu poecilonotus*), and bacteria (*Pseudoalteromonas tetra-odonis*). It decomposes when heated above 437°F. It is soluble in water and dilute acetic acid; slightly soluble in dry alcohol and either. It is practically insoluble in all other organic solvents. When dissolved in water, it is rapidly inactivated by chlorine at 50 ppm. Various salts (solids) have been reported.

This material is hazardous through inhalation, penetration through broken skin, and ingestion. Symptoms include burning, prickling, itching, or tingling sensations (paresthesia) progressing to numbness, headache, excessive sweating (diaphoresis), dizziness, excessive salivation or drooling (ptyalism), nausea, vomiting, diarrhea, a vague feeling of bodily discomfort (malaise), difficulty breathing (dyspnea), incoordination (ataxia), low blood pressure (hypotension), fixed and dilated pupils (mydriasis), coma, seizures, and respiratory arrest.

Used as a research tool in molecular biology.

It is on the Australia Group Core list and the U.S. Select Agents and Toxins list.

Toxicology
$LD_{50\ (Inh)}$: 0.1–0.2 mg
$LD_{50\ (Ing)}$: 2 mg
PAC-1: 6.3×10^{-5} mg/m^3; *PAC-2:* 6.9×10^{-4} mg/m^3; *PAC-3:* 4.1×10^{-3} mg/m^3

C16-A019

Tityustoxin
CAS: 39465-37-7
RTECS: XR3030000
EC: —
UNII: —
ICD-11: XM2HW7

Peptide
Molecular Weight: 8,000

Rapid-acting neurotoxin that binds to sodium channels in nerve tissue leading to an increase in the release of neurotransmitters. It is obtained from the venom of the Brazilian yellow scorpion *Tityus serrulatus*.

This material is hazardous through inhalation and penetration through broken skin. Causes immediate pain in contact with any break in the skin. General symptoms include hyper-excitability, restlessness, salivation, lacrimation, accelerated respiration, convulsions, contractions and muscular twitching, followed by spastic paralysis with rigid limbs.
Used as a research tool in neurobiology.

Toxicology
Human toxicity values have not been established or have not been published. However, based on available information, this material appears to be slightly more toxic than the nerve agent VX (C01-A017, Volume 1).

C16-A020

Veratridine
CAS: 71-62-5; 11076-62-3 (Hydrochloride salt)
RTECS: YX5600000
EC: 200–758-4
UNII: M4BNP1KR7W; 388H04NR0I (Hydrochloride salt)
ICD-11: XM0EH6

$C_{36} H_{51} N O_{11}$
Molecular Weight: 673.8

Neurotoxin that preferentially binds to activated sodium channels and increases the intracellular calcium concentration. It prolongs the action potential duration in the heart. It has the same pharmacological effect as aconitine, grayanotoxins and batrachotoxin. It is obtained from the rootstock of hellebore and sabadilla seeds (*Schoenocaulon officinale*); and is a white to yellowish-white amorphous powder that melts at 356°F. It is soluble ethanol, dimethyl sulfoxide, and chloroform; slightly soluble in ether and insoluble in water. Various salts have been reported.

This material is hazardous through inhalation, penetration through broken skin, and ingestion. Once in the body, the toxin can become stored in body fat and may have a cumulative effect. Symptoms from ingestion include severe nausea, vomiting, burning, prickling, itching, or tingling skin sensation (paresthesia), weakness, low blood pressure (hypotension), slow heart rate (bradycardia), and loss of consciousness (syncope).

Used industrially as an insecticide, for medicinal purposes and in biomedical research.

Toxicology
Human toxicity values have not been established or have not been published. However, based on available information, this material appears to be approximately 100 times less toxic than the nerve agent VX (C01-A017, Volume 1).

C16-A   CYTOTOXINS

C16-A021

Abrins

CAS: 1393-62-0
RTECS: AA5250000
EC: —
UNII: 6YCF89P8OX
ICD-11: XM98Y1

Proteins
Molecular Weight: 63,000–67,000

Ribosome inactivating cytotoxic proteins that irreversibly inhibit protein synthesis in cells causing cell death. They are obtained from the seed of the Jequirity beans plant (*Abrus precatorius*). Typically, yellowish-white powders that are insoluble in distilled water but soluble in salt water. They are fairly heat stable.

This material is hazardous through inhalation, penetration through broken skin, ingestion, ocular absorption, and produces local eye impacts. Exposure may cause sensitization producing severe allergic reactions. Initial signs and symptoms may be for several days. Symptoms include fever, fatigue, weakness, muscle pain (myalgia), joint pain (arthralgia), low body temperature (hypothermia), and low blood pressure (hypotension). Symptoms from inhalation include irritation and pain in the mucous membranes, cough, difficulty breathing (dyspnea), tightness in the chest, excessive sweating (diaphoresis), low blood oxygen (hypoxemia), blue skin and lips (cyanosis), accumulation of fluid in the lungs (pulmonary edema), and multi-system organ dysfunction. Symptoms from ingestion include abdominal pain, vomiting, diarrhea (may be bloody), gastrointestinal hemorrhage, necroses in the liver and kidneys, severe dehydration, limited urine production (oliguria), blood in the urine (hematuria), thirst, burning throat, headache, low blood volume (hypovolemia), seizures; liver, spleen and kidney failure; and shock. Incidental eye exposure can cause tearing (lacrimation), swelling, redness, bleeding (retinal hemorrhage), impaired vision, and blindness. Systemic toxicity through eye exposure is possible.

Used in medical research to study cancer cells.

They on the Australia Group Core list and the U.S. Select Agents and Toxins list.

Toxicology

LCt_{50}: 5 mg-min/m^3 (estimate) $LD_{50\ (Inh)}$: 0.2 mg (estimate) $LD_{50\ (Ing)}$: 0.007–0.07 mg (estimate) PAC-1: 6.60 × 10^{-7} mg/m^3; PAC-2: 7.20 × 10^{-6} mg/m^3; PAC-3: 4.30 × 10^{-5} mg/m^3

C16-A022

Aflatoxins

CAS: 1402-68-2
RTECS: AWQ5950000
ICD-11: XM7U84

Delayed action cytotoxins that inhibits the synthesis of nucleic acids. They are obtained from various molds/fungi (*Aspergillus* spp.). They are colorless to pale-yellow crystalline materials melting above 450°F. The B-series toxins fluoresce blue in the presence of UV light while the G-series toxins fluoresce green. They are only slightly soluble in water, but are soluble in methanol, acetone, dimethyl sulfoxide, dimethylformamide, and chloroform.

These materials are hazardous through inhalation and ingestion. Symptoms include vomiting, abdominal pain, convulsions, pulmonary edema, coma, and death. All aflatoxins are potential carcinogens. Aflatoxin B1 is the most potent natural carcinogen known.

High-level exposure to aflatoxins disrupts and inhibits metabolism of carbohydrates and lipids, as well as the synthesis of proteins, leading to bleeding (hemorrhaging), premature cell death, and tissue necrosis in liver. Symptoms include fever, yellow discoloration of the skin and eyes (jaundice), nausea, vomiting, abdominal pain, enlargement of the liver (hepatomegaly), spontaneous bleeding (hemorrhagic diathesis), swollen limbs (edema), accumulation of fluid in the lungs (pulmonary edema), seizures, and coma.

They are on the Australia Group Core list.

Toxicology
Human toxicity values have not been established or have not been published.

Aflatoxin B1 ($C_{17} H_{12} O_6$)
CAS: 1162-65-8
RTECS: GY1925000
EC: —
UNII: 9N2N2Y55MH
Molecular Weight: 312.3
Most toxic of the aflatoxins; can permeate through intact skin.

Aflatoxin B2 ($C_{17} H_{14} O_6$)
CAS: 7220-81-7
RTECS: GY1722000
EC: 230–618-8
UNII: 7SKR7S646P
Molecular Weight: 314.3
$LD_{50 \text{ (Ing)}}$ 70–700 mg (estimate)

Aflatoxin G1 ($C_{17} H_{12} O_7$)
CAS: 1165-39-5
RTECS: LV1720000
EC: —
UNII: 1DB78J7PUD
Molecular Weight: 328.3

Aflatoxin G2 ($C_{17} H_{14} O_7$)
CAS: 7241-98-7
RTECS: LV1700000
EC: —
UNII: 2MS0D8WA29
Molecular Weight: 330.3

C16-A023

Cardiotoxins
CAS: 11061-96-4
RTECS: FH7482450
EC: 296–300-6; (*Naja nigricollis* Venom) 296–301-1 (*Naja naja* Venom)

UNII: NY80X52FQ2 (*Naja kaouthia* venom); ZZ4AG7L7VM (*Naja naja* venom)
ICD-11: XM1R49

Protein
Molecular Weight: 6,802 (CTX III)

Cytotoxins obtained from cobra venom (*Naja* spp.). Hydrophobic solids that causes irreversible depolarization of cell membrane and cellular destruction as well as contraction of skeletal and smooth muscle, including the heart.

This material is hazardous by penetration through broken skin. Symptoms include heart irregularities and low blood pressure.

Toxicology
$LD_{50\ (Sub)}$: 0.032 g

C16-A024

Cholera Toxin
CAS: 9012-63-9
RTECS: LF3100000
EC: —
UNII: —
ICD-11: —

Protein

Cytotoxin that stimulates the production of the enzyme adenylate cyclase within cells resulting in cellular hypersecretion of chloride and bicarbonate ions, which results in increased fluid secretion and accumulation in the gut resulting in severe diarrhea – up to 1 liter of water per hour – with the subsequent loss of electrolytes and dehydration. It is obtained from bacteria (*Vibrio cholerae*, C17-A037). It is a water-soluble white powder that is stable for years when stored at 39°F.

This material is hazardous by penetration through inhalation, broken skin, and ingestion. Symptoms include vomiting, abdominal cramps, and watery diarrhea. Inhalation or contact with the skin or eyes may cause severe irritation at the site of exposure.

Used as a research tool in molecular biology and in the production of vaccines.

It is on the Australia Group Core list.

Toxicology
$PAC\text{-}1$: 7.7×10^{-5} mg/m^3; $PAC\text{-}2$: 8.5×10^{-4} mg/m^3; $PAC\text{-}3$: 5.1×10^{-3} mg/m^3

Based on available information, this material appears to be less than one-tenth as toxic as the nerve agent VX (C01-A017, Volume 1).

C16-A025

***Clostridium perfringens*Toxins**
CAS: —
RTECS: —
EC: 297–503-2 (Toxin A)
UNII: —
ICD-11:

Proteins

Obtained from the bacteria *Clostridium perfringens* (C17-A015), at least 20 individual exotoxins have been identified. The alpha toxin attacks lipids in the membrane of cells (phospholipase), while the other toxins either attack red blood cells (hemolysins) or cause cell death (necrosis). These exotoxins are heat labile with 90% destruction in 4 minutes at 140°F.

These materials are hazardous through inhalation, penetration through broken skin, and ingestion. Symptoms include confusion, pallor, excessive sweating (diaphoresis), rapid heart rate (tachycardia), low blood pressure (hypotension), renal failure, and septic shock. In open wounds, toxins produce edema and intense pain that is out of proportion to the nature of the wound. The tissue around the wound may become red or bronze and ultimately blackish green. The gas filled vesicles characteristic of typical gas gangrene are unlikely since the carbon dioxide and hydrogen producing this symptom is caused by the *C. perfringens* bacteria and not as a result of the action of the toxins on the cells. Rapid spreading necrosis of muscle tissue (myonecrosis) resulting in fatal systemic toxicity and shock. Ingestion of toxins causes fever, abdominal pain, diarrhea, vomiting, bloody stool, and septic shock.

It is on the Australia Group Core list.

Toxicology
Human toxicity values have not been established or have not been published.

C16-A026

Diamphotoxin
CAS: 87915-42-2
RTECS: —
EC: —
UNII: —
ICD-11: —

Protein
Molecular Weight: 60,000

Obtained from the larva of leaf-cutting beetles (*Diamphidia nigroornata*) and used by bushmen of Southern Africa as an arrow poison. It is destructive to red blood cells (hemolytic), reducing hemoglobin levels by up to 75%, and causing convulsions, paralysis, then death. Dried poison is stable for over a year.

This material is hazardous by penetration through broken skin. Symptoms include partial paralysis, difficulty breathing (dyspnea), blue skin and lips (cyanosis), blood in the urine (hematuria), cardiac arrhythmias, and respiratory failure.

Toxicology
Human toxicity values have not been established or have not been published.

C16-A027

Maitotoxin
CAS: 59392-53-9
RTECS: OM5470000
EC: —

UNII: 9P59GES78D

ICD-11: —

$C_{164} H_{256} O_{68} S_2 Na_2$

Molecular Weight: 3,426

Cyclic polyether that affects the voltage-gated calcium channels causing an increased calcium influx into the cell, ultimately resulting in breakup of the cell's nuclear envelope (blebbing). It is obtained from dinoflagellates (*Gambierdiscus toxicus*). It is a colorless, amorphous solid that is heat stable and soluble in water and alcohols. It is the most lethal non-peptide natural product currently known.

This material is hazardous through inhalation, penetration through broken skin, and ingestion. This toxin, in conjunction with the ciguatoxins (C16-A008), is usually associated with ciguatera fish poisoning (CFP). Although maitotoxin is approximately three times more toxic than ciguatoxin, the symptoms of CFP are dominated by the neurologic presentation from ciguatoxin. The specific symptoms associated with unique exposure to maitotoxin have not been established or have not been published.

Used as a research tool in neurophysiology.

Toxicology

Human toxicity values have not been established or have not been published. However, based on available information, this material appears to be approximately 100 times more toxic than the nerve agent VX (C01-A017, Volume 1).

C16-A028

Microcystin LR (FDF)

CAS: 101043-37-2

RTECS: GT2810000

EC: —

UNII: 4G08121T5U

ICD-11: —

$C_{49} H_{74} N_{10} O_{12}$

Molecular Weight: 995.2

Rapid-acting cytotoxin that disrupts cell membranes in the liver (hepatoxin) causing an accumulation of blood. It is the most toxic of the Microcystins. It is a solid obtained from freshwater blue-green cyanobacteria (*Microcystis* spp.). It is heat stable and soluble in water, alcohols, dimethyl sulfoxide, dimethylformamide, and acetone.

This material is hazardous through inhalation, penetration through broken skin, and ingestion. Symptoms include shivering, rapid breathing (tachypnea), progressing to twitching, convulsions, and gasping respirations. Shock and death occur within a matter of hours.

It is on the Australia Group Core list.

Toxicology

ID_{50}: 1–10 mg (estimate, route unspecified)

C16-A029

Modeccin
CAS: 65988-88-7
RTECS: XW5790000
EC: —
UNII: —
ICD-11: —

Protein
Molecular Weight: 57,000

Ribosome inactivating cytotoxic protein that irreversibly inhibit protein synthesis in cells causing cell death. It is obtained from the roots of *Adenia digitata* plant.

This material is hazardous through inhalation, penetration through broken skin, ingestion, ocular absorption, and produces local eye impacts. Exposure may cause sensitization producing severe allergic reactions. Initial signs and symptoms may be for several days. Symptoms include fever, fatigue, weakness, muscle pain (myalgia), joint pain (arthralgia), low body temperature (hypothermia), and low blood pressure (hypotension). Symptoms from inhalation include irritation and pain in the mucous membranes, cough, difficulty breathing (dyspnea), tightness in the chest, excessive sweating (diaphoresis), low blood oxygen (hypoxemia), blue skin and lips (cyanosis), accumulation of fluid in the lungs (pulmonary edema), and multi-system organ dysfunction. Symptoms from ingestion include abdominal pain, vomiting, diarrhea (may be bloody), gastrointestinal hemorrhage, necroses in the liver and kidneys, severe dehydration, limited urine production (oliguria), blood in the urine (hematuria), thirst, burning throat, headache, low blood volume (hypovolemia), seizures; liver, spleen, and kidney failure; and shock. Incidental eye exposure can cause tearing (lacrimation), swelling, redness, bleeding (retinal hemorrhage), impaired vision, and blindness. Systemic toxicity through eye exposure is possible.

It is on the Australia Group Core list.

Toxicology
PAC-1: 5.4×10^{-6} mg/m^3; *PAC-2:* 5.9×10^{-5} mg/m^3; *PAC-3:* 3.6×10^{-4} mg/m^3

Based on available information, this material appears to be approximately 1,000 times more toxic than the nerve agent VX (C01-A017, Volume 1) when injected.

C16-A030

Pertussis Toxin
CAS: 70323-44-3
RTECS: XW5883750
EC: —
UNII: —
ICD-11: —

Protein
Molecular Weight: 105,000

Cytotoxin that inactivates G proteins involved in cellular metabolism. It is a heat-stable, freeze-dried solid obtained from virulent strains of the bacteria *Bordetella pertussis*, the etiological agent of whooping cough.

This material is hazardous through inhalation and penetration through broken skin. It causes an increased release of insulin, resulting in low blood sugar (hypoglycemia), and sensitivity to histamine that results in increased capillary permeability, low blood pressure (hypotension), and shock.

Used as a research tool in molecular biology and as a component (toxoid) in pertussis vaccines.

Toxicology

Human toxicity values have not been established or have not been published.

C16-A031

Ricin (Agent W)
CAS: 9009-86-3
RTECS: VJ2625000
EC: —
UNII: U6E7R9OE4R
ICD-11: XM2VG9

Protein
Molecular Weight: 65,000

Ribosome inactivating cytotoxic protein that irreversibly inhibit protein synthesis in cells causing cellular death. It is obtained from castor beans (*Ricinus communis*). Waste from production of castor oil contains about 5% ricin by weight. It is a white powder that is soluble in water and relatively heat stable. Aqueous solutions are resistant to chlorine at 10 ppm. It is persistent in the environment.

This material is hazardous through inhalation, penetration through broken skin, ingestion, ocular absorption, and produces local eye impacts. Exposure may cause sensitization producing severe allergic reactions. Initial signs and symptoms may be for several days. Symptoms include fever, fatigue, weakness, muscle pain (myalgia), joint pain (arthralgia), low body temperature (hypothermia), and low blood pressure (hypotension). Symptoms from inhalation include irritation and pain in the mucous membranes, cough, difficulty breathing (dyspnea), tightness in the chest, excessive sweating (diaphoresis), low blood oxygen (hypoxemia), blue skin and lips (cyanosis), accumulation of fluid in the lungs (pulmonary edema), and multi-system organ dysfunction. Symptoms from ingestion include abdominal pain, vomiting, diarrhea (may be bloody), gastrointestinal hemorrhage, necroses in the liver and kidneys, severe dehydration, limited urine production (oliguria), blood in the urine (hematuria), thirst, burning throat, headache, low blood volume (hypovolemia), seizures; liver, spleen, and kidney failure; and shock. Incidental eye exposure can cause tearing (lacrimation), swelling, redness, bleeding (retinal hemorrhage), impaired vision, and blindness. Systemic toxicity through eye exposure is possible.

Used as a research tool in molecular biology, and has been used as a rodenticide.

It is on Schedule 1 of the CWC, the Australia Group Core list, and the U.S. Select Agents and Toxins list.

Toxicology

LCt_{50}: 5 mg-min/m^3 (estimate)
$LD_{50\ (Inh)}$: 0.3 mg (estimate)
$LD_{50\ (Inj)}$: 0.7 mg (estimate)

$LD_{50\ (Ing)}$: 1,400 mg (estimate)

$MEG_{(1hr)}$: Neg: —; Mar: —; Crit: 0.0048 mg/m^3; Cat: —

PAC-1: 8.20 × 10^{-6} mg/m^3; PAC-2: 9.00 × 10^{-5} mg/m^3; PAC-3: 5.4 × 10^{-4} mg/m^3

C16-A032

Shiga Toxins

CAS: 75757-64-1

RTECS: —

EC: —

UNII: —

ICD-11: —

Protein

Molecular Weight: 70,000

Ribosome inactivating cytotoxic protein that irreversibly inhibit protein synthesis in cells and causing cell death. The toxin acts on the vascular endothelium of small blood vessels but not against large vessels such as arteries or major veins. It causes destruction of blood cells (hemolytic anemia) and a low level of platelets in the blood (thrombocytopenia). It is a solid obtained from bacteria (*Shigella dysenteriae* (C17-A036) and some serotypes of *Escherichia coli* (C17-A035).

This material is hazardous through inhalation and ingestion. Symptom include fever, pallor, extreme fatigue, difficulty breathing (dyspnea), high blood pressure (hypertension), swelling (edema), cramping abdominal pain, vomiting, diarrhea (may be bloody), blood in the urine (hematuria), limited urine production (oliguria), and renal failure. Neurological symptoms include double vision (diplopia), cortical blindness, difficulty swallowing (dysphagia), facial palsy, incoordination (ataxia), weakness of one entire side of the body (hemiparesis), confusion, seizures, stroke, and coma. Inhalation of toxin may cause direct impacts to lung tissue including accumulation of fluid (pulmonary edema).

It is on the Australia Group Core list.

Toxicology

PAC-1: 1.5 × 10^{-6} mg/m^3; PAC-2: 1.6 × 10^{-5} mg/m^3; PAC-3: 9.7 × 10^{-5} mg/m^3

Based on available information, this material appears to be several magnitudes more toxic than the nerve agent VX (C01-A017, Volume 1).

C16-A033

Staphylococcal Enterotoxin B (Agent PG)

CAS: 11100-45-1

RTECS: XW5807700

EC: 297–505-3

UNII: —

ICD-11: —

Protein

Molecular Weight: 28,000

Cytotoxic superantigen that activates the immune system to produce a burst of anti-inflammatory cytokines. Originally developed as an incapacitating agent, variation in dose can easily produce lethal effects. It is one of seven enterotoxins obtained from bacteria (*Staphylococcus aureus*). It is a white, fluffy solid that is water soluble and heat stable (not destroyed by cooking or freezing). It resists chlorine in amounts found in potable water. As a freeze-dried powder, it can be stored for more than a year.

This material is hazardous through inhalation, ingestion and may produce red and inflamed whites of the eyes (conjunctivitis) on contact. The latent period is typically 1 to 12 hours after ingestion and 2 to 24 hours after inhalation. Initial symptoms include fever, chills, headache, and muscle pain (myalgia). Additional symptoms from inhalation include nasal irritation, congestion, nonproductive cough, retrosternal chest pain, difficulty breathing (dyspnea), and rapid heart rate (tachycardia); may progress to accumulation of fluid in the lungs (pulmonary edema) and circulatory collapse. Additional symptoms from ingestion include sudden onset of vomiting, abdominal cramps, nausea, explosive watery diarrhea, and severe weakness.

It is on the Australia Group Core list and the U.S. Select Agents and Toxins list.

Toxicology

$ID_{50\ (Inh)}$: 0.00003 mg (estimate)

$LD_{50\ (Inh)}$: 0.0017 mg (estimate)

$LCt_{50\ (Inh)}$: 5 mg-min/m^3 (estimate)

PAC-1: 1.9×10^{-4} mg/m^3; *PAC-2:* 2.1×10^{-3} mg/m^3; *PAC-3:* 1.3×10^{-2} mg/m^3

C16-A034

***Staphylococcus aureus* Alpha Toxin**

CAS: 94716-94-6

RTECS: —

EC: 297–504-8

UNII: —

ICD-11: —

Protein

Molecular Weight: 33,000

Pore-forming cytotoxin that causes cell death and lysis. An exotoxin that adds to the pathogenicity and virulence of the bacteria. Causes respiratory paralysis, vascular and smooth muscle spasms, tissue necrosis with inability to stop bleeding within cells (hemostasis disturbances), low blood platelets and limited clotting with excessive bleeding (thrombocytopenia), and damage to lung tissue (pulmonary lesions). It can be stored as a white freeze-dried powder that is environmentally stable. It is soluble in water and stable over a wide pH range.

This material is hazardous from penetration through broken skin, and ingestion, and produces local skin/eye impacts. Symptoms include skin and eye irritation, promotion of blood coagulation, pulmonary edema, and adult respiratory distress syndrome.

This material is on the Australia Group Core list.

Toxicology

Human toxicity values have not been established or have not been published.

C16-A035

Toxic Shock Syndrome Toxin 1

CAS: —
RTECS: —
EC: —
UNII: —
ICD-11: —

Protein
Molecular Weight: 24,000

Exotoxin that is a superantigen, stimulating the release of large amounts of Interleukin-1 (C20-A005), interleukin-2 and Tumor necrosis factor alpha (C20-A009). It is water soluble, heat stable (>140°F) and unaffected by a wide pH range (2.5–11). It can be stored as a white to off-white freeze-dried powder.

This material is hazardous through inhalation and penetration through broken skin. Symptoms include fever, rash, vomiting, diarrhea, low blood pressure (hypotension), yellow discoloration of the skin and eyes (jaundice), desquamation of the skin over the entire body, with rapid development of multiorgan failure.

This material is on the Australia Group Core list.

Toxicology
Human toxicity values have not been established or have not been published.

C16-A036

Viscumin

CAS: 83590-17-4
RTECS: —
EC: —
UNII: —
ICD-11: —

Protein
Molecular Weight: 119,000 (as a dimer)

Ribosome inactivating cytotoxic protein that irreversibly inhibit protein synthesis in cells causing cell death. It is a solid obtained from the mistletoe plant (*Viscum album*).

This material is hazardous through inhalation, penetration through broken skin, ingestion, ocular absorption, and produces local eye impacts. Exposure may cause sensitization producing severe allergic reactions. Initial signs and symptoms may be for several days. Symptoms include fever, fatigue, weakness, muscle pain (myalgia), joint pain (arthralgia), low body temperature (hypothermia), and low blood pressure (hypotension). Symptoms from inhalation include irritation and pain in the mucous membranes, cough, difficulty breathing (dyspnea), tightness in the chest, excessive sweating (diaphoresis), low blood oxygen (hypoxemia), blue skin and lips (cyanosis), accumulation of fluid in the lungs (pulmonary edema), and multi-system organ dysfunction. Symptoms from ingestion include abdominal pain, vomiting, diarrhea (may be bloody), gastrointestinal hemorrhage, necroses in the liver and kidneys, severe dehydration, limited urine production (oliguria), blood in the urine (hematuria), thirst, burning throat, headache, low blood volume (hypovolemia), seizures; liver, spleen, and kidney failure; and shock. Incidental eye exposure can cause tearing (lacrimation), swelling, redness, bleeding (retinal hemorrhage), impaired vision, and blindness. Systemic toxicity through eye exposure is possible.

It is being tested for medicinal use as an anti-tumor agent.

It is on the Australia Group Core list.

Toxicology

PAC-1: 7.7×10^{-6} mg/m^3; *PAC-2:* 8.5×10^{-5} mg/m^3; *PAC-3:* 5.1×10^{-4} mg/m^3

Based on available information, this material appears to have a comparable toxicity to ricin (C16-A031).

C16-A037

Volkensin
CAS: 91933-11-8
RTECS: —
EC: —
UNII: —
ICD-11: —

Protein
Molecular Weight: 62,000

Ribosome inactivating cytotoxic protein that irreversibly inhibit protein synthesis in cells causing cell death. It is a solid obtained from the roots of the kilyambiti plant (*Adenia volkensii*).

This material is hazardous through inhalation, penetration through broken skin, ingestion, ocular absorption, and produces local eye impacts. Exposure may cause sensitization producing severe allergic reactions. Initial signs and symptoms may be for several days. Symptoms include fever, fatigue, weakness, muscle pain (myalgia), joint pain (arthralgia), low body temperature (hypothermia), and low blood pressure (hypotension). Symptoms from inhalation include irritation and pain in the mucous membranes, cough, difficulty breathing (dyspnea), tightness in the chest, excessive sweating (diaphoresis), low blood oxygen (hypoxemia), blue skin and lips (cyanosis), accumulation of fluid in the lungs (pulmonary edema), and multi-system organ dysfunction. Symptoms from ingestion include abdominal pain, vomiting, diarrhea (may be bloody), gastrointestinal hemorrhage, necroses in the liver and kidneys, severe dehydration, limited urine production (oliguria), blood in the urine (hematuria), thirst, burning throat, headache, low blood volume (hypovolemia), seizures; liver, spleen, and kidney failure; and shock. Incidental eye exposure can cause tearing (lacrimation), swelling, redness, bleeding (retinal hemorrhage), impaired vision, and blindness. Systemic toxicity through eye exposure is possible.

Used as a research tool in neurology.

It is on the Australia Group Core list.

Toxicology

PAC-1: 3.7×10^{-7} mg/m^3; *PAC-2:* 4.0×10^{-6} mg/m^3; *PAC-3:* 2.4×10^{-5} mg/m^3

Based on available information, this material appears to be several magnitudes more toxic than the nerve agent VX (C01-A017, Volume 1).

C16-A   DERMALLY HAZARDOUS CYTOTOXINS

Trichothecene Mycotoxins

ICD-11: XM6X33

Dermally hazardous cytotoxins obtained from various molds and fungi (*Fusarium* spp.) that inhibit protein synthesis in cells. They are colorless, crystalline solids that are heat stable and can be stored for long periods.

These materials are hazardous through inhalation, skin absorption, penetration through broken skin, and ingestion and produces local skin/eye impacts. They are lipophilic and thus easily absorbed through the skin, gut, and pulmonary mucosa. Trichothecenes are radiomimetic and may cause bone marrow suppression, liver failure, and internal bleeding. They are also immunosuppressive.

On the skin, exposure causes irritation, burning, itching, rash, blisters, bleeding, and peeling off of the skin (desquamation). Contact with the eyes can cause tearing (lacrimation), burning sensation, red eyes (conjunctivitis), and blurred vision. Systemic effects include fever, headache, chills, dizziness, incoordination (ataxia), nausea, difficulty breathing (dyspnea), chest pains, vomiting, diarrhea (may be bloody), abdominal pain, loss of appetite (anorexia), excessive bleeding and a lack of clotting (coagulopathy), and low blood pressure (hypotension). Inhalation can also cause respiratory irritation, cough, and necrotic lesions in the nose, mouth, and throat. Once in the body, the toxin can become stored in body fat and may have a cumulative effect.

Toxicology

Human toxicity values for most of these toxins have not been established or have not been published.

C16-A038

Deoxynivalenol ($C_{15} H_{20} O_6$)
CAS: 51481-10-8
RTECS: YD0167000
EC: —
UNII: JT37HYP23V
Molecular Weight: 296.3
Melting point: 304°F.
Soluble in water and ethanol.

Toxicology

Least toxic of the trichothecene mycotoxins, primarily causing vomiting, dizziness, diarrhea, and headache.

C16-A039

Diacetoxyscirpenol ($C_{15} H_{20} O_6$)
It is on the Australia Group Core list and the U.S. Select Agents and Toxins list.
CAS: 2270-40-8
RTECS: YD0112000

EC: 218–873-3

UNII: UYL28I099N

Molecular Weight: 296.3

Melting point: 322°F.

Soluble in chloroform.

Toxicology

PAC-1: 99 x 10^{-4} mg/m^3; *PAC-2:* 1.1 × 10^{-2} mg/m^3; *PAC-3:* 3.4 × 10^{-1} mg/m^3

C16-A040

Diacetylnivalenol ($C_{19} H_{24} O_9$)

CAS: 14287-82-2

RTECS: —

EC: —

UNII: —

Molecular Weight: 396.4

Melting point: —

Soluble in acetone, ethyl acetate, diethyl ether, chloroform, and methylene chloride. It hydrolyzes to form Nivalenol.

C16-A041

Fusarenon X ($C_{17} H_{22} O_8$)

CAS: 23255-69-8

RTECS: —

EC: —

UNII: 0CV8D1DR96

Molecular Weight: 354.4

Melting point: 196° F

Soluble in water, alcohols, chloroform, and ethyl acetate. It hydrolyzes to form Nivalenol.

C16-A042

HT-2 Toxin ($C_{22} H_{32} O_8$)

It is on the Australia Group Core list.

CAS: 26934-87-2

RTECS: YD0050000

EC: —

UNII: NC6C26RM46

Molecular Weight: 424.5

Melting point: —

Soluble in acetone, ethyl acetate, diethyl ether, chloroform, and methylene chloride.

C16-A043

Neosolaniol ($C_{19} H_{26} O_8$)
CAS: 36519-25-2
RTECS: YD0080000
EC: —
UNII: PLZ86LH7A6
Molecular Weight: 382.4
Melting point: 325°F
Soluble in acetone, ethyl acetate, diethyl ether, chloroform, and methylene chloride.

C16-A044

Nivalenol ($C_{15} H_{20} O_7$)
CAS: 23282-20-4
RTECS: YD0165000
EC: —
UNII: 5WOP02RM1U
Molecular Weight: 312.3
Melting point: 432°F (decomposes)
Slightly soluble in water; soluble in methanol, ethanol, and acetonitrile.

C16-A045

T2 Toxin ($C_{24} H_{34} O_9$)
It is on the Australia Group Core list and the U.S. Select Agents and Toxins list.
CAS: 21259-20-1
RTECS: YD0100000
EC: 244–297-7
UNII: I3FL5NM3MO
Molecular Weight: 466.5
Melting point: 306°F
Slightly soluble in water; soluble in ethyl acetate, ethanol, chloroform, dimethyl sulfoxide, and dimethylformamide.

Toxicology

Most toxic member of the trichothecene mycotoxins.
$LC_{50\ (Inh)}$: 200–5,800 mg /m^3
$LD_{50\ (Per)}$: 140–840 mg (in a percutaneous solvent)
$ID_{50\ (Ing)}$: 0.78 mg/L (drinking water)
Nausea & Vomiting: 0.05 mg/L (drinking water)
$MEG_{(1hr)}$: Neg: 0.004 mg/m^3; Mar: 0.03 mg/m^3; Crit: 0.4 mg/m^3; Cat: —
PAC-1: 0.0014 mg/m^3; *PAC-2:* 0.015 mg/m^3; *PAC-3:* 0.093 mg/m^3

Macrocyclic Trichothecene Mycotoxins
ICD-11: XM6X33

Dermally hazardous cytotoxins obtained from various molds and fungi (*Stachybotrys* spp., *Myrothecium* spp., *Podostroma* spp.). Many of these toxins have been associated with poor indoor air quality resulting from the growth of molds. They are colorless to white, crystalline solids that are heat stable, insoluble in water, and can be stored for long periods.

These materials are hazardous through inhalation, skin absorption, penetration through broken skin, ingestion, and produces local skin/eye impacts. Trichothecenes are radiomimetic and may cause bone marrow suppression, liver failure and internal bleeding. They are also immunosuppressive.

On the skin, exposure causes irritation, burning, itching, rash, and blisters. Contact with the eyes can cause tearing (lacrimation), burning sensation, red eyes (conjunctivitis), and blurred vision. Systemic effects include fever, headache, fatigue, chills, dizziness, vertigo, nausea, difficulty breathing (dyspnea), chest pains, vomiting, diarrhea (may be bloody), abdominal pain, and loss of appetite (anorexia). Inhalation can also cause respiratory irritation, cough, and burning in the nose and nasal passages nose bleeds (epistaxis). Once in the body, the toxin can become stored in body fat and may have a cumulative effect.

Toxicology

Human toxicity values have not been established or have not been published. However, they are generally considered to be more toxic than mycotoxin T-2 (C16-A045)

C16-A046

Roridin A ($C_{29} H_{40} O_9$)
CAS: 14729-29-4
RTECS: VL0355000
EC: 238–783-8
UNII: U28899D1U2
Molecular Weight: 532.6
Melting point: 388°F
Soluble in chloroform.

C16-A047

Roridin E ($C_{29} H_{38} O_8$)
CAS: 16891-85-3
RTECS: YX9821500
EC: —
UNII: 98826FBF79
Molecular Weight: 514.6
Melting point: 361°F
Soluble in alcohols and dimethyl sulfoxide.

C16-A048

Satratoxin G ($C_{29} H_{36} O_{10}$)
CAS: 53126-63-9
RTECS: —
EC: —
UNII: —
Molecular Weight: 544.6
Melting point: 333°F
Soluble in dimethyl sulfoxide.

C16-A049

Satratoxin H ($C_{29} H_{36} O_9$)
CAS: 53126-64-0
RTECS: VQ5950000
EC: —
UNII: —
Molecular Weight: 528.6
Melting point: 327°F
Soluble in alcohols, acetone, and chloroform.

Toxicology
PAC-1: 0.0046 mg/m^3; *PAC-2:* 0.051 mg/m^3; *PAC-3:* 0.3 mg/m^3

C16-A050

Verrucarin A ($C_{27} H_{34} O_9$)
CAS: 3148-09-2
RTECS: WH1314900
EC: —
UNII: OL62X66O4I
Molecular Weight: 502.6
Melting point: >680°F (decomposes)
Soluble in chloroform.

C16-A051

Zearalenone
CAS: 17924-92-4
RTECS: DM2550000
EC: 241–864-0
UNII: 5W827M159J
ICD-11: XM4E53

$C_{18} H_{22} O_5$
Molecular Weight: 318.4

Dermally hazardous cytotoxins obtained from various molds and fungi (*Gibberella* spp. and *Fusarium* spp.). It is a white, crystalline solid that melts at 322°F. It is slightly soluble in water, but soluble in benzene, acetonitrile, methylene chloride, methanol, ethanol, and acetone. It is heat stable and can be stored for long periods.

This material is hazardous through inhalation, skin absorption, penetration through broken skin, ingestion, and produces local skin/eye impacts. It is known to mimic the body's production of estrogen (hyperestrogenism). In animals, it has caused feminization of male animals and interfered with conception, ovulation, and fetal development in female animals. Specific signs and symptoms of acute high-dose exposure to zearalenone in humans have not been established or have not been published.

Toxicology
Human toxicity values have not been established or have not been published.

C16   CHEMICAL STRUCTURES

C16-A001
Aconitine

C16-A002
Anatoxin A

C16-A003
Anatoxin-A(S)

C16-A004
Batrachotoxin

C16-A006
Brevetoxin PbTx-1

C16-A006
Brevetoxin PbTx-2

C16-A006
Brevetoxin PbTx-3

C16-A007
α-Bungarotoxin

C16-A007
β-Bungarotoxin

C16-A008
Ciguatoxin-1

C16-A008
Ciguatoxin-2

C16-A008
Ciguatoxin-3

C16-A011
Gonyautoxin-1

C16-A011
Gonyautoxin-2

C16-A011
Gonyautoxin-3

C16-A011
Gonyautoxin-4

C16-A011
Gonyautoxin-5

C16-A011
Gonyautoxin-6

C16-A011
Gonyautoxin-8

C16-A013
Neosaxitoxin

C16-A014
Palytoxin

C16-A015
Saxitoxin

C16-A018
Tetrodotoxin

C16-A020
Veratridine

C16-A022
Aflatoxin B1

C16-A022
Aflatoxin B2

C16-A022
Aflatoxin G1

C16-A022
Aflatoxin G2

C16-A027
Maitotoxin

C16-A028
Microcystin LR

C16-A038
Deoxynivalenol

C16-A039
Diacetoxyscirpenol

C16-A040
Diacetylnivalenol

C16-A041
Fusarenon X

C16-A042
HT-2 Toxin

C16-A043
Neosolaniol

C16-A044
Nivalenol

C16-A045
T2 Toxin

C16-A046
Roridin A

C16-A047
Roridin E

C16-A048
Satratoxin G

C16-A049
Satratoxin H

C16-A050
Verrucarin A

C16-A051
Zearalenone

BIBLIOGRAPHY

American Society for Microbiology. *Sentinel Level Clinical Laboratory Guidelines for Suspected Agents of Bioterrorism and Emerging Infectious Diseases: Staphylococcal Enterotoxin B*. Revised June 2013.

American Society for Microbiology. *Sentinel Level Clinical Laboratory Guidelines for Suspected Agents of Bioterrorism and Emerging Infectious Diseases: Botulinum Toxin*. Revised June 2013.

Arnon, Stephen S., Robert Schechter, Thomas V. Inglesby, Donald A. Henderson, John G. Bartlett, Michael S. Ascher, Edward Eitzen, Anne D. Fine, Jerome Hauer, Marcelle Layton, Scott Lillibridge, Michael T. Osterholm, Tara O'Toole, Gerald Parker, Trish M. Perl, Philip K. Russell, David L. Swerdlow and Kevin Tonat. "Botulinum Toxin as a Biological Weapon: Medical and Public Health Management." *Journal of the American Medical Association* 285 (2001): 1059–1070.

Argonne National Laboratory. *Australia Group Common Control List Handbook Volume II: Biological Weapons-Related Common Control Lists*, Revision 4. Washington, D.C.: US Government Printing Office, February 2018.

The Australia Group. *Australia Group Common Control Lists*. February 28, 2020 [https://www.dfat.gov.au/publications/minisite/theaustraliagroupnet/site/en/human_animal_pathogens.html]. December 31, 2020.

Burrows, W. Dickinson and Sara E. Renner. "Biological Warfare Agents as Threats to Potable Water." *Environmental Health Perspectives* 107 (1999): 975–984.

Centers for Disease Control and Prevention. "Biological and Chemical Terrorism: Strategic Plan for Preparedness and Response. Recommendations of the CDC Strategic Planning Workgroup." *MMWR Recommendations and Reports* 49 (RR04) (April 21, 2000): 1–14.

Centers for Disease Control and Prevention. "Case Definitions for Chemical Poisoning." *MMWR Recommendations and Reports* 54 (RR01) (January 14, 2005): 1–24.

Centers for Disease Control and Prevention. *Case Definitions for Chemical Poisoning*. April 4, 2018, [https://emergency.cdc.gov/chemical/casedef.asp]. December 31, 2020.

Centers for Disease Control and Prevention. *Chemical Agent*. 2018 [https://emergency.cdc.gov/agent/agentlistchem.asp]. December 31, 2020.

Chaboo C.S., Biesele M., Hitchcock R.K. and Weeks A. "Beetle and Plant Arrow Poisons of the Jul'hoan and Haillom San Peoples of Namibia (*Insecta, Coleoptera, Chrysomelidae; Plantae, Anacardiaceae, Apocynaceae, Burseraceae*)." *Zookeys* 558 (February 1, 2016): 9–54.

Chandrasekhar, J. "Maitotoxin – Holder of Two World Records." *Resonance* 5 (May 1996): 68–70.

Chorus, Ingrid and Jamie Bartram, ed. *Toxic Cyanobacteria in Water: A Guide to Their Public Health Consequences, Monitoring and Management*. London, United Kingdom: E & FN Spon, 1999.

Compton, James A.F. *Military Chemical and Biological Agents: Chemical and Toxicological Properties*. Caldwell, New Jersey: The Telford Press, 1987.

Cope, Arthur C. and Joseph Dec "Ricin," chapter 12 in *Chemical Warfare Agents, and Related Chemical Problems, Parts I – II*. Office of Scientific Research and Development, National Defense Research Committee. Washington, DC: Government Printing Office, 1946.

Cope, Author, John Dec, Keith Cannan, Birdsey Renshaw and Stanford Moore. "Ricin," chapter 12 in *Chemical Warfare Agents, and Related Chemical Problems, Parts I – II*. Office of Scientific Research and Development, National Defense Research Committee. Washington, DC: Government Printing Office, 1946.

Department of Agriculture. 9 CFR Part 121 – "Possession, Use, and Transfer of Biological Agents and Toxins," 2005.

Donohue-Rolfe, Arthur, David W. K. Acheson, Anne V. Kane and Gerald T. Keusch. "Purification of Shiga Toxin and Shiga-Like Toxins I and II by Receptor Analog Affinity Chromatography with Immobilized P1 Glycoprotein and Production of Cross-Reactive Monoclonal Antibodies." *Infection and Immunity* 57 (December 1989): 3,888–3,893.

Estacion, Mark and William P. Schilling. "Maitotoxin-Induced Membrane Blebbing and Cell Death in Bovine Aortic Endothelial Cells." *BMC Physiology* (2001): 1:2 [http://www.biomedcentral.com/1472–6793/1/2]. December 31, 2020.

Food and Agriculture Organization of the United Nations, and the World Health Organization. *Technical Paper on Toxicity Equivalency Factors for Marine Biotoxins Associated with Bivalve Molluscs*. Rome, 2016.

Franz, David R. *Defense Against Toxin Weapons*. Revised Edition. Fort Detrick, Maryland: United States Army Medical Research and Materiel Command, 1997.

Friedman, Melissa A. and Bonnie E. Levin. "Neurobehavioral Effects of Harmful Algal Bloom (HAB) Toxins: A Critical Review." *Journal of the International Neuropsychological Society* 11 (2005): 331–338.

Gasperi-Campani, Anna, Luigi Barbieri, Enzo Lorenzoni, Lucio Montanaro, Simonetta Sperti, Eugenio Bonetti and Fiorenzo Stirpe. "Modeccin, the Toxin of Adenia digitata: Purification, Toxicity and Inhibition of Protein Synthesis in Vitro." *Biochemical Journal* 174 (1978): 491–496.

Griffiths, Gareth. "Understanding Ricin from a Defensive Viewpoint." *Toxins* 3 (2011): 1373–1392.

Gupta, Ramesh C., ed. *Handbook of Toxicology of Chemical Warfare Agents. 2nd Edition.* London, United Kingdom: Elsevier, 2015.

Harpe, Jon de la, E. Reich, Karl A. Reich and Eugene B. Dowdle. "Diamphotoxin: The Arrow Poison of the!Kung Bushmen." *The Journal of Biological Chemistry* 258 (1983): 11,924–11,931.

Henderson, Donald A., Thomas V. Inglesby, and Tara O'Toole, ed. *Bioterrorism: Guidelines for Medial and Public Health Management.* Chicago, Illinois: AMA Press, 2002.

International Programme on Chemical Safety. *Environmental Health Criteria Monographs.* [http://www.inchem.org/pages/ehc.html]. December 31, 2020.

International Programme on Chemical Safety. *Joint Expert Committee on Food Additives (JECFA) – Monographs and Evaluations.* [http://www.inchem.org/pages/jecfa.html]. December 31, 2020.

Ishiguro, Masatsune, Takao Takahashi, Gunki Funatsu, Katsuya Hayashi and Masaru Funatsu. "Biochemical Studies on Ricin. I. Purification of Ricin." *Journal of Biochemistry* (Tokyo, Japan) 55 (1964) 587–592.

Kabat, Elvin E., Michael Heidelberger and Ada E. Bezer. "A Study of the Purification and Properties of Ricin." *Journal of Biological Chemistry* 168 (1947): 629–639.

Katircioglu, Hikmet, Beril S. Akin and Tahir Atici. "Microalgal Toxin(s): Characteristics and Importance." *African Journal of Biotechnology* 3 (2004): 667–674.

Kishi, Yoshito. "Complete Structure of Maitotoxin." *Pure & Applied Chemistry* 70 (1998): 339–344.

Richard Lewis, Sebastien Dutertre, Irina Vetter and MacDonald Christie. "Conus Venom Peptide Pharmacology." *Pharmacological Reviews* 64 (2012):259–298.

Lindler, Luther E., Frank J. Lebeda and George W. Korch, ed. *Biological Weapons Defense: Infectious Diseases and Counterbioterrorism.* Totowa, New Jersey: Humana Press, Inc., 2005.

List Biological Laboratories, Inc. *Safety Data Sheets (SDSs) for Bacterial Toxins.* [http://www.listlabs.com]. December 31, 2020.

Mankiewicz, Joanna, Malgorzata Tarczynska, Zofia Walter and Maciej Zalewski. "Natural Toxins from Cyanobacteria." *Acta Biologica Cracoviensia Series Botanica* 45 (2003): 9–20.

Mebs, Dietrich. *Venomous and Poisonous Animals: A Handbook for Biologists, Toxicologists and Toxinologists, Physicians and Pharmacists.* Stuttgart: Medpharm Scientific Publishers, 2002.

Melton-Celsa, Angela R. *Shiga Toxin (Stx) Classification, Structure, and Function.* Microbiol Spectr. 2014; 2(2): doi:10.1128/microbiolspec.EHEC-0024-2013.

Merck & Co. *Merck Manual Professional Version. Medical Topics and Chapter: Fish Poisoning and Shellfish Poisoning.* April 2020 [https://www.merckmanuals.com/professional/injuries-poisoning/poisoning/fish-poisoning-and-shellfish-poisoning]. December 31, 2020.

Merck & Co. *Merck Manual Professional Version. Medical Topics and Chapter: Plant Poisoning.* April 2020 [https://www.merckmanuals.com/professional/injuries-poisoning/poisoning/plant-poisoning]. December 31, 2020.

Merck & Co. *Merck Manual Professional Version. Medical Topics and Chapter: Toxins as Mass-Casualty Weapons.* May 2019 [https://www.merckmanuals.com/professional/injuries-poisoning/mass-casualty-weapons/toxins-as-mass-casualty-weapons]. December 31, 2020.

Merck & Co. *Merck Veterinary Manual.* 2020 [https://www.merckvetmanual.com/]. December 31, 2020.

Moriyama, Hideo. "Purification and Properties of Ricin." *Igaku to Seibutsugaku* 10 (1947): 163–166.

Moshiri, Mohammad, Fatemeh Hamid and Leila Etemad. "Ricin Toxicity: Clinical and Molecular Aspects" *Reports of Biochemistry & Molecular Biology* 4 (April 2016): 60–65.

Olsnes, Sjur, Fiorenzo Stirpe, Kirsten Sandvig and Alexander Pihl. "Isolation and Characterization of Viscumin, a Toxic Lectin from Viscum album L. (Mistletoe)." *The Journal of Biological Chemistry* 257 (1982): 13,263–13,270.

Olson, Kent R., ed. *Poisoning & Drug Overdose.* 4th Edition. New York, New York: Lange Medical Books/ McGraw-Hill, 2004.

Patocka Jiri and Ladislav Streda. "Plant Toxic Proteins and Their Current Significance for Warfare and Medicine." *Journal of Applied Biomedicine* 1 (2003): 141–147.

Public Health Service. 42 CFR Part Patocka Jiri and Ladislav Streda. "Plant Toxic Proteins and Their Current Significance for Warfare and Medicine." *Journal of Applied Biomedicine* 1 (2003): 141–147 73 – "Select Agents and Toxins," 2004: 443-458.

Schmitt, Clare K., Karen C. Meysick and Alison D. O'Brien. "Bacterial Toxins: Friends or Foes?" *Emerging Infectious Diseases* 5, no. 2, (April/June 1999): 224–234.

Sidell, Fredrick R., Ernest T. Takafuji and David R. Franz, ed. *Medical Aspects of Chemical and Biological Warfare, Textbook of Military Medicine Series, Part 1, Warfare, Weaponry, and the Casualty.* Washington, D.C.: Office of the Surgeon General, Department of the Army, 1997.

Sifton, David W. ed. *PDR Guide to Biological and Chemical Warfare Response.*Montvale, New Jersey: Thompson/Physicians' Desk Reference, 2002.

Singh, Bal Ram and Anthony T. Tu, ed. *Natural Toxins 2: Structure, Mechanism of Action, and Detection. Volume 391 of Advances in Experimental Medicine and Biology.* New York, New York: Plenum Press, 1996.

Somani, Satu M. and James A. Romano, Jr., ed. *Chemical Warfare Agents: Toxicity at Low Levels.* Boca Raton, Florida: CRC Press, 2001.

Somani, Satu M., ed. *Chemical Warfare Agents.* New York, New York: Academic Press, 1992.

Stirpe, Fiorenzo, Kirsten Sandvig, Sjur Olsnes and Alexander Pihl. "Action of Viscumin, A Toxic Lectin from Mistletoe, on Cells in Culture." *The Journal of Biological Chemistry*, 257 (1982): 13,271–13,277.

Stirpe, Fiorenzo, Luigi Barbieri, Ada Abbondanza, Anna Ida Falasca, Alex N. F. Brown, Kirsten Sandvig, Sjur Olsnes and Alexander Pihl. "Properties of Volkensin, a Toxic Lectin from Adenia volkensii." *The Journal of Biological Chemistry* 260 (1985): 14,589–14,595.

Sweeney, Edel C., Alexander G. Tonevitsky, Rex A. Palmer, Hidie Niwa, Uwe Pfueller, Juergen Eck, Hans Lentzen, Igor I. Agapov and Mikhail P. Kirpichnikov. "Mistletoe Lectin I Forms a Double Trefoil Structure." *FEBS Letters* 431 (1998): 367–370.

True, Bey-Lorraine and Robert H. Dreisbach. *Dreisbach's Handbook of Poisoning: Prevention, Diagnosis and Treatment.* 13th Edition. London, United Kingdom: The Parthenon Publishing Group, 2002.

United States Army Headquarters. *Potential Military Chemical/Biological Agents and Compounds, Field Manual No. 3-11.9.* Washington, D.C.: Government Printing Office, January 10, 2005.

United States Department of Health and Human Services Centers for Disease Control and Prevention and National Institutes of Health. *Biosafety in Microbiological and Biomedical Laboratories (BMBL).* 5th Edition. Washington, D.C.: US Government Printing Office, 2009.

United State Environmental Protection Agency. *Comptox.* 2020 [https://comptox.epa.gov/dashboard]. December 31, 2020.

United State Environmental Protection Agency. "Cyanobacteria and Cyanotoxins: Information for Drinking Water Systems." Factsheet EPA-810F11001, September 2014.

United States Food & Drug Administration, Center for Food Safety & Applied Nutrition. *Foodborne Pathogenic Microorganisms and Natural Toxins Handbook, 2nd Edition.* October 24, 2017 [https://www.fda.gov/media/83271/download]. December 31, 2020.

United States National Institute for Occupational Safety and Health. "Emergency Response Card for Abrin." Interim Document, March 24, 2003.

United States National Institute for Occupational Safety and Health. "Emergency Response Card for Ricin." Interim Document, March 24, 2003.

United States National Institute of Health, National Library of Medicine. *PubChem.* 2020 [https://pubchem.ncbi.nlm.nih.gov/]. December 31, 2020.

Withers, Mark, ed. *Medical Management of Biological Casualties Handbook.* 8th Edition. Fort Detrick, Maryland: United States Army Medical Research Institute of Infectious Diseases, September 2014.

World Health Organization. *Health Aspects of Chemical and Biological Weapons: Report of A WHO Group of Consultants.* Geneva, Switzerland: World Health Organization, 1970.

World Health Organization. *Public Health Response to Biological and Chemical Weapons: WHO Guidance.* Geneva Switzerland: World Health Organization, 2004.

Yang, Chen-Chung. "Crystallization and Properties of the Cobrotoxin from Formosan Cobra Venom." *The Journal of Biological Chemistry* 240 (1965): 1,616–1,618.

17 Bacterial Pathogens

GENERAL INFORMATION

Bacteria are single-celled microorganisms. They are easy to grow but production, isolation, harvesting, and storage of large quantities of these organisms is difficult. Bacteria come in various shapes including rods (bacilli), spheres (cocci), and commas or spirals (spirilla). Individual organisms range in size from less than 1 micron to tens of microns. They may or may not be able to move on their own (i.e., motile). Bacteria can be aerobic; that is, they can live and grow in the presence of oxygen, or they can be anaerobic and live without oxygen.

This class of agents includes rickettsiae; which, because of their small size – approximately 0.3 microns in diameter and ranging from 0.3 to 0.5 microns in length, which is intermediate between other bacteria and viruses – and being strict obligate parasites (i.e., can only grow inside living cells), was historically considered a separate category of biological agents by the military. Rickettsiae are Gram-negative, non-motile, and non-sporing, and occur in the form of rods or spheres. They are more difficult to produce in quantity than other bacteria and are easily killed by heat, dehydration, or common disinfecting agents. They are normally transmitted by an arthropod vector (i.e. ticks, lice, fleas, mites), which also serves as either the primary or intermediate host.

In adverse conditions, some bacteria can enter a dormant state known as a spore. Spores can remain dormant for decades and can survive under extreme temperatures and other adverse environmental conditions. Unlike fungi (C19), formation of spores is not related to reproduction and is done strictly as a protective mechanism. Upon reactivation, each spore produces a single active bacterium. Spores are normally spherical or oval and are only a fraction of the size of the active (i.e., vegetative) cell.

Pathogens employed as biological weapons can be used for both lethal and incapacitating purposes. They cause disease by invading tissue or by producing toxins (C16) that are detrimental to the infected individual. Pathogens can be selected to target a specific host (e.g., humans, cows, fowl) or they may pose a broad threat to both humans and animals.

Pathogens deployed to target animals are generally used to produce lethal effects in an agriculturally significant species such as cows, pigs, or chickens. Although these pathogens are selected to target a specific animal species, there is the possibility that the disease may migrate to humans. The diseases produced by these crossover pathogens may be difficult for medical personnel not trained in exotic pathology to diagnose.

Other pathogens are selected to produce lethal effects in an agriculturally significant crop species such as wheat, corn, or rice. There is little potential for migration of these pathogens to either humans or animals.

A final group of biological warfare pathogens are those used as simulants to model the release of other, more hazardous agents. Pathogens employed as biological warfare simulants do not generally pose a significant risk to healthy people, animals, or plants. However, individuals with respiratory illness or suppressed immune systems may be at risk should they be exposed to an infectious dose of the agent.

Bacteria can be stored as either liquids (e.g., organisms in concentrated growth media) or powders (e.g., spores or freeze-dried mixtures of agent and growth media) and are easy to disperse. However, because they are living organisms and can be killed during the dispersal process there are limitations to the methods that can be used. Unlike chemical or toxin agents, pathogens can also be stored and dispersed via infected vectors (e.g., mosquitoes, fleas). For more information on methods of dispersal, see appendix 1. In most cases, large-scale attacks will be clandestine and only detected through epidemiological analysis of resulting disease patterns. Even in the case of small-scale incidents or attacks

DOI: 10.4324/9781003230564-8

directed at specific individuals (e.g., white-powder letters), without the inclusion of a threat the attack may go unrecognized until the disease appears in exposed individuals (e.g., the initial 2001 anthrax attack at American Media Inc., which claimed the life of Robert Stevens).

In general, unless a local reservoir (i.e., intermediate host that may or may not be affected by the bacteria) is established, pathogens are easily killed by unfavorable environmental factors such as fluctuations in temperature, humidity, food sources, or ultraviolet light. For this reason, their persistency is generally limited to days. However, bacterial spores are highly resistant to impacts from changes in environmental factors. Agents that can form spores can survive in this state for decades and then become active again under the proper conditions. In addition, pathogens can be freeze-dried (i.e., lyophilized) and remain in a preserved state almost indefinitely. Freeze-dried pathogens are reactivated when exposed to moisture.

Incubation times for diseases resulting from infection vary depending on the specific pathogen, but are generally on the order of days to weeks. Exposures to extremely high doses of some pathogens may reduce the incubation period to as short as several hours. The pathway of exposure (e.g., inhalation, ingestion) can also cause a significant change in the incubation time required as well as the clinical presentation of the disease.

Some diseases caused by bacteria are communicable and easily transferred from an infected individual to anyone in close proximity. Typically, this occurs when the infected individual coughs or sneezes creating an infectious aerosol. These aerosols enter the body of a new host through inhalation and/or contact of the aerosol with the mucous membranes (e.g., eyes, nose, mouth). In addition, although intact skin is an effective barrier against most pathogens, abrasions, or lacerations circumvents this protective barrier and allows entry of the pathogen into the body.

Along with direct aerosol exposure, some bacteria are potentially associated with exposure through ingestion of contaminated food and/or water. Pathogens that normally infect individuals through an ingestion pathway also often pose a significant risk of secondary infections from the fecal/oral cycle. Individuals or animals may become asymptomatic carriers of these types of diseases and become capable of spreading the disease long after their apparent recovery (e.g., Typhoid Mary). Mechanical vectors, (e.g., flies, roaches) can also transport these types of pathogens to remote food items.

RESPONSE

PERSONAL PROTECTIVE REQUIREMENTS

Responding to the Scene of a Release

A number of conditions must be considered when selecting protective equipment for individuals at the scene of a release. For instances such as white-powder letters when the mechanism of release is known and it does not involve an aerosol generating device, then responders can use Level C with N95 or higher-level filters.

If an aerosol generating device is employed (e.g., sprayer), or the dissemination method is unknown and the release is ongoing, then responders should wear a Level A protective ensemble. Once the device has stopped generating the aerosol or has been rendered inoperable, and the aerosol has settled, then responders can downgrade to Level B.

In all cases, there is a significant hazard posed by contact of contaminated material with skin that has been cut or lacerated, or through injection of pathogens by contact with debris. Appropriate protection to avoid any potential abrasion, laceration, or puncture of the skin is essential. Individuals with damaged or open skin should not be allowed to enter the contaminated area.

Working with Infected Individuals

For most bacteria, use infection control guidelines standard precautions. If appropriate, or the identity of the bacteria is unknown, use additional droplet and airborne precautions. Avoid direct contact with all body fluids, wounds, or wound drainage.

Standard Precautions

Include hand hygiene, the use of gloves and eye protection, wearing a gown and mask as appropriate, and using care when handling sharps. If appropriate, have patients employ respiratory hygiene/cough etiquette practices (i.e., patients covering the mouth/nose with a tissue when coughing and prompt disposal of used tissues; using surgical masks on the coughing person when tolerated and appropriate; hand hygiene after contact with respiratory secretions; and spatial separation – ideally greater than 3 feet).

Contact Precautions

Isolate infected individuals in single-patient rooms. If this is not possible, ensure greater than 3 feet spatial separation between beds to reduce the opportunities for inadvertent sharing of items between patients.

   Employ standard precaution and wear gowns. Protective garments are donned upon room entry and discarded before exiting the patient room.

Droplet Precautions

Isolate infected individuals in single-patient rooms. If this is not possible, ensure greater than 3 feet spatial separation with drawing curtain between patient beds.

   Employ standard precaution and wear gowns. Wear a mask for close contact with infectious patient (i.e., within 3 feet of the patient). Protective garments are donned upon room entry and discarded before exiting the patient room.

   Patients who must be transported outside of the room should wear a mask if tolerated and follow respiratory hygiene/cough etiquette practices (i.e., patients covering the mouth/nose with a tissue when coughing and prompt disposal of used tissues, hand hygiene after contact with respiratory secretions, and spatial separation – ideally greater than 3 feet).

Airborne Precautions

Isolate infected individuals in in an airborne infection isolation room with the door closed except during entry and exit. If such an isolation room is unavailable, mask the patient and place them in a private room with the door closed. Cohort patients who are presumed to have the same infection in areas of the facility that are away from other patients. Use portable exhaust fans to create a negative pressure environment in the converted area of the facility. Discharge air directly to the outside, away from people and air intakes, or direct exhaust through HEPA filters before it is introduced to other air spaces.

   Employ standard precaution and wear gowns. Wear an N95 or higher-level mask or respirator. Masks and respirators must be fit-tested to each individual and users must perform a seal check each time they don the facepiece. Respiratory protection is donned prior to room entry. It is removed only after exiting the room and performing hand hygiene. Protective garments are donned upon room entry and discarded before exiting the patient room.

   Patients who must be transported outside of the isolation area should wear a mask if tolerated and follow respiratory hygiene/cough etiquette practices (i.e., patients covering the mouth/nose with a tissue when coughing and prompt disposal of used tissues, hand hygiene after contact with respiratory secretions, and spatial separation – ideally greater than 3 feet).

ADDITIONAL HAZARDS

It is possible that local insects can become both a reservoir and a vector for the pathogen. Under these circumstances, the pathogen can survive well after the initial release and can rapidly spread beyond the immediately affected area. A host-vector cycle (i.e., an infected individual/animal/plant is fed on by multiple parasitic insects that then move on to infect other individuals/animals/plants) may be established that rapidly leads to vector amplification and an expediential expansion of the outbreak. In many cases, once a vector is infected, it is capable of transmitting the disease throughout its life span. Some pathogens are also transmitted directly to the young of the vector so that the next generation is born infected. Response activities must also include efforts to contain and eliminate these vectors (e.g., barriers to prevent exposures, application of pesticides).

DECONTAMINATION

Casualties/Personnel

Infected individuals

Unless the individual is reporting directly from the scene of an attack (e.g., white-powder letter, aerosol release, etc.), then decontamination is not necessary. Use standard protocols for individuals that may be infected with a communicable disease transmissible via an aerosol.

Direct Exposure

In the event that an individual is at the scene of a known or suspected attack (e.g., white-powder letter, aerosol release, etc.), have them wash their hands and face thoroughly with antimicrobial soap and water as soon as possible. If antimicrobial soap is not available, use any available soap or shampoo. They should also blow their nose to remove any agent particles that may have been captured by nasal mucous. If the release involved a powdered agent and it is practicable, dampen the agent with a water mist to help prevent aerosolization. Remove all clothing and seal in a plastic bag. To avoid further exposure of the head, neck, and face to the agent, cut off potentially contaminated clothing that must be pulled over the head. Shower using copious amounts of antimicrobial soap (if available) and water. Ideally, showers will be high-volume with low-pressure. Ensure that the hair has been washed and rinsed to remove potentially trapped agent. The CDC does not recommend that individuals use bleach or other disinfectants directly on their skin.

Animals

Unless the animals are at the actual scene of a release then decontamination is not necessary.

Apply universal decontamination procedures using antimicrobial soap and water. If antimicrobial soap is not available, use any available soap, shampoo, or detergent. In some cases, severe infection may require euthanasia of animals and/or herds. Consult local/state veterinary assistance office. If the pathogen has not been identified, then wear a fitted N95 protective mask, eye protection, disposable protective coveralls, disposable boot covers, and disposable gloves when dealing with infected animals.

Plants

May require removal and destruction of infected plants. Incineration of impacted fields may be required. Consult local/state agricultural assistance office. Some bacteria are easily spread by mechanical vectors (e.g., farm implements, track out, running water) and extreme care must be taken to avoid further contamination. Wear disposable protective coveralls, disposable boot covers, and disposable gloves to prevent spread of contamination. Many plant pathogens are spread via insects and in these situations response activities must also include efforts to contain and eliminate these vectors.

Food

All foodstuffs in the area of a release should be considered contaminated. Unopened items may be used after decontamination of the container. Opened or unpackaged items should be destroyed. Fruits and vegetables should be washed thoroughly with antimicrobial soap and water. Many pathogens can survive in food containers for extended periods.

Property

Surface Disinfectants

Compounds containing phenolics, chlorhexidine (not effective against bacteria spores), quaternary ammonium salts (additional activity if bis-n-tributyltin oxide present), hypochlorites such as household bleach, alcohols such as 70 to 90% ethanol and isopropanol (not effective against bacteria spores), potassium peroxymonosulfate, hydrogen peroxide, iodine/iodophors, and triclosan.

Sandia decontamination foam formulations containing high pH aqueous solutions of hydrogen peroxide, Canadian Aqueous System for Chemical-Biological Agent Decontamination (CASCAD) containing aqueous solutions of chloroisocyanurates and complexing agents, as well as Dahlgren decontamination formulations containing aqueous solutions of peracetyl borate and surfactants are all effective against bacterial agents, although with varying contact times.

The military also identifies the following nonstandard decontaminants: Detrochlorite (thickened bleach mixture of diatomaceous earth, anionic wetting agent, calcium hypochlorite, and water), 3% aqueous peracetic acid solution, 1% aqueous hyamine solution, and 10% aqueous sodium or potassium hydroxide solution.

Large Area Fumigants

Gases, including formaldehyde, ozone, ethylene oxide, or chlorine dioxide are effective against many bacteria. However, these materials are highly toxic to humans and animals, and fumigation operations must be adequately controlled to prevent unnecessary exposure. Additional methods include vaporized hydrogen peroxide and an ionized hydrogen peroxide aerosol.

Fomites

Some pathogens may be adsorbed onto clothing or bedding causing these items to become infectious and capable of transmitting the disease. Others may contain vectors (e.g., lice, ticks) that pose a transmission hazard. Deposit items in an appropriate biological waste container and send to a medical waste disposal facility.

Alternatively, cotton or wool articles can be boiled in water for 30 minutes, autoclaved at 253°F for 45 minutes, or immersed in a 2% household-bleach solution (i.e., 1 liter of bleach in 2 liters of water) for 30 minutes followed by rinsing.

FATALITY MANAGEMENT

Unless the cadaver is coming directly from the scene of an attack (e.g., white-powder letter, aerosol release), then process the body according to established procedures for handling potentially infectious remains.

Because of the nature of biological warfare agents, it is highly unlikely that a contaminated cadaver will be recovered from the scene of an attack unless it is from an individual who died from trauma or other complications while the attack was ongoing. If a fatality is grossly contaminated with a biological agent, wear disposable protective clothing with integral hood and booties, disposable gloves, eye protection, and an N95 respirator or powered air-purifying respirator (PAPR) equipped with N95 or high-efficiency particulate air (HEPA) filters.

Remove all clothing and personal effects. Items that will be retained for further processing should be double sealed in impermeable containers, ensuring that the inner container is decontaminated before placing it in the outer one. Otherwise, dispose of contaminated articles at an appropriate medical waste disposal facility.

Thoroughly wash the remains with antimicrobial soap and water. Pay particular attention to areas where agent may get trapped, such as hair, scalp, pubic areas, fingernails, folds of skin, and wounds. If deemed appropriate, the body can be washed with a 2% sodium hypochlorite bleach solution (i.e., 2 liters of water for every liter of household bleach), ensuring the solution is introduced into the ears, nostrils, mouth, and any wounds. This concentration of bleach will not affect remains but will disinfect the offending agent. Higher concentrations of bleach can harm remains. The bleach solution should remain on the cadaver for a minimum of 15 minutes. Wash with soap and water. If the body is to be embalmed (not recommended by the CDC), ensure that all the bleach solution is removed as it will react with embalming fluid.

If there is a potential that vectors may be involved, care must be taken to kill any vectors (e.g., lice, fleas) remaining either on the cadaver or residing in fomites. Remove all potentially infested clothing depositing it in a container that will trap and eliminate vectors. Dispose of contaminated articles at an appropriate medical waste disposal facility.

Once the remains have been thoroughly decontaminated, process the body according to established procedures for handling potentially infectious bodies. Use appropriate burial procedures.

The CDC has determined that the risks of occupational exposure to biological terrorism agents outweighs the advantages of embalming fatalities and recommends that bodies should be buried without embalming, unless the bacterium forms spores (e.g., *Bacillus anthracis* (C17-A001)). Fatalities due to spore forming bacteria should be cremated without embalming.

C17-A AGENTS

C17-A001

Bacillus anthracis (Agent N)
Anthrax
ICD-11: XN94F
UNII: IT7Z319PFY
EPPO: —

It is an aerobic, Gram-positive, spore-forming, rod-shaped bacterium. It is endemic in many countries of the world, particularly in tropical and sub-tropical areas. Dry spores are stable for decades. Spores are stable in water for up to two years and are resistant to chlorine at purification concentrations. This is a biosafety level 2 agent for clinical samples. Biosafety level 3 should be used for activities with a high potential for aerosol or droplet production or high concentrations of agent.

It is on the Australia Group Core list, a Tier-1 agent on the U.S. Select Agents and Toxins list, and the OIE list of Notifiable Diseases for 2020.

In People:
CDC Case Definition: Clinical criteria is 1) an illness with at least one specific or two non-specific symptoms and signs that are compatible with cutaneous, ingestion, inhalation, or injection anthrax; systemic involvement; or anthrax meningitis; OR 2) a death of unknown cause and organ involvement consistent with anthrax.

Presumptive laboratory criteria is gram stain demonstrating Gram-positive rods, square-ended, in pairs or short chains.

Confirmatory laboratory criteria is 1) Culture and identification from clinical specimens by Laboratory Response Network; OR 2) Demonstration of *B. anthracis* antigens in tissues by immunohistochemical staining using both *B. anthracis* cell wall and capsule monoclonal antibodies; OR 3) Evidence of a four-fold rise in antibodies to protective antigen between acute and convalescent sera or a fourfold change in antibodies to protective antigen in paired convalescent sera using CDC quantitative anti-PA immunoglobulin G ELISA testing in an unvaccinated person; OR 4) Detection of *B. anthracis* or anthrax toxin genes by polymerase chain reaction and/ or sequencing in clinical specimens collected from a normally sterile site (such as blood or CSF) or lesion of other affected tissue (skin, pulmonary, reticuloendothelial, or gastrointestinal); OR 5) Detection of lethal factor in clinical serum specimens by mass spectrometry.

Communicability: Direct person-to-person transmission is rare. When dealing with infected individuals, use standard precautions. Avoid direct contact with wounds or wound drainage.

Normal Routes of Exposure: Inhalation; ingestion; abraded skin; mucous membranes; possibly vectors (biting flies).

Infectious Dose: 2,500–55,000 spores (Inhalation).

Secondary Hazards: Spores; blood; body tissue; fomites.

Incubation: Hours (high dose) – 7 days. May be prolonged up to 2 months.

Signs & Symptoms: Inhalation: often biphasic, but symptoms may progress rapidly. Mild and nonspecific (i.e., flu-like) with fever, vague feeling of bodily discomfort (malaise), and fatigue accompanied by a nonproductive cough and chest discomfort. This progresses to severe respiratory distress with difficulty breathing (dyspnea), rapid breathing (tachypnea), high-pitched whistling respiration (stridor), blue lips (cyanosis), excessive sweating (diaphoresis), and chest pain that may be severe enough to mimic a heart attack. Additional symptoms may include nausea, vomiting, abdominal pain, headache, and altered mental status.

Skin: usually begins as a small, painless, itchy pimple (papule) on an exposed surface, which turns into a blister (vesicle) and then progresses into a depressed black scab (eschar) that is often surrounded by swelling or redness and may be accompanied by swollen lymph glands. Fever is also common.

Ingestion: if the infection occurs in the mouth and throat (oropharynx), a lesion may be observed in the mouth or back of the throat. Symptoms include sore throat, difficulty swallowing (dysphagia), and swelling of the neck. Additional flu-like symptoms could include fever, headache, fatigue, difficulty breathing (dyspnea), abdominal swelling with pain, and vomiting or diarrhea, either of which may contain blood.

Suggested Alternatives for Differential Diagnosis: Early diagnosis is difficult. Potential alternatives include abdominal aneurysm, aortic dissection, bacterial mediastinitis, bronchitis, central nervous system infection, coccidioidomycosis, diphtheria, dysentery, ecthyma, fulminate mediastinal tumors, gastroenteritis, glanders, hantavirus pulmonary syndrome, histoplasmosis, influenza, leprosy, meningitis, nonspecific viral syndrome, orf, plague, pleural effusion, pneumonia, primary syphilis, psittacosis, rat bite fever, rickettsial pox, septicemia, spider bite, subarachnoid hemorrhage, superior vena cava syndrome, tularemia, typhoid fever, and valley fever.

Mortality Rate (untreated): ≤20% (cutaneous); 25%–75% (gastrointestinal); ≤100% (respiratory).

In Domestic Animals:

Susceptible species: Cattle, sheep, goats, horses.

Communicability: Direct transmission does not occur.

Normal Routes of Exposure: Inhalation; Ingestion; abraded Skin; mucous membranes; vectors (biting flies).

Secondary Hazards: Spores; blood; body tissue; animal products (e.g., hides, hair, wool, bones, feedstuffs, handicrafts).

Incubation: 1–14 days.

Signs: Often, the course of disease is so rapid that illness is not observed and animals are found dead. Symptoms include abrupt fever, chills, severe colic, loss of appetite (anorexia), depression, staggering, difficulty breathing (dyspnea), a period of excitement followed by depression, stupor, trembling, collapse, convulsions, and death. Rigor mortis is frequently absent or incomplete. Dark blood may ooze from the mouth, nostrils, and anus with marked bloating and rapid body decomposition. The blood is dark and thickened and fails to clot readily.

Suggested Alternatives for Differential Diagnosis: Conditions that cause sudden death such as acute classical swine fever, acute infectious anemia, acute leptospirosis, African swine fever, anaplasmosis, bacillary hemoglobinuria, bloat, clostridial infections, colic, pharyngeal malignant edema, purpura; acute poisonings by bracken fern, sweet clover, and lead; lightning strike and sunstroke. *Mortality Rate (untreated):* Very high.

C17-A002

Bacillus cereu sbv. anthracis
Clinically compatible with anthrax
ICD-11: XN8PY
UNII: IT7Z319PFY
EPPO: —

It is an aerobic, Gram-positive, spore-forming, rod-shaped bacterium. Unlike *B. anthracis*, most *B. cereus* bv. *anthracis* strains are motile. It has only been identified as endemic in West and Central Africa. All are gamma-phage resistant and many are resistant to penicillin. Dry spores are likely to be stable for decades. This is a biosafety level 2 agent for clinical samples. Biosafety level 3 should be used for activities with a high potential for aerosol or droplet production or high concentrations of agent.

It is a Tier-1 agent on the U.S. Select Agents and Toxins List.

In People:
There have not been any reported cases of this disease in humans.

CDC Case Definition: Clinical criteria is 1) an illness with at least one specific or two non-specific symptoms and signs that are compatible with cutaneous, ingestion, inhalation, or injection anthrax; systemic involvement; or anthrax meningitis; OR 2) a death of unknown cause and organ involvement consistent with anthrax.

Presumptive laboratory criteria is gram stain demonstrating Gram-positive rods, square-ended, in pairs or short chains.

Confirmatory laboratory criteria is 1) Culture and identification from clinical specimens by Laboratory Response Network; OR 2) Demonstration of *B. anthracis* antigens in tissues by immunohistochemical staining using both *B. anthracis* cell wall and capsule monoclonal antibodies; OR 3) Evidence of a four-fold rise in antibodies to protective antigen between acute and convalescent sera or a fourfold change in antibodies to protective antigen in paired convalescent sera using CDC quantitative anti-PA immunoglobulin G ELISA testing in an unvaccinated person; OR 4) Detection of *B. anthracis* or anthrax toxin genes by polymerase chain reaction and/ or sequencing in clinical specimens collected from a normally sterile site (such as blood or CSF) or lesion of other affected tissue (skin, pulmonary, reticuloendothelial, or gastrointestinal); OR 5) Detection of lethal factor in clinical serum specimens by mass spectrometry.

Communicability: Based on similarities to *B. anthracis*, direct person-to-person transmission should be rare or not at all. When dealing with infected individuals, use standard precautions. Avoid direct contact with wounds or wound drainage.

Normal Routes of Exposure: Inhalation; Ingestion; abraded skin; mucous membranes; possibly vectors (biting flies); possibly mechanical vectors (carrion flies).

Infectious Dose: Unknown.

Secondary Hazards: Spores; blood; body tissue; fomites.

Incubation: Unknown, but evidence suggests it is highly virulent.

Signs & Symptoms: No human cases reported, see anthrax (C17-A001).

Suggested Alternatives for Differential Diagnosis: Potential alternatives include abdominal aneurysm, aortic dissection, bacterial mediastinitis, bronchitis, central nervous system infection, coccidioidomycosis, diphtheria, dysentery, ecthyma, fulminate mediastinal tumors, gastroenteritis, glanders, hantavirus pulmonary syndrome, histoplasmosis, influenza, leprosy, meningitis, nonspecific viral syndrome, orf, plague, pleural effusion, pneumonia, primary syphilis, psittacosis, rat bite fever, rickettsial pox, septicemia, spider bite, subarachnoid hemorrhage, superior vena cava syndrome, tularemia, typhoid fever, and valley fever.

Mortality Rate (untreated): Unknown. Evidence suggests that among animals, systemic infections are generally fatal.

In Domestic Animals:

Susceptible species: Goats; likely cattle, sheep, horses.

Communicability: Direct transmission does not occur.

Normal Routes of Exposure: Inhalation; Ingestion; abraded skin; mucous membranes; vectors (biting flies).

Secondary Hazards: Spores; blood; body tissue; animal products such as hides, hair, wool, bones, feedstuffs, or handicrafts.

Incubation: Unknown.

Signs: Often, the course of disease is so rapid that illness is not observed and animals are found dead. Symptoms include abrupt fever, chills, severe colic, loss of appetite (anorexia), depression, staggering, difficulty breathing (dyspnea), a period of excitement followed by depression, stupor, trembling, collapse, convulsions, and death. Rigor mortis is frequently absent or incomplete. Dark blood may ooze from the mouth, nostrils, and anus with marked bloating and rapid body decomposition. The blood is dark and thickened and fails to clot readily.

Suggested Alternatives for Differential Diagnosis: Conditions that cause sudden death such as acute classical swine fever, acute infectious anemia, acute leptospirosis, African swine fever, anaplasmosis, bacillary hemoglobinuria, bloat, clostridial infections, colic, pharyngeal malignant edema, purpura; acute poisonings by bracken fern, sweet clover, and lead; lightning strike and sunstroke.

Mortality Rate (untreated): Generally fatal.

C17-A003

Bacillus subtilis (Agent BG)
BW Simulant
ICD-11: XM4SG9
UNII: 8CF93KW41W
EPPO: BACISU

It is a Gram-positive, spore-forming, rod-shaped, facultative anaerobe (can live with or without oxygen). It is very common in soil, water and air.

No significant harmful health effects to humans or animals are expected from exposure to this pathogen unless the individual has a compromised respiratory system or suppressed immune system. Direct contact with large quantities of B. subtilis spores may cause redness or irritation of the skin.

Some strains have been used as a medicinal product to treat dysentery and other intestinal problems, and as a pesticide to control plant diseases and fungal pathogens.

C17-A004

Bacillus thuringiensis (Agent BT)
BW Simulant
ICD-11: —
UNII: 3TK3LQP1N7
EPPO: BACITH

B. thuringiensis is a Gram-positive, spore-forming, rod-shaped, facultative anaerobe (can live with or without oxygen). Vegetative cells are motile and approximately 1 micron wide and 5 microns long. It is very common in soil and on plants.

No significant harmful health effects to humans, animals or plants are expected from exposure to this pathogen. Some strains have been used as pesticides to control crop damaging insects.

C17-A005

Bartonella quintana
Trench Fever
ICD-11: XN14D
UNII: 9R2271TA4J
EPPO: BARNQU

It is an aerobic, Gram-negative, non-motile, rod-shaped bacterium that can only reproduce within a host's cells (obligate intracellular). It is endemic throughout the world. Can survive for only short period in the environment outside a host. The natural reservoir is humans. This is a biosafety level 2 agent. It was removed from the Australia Group list in June 2011 and is no longer regulated for export control.

In People:
CDC Case Definition: None established.

Communicability: Direct person-to-person transmission does not occur. When dealing with infected individuals, use standard precautions and protect from vectors.

Normal Routes of Exposure: Vectors (lice).

Infectious Dose: Unknown.

Secondary Hazards: Fomites; Vector amplification.

Incubation: 3–48 days.

Signs & Symptoms: Abrupt onset of fever (up to 104°F), chills, excessive sweating (diaphoresis), headache with pain behind the eyes, eye infection (pink eye – conjunctivitis), sensitivity to light (photophobia), difficulty breathing (dyspnea), a vague feeling of bodily discomfort (malaise), weakness, depression, restlessness, insomnia, and pain in the bones of the shins, neck, and back. May produce a transient rash of flat (macular) red spots. In rare cases, it may progress to infection of the heart (endocarditis).

Suggested Alternatives for Differential Diagnosis: AIDS, babesiosis, bacillary angiomatosis, brucellosis, cat scratch disease, chronic fatigue syndrome, cryptococcosis, culture-negative endocarditis, epidemic typhus, Epstein-Barr virus infection, histoplasmosis, Lyme disease, non-Hodgkin lymphoma, Q fever, relapsing fever, Rocky Mountain spotted fever, tuberculosis, and typhoid fever.

Mortality Rate (untreated): 0%; 12% (endocarditis).

C17-A006

***Brucella* species** including *B. abortus*; *B. melitensis*; *B. suis* (Agent US))
Brucellosis
ICD-11: XN22N; XN7ZW (*B. melitensis*); XN7A8 (*B. abortus*); XN3UP (*B. suis*)
UNII: OQF85710LZ (*B. melitensis*); 492LCM0TUL (*B. abortus*)
EPPO: BRULME (*B. melitensis*)

They are Gram-negative, non-motile, oval-shaped, facultative anaerobes (can live with or without oxygen), that can only reproduce within a host's cells (obligate intracellular). It is endemic throughout the world. The natural reservoirs are cattle (*B. abortus);* sheep and goats (*B. melitensis*); and pigs (*B. suis*). *Brucella* spp. are stable in water for 20 to 72 days and in fecal matter for over 60 days (in a cool environment). Direct exposure to sunlight kills the organism within a few hours. The most pathogenic of the species for man is *B. melitensis,* followed by *B. suis* and then *B. abortus.* This is a biosafety level 2 agent for clinical samples. Biosafety level 3 should be used for activities with a high potential for aerosol or droplet production or high concentrations of agent.

They are on the Australia Group Core list, a Tier-1 agent on the USU.S. Select Agents and Toxins list, and the OIE list of Notifiable Diseases for 2020.

In People:

CDC Case Definition: Clinical criteria is an illness characterized by acute or insidious onset of fever and one or more of the following: night sweats, arthralgia, headache, fatigue, anorexia, myalgia, weight loss, arthritis/spondylitis, meningitis, and/or focal organ involvement (endocarditis, orchitis/ epididymitis, hepatomegaly, splenomegaly).

Presumptive laboratory criteria is 1) *Brucella* total antibody titer of greater than or equal to 160 by standard tube agglutination test or *Brucella* microagglutination test in one or more serum specimens obtained after onset of symptoms; OR 2) detection of *Brucella* DNA in a clinical specimen by PCR assay.

Confirmatory laboratory criteria is 1) culture and identification of *Brucella* spp. from clinical specimens; OR 2) evidence of a fourfold or greater rise in *Brucella* antibody titer between acute- and convalescent-phase serum specimens obtained greater than or equal to 2 weeks apart.

Communicability: Direct person-to-person transmission does not occur except rarely during sexual contact. When dealing with infected individuals who have open skin lesions, use standard precautions. Avoid direct contact with wounds or wound drainage.

Normal Routes of Exposure: Inhalation; ingestion; abraded skin; mucous membranes.

Infectious Dose: 10–100 organisms.

Secondary Hazards: Blood and body fluids; body tissue.

Incubation: 3–180 days.

Signs & Symptoms: Are non-specific and consist of irregular fever, vague feeling of bodily discomfort (malaise), headache, profound weakness and fatigue, chills and sweating, generalized severe joint pain (arthralgia) and muscle pain (myalgia), loss of appetite (anorexia), weight loss, and depression. Joint complications are common.

Suggested Alternatives for Differential Diagnosis: Potential alternatives include abortion complications, acute epididymitis, ankylosing spondylitis, brain abscess in emergency medicine, bronchitis, cat scratch fever, chronic fatigue syndrome, collagen-vascular disease, cryptococcosis, cystitis, depression and suicide, emergent management of subarachnoid hemorrhage, emergent treatment of gastroenteritis, Epstein-Barr virus, erythema nodosum, hepatitis, hepatitis, histiocytosis, histoplasmosis, infectious mononucleosis, infective endocarditis, influenza, leptospirosis, lumbar disk disorders, malaria, malignancy, mechanical back pain, meningitis, osteomyelitis, pneumonia, rickettsial diseases, sacroiliitis, spontaneous bacterial peritonitis, thrombotic thrombocytopenic purpura, tuberculosis, tularemia, typhoid fever, undifferentiated spondyloarthropathy, urinary tract infection, and vasculitis.

Mortality Rate (untreated): ≤2%.

In Domestic Animals:

Target species: Cattle, pigs, sheep, goats.

Communicability: Direct transmission is possible.

Normal Routes of Exposure: Ingestion; mucous membranes; mating.

Secondary Hazards: Blood and body fluids; body tissue; mechanical vectors.

Incubation: 14–365 days.

Signs: There is no effective way to detect infected animals by their appearance. Some potential signs and symptoms include fever, increased respiration and depression; inflammation of testes and epididymis; swelling of the scrotum; atrophic testicles; edematous placenta and fetus; infertility; abortion in the last 3 to 4 months of pregnancy; and lymph-filled cystic cavities (hygromas) on the knees, stifles, hock, and angle of the haunch.

Suggested Alternatives for Differential Diagnosis: In cattle: Other causes of abortion including infectious bovine rhinotracheitis, vibriosis, leptospirosis, trichomoniasis, mycoplasma infections, mycosis, nutritional and physiological causes. Brucellosis should always be suspected when there are multiple late-term abortions in a herd. In pigs: Other common diseases causing abortion are Aujeszky's disease, leptospirosis, salmonellosis, streptococcidiosis, classical Swine Fever, and parvorisosis. In sheep, goats: Abortions due to chlamydiosis and coxiellosis.

Mortality Rate (untreated): Not normally fatal.

C17-A007

Burkholderia mallei (Agent LA)
Glanders
ICD-11 XN52E
UNII: IDH134C231
EPPO: —

It is an aerobic, Gram-negative, non-motile, rod-shaped bacterium. It is endemic in parts of South America, Africa, the Middle East and Asia. The natural reservoirs are horses, donkeys and mules. It is normally a disease of horses but is transmissible to humans by direct contact with sick animals or fomites. It can survive in the environment for 3–5 weeks in wet, humid, or dark conditions at temperatures between 68°F and 77°F, but is inactivated by heat or direct sunlight. This is a biosafety level 2 agent for clinical samples. Biosafety level 3 should be used for activities with a high potential for aerosol or droplet production or high concentrations of agent.

It is on the Australia Group Core list, a Tier-1 agent on the U.S. Select Agents and Toxins list, and the OIE list of Notifiable Diseases for 2020.

In People:

CDC Case Definition: Has not been developed.

Communicability: Direct person-to-person transmission is rare. When dealing with infected individuals, use standard precautions.

Normal Routes of Exposure: Inhalation; abraded skin; mucous membranes; ingestion.

Infectious Dose: Unknown, but very few organisms.

Secondary Hazards: Aerosols; body fluids; body tissue; fomites.

Incubation: 1–14 days.

Signs & Symptoms: Symptoms will vary depending on the type of infection but generally include fever with chills and sweating, chest pain, headache, sensitivity to light (photophobia), muscle tightness and pain (myalgia). Lymph glands may be swollen, and mucous membranes may produce a discharge of mucus and pus (mucopurulent) that may be bloody.

Pulmonary infection: Include pneumonia, abscesses in the lungs, fluid in the chest cavity (pleural effusion) with coughing, and difficulty breathing (dyspnea).

Systemic infection: Include a sharp, stabbing, or burning (pleuritic) chest pain with flushing (erythroderma), rash, necrotic lesions, jaundice, tearing (lacrimation), and diarrhea. May also develop rapid heart rate (tachycardia), and enlargement of both the liver and the spleen (hepatosplenomegaly). Multi-organ failure is common.

Skin infection: Include localized ulceration of the skin or mucous membranes. There may be a bumpy (papular) or blister (pustular) rash.

Suggested Alternatives for Differential Diagnosis: Anthrax, malaria, plague, pneumonia, smallpox, typhoid fever.

Mortality Rate (untreated): ≤ 95%.

In Domestic Animals:

Target species: Horses, mules, donkeys.

Communicability: Direct transmission is possible.

Normal routes of exposure: Inhalation; ingestion; mucous membranes.

Secondary Hazards: Blood and body fluids; body tissue; fomites (including water).

Incubation: 3–180 days.

Signs: High fever, cough, and difficulty inhaling (inspiratory dyspnea). There is a thick, sticky, yellowish nasal discharge of mucus and pus (mucopurulent) from one or both nostrils that may be bloody. Animals may develop bronchopneumonia or nodules and abscesses in the lungs. Nasal mucosa develop deep, rapidly spreading ulcers that become star-shaped scars when healed. Lymph nodes are usually swollen and painful, and multiple nodules may develop in the skin along the course of lymphatic vessels. These nodules often rupture and ulcerate, discharging an oily, thick yellow pus.

Postmortem findings include ulcers, nodules and star-shaped scars in the nasal cavity, trachea, pharynx, larynx, skin and subcutaneous tissues. Lymph nodes are typically enlarged, fibrotic and abscessed with focal abscesses. There may be evidence of severe bronchopneumonia as well as necrosis noted in the internal organs and testes. The lungs, liver, spleen and kidneys may contain firm, rounded nodules with greyish centers that are approximately 1 cm in diameter and surrounded by areas of inflammation.

Suggested Alternatives for Differential Diagnosis: Allergy, botryomycosis, dourine, epizootic lymphangitis, fungal infections, horsepox, melioidosis, pseudotuberculosis, sporotrichosis, strangles, trauma, tuberculosis, ulcerative lymphangitis.

Mortality Rate (untreated): High.

C17-A008

Burkholderia pseudomallei (Agent HI)
Melioidosis
ICD-11: XN3LD
UNII: 531JR2TJ2Z
EPPO: —

It is an aerobic, Gram-negative, motile, rod shaped bacterium. It is endemic in Southeast Asia and Australia, as well as parts of India, and China. The natural reservoir is soil and water. The optimal temperature range for growth is 75°F to 90°F, but it can grow in temperatures up to 108°F. This is a biosafety level 2 agent for clinical samples. Biosafety level 3 should be used for activities with a high potential for aerosol or droplet production or high concentrations of agent.

It is on the Australia Group Core list and a Tier-1 agent on the U.S. Select Agents and Toxins list.

In People:

CDC Case Definition: Has not been developed.

Communicability: Direct person-to-person transmission does not occur except rarely during sexual contact. When dealing with infected individuals, use standard precautions.

Normal Routes of Exposure: Inhalation; ingestion; abraded skin; mucous membranes.

Infectious Dose: Unknown.

Secondary Hazards: Aerosols (body fluids); blood and body fluids; body tissue.

Incubation: Unclear, 1 day to years.

Signs & Symptoms: Great clinical diversity; onset of symptoms may be sudden or gradual and include fever; headache; muscle pain (myalgia); joint pain (arthralgia); rigors; chest pain; respiratory distress ranging from mild bronchitis to severe pneumonia with difficulty breathing (dyspnea); pulmonary abscesses; fluid in the chest cavity (pleural effusion); cough; abdominal pain or discomfort; enlargement of the lymph glands, particularly of the neck (cervical adenopathy); enlargement of the liver (hepatomegaly) and/or the spleen (splenomegaly); jaundice; difficult or painful urination (dysuria); loss of appetite (anorexia); sensitivity to light (photophobia); tearing (lacrimation); and disorientation. There can be localized subcutaneous skin infection (cellulitis) with pain or swelling, ulceration, and abscess.

Suggested Alternatives for Differential Diagnosis: Anthrax, chronic abscesses, empyema, fungal nodules, invasive pneumococcal disease, Leptospirosis, lung carcinoma, osteomyelitis, plague, pneumonia, scrub typhus, smallpox, staphylococcal sepsis, tuberculosis, typhoid fever.

Mortality Rate (untreated): ≤90%.

In Domestic Animals:

Target species: Cattle, pigs, sheep, goats, horses.

Communicability: Direct transmission does not occur except rarely during sexual contact.

Normal Routes of Exposure: Inhalation; ingestion; abraded skin; mucous membranes.

Secondary Hazards: Aerosols; blood and body fluids; body tissue.

Incubation: Variable.

Signs: Disease most likely due to percutaneous inoculation often develops at distant sites without evidence of active infection at the point of inoculation. The organs most commonly affected include the lungs, spleen, liver, and associated lymph nodes. Great clinical diversity; onset of symptoms may be sudden or gradual and include high fever (106°F), discharge of mucus and pus (mucopurulent) from the nose and eyes, severe coughing, pneumonia with respiratory distress, difficulty breathing (dyspnea), red streaking with swollen lymph nodes (lymphangitis), edema of the limbs, skin abscesses, blindness or rapid involuntary movements of the eyes (nystagmus), spasms, circling, incoordination, partial paralysis of the hind legs (posterior paresis), colic and diarrhea, loss of appetite (anorexia), abortion, and still births.

Postmortem findings include multiple abscesses in most organs especially in the regional lymph nodes, spleen and liver. Abscesses contain a thick, caseous greenish-yellow or off-white pus. There is usually no calcification. Additional findings may include pneumonic changes in the lungs, suppurative polyarthritis with the joint capsules containing fluid and large masses of greenish-yellow pus, bulges in major blood vessels (aortic aneurysms), and meningoencephalitis.

Suggested Alternatives for Differential Diagnosis: Clinical presentation is not characteristic and diagnosis is difficult. Consider actinobacillosis, caseous lymphadenitis, fungal eczema, glanders, non-specific purulent conditions, strangles, tuberculosis.

Mortality Rate (untreated): ≤90%.

C17-A009

Chlamydophila psittaci (Agent SI)
Psittacosis
ICD-11: XN4S7
UNII: A16IX59JOH
EPPO: —

It is a non-motile Gram-negative, spherical bacterium that can only reproduce within a host's cells (obligate intracellular). It is endemic throughout the world. The natural reservoir is birds. Survival of the bacteria outside the host depends on the source: Infected fluid from eggs-52 hours; bird droppings–a few days; bird feed-2 months; glass-15 days; straw-20 days. This is a biosafety level 3 agent. It is highly communicable from infected birds to people.

It is on the Australia Group Core list and the OIE list of Notifiable Diseases for 2020.

In People:
CDC Case Definition: Clinical criteria is an illness characterized by fever, chills, headache, myalgia, and a dry cough with pneumonia often evident on chest x-ray. Severe pneumonia requiring intensive-care support, endocarditis, hepatitis, and neurologic complications occasionally occur.

Laboratory criteria for diagnosis are 1) isolation of *C. psittaci* from respiratory specimens (e.g., sputum, pleural fluid, or tissue), or blood, OR 2) fourfold or greater increase in antibody against *C. psittaci* by complement fixation or microimmunofluorescence between paired acute- and convalescent-phase serum specimens obtained at least 2–4 weeks apart, OR 3) supportive serology, OR 4) detection of *C. psittaci* DNA in a respiratory specimen via amplification of a specific target by polymerase chain reaction assay.

Communicability: Direct person-to-person transmission is rare. When dealing with infected individuals, use standard precautions.

Normal Routes of Exposure: Inhalation; abraded skin; mucous membranes.
Infectious Dose: Unknown.

Secondary Hazards: Aerosols (respiratory secretions); blood and body fluids; fecal (from birds); fomites (from birds).

Incubation: 5–28 days.

Signs & Symptoms: Diagnosis of psittacosis can be difficult. There is a variable clinical presentation with flu-like symptoms including abrupt onset of fever, headache, muscle pain (myalgia), chills, upper or lower respiratory tract disease, and dry non-productive cough. Pneumonia is often evident in chest x-rays.

Suggested Alternatives for Differential Diagnosis: Brucellosis, chlamydial pneumonias, fungal pneumonia, infective endocarditis, legionnaires disease, mycoplasma pneumoniae, Q fever, tuberculosis, tularemia, typhoid fever, and viral pneumonia.

Mortality Rate (untreated): ≤20%.

In Domestic Animals:

Target species: Poultry.

Communicability: Direct transmission is possible.

Normal routes of exposure: Inhalation; ingestion; mucous membranes.

Secondary Hazards: Aerosols; body fluids; fecal matter; fomites.

Incubation: Varies, most common 3–10 days.

Signs: Asymptomatic infections are common. Otherwise, nasal and ocular discharge, eye (pink eye – conjunctivitis) and sinus (sinusitis) infections, green to yellow-green fecal matter, fever, loss of condition, inactivity, ruffled feathers, weakness, loss of appetite (anorexia), and weight loss.

Postmortem findings include inflammation of the lungs, air sacs, liver, heart, spleen, kidneys, and the lining of the abdomen (peritoneum).

Suggested Alternatives for Differential Diagnosis: Aspergillosis, avian influenza, Enterobacteriaceae, fowl cholera, herpesviruses, *Mycoplasma gallisepticum* infections, paramyxoviruses, pasteurellosis, salmonellosis, as well as other viral and bacterial infections.

Mortality Rate (untreated): ≥30% (turkeys); varies greatly on serotype and bird species.

C17-A010

Clavibacter michiganensis **subsp.***sepedonicus*
Ring Rot of Potatoes
ICD-11: —
UNII: —
EPPO: CORBSE

It is a Gram-positive, non-motile, rod-shaped bacterium. Development is most rapid at temperatures of 70°F to 75°F. It is endemic to North America, Europe, and China. In wet soil, the bacterium can survive for several months; in drier soil it survives for more than a year. Ring rot infection can pass through one or more field generations without causing symptoms in stems and tubers. It can spread easily and adheres indefinitely on all types of surfaces that it contacts. It is capable of surviving two to five years in dried slime on surfaces of crates, bins, burlap sacks, or harvesting and grading machinery, even if exposed to temperatures well below freezing.

It is on the Australia Group Core list.

In Plants:

Target species: Potatoes; can colonize roots of sugar beets.

Normal routes of transmission: Contact (wounds in tubers).

Secondary Hazards: Crop debris; mechanical vectors (contaminated containers, farm implements, wash water); vectors (Colorado potato beetle, flea beetle, leafhoppers, sucking insects such as aphids).

Signs: Often persists latently without visible signs while still being able to spread. It can cause wilting of the leaf margins, especially on the lower leaves and often on only one side of the plant. Leaves curl and progressively lose their shine (appearing dull and greasy) with the onset of yellowing, browning, and eventual necrosis.

In the infected tuber there is a creamy yellow to light brown rot evident in the area approximately 0.6 cm below the skin (vascular ring). Later, this area disintegrates and brown cavities form. In advanced stages these cavities involve much of the center of the potato and leave only a hollow shell of tissue. Often, the tuber will have external cracking and swelling.

Crop Losses: ≤50%

C17-A011

Clostridium argentinense
Produces Botulinum Toxin Type G (C16-A005)

ICD-11: —
UNII: —
EPPO: —

It is an anaerobic, Gram-positive, spore-forming, rod-shaped bacterium that produces neurotoxins (see C16-A005). It is endemic to Argentina. The natural reservoir is soil. This is a biosafety level 2 agent.

It is on the Australia Group Core list and a Tier-1 agent on the U.S. Select Agents and Toxins list.

In People:

CDC Case Definition: See *Clostridium botulinum* (C17-A013).

Communicability: Direct person-to-person transmission does not occur. When dealing with infected individuals, use standard precautions.

Normal Routes of Exposure: Danger is from exposure to the toxin (inhalation, ingestion, abraded skin).

Infectious Dose: Not applicable.

Secondary Hazards: Spores; residual toxin.

Incubation: Not applicable.

Signs & Symptoms: See Botulinum toxin (C16-A005).

Suggested Alternatives for Differential Diagnosis: See *Clostridium botulinum* (C17-A013).

Mortality Rate (untreated): See *Clostridium botulinum* (C17-A013).

C17-A012

Clostridium baratii
Produces Botulinum Toxin (C16-A005)

ICD-11: —
UNII: —
EPPO: —

It is an anaerobic, Gram-positive, spore-forming, rod-shaped bacterium that produces neurotoxins (see C16-A005). It is endemic throughout the world. This is a biosafety level 2 agent.

It is on the Australia Group Core list and a Tier-1 agent on the U.S. Select Agents and Toxins list.

In People:

CDC Case Definition: See Botulinum toxin (C16-A005).

Communicability: Direct person-to-person transmission does not occur. When dealing with infected individuals, use standard precautions.

Normal Routes of Exposure: Danger is from exposure to the toxin (inhalation, ingestion, abraded skin).

Infectious Dose: Not applicable.

Secondary Hazards: Spores; residual toxin.

Incubation: Not applicable.

Signs & Symptoms: See Botulinum toxin (C16-A005).

Suggested Alternatives for Differential Diagnosis: See Botulinum toxin (C16-A005).

Mortality Rate (untreated): See Botulinum toxin (C16-A005).

C17-A013

Clostridium botulinum
Produces Botulinum Toxin (C16-A005)
ICD-11: XN2JN
UNII: 0296055VE0
EPPO: —

It is an anaerobic, Gram-positive, spore-forming, rod-shaped bacterium that produces neurotoxins (see C16-A005), especially in low-acid foods. It is endemic throughout the world. The natural reservoir is soil. It is also found in honey. This is a biosafety level 2 agent.

It is on the Australia Group Core list and a Tier-1 agent on the U.S. Select Agents and Toxins list.

In People:

CDC Case Definition: Clinical description incudes symptoms such as diplopia, blurred vision, and bulbar weakness. Symmetric paralysis may progress rapidly.

Laboratory criteria for diagnosis is 1) detection of botulinum toxin in serum, stool, or patient's food; OR 2) isolation of *Clostridium botulinum* from stool, wound, or other clinical specimen.

Case confirmation is a clinically compatible case that is laboratory confirmed in a patient aged greater than or equal to 1 year; or that occurs among persons who ate the same food as persons who have laboratory-confirmed botulism; or who has a history of a fresh, contaminated wound during the 2 weeks before onset of symptoms.

Communicability: Direct person-to-person transmission does not occur. When dealing with infected individuals, use standard precautions.

Normal Routes of Exposure: Danger is from exposure to the toxin (inhalation, ingestion, abraded skin).

Infectious Dose: Not applicable.

Secondary Hazards: Spores; Residual toxin.

Incubation: Not applicable.

Signs & Symptoms: See Botulinum toxin (C16-A005).

Suggested Alternatives for Differential Diagnosis: Basilar artery stroke, cerebrovascular disease of the brainstem, congenital neuropathy or myopathy, diphtheria, encephalitis, Guillain-Barré syndrome, hypermagnesemia, hyperthyroidism, intracranial mass lesions, Lambert-Eaton syndrome, Mediterranean fever, myasthenia gravis, neurasthenia, poliomyelitis, progressive external ophthalmoplegia, thyrotoxicosis, tick paralysis, and poisonings by amanita mushrooms, atropine, carbon monoxide, methyl alcohol, methyl chloride, organophosphates, scopolamine, shellfish, sodium fluoride, and drugs (e.g., penicillamine, aminoglycosides).

Mortality Rate (untreated): ≤70%.

C17-A014

Clostridium butyricum
Producing Botulinum Toxin (C16-A005)
ICD-11: —
UNII: —
EPPO: —

It is an anaerobic, Gram-positive, spore-forming, rod-shaped bacterium that produces neurotoxins (see C16-A005). It is endemic throughout the world. Non-toxic strains are a common human and animal gut commensal bacterium and widely used as a probiotic in Asia. This is a biosafety level 2 agent.

It is on the Australia Group Core list and a Tier-1 agent on the U.S. Select Agents and Toxins list.

In People:
CDC Case Definition: See *Clostridium botulinum* (C17-A013).

Communicability: Direct person-to-person transmission does not occur. When dealing with infected individuals, use standard precautions.

Normal Routes of Exposure: Danger is from exposure to the toxin (inhalation, ingestion, abraded skin).

Infectious Dose: Not applicable.

Secondary Hazards: Spores; residual toxin.

Incubation: Not applicable.

Signs & Symptoms: See Botulinum Toxin (C16-A005).

Suggested Alternatives for Differential Diagnosis: See *Clostridium botulinum* (C17-A013).

Mortality Rate (untreated): See *Clostridium botulinum* (C17-A013).

C17-A015

Clostridium perfringens Epsilon Toxin Producing (Agent G)
Gas Gangrene
ICD-11: XN7J5
UNII: OVP6XX033E
EPPO: —

It is an anaerobic, Gram-positive spore-forming rod-shaped bacterium that produces cytotoxins (see C16-A025). It is endemic throughout the world. It is widely distributed in the environment and frequently occurs in the intestines of humans and many animals. It is stable in water (sewage) and resistant to chlorine at purification concentrations. It grows between 54°F and 140°F, with very rapid reproduction between 109°F and 117°F. Generation times can be as short as 10minutes. Causes gas gangrene, enteritis necroticans, food poisoning, and nonfood-borne enterotoxemic infections. Gas gangrene results from wound contamination with spores. Only rare cases of pulmonary infections, and no apparent disease caused by inhalation of spores. This is a biosafety level 2 agent.

It is on the Australia Group Core list and the OIE list of Notifiable Diseases for 2020.

In People:

CDC Case Definition: None established.

Communicability: Direct person-to-person transmission does not occur. When dealing with infected individuals, use standard precautions. Avoid direct contact with wounds or wound drainage. If drainage is extensive, use contact precautions.

Normal Routes of Exposure: Ingestion; abraded skin; mucous membranes.

Infectious Dose: >100,000,000 vegetative cells via ingestion.

Secondary Hazards: Body fluids (wound drainage); fecal matter (if intestinal).

Incubation: 8–22 hours (enteritis); 6 hours–several days (gas gangrene).

Signs & Symptoms: Gas Gangrene: low-grade fever, rapid heart rate (tachycardia), rapid breathing (tachypnea), excessive perspiration (diaphoresis), and possibly altered mental status. In later stages blood pressure may drop (hypotension). Affected area may have crepitance or subcutaneous air, large blisters (bullae) surrounded by tense edema. Ruptured blisters produce a profuse, discolored discharge that may have a sweet mousy odor. Because gas gangrene affects the deep muscle tissue, the superficial skin often appears normal early in the disease course but eventually turns pale and then becomes grey, bronze, purplish-red or even black. Pain is often out of proportion to physical findings. Decreased pain or anesthesia at the site of infection can indicate that cutaneous nerve endings are being destroyed and that the disease is advanced. Progression to toxemia and shock can be rapid.

Enteritis: acute onset of abdominal pain and distention, diarrhea (sometimes bloody), and vomiting.

Suggested Alternatives for Differential Diagnosis: Gas Gangrene – abdominal abscess, acute gout, bacterial sepsis, carcinoma erysipeloides, cellulitis, cutaneous anthrax, deep venous thrombosis and thrombophlebitis, elective abortion, emphysematous cholecystitis, familial Mediterranean fever, fixed drug reaction, group a streptococcal infection, necrotizing fasciitis, other causes of necrotizing myositis, penetrating abdominal trauma, pyoderma gangrenosa, pyomyositis, septic arthritis, septic shock, Sweet syndrome, toxic shock syndrome, vaccinia vaccination, vibrio infections, water-borne skin infections, Wells syndrome, other causes of soft tissue gas (e.g., pneumomediastinum, pneumothorax, fractured larynx, fractured trachea).

Mortality Rate (untreated): >25%.

In Domestic Animals:

Susceptible species: All species.

Communicability: Direct transmission does not occur.

Normal routes of exposure: Ingestion; abraded skin.

Secondary Hazards: None.

Incubation: Varies, as short as 2 hours for intestinal.

Signs: Causes enteritis, dysentery, and toxemia in horses, sheep, cattle and pigs without fever (afebrile). Vomiting is rare. Mortality may be high in lambs, calves, pigs, and foals. In birds, typically the only sign is a sudden increase in mortality (≤50%). However, birds with depression, ruffled feathers, and diarrhea may also be seen. Gangrenous dermatitis is characterized by gangrenous necrosis of the skin and a sharp increase in mortality (≤60%).

Suggested Alternatives for Differential Diagnosis: Acute rumen impaction, enterotoxemia due to *E. coli*, hypocalcemia, hypomagnesemia, louping-ill, monocytic ehrlichiosis, pasteurellosis, polioencephalomalatia, Potomac horse fever, pregnancy toxemia, rabies, salmonellosis, and acute lead poisoning.

Mortality Rate (untreated): Variable depending on the species and location of infection.

C17-A016

Clostridium tetani
Tetanus
ICD-11: XN5NQ
UNII: 751E8J54VM
EPPO: —

It is an anaerobic, Gram-positive spore-forming rod-shaped bacterium that produces neurotoxins (see C16-A017). It is endemic throughout the world. It is found in found in soil and intestinal tracts of animals. Spores may remain viable for years if protected from light and heat. They can be destroyed by boiling water. Spores are unable to grow in normal tissue or even in wounds unless necrosis is present. The bacteria remain localized in the necrotic tissue at the original site of infection and multiply. This is a biosafety level 2 agent.

It is on the Australia Group Warning list.

In People:
CDC Case Definition: Clinical criteria is acute onset of hypertonia and/or painful muscular contractions (usually of the muscles of the jaw and neck) and generalized muscle spasms without other apparent medical cause.

Communicability: Direct person-to-person transmission does not occur. When dealing with infected individuals who have open skin lesions, use standard precautions.

Normal Routes of Exposure: Abraded skin; wounds (e.g., punctures, lacerations, burns, insect bites).
Infectious Dose: Unknown.

Secondary Hazards: Spores; body fluids (wound drainage).

Incubation: 3–21 days, but may extend to 60 days. In general, the shorter the incubation the worse the prognosis.

Signs & Symptoms: Tetanus is a clinical syndrome without confirmatory laboratory tests. There is a direct relationship between the distance from the inoculation wound to the central nervous system and the onset of symptoms.

Headache, fever, and sweating. Trismus (i.e., lockjaw), stiffness, neck rigidity, difficulty swallowing (dysphagia), restlessness, and reflex spasms. Subsequently, muscle rigidity becomes the major manifestation. Sustained contraction of facial musculature produces a sneering grin expression known as risus sardonicus. In rare cases, tetanus associated with head wounds may result in flaccid cranial nerve palsies rather than spasm. Muscle rigidity spreads from the jaw and facial muscles to the muscle groups in the neck, trunk, and extremities. Patients with tetanus may present with abdominal tenderness and guarding, mimicking an acute abdomen. Reflex spasms develop in most patients and can be triggered by minimal external stimuli such as noise, light, or touch (hyperesthesia) leading to the classic backward bridging position (opisthotonos). Tetanic seizures may occur and portend a poor prognosis.

Complications of tetanus include high blood pressure (hypertension), fractures, pulmonary embolism, and aspiration pneumonia. Death results from exhaustion, respiratory failure, or cardiac arrest.

Suggested Alternatives for Differential Diagnosis: Strychnine (see C11-A159) poisoning is the only true mimic. Otherwise acute abdomen, arthrogryposis, conversion disorder, dental infections, dystonic drug reactions (e.g., phenothiazines, metoclopramide), encephalitis, globus hystericus, hemorrhagic stroke, hepatic encephalopathy, hypocalcemia, hysteria, intracranial hemorrhage, intraoral disease, local infections, malignant hyperthermia, mandible dislocation, medication-induced dystonic reactions, meningitis, neoplasms, neuroleptic malignant syndrome, odontogenic infections, peritonsillar abscess, rabies, seizure disorder, serotonin syndrome, stimulant use, stroke, subarachnoid hemorrhage, tardive dystonia, widow spider envenomation (i.e., genus Latrodectus). *Mortality Rate (untreated):* ≤90%.

In Domestic Animals:

Susceptible species: Cattle, sheep, goats, horses.

Communicability: Direct transmission does not occur.

Normal routes of exposure: Wounds (punctures); birthing.

Secondary Hazards: Spores; fecal matter; necrotic animal tissues.

Incubation: 7 days–4 months.

Signs: Initially display localized stiffness, often involving the jaw and neck muscles, the hindlimbs, and the region of the infected wound. Shortly thereafter the animal shows pronounced general stiffness accompanied by tonic spasms and sensitivity to external stimuli (hyperesthesia). Sweating, rapid heart rate (tachycardia), rapid breathing (tachypnea) and congestion of mucous membranes are common. Spasms of the head muscles lead to lockjaw. Horses will often assume a sawhorse stance while sheep, goats, and pigs will often fall to the ground and assume a backward arch (opisthotonos). Consciousness is not affected.

Suggested Alternatives for Differential Diagnosis: Cerebrospinal meningitis, enterotoxaemia of lambs, enzootic muscular dystrophy, hypocalcemia of mares, polioencephalomalacia, poisoning by strychnine.

Mortality Rate (untreated): ≤80%.

C17-A017

Coxiella burnetii (Agent OU)
Q Fever
ICD-11: XN5H6
UNII: GRY5SDU86N
EPPO: COXIBU

It is an aerobic, Gram-negative, oval-shaped bacterium that can only reproduce within a host's cells (obligate intracellular). It is endemic throughout the world. It exists in two vegetative forms: a large one that is found in infected cells, and a small, dense spore-like one that exists outside a host and is resistant to heat and desiccation. It can survive in the environment for up to ten months and can also withstand exposure to many standard disinfectants. Although the disease is usually subclinical in ruminants (e.g., cattle, sheep and goats), it can cause loss of appetite (anorexia) and/or abortions. Inhalation of dust contaminated with placental tissue or birth fluids is a common way human contract the disease. Domestic animals, (e.g., cats, dogs, rodents, birds), are also susceptible to infection and can act as a source of infection for humans. Ticks may transmit the disease among domestic ruminants, but are not thought to play an important role in transmission of the disease to humans. This is a biosafety level 2 agent for clinical samples. Biosafety level 3 should be used for activities with a high potential for aerosol or droplet production or high concentrations of agent.

It is on the Australia Group Core list, the HHS Select Agents and Toxins list and the OIE list of Notifiable Diseases for 2020.

In People:

CDC Case Definition: Clinical criteria is acute fever and one or more of the following: rigors, severe retrobulbar headache, acute hepatitis, pneumonia, and/or elevated liver enzyme levels.

Confirmatory laboratory criteria is 1) serological evidence of a fourfold change in immunoglobulin G-specific antibody titer to *C. burnetii* phase II antigen by indirect immunofluorescence assay between paired serum samples ideally taken 3–6 weeks apart; OR 2) detection of *C. burnetii* DNA in a clinical specimen via amplification of a specific target by polymerase chain reaction assay; OR 3) demonstration of *C. burnetii* in a clinical specimen by immunohistochemical methods; OR 4) isolation of *C. burnetii* from a clinical specimen by culture.

Communicability: Direct person-to-person transmission is rare. When dealing with infected individuals, use standard precautions.

Normal Routes of Exposure: Inhalation; ingestion; mucous membranes; sexual contact (rarely).

Infectious Dose: 1–10 organisms (inhalation).

Secondary Hazards: Aerosols (contaminated dust; respiratory secretions from individuals with pneumonia); body fluids (e.g., milk, urine from ruminants); fecal (from ruminants); fomites (e.g., straw, bedding); animal products (e.g., hides, wool); vectors (ticks).

Incubation: 14–39 days (can be less with high doses).

Signs & Symptoms: Clinical presentation of acute Q fever varies considerably cannot be diagnosed based on clinical presentation alone. Symptoms are typically flu-like and include fever, chills, fatigue, sweating, weakness, vague feeling of bodily discomfort (malaise), severe headache behind the eyes, muscle aches (myalgia), joint aches (arthralgia), non-productive cough with minimal auscultatory abnormalities on exam. It may progress to pneumonia; inflammation of the liver (hepatitis) with nausea, vomiting, and diarrhea; inflammation around the brain and spinal cord (meningitis); inflammation of the heart muscle (myocarditis); inflammation of the gall bladder (cholecystitis); or infection of the lymph system (lymphadenitis). Total recovery may be prolonged.

Suggested Alternatives for Differential Diagnosis: Acute interstitial pneumonitis, arbovirus ence-phalitis, aseptic meningitis, atypical pneumonias, bacterial pericarditis, brucellosis, chlamydial infections, chronic fatigue syndrome, connective-tissue diseases, cytomegalovirus, elective abortion, endocarditis, Epstein-Barr Virus, hepatitis, Hodgkin lymphoma, influenza, legionella infection, leptospirosis, Mediterranean fever, meningoencephalitis, myocarditis, non-Hodgkin lymphoma, osteoarticular infection, osteomyelitis, placentitis, rickettsial infection, sarcoidosis, tick-borne diseases, torch syndrome (e.g., toxoplasmosis, rubella, cytomegalovirus, herpes simplex infections), vascular graft infections, visceral leishmaniasis.

Mortality Rate (untreated): ≤3%.

C17-A018

Ehrlichia ruminantium
Heartwater
ICD-11: —
UNII: —
EPPO: COWDRU

It is an aerobic, Gram-negative, non-motile, spherical bacterium that can only reproduce within a host's cells (obligate intracellular). It is endemic to sub-Saharan Africa and the Caribbean. The natural reservoir is ruminants (e.g., cattle, sheep and goats) and ticks. Ticks remain infected for over 15 months. Infected animals that recover can become chronic carriers. It is extremely fragile and cannot persist outside of a host for more than a few hours.

It is on the OIE list of Notifiable Diseases for 2020.

In People:
Does not occur in humans.

In Domestic Animals:
Target species: Cattle, sheep, goats.

Communicability: Direct transmission does not occur.

Normal routes of exposure: Vectors (ticks).

Secondary Hazards: Blood (during febrile stage); fomites (with vectors); vector amplification.

Incubation: 10–30 days.

Signs: In the peracute form, animals may drop dead within a few hours of developing a fever without apparent clinical signs. Otherwise there is a brief interval of high fever, severe respiratory distress, spasms triggered by minimal external stimuli such as noise, light, or touch (hyperesthesia), tearing (lacrimation), terminal convulsions and sudden death. In the acute form, there is sudden high fever, difficulty breathing (dyspnea), rapid breathing (tachypnea), loss of appetite (anorexia), diarrhea, depression and listlessness. Nervous signs may include extending the tongue, sucking movements, clamping the jaw, spasms triggered by minimal external stimuli such as noise, light, or touch (hyperesthesia), high-stepping stiff gait, walking in circles, or standing rigidly with tremors of the superficial muscles. Animals may become recumbent and exhibit pedaling with legs, twitching eyelids (nystagmus), chewing movements, frothing at the mouth, spasms, and convulsions.

Postmortem findings include excessive straw-colored to reddish fluid in the sac surrounding the heart (hydropericardium), in the abdomen (ascites), and in the chest cavity (hydrothorax); pulmonary and mediastinal edema; inflammation and hemorrhage in the stomach and intestines (hemorrhagic gastroenteritis); enlargement of the liver (hepatomegaly), spleen (splenomegaly) and lymph nodes; swelling (edema) and hemorrhage of the brain.

Suggested Alternatives for Differential Diagnosis: Anthrax, cerebral trypanosomiasis, encephalitis, listeriosis, meningitis, parasitism, piroplasmosis, rabies, tetanus, theileriosis; poisoning by strychnine, lead, organophosphates, arsenic, and various plants that affect the central nervous system.

Mortality Rate (untreated): ≤90%; varies widely by species.

C17-A019

Escherichia coli (Agent EC)
BW Simulant
ICD-11: XN6P4
UNII: 514B9K0L10
EPPO: ESCHCO

It is a Gram-negative, motile, rod-shape, facultative anaerobe (can live with or without oxygen). It can survive outside a host for extended periods. It is part of the common bacterium that normally inhabits the intestinal tracts of humans and animals. It helps protect the intestinal tract from bacterial infection and aids in digestion.

Although used as a simulant, it can cause acute bacterial meningitis, pneumonia, intra-abdominal infections, enteric infections, urinary tract infections, septic arthritis, endophthalmitis, suppurative thyroiditis, sinusitis, osteomyelitis, endocarditis, and skin and soft tissue infections. There are also strains of *E. coli* (C17-A035) that produce lethal cytotoxins (C16-A032).

C17-A020

Francisella tularensis (Agent UL)
Tularemia
ICD-11: XN0BX
UNII: NNR1301B0H
EPPO: FRNSTU

It is an aerobic, Gram-negative, non-motile, rod-shaped bacterium. It is endemic to North America and parts of Europe and Asia. The natural reservoir is rodents, particularly rabbits and hares. However, domestic cats are also very susceptible to tularemia and have been known to transmit the bacteria to humans. The organism can remain viable for weeks in soil, water, carcasses and hides. It remains viable for months at temperatures at or below freezing. This is a biosafety level 2 agent for clinical samples. Biosafety level 3 should be used for activities with a high potential for aerosol or droplet production or high concentrations of agent.

It is on the Australia Group Core list, a Tier-1 agent on the U.S. Select Agents and Toxins list, and the OIE list of Notifiable Diseases for 2020.

In People:

CDC Case Definition: Clinical criteria is an illness characterized by several distinct forms, including the following: 1) Ulceroglandular: cutaneous ulcer with regional lymphadenopathy; 2) Glandular: regional lymphadenopathy with no ulcer; 3) Oculoglandular: conjunctivitis with preauricular lymphadenopathy; 4) Oropharyngeal: stomatitis or pharyngitis or tonsillitis and cervical lympha-denopathy; 5) Pneumonic: primary pulmonary disease; 6) Typhoidal: febrile illness without localizing signs and symptoms.

Confirmatory laboratory criteria is 1) Isolation of *F. tularensis* in a clinical or autopsy specimen; OR 2) Fourfold or greater change in serum antibody titer to *F. tularensis* antigen between acute and convalescent specimens.

Communicability: Direct person-to-person transmission does not occur. When dealing with infected individuals, use standard precautions.

Normal Routes of Exposure: Inhalation; ingestion; abraded skin; mucous membranes; vectors (mosquitoes; ticks; biting flies).

Infectious Dose: 10–50 organisms.

Secondary Hazards: Blood and body fluids; body tissue; fomites (with vectors).

Incubation: 1–21 days (dose dependent).

Signs & Symptoms: Can be mistaken for common illnesses and presentation can vary depending on the route of entry. Common clinical findings for all forms include abrupt onset of fever (100°–104°F) with pulse-temperature disassociation, chills, headache, loss of appetite (anorexia), vague feeling of bodily discomfort (malaise), fatigue, muscle pain (myalgia), cough, vomiting, sore throat and abdominal pain. From an injection (e.g., insect bite) there will be swelling of regional lymph glands, usually in the armpit or groin. There may or may not be an ulcerated skin lesion at the point of entry through the skin. In the eye it causes unilateral irritation and inflammation with tearing (lacrimation), sensitivity to light (photophobia), with swelling of lymph glands in front of the ear.

Ingestion causes mouth ulcers, sore throat, tonsillitis, swelling of lymph glands in the neck, abdominal pain, nausea, diarrhea, and gastrointestinal bleeding. Inhalation causes pneumonia with dry cough, difficulty breathing (dyspnea), and chest pain. There may be hemorrhagic inflammation of the airways early in the course of illness. If untreated, the pathogen can also migrate to the lungs through any of the other routes of entry and also cause pneumonia.

Suggested Alternatives for Differential Diagnosis: Anthrax, bacterial pericarditis, brucellosis, cat scratch disease, chlamydial infections, diphtheria, disseminated fungal disease, disseminated mycobacterial disease, endocarditis, Epstein-Barr virus, influenza, legionella infection, leishmaniasis, lymphogranuloma venereum, malaria, mononucleosis, mumps, mycoplasma infections, nontuberculous mycobacterial infections, parainfluenza virus infections, pharyngitis, plague, pneumonia, psittacosis, Q fever, rat-bite fever, rhabdomyolysis, rickettsial infection, salmonella infection, sporotrichosis, syphilis, tick-borne diseases (e.g., Lyme disease, Rocky Mountain spotted fever), toxoplasmosis, tuberculosis, tuberculous cervical lymphadenitis, viral pericarditis.

Mortality Rate (untreated): ≤15%; >50% (respiratory).

In Domestic Animals:

Susceptible species: Pigs, sheep, horses

Communicability: Direct transmission does not occur.

Normal Routes of Exposure: Inhalation; ingestion; abraded skin; mucous membranes; vectors (mosquitoes; ticks; biting flies).

Secondary Hazards: Blood and body fluids; body tissue; fomites (with vectors).

Incubation: 1–10 days.

Signs: Characterized by sudden onset of high fever, lethargy, loss of appetite (anorexia), stiffness, and incoordination (ataxia). Pulse and respiratory rates are increased. Coughing, diarrhea, dehydration, with frequent urination (pollakiuria), swollen lymph glands, and simultaneous enlargement of both the liver and the spleen (hepatosplenomegaly). Prostration and death may occur in a few hours or days. Subclinical cases may be common.

Suggested Alternatives for Differential Diagnosis: Acute pneumonia, plague, pseudotuberculosis, and other septicemic diseases, tick paralysis.

Mortality Rate (untreated): ≤15% (lambs).

C17-A021

Legionella pneumophila
Legionellosis
ICD-11: XN9YS
UNII: TJR6ZFY0F0
EPPO: LEGIPN

It is an aerobic, Gram-negative (faintly staining), bacterium. It is endemic throughout the world. It can appear in different forms: in tissue and fluid samples it is rod-shaped, but in cultures it is long and filamentous. They are normally intracellular parasites of protozoa in aquatic environments all over the world. It is also found in cooling towers, plumbing systems, water heaters, and warm water spas. They have a preference for hot environments. It can survive for months in tap or distilled water. It does not naturally occur In domestic animals. This is a biosafety level 2 agent for clinical samples. Biosafety level 3 should be used for activities with a high potential for aerosol or droplet production or high concentrations of agent.

It is on the Australia Group Warning list.

In People:

CDC Case Definition: Clinical Description is 1) Legionnaires' disease is characterized by fever, myalgia, cough, and clinical or radiographic pneumonia; 2) Pontiac fever is a milder illness without pneumonia.

Confirmatory laboratory criteria is 1) Isolation of any *Legionella* organism from respiratory secretions, lung tissue, pleural fluid, or other normally sterile fluid; OR 2) By detection of *Legionella pneumophila* serogroup 1 antigen in urine using validated reagents; OR 3) By fourfold or greater rise in specific serum antibody titer to *Legionella pneumophila* serogroup 1 using validated reagents.

Communicability: Direct person-to-person transmission does not occur. When dealing with infected individuals, use standard precautions.

Normal Routes of Exposure: Inhalation.

Infectious Dose: Unknown but low

Secondary Hazards: None.

Incubation: 2–26 days (Legionnaire's disease); 24–72 hours (Pontiac fever).

Signs & Symptoms: Legionnaires' disease: fever, chills, confusion, headache, diarrhea, abdominal pain, muscle pain (myalgia), pneumonia with a nonproductive cough. Blood-streaked phlegm may be present.

Pontiac fever: Flu-like symptoms including fever, tiredness, muscle pain (myalgia), headache, sore throat, nausea, with or without coughing.

Suggested Alternatives for Differential Diagnosis: Aspiration pneumonia, bronchitis, congestive heart failure, costochondritis, HIV disease, influenza, meningitis, pleural effusion, pneumonia, prostatitis, psittacosis, Q Fever, respiratory distress syndrome, septic shock, tularemia.

Mortality Rate (untreated): ≤50% (Legionnaire's disease); 0% (Pontiac fever).

C17-A022

Liberibacter africanus
African Greening
ICD-11: —
UNII: —
EPPO: LIBEAF

It is endemic to Africa. Normally found at altitudes above 915 meters above sea level. It is temperature sensitive does not cause symptoms at temperatures above 80°F. Although extended periods of higher temperatures will suppress symptom development, they do not suppress infection of the plant. Disease is generally milder than *L. asiaticus* (C17-A023).

Once infected, vectors remain capable of transmitting the disease for up to 3 months, but progeny are free of the bacterium.

In Plants:

Target species: Citrus plants.

Normal routes of transmission: Vectors (psyllids).

Secondary Hazards: Fruit; crop debris.

Signs: It is difficult to recognize due to symptoms of the disease resembling those of other citrus disorders. The first symptom is usually the appearance of a yellow shoot on a tree. This is followed by progressive yellowing of the entire canopy. Leaves display blotchy mottling and are reduced in size without xylem dysfunction or wilting. Fruits are often small, lopsided, poorly colored, and contain aborted seeds.

Crop Losses: ≤100%

C17-A023

Liberibacter asiaticus
Asian Greening
ICD-11: —
UNII: —
EPPO: LIBEAS

It is endemic to southern Asia, South America, Central America, and the southern United States. It is more temperature tolerant than *L. africanus* (C17-A022) and can still produce symptoms at temperatures up to 95°F. Although extended periods of higher temperatures will suppress symptom development, they do not suppress infection of the plant. It is the most severe form of citrus greening and causes greater damage than *L. africanus* (C17-A022)

Once infected, vectors remain capable of transmitting the disease for up to 3 months, but progeny are free of the bacterium.

In Plants:

Target species: Citrus plants.

Normal routes of transmission: Vectors (psyllids).

Secondary Hazards: Fruit; crop debris.

Signs: It is difficult to recognize due to symptoms of the disease resembling those of other citrus disorders. The first symptom is usually the appearance of a yellow shoot on a tree. This is followed by progressive yellowing of the entire canopy. Leaves display blotchy mottling and are reduced in size without xylem dysfunction or wilting. Fruits are often small, lopsided, poorly colored, and contain aborted seeds.

Crop Losses: ≤100%

C17-A024

Mycoplasma capricolum subsp. capripneumoniae
Contagious Caprine Pleuropneumonia
ICD-11: —
UNII: —
EPPO: —

It is a Gram-negative, non-motile, spherical bacterium. It is endemic in Africa, the Middle East and parts of Asia. It is among the smallest of bacterial organisms and lacks a true cell wall. Survives outside the host for up to 3 days in tropical areas and up to 2 weeks in temperate zones. Can survive over 10 years in frozen pleural fluid. Inactivated within minutes by ultraviolet light. Chronic carriers may exist, but this has not been proven.

It is on the Australia Group Core list, the U.S. Select Agents and Toxins list, and the OIE list of Notifiable Diseases for 2020.

In People:
Does not occur in humans.

In Domestic Animals:
Target species: Goats; sheep (rarely).

Communicability: Direct transmission is possible.

Normal routes of exposure: Inhalation.

Secondary Hazards: Aerosols (respiratory secretions).

Incubation: 3–41 days.

Signs: Strictly a respiratory disease; lesions are produced only in the lungs with no dissemination to other organs. In the peracute form, the animal dies within 1–3 days showing minimal clinical signs. In the acute form, symptoms include high fever (104°F to 106°F), weakness and loss of appetite (anorexia); followed in 2–3 days by coughing with rapid, labored respiration. Coughing is frequent, violent, and productive. There is excess salivation with frothy nasal discharge. Respiration is accelerated and painful, and may be accompanied by a grunt. In the terminal stage animals are unable to move. In the chronic form, there is cough, nasal discharge, and debilitation.

Postmortem findings in acute cases include unilateral pneumonia with varying degrees of lung consolidation, and swollen bronchial lymph nodes. There may be pea-sized, yellow nodules surrounded by areas of congestion. Cuts in the surface of the lung have a granular appearance and produce straw-colored exudate. There is serofibrinous pleuritis with straw-colored fluid in the thorax.

Suggested Alternatives for Differential Diagnosis: Peste des petits ruminants, contagious agalactia syndrome, caseous lymphadenitis, pasteurellosis, and other causes of pneumonia.

Mortality Rate (untreated): ≤100% (acute).

C17-A025
Mycoplasma mycoides
Contagious Bovine Pleuropneumonia
ICD-11: —
UNII: —
EPPO: —

It is a Gram-negative, non-motile, spherical bacterium. It is endemic in sub-Sharan Africa. It is among the smallest of bacterial organisms and lack a true cell wall. Survives outside the host for up to 3 days in tropical areas and up to 2 weeks in temperate zones. Can survive over 10 years in frozen fluid. Chronic nonclinical carriers are capable of spreading the disease.

It is on the Australia Group Core list, the U.S. Select Agents and Toxins list, and the OIE list of Notifiable Diseases for 2020.

In People:
Does not occur in humans.

In Domestic Animals:
Target species: Cattle, water buffalo, bison, yak.

Communicability: Direct transmission is possible.

Normal routes of exposure: Inhalation. Transplacental infection of unborn calves can also occur.

Secondary Hazards: Aerosols (respiratory secretions).

Incubation: 3–24 weeks.

Signs & Symptoms: Calves up to 6 months of age: the primary sign may be painful arthritis appearing in multiple joints simultaneously (polyarthritis) and lameness. There is often no sign of respiratory involvement.

Adults: in the peracute form, the animal dies within 1–3 days showing no clinical signs other than fever. In the acute form, symptoms include fever (up to 107°F), loss of appetite (anorexia), depression, a reduced milk production, and respiratory signs (e.g., rapid breathing (tachypnea), difficult breathing (dyspnea), cough, nasal discharge). Animals may have nose bleeds (epistaxis) or diarrhea, and pregnant animals may abort or give birth to stillborn calves. Respiration can become labored and painful; animals may grunt when exhaling and react intensely if pressed between the ribs. Severely affected cattle may stand with their head and neck extended and forelegs apart, breathing through the mouth. Animals become recumbent. In severe cases they often die, typically within 1–3 weeks.

Postmortem findings for calves: With poly-arthritis, affected joints are filled with fluid and abundant fibrin. Infarcts, appearing as chronic fibrotic foci, may be found in the kidneys of these animals. Adults: typically exhibit unilateral pneumonia with of lung consolidation and marbling (i.e., areas of gray, yellow or red separated by a network of pale bands). Bits of dead lung tissue can become encapsulated, forming hardened nodules (sequestra) 2–25 cm in diameter that are surrounded by a fibrous connective tissue capsule up to 1 cm thick. Such nodules deep in the lung may not be seen, but can be still be palpated. These nodules are still evident in recovered animals. Bronchial lymph nodes are swollen (edema). There is fibrinous pleuritis with turbid or straw-colored fluid in the thorax and pericardial sac. The fibrin is replaced by fibrous connective tissue over time.

Suggested Alternatives for Differential Diagnosis: Abscesses, actinobacillosis, acute bovine pasteurellosis, bovine ephemeral fever, bovine farcy, bronchopneumonia, East coast fever, echinococcosis, foreign body pneumonia, hemorrhagic septicemia, infectious bovine rhinotracheitis, lungworms, pleuropneumonia, traumatic pericarditis, tuberculosis.

Mortality Rate (untreated): ≤80%.

C17-A026

Orientia tsutsugamushi
Scrub Typhus
ICD-11: 1C30.3
UNII: YQX511D71D
EPPO: —

It is an aerobic, Gram-negative, non-motile, oval-shaped bacterium that can only reproduce within a host's cells (obligate intracellular). It is endemic to Asia, Australia and the Pacific. The natural reservoir is mites. Mites remain infected for life and transfer pathogens directly to their eggs. Can survive for at least 10 days in blood, and at least 45 days if frozen in blood. This is a biosafety level 2 agent for clinical samples. This is a biosafety level 3 practices for activities with a high potential for aerosol or droplet production, high concentrations of agent, or working with infected arthropods.

In People:

CDC Case Definition: None established.

Communicability: Direct person-to-person transmission does not occur. When dealing with infected individuals, use standard precautions.

Normal Routes of Exposure: Vectors (mites).

Infectious Dose: Unknown.

Secondary Hazards: Aerosols (insect parts and fecal matter); fomites (with vectors); vector amplification.

Incubation: 6–21 days.

Signs & Symptoms: At the bite location, a painless, firm, gradually enlarging pimple (papule) forms that ulcerates before forming a scab (eschar). Approximately 10 days after infection, sudden onset of headache, chills, muscle pain (myalgia), vague feeling of bodily discomfort (malaise), and loss of appetite (anorexia). Progressing to high fever (104°F to 105°F), chills, painful lymph nodes (lymphadenopathy), cough, eye pain with bloodshot eyes (conjunctival injection), swollen, and inflammation of lung (pneumonitis) and/or brain (encephalitis) tissue. A dull red to dark purple rash of mixed flat and raised (maculopapular) spots appears first on the trunk then spreads to the arms and legs. Rash may be transient.

Suggested Alternatives for Differential Diagnosis: Anthrax, brucellosis, dengue, ehrlichiosis, hemorrhagic fever with renal failure syndrome, Kawasaki disease, leptospirosis, malaria, measles, mononucleosis, relapsing fever, rickettsial infections, Rocky Mountain spotted fever, rubella, severe fever with thrombocytopenia syndrome, syphilis, toxic shock syndrome, toxoplasmosis, tularemia, typhoid fever, typhus.

Mortality Rate (untreated): ≤60%.

C17-A027

Ralstonia solanacearum race 3, bv. 2
Bacterial Wilt
ICD-11: —
UNII: —
EPPO: PSDMS3

It is a Gram-negative bacterial pathogen that causes common wilt or brown rot. There are five races of *R. solanacearum* with different hosts and geographic distributions. Race 3 is endemic throughout the world except in the United States and Canada. Bacteria can remain viable in soil and water for extended periods. Growth is most severe between 75°F and 95°F, but the pathogen can survive in fallow soils with temperatures in the winter as low as 39°F.

It is on the Australia Group Core list and the U.S. Select Agents and Toxins list.

In Plants:

Target species: Tomatoes, potatoes, eggplant, peppers, bananas, geraniums.

Normal routes of transmission: Contact (soil, irrigation water).

Secondary Hazards: Shared Irrigation; mechanical vectors (cultivation, track out); vectors (suspected). There is little or no risk of transmission aerially, through leaf contact or splashing of water from leaf-to-leaf.

Signs: The primary symptom is wilted and/or yellowed leaves similar to wilting symptoms caused by other pathogens. However, leaf spots are rarely present. Initially, leaves wilt during the day, often curling upwards, but recover during the nighttime. Plants may become stunted and slightly yellowed (chlorotic). Vascular discoloration of the stem is common, sometimes seen as dark narrow streaks visible through the epidermis, and roots may turn brown. Gray-white bacterial ooze may be present on the stem and bacterial streaming may be seen if cut stem sections are placed into water. Stems may also blacken. In late stages, stems may collapse and the entire plant may desiccate and die.

Infected potato tubers become grayish-brown. Cross sections usually show a ring of distinct grayish-brown discoloration in the area approximately 0.6 cm below the skin (vascular ring) that may extend into the pith. Grayish-white droplets of bacterial slime ooze out with light pressure. Bacterial ooze may collect in the tuber eyes and soil may stick to secretions. Can be confused with *X. campestris* pv. *pelargonii* and *C. michiganensis* subsp. *sepedonicus* (see C17-A010)

Crop Losses: ≤100% (potatoes); ≤90% (tomatoes)

C17-A028

Rathayibacter toxicus
Gumming Disease
ICD-11: —
UNII: —
EPPO: RATHTO

It is a Gram-positive, non-motile, aerobic bacteria experiencing optimum growth at 79°F. It is endemic to Australia and South Africa. It overseasons in the soil and relies on nematode vectors (*Anguina* spp.) to transmit it into the stems and seed heads of grasses. The bacteria infest the galls formed by the nematodes as they feed on the plant. Bacteria reenter the soil when the infected galls fall to the ground. The bacteria can survive in the soil in a dried state for many years.

R. toxicus produces highly potent corynetoxins, heat-stable glycolipid neurotoxins, that affects grazing animals feeding on the grasses in pasture or on hay harvested from infected plants. The poison is cumulative in feeding animals and effects may be delayed if toxins are present in low concentrations with repeated exposures. The toxins could also potentially affect humans who ingest contaminated grain use for food.

It is on the U.S. Select Agents and Toxins list.

In Plants:
Target species: Pastoral and cereals grains.

Normal routes of transmission: Vectors (nematodes)

Secondary Hazards: Crop debris (hay; seeds); mechanical vectors (harvesting, track out, wind, run-off water); toxins.

Signs: Most infected plants do not show any visible signs or the disease is misidentified.

Seeds may be twisted, deformed and discolored. Bacterial growth may produce a yellow slime that covers the seed heads (gummosis). The slime turns orange and then brown as it dries. However, this slime does not always appear and may also be washed off by rain. *Crop Losses:* 0% (from pathogen).

In Domestic Animals:
Target species: Cattle, sheep, goats, horses, bison.

Communicability: Direct transmission is possible.

Normal routes of exposure: Ingestion.

Secondary Hazards: None.

Incubation: Cumulative toxin; symptoms may be delayed based on exposure.

Signs: Abrupt onset with tremors, nystagmus, incoordination (ataxia), rigidity, adoption of a wide-based stance, collapse, convulsions while recumbent. Convulsions can be precipitated by either forced exercise or high ambient temperatures. Animals may appear to recover between episodes of nervous spasms.

Postmortem findings include congestion, edema, hemorrhage of the brain and lungs, degeneration of the liver and kidneys, with small hemorrhages (petechia) in the gallbladder rumen, small intestine, kidney, and lymph nodes throughout the body.

Suggested Alternatives for Differential Diagnosis: Bovine spongiform encephalitis, enterotoxemia, grass tetany, perennial ryegrass staggers, polioencephalomalacia; poisoning by *Phalaris* spp., ergot, or botulism.

Mortality Rate (untreated): ≤50%.

C17-A029

Rickettsia akari
Rickettsialpox
ICD-11: XN7WV
UNII: 0M6MOQ3BOU
EPPO: —

It is an aerobic, Gram-negative, non-motile, rod-shaped bacterium that can only reproduce within a host's cells (obligate intracellular). It is endemic throughout the world, most commonly in urban areas. The natural reservoir is house mice, rats and mites. Mites remain infected for life and transfer pathogens directly to their eggs. Does not produce disease In Domestic Animals. This is a biosafety level 2 agent for clinical samples. This is a biosafety level 3 practices for activities with a high potential for aerosol or droplet production, high concentrations of agent, or working with infected arthropods.

In People: *CDC Case Definition:* None established.

Communicability: Direct person-to-person transmission does not occur. When dealing with infected individuals, use standard precautions.

Normal Routes of Exposure: Vectors (mites).

Infectious Dose: Unknown.

Secondary Hazards: Aerosols (blood, body fluids); blood and body tissue (from rodents).
Incubation: 10–21 days.

Signs & Symptoms: Mild, self-limiting disease. At the bite location, a painless, firm, red pimple (papule) forms that progresses to a blister (vesicle), which ulcerates before forming a scab (eschar) that slowly heals. Approximately 3–7 days after the initial skin lesion appears, there is a sudden onset of high fever (101°F to 104°F), sore throat, chills, headaches, neck and back stiffness, muscle pain (myalgia), and sensitivity to light (photophobia). Shortly after symptoms appear, a generalized pimple and blister (papulovesicular) rash erupts on the face, trunk, and extremities with no particular sequence of involvement. Scabs form but do not leave scars.

Suggested Alternatives for Differential Diagnosis: African tick bite fever, anthrax, chickenpox, Mediterranean spotted fever, *Rickettsia parkeri* rickettsiosis, scrub typhus, smallpox.

Mortality Rate (untreated): 0%.

C17-A030

Rickettsia prowazekii (Agent YE)
Typhus
ICD-11: XN8SY
UNII: TVS414L9M5
EPPO: RICKPR

It is an aerobic, Gram-negative, non-motile, rod-shaped bacterium that can only reproduce within a host's cells (obligate intracellular). It is endemic throughout the world. It is stable in lice feces for up to 100 days and can be viable in blood for extended periods if frozen. The natural reservoirs are humans, and within the United States, flying squirrels. Lice acquire the pathogen when feeding on humans during the fibril phase of the disease. *R. prowazekii* is not transmitted by the bite of the louse, but rather by contamination of the bite, or other open wounds, with infected feces from the louse. Lice begin releasing rickettsia in their feces within six days after becoming infected. Lice die within two weeks after becoming infected. This is a biosafety level 2 agent for clinical samples. This is a biosafety level 3 practices for activities with a high potential for aerosol or droplet production, high concentrations of agent, or working with infected arthropods.

It is on the Australia Group Core list, and the HHS Select Agents list.

In People:

CDC Case Definition: None established.

Communicability: Direct person-to-person transmission does not occur. When dealing with infected individuals, use standard precautions and protect from vectors.

Normal Routes of Exposure: Vectors (lice).

Infectious Dose: <10 organisms.

Secondary Hazards: Aerosols (lice feces); fomites (with vectors); vector amplification.
Incubation: 10–16 days.

Signs & Symptoms: Generally non-specific. Sudden onset of fever and chills, headache, rapid breathing (tachypnea), rapid heart rate (tachycardia), muscle pain (myalgia), cough, nausea, vomiting, confusion. A rash of flat and raised (maculopapular) spots that starts on the trunk of the body and spreads to the extremities occurs in some cases. May progress to drowsiness, confusion, delirium, seizures, and coma.
Suggested Alternatives for Differential Diagnosis: Anthrax, brucellosis, dengue, ehrlichiosis, Epstein-Barr virus, infectious mononucleosis, Kawasaki disease, leptospirosis, malaria, measles, meningitis, meningococcemia, relapsing fever, Rocky Mountain spotted fever, rubella, syphilis, toxic shock syndrome, toxoplasmosis, tularemia, typhoid fever.

Mortality Rate (untreated): ≤40%.

C17-A031

Rickettsia rickettsii
Rocky Mountain Spotted Fever
ICD-11: XN33Q
UNII: V6ZO3M43L5
EPPO: RICKRI

It is an aerobic, Gram-negative, non-motile, rod-shaped bacterium that can only reproduce within a host's cells (obligate intracellular). It is endemic to North and South America. The natural reservoir is ticks. Ticks remain infected for life and transfer pathogens directly to their eggs. Can survive for up to 1 year in tick tissue or blood; however, quickly inactivated by drying. Rickettsia not transmitted unless the tick attached from 4 to 6 hours. This is a biosafety level 2 agent for clinical samples. This is a biosafety level 3 practices for activities with a high potential for aerosol or droplet production, high concentrations of agent, or working with infected arthropods.

It was removed from the Australia Group list in June 2011 and is no longer regulated for export control.

In People:

CDC Case Definition: Clinical criteria is any reported fever and one or more of the following: rash, headache, myalgia, anemia, thrombocytopenia, and/or any hepatic transaminase elevation. Confirmatory laboratory criteria is 1) serological evidence of a fourfold change in immunoglobulin G specific antibody titer reactive with *R. rickettsii* antigen by indirect immunofluorescence assay between paired serum specimens (one taken in the first week of illness and a second 2–4 weeks later); OR 2) detection of *R. rickettsii* DNA in a clinical specimen via amplification of a specific target by polymerase chain reaction assay; OR 3) demonstration of spotted fever group antigen in a biopsy or autopsy specimen by immunohistochemical methods; OR 4) isolation of *R. rickettsii* from a clinical specimen in cell culture.

Communicability: Direct person-to-person transmission does not occur. When dealing with infected individuals, use standard precautions.

Normal Routes of Exposure: Abraded skin; mucous membranes; vectors (ticks).

Infectious Dose: <10 organisms.

Secondary Hazards: Aerosols; blood and body fluids (from ticks); fecal (from ticks); fomites (with vectors).

Incubation: 2–14 days.

Signs & Symptoms: Rapidly progressive disease, can be fatal within days. Sudden onset of fever, headache, nausea, vomiting, loss of appetite (anorexia), abdominal pain that may mimic appendicitis, muscle pain (myalgia), swelling (edema) around the eyes and on the back of hands. Symptoms may progress to altered mental status, coma, cerebral edema, pulmonary edema, and acute respiratory distress syndrome.

A rash usually develops several days after the onset of fever presenting as small, pink, flat spots (macules) on the wrists, forearms, and ankles; spreading to include the trunk and sometimes the palms of hands and soles of feet. Spots may become brown-purple due to bleeding under the skin (petechia). Necrosis of extremities may require amputation.

Suggested Alternatives for Differential Diagnosis: Allergic vasculitis, anterior uveitis, babesiosis, bacterial pneumonia, Brill-Zinsser disease, bronchitis, conjunctivitis, dengue, diffuse unilateral subacute neuroretinitis, disseminated gonococcal infection, drug hypersensitivity, drug reactions, Eales disease, ehrlichiosis, enteroviruses, Epstein-Barr virus, gastroenteritis, Henoch-Schönlein purpura, hepatitis, idiopathic thrombocytopenic purpura, immune complex vasculitis, influenza, Kawasaki disease, leptospirosis, luetic exanthem, malaria, measles, meningitis, meningococcemia, mononucleosis, murine typhus, Mycoplasma pneumonia, ocular ischemic syndrome, optic neuritis, papilledema, pneumonia, Q fever, relapsing fever, retinal artery/vein occlusion, rubella, rubeola, staphylococcal sepsis, syphilis, thrombocytopenic purpura, tick-borne diseases, toxic shock syndrome, toxic shock syndrome, tularemia, typhoid fever, typhus.

Mortality Rate (untreated): ≤25%.

C17-A032

Rickettsia typhi
Endemic Typhus
ICD-11: XN2AR
UNII: TC0HGR44VW
EPPO: RICKTY

It is an aerobic, Gram-negative, non-motile, rod-shaped bacterium that can only reproduce within a host's cells (obligate intracellular). It is endemic throughout the world. The natural reservoirs are rats and fleas. Fleas remain infected for life. *R. typhi* is not transmitted by the bite of the flea, but rather by contamination of the bite, or other open wounds, with infected feces from the flea. It is resistant to drying and can remain infectious for up to 100 days in rat feces. This is a biosafety level 2 agent for clinical samples. This is a biosafety level 3 practices for activities with a high potential for aerosol or droplet production, high concentrations of agent, or working with infected arthropods.

In People:

CDC Case Definition: None established.

Communicability: Direct person-to-person transmission does not occur. When dealing with infected individuals, use standard precautions.

Normal Routes of Exposure: Inhalation; abraded skin; mucous membranes; vectors (fleas).
Infectious Dose: <10 organisms (estimate).

Secondary Hazards: Aerosols (flea feces); fecal (from fleas); fomites (with vectors).
Incubation: 3–14 days.

Signs & Symptoms: Symptoms include fever, severe headache, pain in the muscles (myalgia) and joints (arthralgia), cough, nervousness, nausea, and vomiting. A rash of flat (macular) spots appears first on the trunk then spreads to the arms and legs; but does not appear on the palms of the hands, the soles of the feet, or the face. Rash is often transient and difficult to observe. Can progress to inflammation around the brain and spinal cord (meningitis), seizures, and multiorgan failure including the liver, kidneys, heart, and lungs.

Suggested Alternatives for Differential Diagnosis: Babesiosis, brucellosis, dengue, ehrlichiosis, infectious mononucleosis, Kawasaki disease, Leptospirosis, Lyme disease, malaria, measles, relapsing fever, Rocky Mountain spotted fever, rubella, syphilis, toxic shock syndrome, toxoplasmosis, tularemia, typhoid fever.

Mortality Rate (untreated): ≤4%.

C17-A033

Salmonella enterica subsp. *enterica* ser. **Typhi**
Typhoid Fever
ICD-11: XN4AM
UNII: 760T5R8B3O
EPPO: SALLTP

It is a Gram-negative, motile, rod-shaped, facultative anaerobe (can live with or without oxygen). It is endemic throughout the world but primarily found in developing countries. Humans are the natural reservoir. It can survive outside a host on the skin for up to 20 minutes, in dust for up to thirty days, in water or ice for many weeks, and in feces for up to sixty-two days. The optimal temperature for growth is 99°F. This is a biosafety level 2 agent for clinical samples. Biosafety level

3 should be used for activities with a high potential for aerosol or droplet production or high concentrations of agent.

It is on the Australia Group Core list and the OIE list of Notifiable Diseases for 2020.

In People:

CDC Case Definition: Clinical criteria is an illness caused by *Salmonella enterica* serotype Typhi that is often characterized by insidious onset of sustained fever, headache, malaise, anorexia, relative bradycardia, constipation or diarrhea, and nonproductive cough. However, many mild and atypical infections occur. Carriage of serotype Typhi may be prolonged.

Confirmatory laboratory criteria is isolation of serotype Typhi from blood, stool, or other clinical specimen.

Communicability: Direct person-to-person transmission is possible through fecal/oral. When dealing with infected individuals, use standard precautions. Wash hands frequently. If working directly with fecal matter, use contact precautions.

Normal Routes of Exposure: Ingestion.

Infectious Dose: 100,000 organisms.

Secondary Hazards: Blood and body fluids; fecal matter; mechanical vectors.

Incubation: 6–30 days.

Signs & Symptoms: There are few clinical features that reliably distinguish it from a variety of other infectious diseases. Clinical presentation includes insidious onset of sustained high fever (103°F to 104°F), severe dull frontal headache, vague feeling of bodily discomfort (malaise), chills, dry cough, nausea, vomiting, abdominal pain, transient rash of mixed flat and raised (maculopapular) rose-colored spots on the trunk, muscle pain (myalgia), loss of appetite (anorexia), usually constipation (although it may cause diarrhea), and bloody stools. The abdomen becomes distended, an enlarged spleen (splenomegaly) is common. Severe cases produce confusion, delirium, intestinal perforation, and death. The illness may last for 3 to 4 weeks. Individuals may become asymptomatic carriers capable of spreading the disease (e.g., Typhoid Mary).

Suggested Alternatives for Differential Diagnosis: Abdominal abscess, amebic liver, appendicitis, brucellosis, campylobacteriosis, cryptosporidiosis, cyclosporiasis, dengue, *E. coli* infections, hepatic abscesses, influenza, leishmaniasis, *Listeria* monocytogenes, malaria, rickettsial diseases, shigellosis, toxoplasmosis, tuberculosis, tularemia, typhus, *Vibrio* infections, yersiniosis, ingestion of bacterial toxins such as staphylococcal enterotoxins or botulinum toxin.

Mortality Rate (untreated): ≤30%.

C17-A034

Serratia marcescens (Agent SM)
BW Simulant
ICD-11: —
UNII: FB3Y68CF2Y
EPPO: SERRMA

It is a Gram-negative, motile, rod-shaped, facultative anaerobe (can live with or without oxygen) with a blood red color. It is endemic throughout the world. S. marcescens can survive from 3 days to 2 months on dry, inanimate surfaces and up to 5 weeks on a dry floor. The main risk factor for exposure to *S. marcescens* is hospitalization where it causes opportunistic infections of the heart lining (endocardium), eyes, blood, wounds, urinary and respiratory tracts. Until the 1960s, medical researchers used *S. marcescens* as a biological marker for studying the transmission of microorganisms because it was considered a harmless organism that only fed on dead or decaying organic matter (saprophyte).

In People:

CDC Case Definition: None established.

Communicability: Direct person-to-person transmission does not occur.

Normal Routes of Exposure: Inhalation; abraded skin; mucous membranes.

Infectious Dose: Unknown.

Secondary Hazards: None.

Incubation: Unknown.

Signs & Symptoms: Depend on the site of infection. In the hospital (nosocomial), Serratia species tend to colonize the respiratory tract (fever, chills, cough, low blood pressure (hypotension), difficulty breathing (dyspnea), and/or chest pain); urinary tract (fever, frequent urination, difficult or painful urination (dysuria), pus in the urine (pyuria), or inability to urinate); surgical wounds (drainage, abscesses, and oozing (exudate)); and soft tissues (skin infections, subcutaneous skin infections (cellulitis), vein inflammation (phlebitis)).

Suggested Alternatives for Differential Diagnosis: Infections from other bacteria (e.g., *Enterobacter*, *E. coli*, *Klebsiella*, *Proteus*, or *Providencia*) producing meningitis, pneumonia, or sepsis.

Mortality Rate (untreated): ≤40% (vulnerable population).

C17-A035

Shiga Toxin Producing *Escherichia coli*
Enterohemorrhagic *Escherichia coli*
ICD-11: XN5N5
UNII: —
EPPO: —

It is a facultative (capable of growing with or without the presence of oxygen) anaerobic, Gram-negative, motile, rod-shaped bacterium. It is endemic throughout the world. Although *E. coli* is part of the common bacterium that normally inhabits the intestinal tracts of humans and animals, these serotypes produce Shiga toxins (C16-A032) that causes bleeding (hemorrhage) in the intestines. Cattle, including calves, are one of the main reservoirs for infection in humans. It can survive outside a host for extended periods. This is a biosafety level 2 agent.

E. coli can also cause acute bacterial meningitis, pneumonia, intra-abdominal infections, enteric infections, urinary tract infections, septic arthritis, endophthalmitis, suppurative thyroiditis, sinusitis, osteomyelitis, endocarditis, and skin and soft tissue infections.

It is on the Australia Group Core list and the OIE list of Notifiable Diseases for 2020.

In People:

CDC Case Definition: Clinical criteria is an infection of variable severity characterized by diarrhea (often bloody) and/or abdominal cramps. Illness may be complicated by hemolytic-uremic syndrome (HUS).

Confirmatory laboratory criteria is 1) Isolation of *E. coli* O157: H7 from a clinical specimen; OR 2) Isolation of *E. coli* from a clinical specimen with detection of Shiga toxin or Shiga toxin genes.

Communicability: Direct person-to-person transmission is possible through fecal/oral. When dealing with infected individuals, use standard precautions. Wash hands frequently. If working directly with fecal matter, use contact precautions.

Normal Routes of Exposure: Ingestion.

Infectious Dose: 10 organisms (estimate).

Secondary Hazards: Aerosols (possibly); fecal matter; mechanical vectors (flies, roaches).
Incubation: 2–8 days.

Signs & Symptoms: Severe cramping with bowel movements ranging from non-bloody diarrhea to stools that are almost pure blood. Vomiting occurs in approximately half the cases. There is generally no fever associated with the infection. Usually lasts 8 days.

After approximately 7 days from symptom onset, may progress to HUS resulting from the toxin destroying red blood cells. Symptoms include high blood pressure (hypertension), pallor, skin and mucous membranes may be dry, decreased urination, exhaustion, and ultimately kidney failure. Neurologic findings such as blindness, incoordination (ataxia), weakness of one entire side of the body (hemiparesis), and coma indicate a poorer prognosis.

Suggested Alternatives for Differential Diagnosis: Disseminated intravascular coagulation, malignant hypertension, neurological manifestations of uremic encephalopathy, neurological sequelae of infectious endocarditis, neurosarcoidosis, polyarteritis nodosa, primary CNS lymphoma, Scleroderma renal crisis, shigellosis, systemic lupus erythematosus, thrombotic thrombocytopenic purpura, viral encephalitis, viral meningitis.

Mortality Rate (untreated): ≤90% (hemolytic uremic syndrome).

C17-A036

Shigella dysenteriae (Agent Y)
Shigellosis
ICD-11: XN285
UNII: 1EP6R5562J
EPPO: SHIGDY

It is a Gram-negative, non-motile, rod-shaped, facultative anaerobe (can live with or without oxygen). It is endemic throughout the world but primarily found in developing countries. Humans and other large primates are the only natural reservoirs. It can survive outside a host on dry surfaces for months; on metal utensils for up to 28 days at 60°F or up to 13 days at 99°F; in feces for up to 12 days at 77°F; in citric juices and carbonated soft drinks for up to 10 days; on vegetables for several days; and on hands for over 3 hours. Flies can carry Shigella for up to 24 days. It grows between 77°F and 99°F. The organism is frequently found in water polluted with human feces where it can survive for approximately 3 days. This is a biosafety level 2 agent.

It is on the Australia Group Core list.

In People:
CDC Case Definition: Clinical criteria is an illness of variable severity commonly manifested by diarrhea, fever, nausea, cramps, and tenesmus. Asymptomatic infections may occur.

Confirmatory laboratory criteria is Isolation of *Shigella* spp. from a clinical specimen.

Communicability: Direct person-to-person transmission is possible through fecal/oral. When dealing with infected individuals, use standard precautions. Wash hands frequently. If working directly with fecal matter, use contact precautions.

Normal Routes of Exposure: Ingestion.

Infectious Dose: 10–200 organisms.

Secondary Hazards: Fecal matter; blood (rarely); mechanical vectors (flies, roaches).

Incubation: 12 hours–7 days.

Signs & Symptoms: Acute bloody diarrhea, abdominal pain (cramping or tenderness), cramping rectal pain with a feeling of the need to pass stool even when the bowels are empty (tenesmus), and fever; occasional vomiting and dehydration. Symptoms usually last 5 to 7 days, but some people may experience symptoms for more than 4 weeks. Serious less frequent complications include sepsis, seizures, convulsions, renal failure, and hemolytic uremic syndrome. Individuals may become asymptomatic carriers capable of spreading the disease.

Suggested Alternatives for Differential Diagnosis: Amebiasis, bacterial gastroenteritis, campylo-bacter infections, cholera, *Clostridioides difficile* colitis, colon cancer, Crohn's disease, cryptos-poridiosis, *Escherichia coli* infections, pseudomembranous colitis, salmonella infections, schistosomiasis, ulcerative colitis, viral gastroenteritis, *Yersinia enterocolitica.*

Mortality Rate (untreated): ≤15%; ≤50% (hemolytic uremic syndrome).

C17-A037

Vibrio cholerae (Agent HO)
Cholera
ICD-11: XN7N1; XN62R (El Tor)
UNII: 4M0784008H; GDV6F3DW7S (El Tor)
EPPO: VIBRCH

It is a Gram-negative, motile, rod- or curved rod-shaped, facultative anaerobe (can live with or without oxygen). Although it has occurred worldwide, it is currently endemic in Africa and India. The natural reservoir is humans, although marine shellfish and plankton can also serve as hosts. *V. cholerae* releases a potent enterotoxin in the small intestine that causes a dramatic efflux of ions and water leading to watery diarrhea. The El Tor strain can survive in fresh water and salt water (up to 3% saline). It is easily killed by heating (>115°F), freezing, chlorine at purification concentrations. This is a biosafety level 2 agent.

It is on the Australia Group Core list.

In People:
CDC Case Definition: Clinical criteria is an illness characterized by diarrhea and/or vomiting; severity is variable.

Confirmatory laboratory criteria is 1) isolation of toxigenic (i.e., cholera toxin-producing) *Vibrio cholerae* O1 or O139 from stool or vomitus, OR 2) serologic evidence of recent infection.

Communicability: Direct person-to-person transmission is possible through fecal/oral. When dealing with infected individuals, use standard precautions. Wash hands frequently. If working directly with fecal matter, use contact precautions.

Normal Routes of Exposure: Ingestion.

Infectious Dose: 1,000,000 organisms.

Secondary Hazards: Fecal matter; vomit; mechanical vectors (flies).

Incubation: 6 hours–5 days.

Signs & Symptoms: Vomiting is a prominent early manifestation of the disease followed rapidly by abdominal cramps with profuse watery diarrhea (opaque white liquid that does not have a bad odor; often described as resembling rice water). Bowel movements are frequent and often uncontrolled. Stool volume is more than that from any other infectious diarrhea. Diarrhea and vomiting can lead to severe dehydration accompanied by rapid heart rate (tachycardia), rapid breathing (tachypnea), low blood pressure (hypotension), vascular collapse, shock, and death. Dehydration can develop within hours after the onset of symptoms.

Suggested Alternatives for Differential Diagnosis: Clinical picture is unlikely to be confused with any other disease. Consider Escherichia coli infections, rotavirus.

Mortality Rate (untreated): ≤70%.

C17-A038

Xanthomonas albilineans
Leaf Scald of Sugarcane
ICD-11: —
UNII: —
EPPO: XANTAB

It is an aerobic, Gram-negative, motile rod-shaped bacterium. It is endemic to Central and South America, Africa, India, Southeast Asia, and Australia. *X. albilineans* produces the toxin albicidin that blocks chloroplast differentiation and results in leaf scald disease symptoms. It does not survive long in soil.

It is on the Australia Group Core list.

In Plants:

Target species: Sugarcane.

Normal routes of transmission: Contact.

Secondary Hazards: Airborne (rain); crop debris; mechanical vectors (cultivation, track out).

Signs: Infected plants can remain in a latency phase for several months without showing symptoms. On young leaves, the disease presents a single, white to yellow (chlorotic), pencil-line extending nearly the entire length of the upper side of the leaf (lamina). As leaves become older, the width of the white line increases and turns into a diffused white stripe. The stripe eventually starts drying from the tip, leaves develop an inward curl, wilt, and take on a partially burnt appearance. A chronic version of the disease causes stunting of cane stalks with the development of side shoots in a bottom to top (acropetal) fashion, giving a bushy appearance to the affected plant.

Crop Losses: ≤100%

C17-A039

Xanthomonas citri subsp. *citri*
Citrus Canker
ICD-11: —
UNII: —
EPPO: XANTCI

It is a Gram-negative, motile, rod-shaped bacterium that thrives in humid subtropical regions of the world.

It is on the Australia Group Core list.

In Plants:

Target species: Many varies of citrus including grapefruits, oranges, lemons, limes, and tangerines.

Normal routes of transmission: Contact.

Secondary Hazards: Airborne (rain, splashing water); crop debris; fomites (mulch, soil); mechanical vectors (cultivation, track out).

Signs: Initially, leaves develop bright yellow spots on the underside that are followed by erumpent brownish lesions on both sides. Lesions are rough, cracked, and corky. They may be surrounded by a water-soaked yellow or chlorotic halo margin. Affected areas of leaves may fall out in time, leaving holes. Fruit has characteristic scabs or crater-like lesions on the surface that do not extend through the rind. In dry conditions, branches develop a raised, corky, or spongy canker with a ruptured surface. In moist conditions, lesions on branches remain unruptured with an oily margin. Older leaf tissue and mature fruit are more resistant to citrus canker infection. The disease causes trees to prematurely drop leaves and fruit. Ultimately, fruit production declines then stops.

Crop Losses: ≤100%

C17-A040

Xanthomonas oryzae pv. oryzae
Rice Leaf Blight
ICD-11: —
UNII: —
EPPO: XANTOR

It is a Gram-negative, motile, rod-shaped bacterium. Endemic to Japan and rice producing countries of Asia. The optimal temperature range for growth is 68°F to 86°F.

It is on the Australia Group Core list and the HHS Select Agents and Toxins list.

In Plants:

Target species: Rice.

Normal routes of transmission: Contact.

Secondary Hazards: Airborne (wind, rain, storm); mechanical vectors (irrigation); crop debris; weed hosts.

Signs: Symptoms appear on the leaves of young plants as pale-green to grey-green, water-soaked streaks near the leaf tip and margins. These lesions coalesce and become yellowish-white with wavy edges. The whole leaf may be affected, becoming whitish or greyish, and then die. Systemic infection results in wilting, desiccation of leaves and death.

Crop Losses: ≤23%

C17-A041

Xylella fastidiosa
Pierce's Disease
ICD-11: —
UNII: —
EPPO: XYLEFA

It is a Gram-negative, non-motile, rod-shaped bacterium that is limited to the xylem of plants. It is endemic to North and South America.

It is on the Australia Group Awareness list.

In Plants:

Target species: Grapes, almonds, olives, citrus, coffee.

Normal routes of transmission: Vectors (sharpshooters, spittlebugs).

Secondary Hazards: Crop debris; mechanical vectors (cultivation).

Signs: Can be confused with other diseases, environmental stresses, water deficiency, air pollutants, or nutritional problems. Leaves develop yellow (chlorotic) spots or patches, then take on a scorched appearance – becoming brown and necrotic – before dropping prematurely. Symptoms may progress to dieback of shoots, twigs, and branches. The overall plant may become stunted. Fruit size and yield are generally reduced. Depending on the affected species, individual plants may die rapidly, survive for several years, or simply suffer a general decline in vigor with severely reduced productivity.

Crop Losses: ≤50%

C17-A042

Yersinia pestis (Agent LE)
Plague
ICD-11: XN6QS; 1B93.2 (Pneumonic); 1G40 (Sepsis); 1B93.Z (Unspecified)
UNII: S6JJH3XV1D
EPPO: YERSPE

It is a Gram-negative, non-motile, rod-shaped, facultative anaerobe (can live with or without oxygen). It is endemic to the western United States, South America, and parts of Africa and Asia. The natural reservoir is numerous mammalian species. Cattle, horses, sheep, and pigs are not known to develop symptomatic illness from plague. It does cause illness in cats and dogs.

Y. pestis can survive for up to 2 months in infected fleas, 5–6 months on clothing, up to 7 months in soil, and for weeks in water or damp food and grains. At or near freezing temperatures it can remain alive for years. It survives for long periods in dried sputum, blood, flea feces, or dead animals. It is killed by exposure to several hours of sunlight. This is a biosafety level 2 agent for clinical samples. Biosafety level 3 should be used for activities with a high potential for aerosol or droplet production, high concentrations of agent, or work involving infected arthropods.

It is on the Australia Group Core list and a Tier-1 agent on the U.S. Select Agents and Toxins list.

In People:

CDC Case Definition: Clinical criteria: Plague is transmitted to humans by fleas or by direct exposure to infected tissues or respiratory droplets; the disease is characterized by fever, chills, headache, malaise, prostration, and leukocytosis that manifests in one or more of the following principal clinical forms: 1) Regional lymphadenitis (bubonic plague); 2) Septicemia without an evident bubo (septicemic plague); 3) Plague pneumonia, resulting from hematogenous spread in bubonic or septicemic cases (secondary pneumonic plague) or inhalation of infectious droplets (primary pneumonic plague); 4) Pharyngitis and cervical lymphadenitis resulting from exposure to larger infectious droplets or ingestion of infected tissues (pharyngeal plague).

Presumptive laboratory criteria is 1) elevated serum antibody titer(s) to *Yersinia pestis* fraction 1 antigen (without documented fourfold or greater change) in a patient with no history of plague vaccination; OR 2) detection of fraction 1 antigen in a clinical specimen by fluorescent assay. Confirmatory laboratory criteria is 1) isolation of *Y. pestis* from a clinical specimen; OR 2) fourfold or greater change in serum antibody titer to *Y. pestis* fraction 1 antigen.

Communicability: Direct person-to-person transmission is possible for pneumonic plague but not bubonic plague. When dealing with individuals displaying signs of pneumonic plague, use droplet precautions. When dealing with individuals displaying signs of bubonic plague, use standard precautions. Avoid direct contact with wounds or wound drainage.

Normal Routes of Exposure: Inhalation; abraded skin; vectors (fleas).

Infectious Dose: 100 organisms.

Secondary Hazards: Aerosols (respiratory secretions); blood and body fluids; body tissue; fomites (with vectors); vector amplification.

Incubation: 1–6 days.

Signs & Symptoms: Initial signs and symptoms may be nonspecific and include fever, chills, headache, a vague feeling of bodily discomfort (malaise), muscle pain (myalgia), nausea, sore throat, and headache.

Bubonic plague: abdominal pain and diarrhea. There may be blisters (vesicles) at the site of the flea bite. As the disease progresses, there may be pimples (papules), small pus-filled blisters (pustules), boil-like eruptions (carbuncles), or scabs (eschars) in the area of the bite and a generalized rash on the hands and feet. Painful, swollen lymph glands (buboes) arise, usually in the groin, armpit, or neck. Buboes are unilateral, oval, extremely tender and can vary from 2–10 cm in size. Enlargement of the buboes leads to rupture and discharge of malodorous pus. Fluid from the bubo is infectious. Swollen livers (hepatomegaly) and/or spleens (splenomegaly) often occur and may be tender.

Septicemic plague: occurs when the pathogen moves into the blood. There is a rapid onset of symptoms including rapid heart rate (tachycardia), rapid breathing (tachypnea), and low blood pressure (hypotension). Hypothermia is common. Patients experience nausea, vomiting, abdominal pain, and diarrhea. Buboes are uncommon. There may be generalized dermal hemorrhages (purpura) that can progress to necrosis and gangrene of the distal extremities. There is a high mortality rate associated with disseminated intravascular coagulation, multiorgan failure, and profound low blood pressure (hypotension).

Pneumonic plague: occurs when the pathogen moves to the lungs. Symptoms include cough, chest pain, difficulty breathing (dyspnea), sputum containing pus, or coughing up blood (hemoptysis). Pneumonic plague is highly contagious and poses significant hazard of person-to-person transmission through aerosol generated by coughing. If the initial infection is due to inhaling aerosolized pathogen, buboes may or may not be seen.

Suggested Alternatives for Differential Diagnosis: Acute respiratory distress syndrome, anthrax, bacterial pharyngitis, bacterial pneumonia, bacterial sepsis, B-cell lymphoma, brucellosis, cat scratch fever, cellulitis, chancroid, community-acquired pneumonia, dengue, disseminated intravascular coagulation, gas gangrene, lymphadenitis, lymphadenopathy, lymphogranuloma venereum, malaria, necrotizing fasciitis, *Pasteurella multocida* infection, peritonsillar abscess, pneumonia, scarlet fever, septic shock, syphilis, systemic inflammatory response syndrome, tick-borne diseases (e.g., Lyme disease, Rocky Mountain spotted fever), tonsillitis, tularemia, typhus.

Mortality Rate (untreated): ≤60% (bubonic); ≤100% (septicemic; pneumonic).

C17-A043

Yersinia pseudotuberculosis
Yersiniosis
ICD-11: 1B9A
UNII: W9AS1Z91Q7
EPPO: YERSPS

It is a Gram-negative, motile, rod-shaped, facultative anaerobe (can live with or without oxygen). It is endemic throughout the world. It is primarily a zoonotic disease of birds and mammals, with humans as incidental hosts. It can survive for up to 15 days in fresh water, 4 days in seawater, and 9 months in soil. It is tolerant to cold temperatures. This is a biosafety level 2 agent.

It is on the Australia Group Warning list and the OIE list of Notifiable Diseases for 2020.

In People:

CDC Case Definition: None established.

Communicability: Direct person-to-person transmission is possible through fecal/oral. When dealing with infected individuals, use standard precautions. Wash hands frequently. If working directly with fecal matter, use contact precautions.

Normal Routes of Exposure: Ingestion.

Infectious Dose: 100,000,000 organisms.

Secondary Hazards: Fecal matter; fomites.

Incubation: 3–21 days.

Signs & Symptoms: Symptoms include pain in the lower-right abdominal area resembling appendicitis, as well as fever, headache, sore throat (pharyngitis), and loss of appetite (anorexia). In some cases, there may be vomiting and watery diarrhea. May also produce arthritis, inflammation of the iris (iritis), or skin rash resembling scarlet fever. May also cause infections of other sites such as wounds or the urinary tract. Individuals may become asymptomatic carriers capable of spreading the disease.

Suggested Alternatives for Differential Diagnosis: Acute pancreatitis, appendicitis, bacterial gastroenteritis, bacterial sepsis, *Clostridioides difficile* colitis, Crohn's disease, enterocolitis, enteropathic arthropathies, erythema multiforme, Kawasaki disease, leptospirosis, mesenteric lymphadenitis, neutropenic enterocolitis, sarcoidosis, staphylococcal infections, toxic shock syndrome, typhoid fever, ulcerative colitis, Whipple disease.

Mortality Rate (untreated): ≤11%.

BIBLIOGRAPHY

Acha, Pedro N. and Boris Szyfres. *Zoonoses and Communicable Diseases Common to Man and Animals, Scientific and Technical Publication No. 580. 3rd Edition. Volume 1, Bacterioses and Mycoses.* Washington, D.C.: Pan American Health Organization, 2003.

Acha, Pedro N. and Boris Szyfres. *Zoonoses and Communicable Diseases Common to Man and Animals, Scientific and Technical Publication No. 580. 3rd Edition. Volume 2, Chlamydioses, Rickettsioses, and Viroses.* Washington, D.C.: Pan American Health Organization, 2003.

American Committee of Medical Entomology; American Society of Tropical Medicine and Hygiene. *Arthropod Containment Guidelines, Version 3.2.* Vector Borne and Zoonotic Diseases 19, #3, (2019): 152–173.

American Society for Microbiology. *Sentinel Level Clinical Laboratory Guidelines for Suspected Agents of Bioterrorism and Emerging Infectious Diseases: Bacillus anthracis and Bacillus cereus biovar anthracis.* Revised September 2017.

American Society for Microbiology. *Sentinel Level Clinical Laboratory Guidelines for Suspected Agents of Bioterrorism and Emerging Infectious Diseases: Botulinum Toxin.* Revised June 2013.

American Society for Microbiology. *Sentinel Level Clinical Laboratory Guidelines for Suspected Agents of Bioterrorism and Emerging Infectious Diseases: Brucella species.* Revised March 2016.

American Society for Microbiology. *Sentinel Level Clinical Laboratory Guidelines for Suspected Agents of Bioterrorism and Emerging Infectious Diseases: Coxiella burnetti*. Revised March 2016.

American Society for Microbiology. *Sentinel Level Clinical Laboratory Guidelines for Suspected Agents of Bioterrorism and Emerging Infectious Diseases: Francisella tularensis*. Revised March 2016.

American Society for Microbiology. *Sentinel Level Clinical Laboratory Guidelines for Suspected Agents of Bioterrorism and Emerging Infectious Diseases: Melioidosis: Burkholderia pseudomallei*. Revised March 2016.

American Society for Microbiology. *Sentinel Level Clinical Laboratory Guidelines for Suspected Agents of Bioterrorism and Emerging Infectious Diseases: Yersinia pestis*. Revised March 2016.

Argonne National Laboratory. *Australia Group Common Control List Handbook Volume II: Biological Weapons-Related Common Control Lists*, Revision 4. Washington, D.C.: U.S. Government Printing Office, February 2018.

The Australia Group. *Australia Group Common Control Lists*. February 28, 2020 [https://www.dfat.gov.au/publications/minisite/theaustraliagroupnet/site/en/human_animal_pathogens.html]. December 31, 2020.

Bozue, Joel, Christopher K. Cote and Pamela J. Glass, ed. *Medical Aspects of Biological Warfare, Textbooks of Military Medicine Series*. Washington, D.C.: Office of the Surgeon General, Department of the Army, 2018.

Brunette, Gary W. and Jeffrey B. Nemhauser, ed. *CDC Yellow Book 2020: Health Information for International Travel*. July 18, 2019 [https://wwwnc.cdc.gov/travel/page/yellowbook-home-2014]. December 31, 2020.

Burrows, W. Dickinson and Sara E. Renner. "Biological Warfare Agents as Threats to Potable Water." *Environmental Health Perspectives* 107 (1999): 975–984.

California Department of Food and Agriculture. Animal Health and Food Safety Services. Animal Health Branch. *Biosecurity: Selection and Use of Surface Disinfectants*, Revision June 2002.

Canada Centre for Biosecurity. *Pathogen Safety Data Sheets*. December 6, 2019 [https://www.canada.ca/en/public-health/services/laboratory-biosafety-biosecurity/pathogen-safety-data-sheets-risk-assessment.html]. December 31, 2020.

Canada Centre for Biosecurity. *Canadian Biosafety Standard*, 2nd Edition, Ottawa, Canada, March 2015.

Centers for Disease Control and Prevention. *National Notifiable Diseases Surveillance System: Surveillance Case Definitions for Current and Historical Conditions*. August 2, 2017 [https://wwwn.cdc.gov/nndss/conditions/]. December 31, 2020.

Centers for Disease Control and Prevention. "Biological and Chemical Terrorism: Strategic Plan for Preparedness and Response. Recommendations of the CDC Strategic Planning Workgroup." *Morbidity and Mortality Weekly Report* 49 (RR-4) (2000): 1–14.

Centers for Disease Control and Prevention. "Medical Examiners, Coroners, and Biologic Terrorism: A Guidebook for Surveillance and Case Management." *Morbidity and Mortality Weekly Report* 53 (RR-8) (2004).

Centre for Agriculture and Bioscience International. *Invasive Species Compendium*. 2020 [https://www.cabi.org/isc]. December 31, 2020.

Centre for Agriculture and Bioscience International. *Plantwise Knowledge Bank*. 2020 [https://www.plantwise.org/KnowledgeBank]. December 31, 2020.

Chosewood, L. Casey and Deborah E. Wilson, ed. *Biosafety in Microbiological and Biomedical Laboratories*. 5th Edition. Washington, D.C.: U.S. Government Printing Office, 2009.

Committee on Foreign Animal Diseases of the United States Animal Health Association. *Foreign Animal Diseases*. 7th Edition. Revised 2008. Boca Raton, Florida: Boca Publishing Group, Inc., 2008.

Compton, James A.F. *Military Chemical and Biological Agents: Chemical and Toxicological Properties*. Caldwell, New Jersey: The Telford Press, 1987.

Dennis, David T., Thomas V. Inglesby, Donald A. Henderson, John G. Bartlett, Michael S. Ascher, Edward Eitzen, Anne D. Fine, Arthur M. Friedlander, Jerome Hauer, Marcelle Layton, Scott R. Lillibridge, Joseph E. McDade, Michael T. Osterholm, Tara O'Toole, Gerald Parker, Trish M. Perl, Philip K. Russell and Kevin Tonat. "Tularemia as a Biological Weapon: Medical and Public Health Management." *Journal of the American Medical Association* 285 (2001): 2763–2773.

Department of Agriculture. 9 CFR Part 121 – "Possession, Use, and Transfer of Biological Agents and Toxins," 2005: 817–832.

European and Mediterranean Plant Protection Organization. *EPPO Global Database*. 2020 [https://gd.eppo.int/]. December 31, 2020.

European Association of Zoo and Wildlife Veterinarians. *Transmissible Disease Handbook, 5th Edition*. 2020 [https://www.eazwv.org/page/inf_handbook]. December 31, 2020.

European Centre for Disease Prevention and Control. *All Topics*. 2020 [https://www.ecdc.europa.eu/en/all-topics]. December 31, 2020.

Food and Agriculture Organization of the United Nations (FAO). International Plant Protection Convention. "International Standard for Phytosanitary Measures 27, Diagnostic Protocols for Regulated Pests, DP 25: *Xylella fastidiosa*." August 2018.

Food and Agriculture Organization of the United Nations (FAO) International Plant Protection Convention. "International Standard for Phytosanitary Measures 27, Diagnostic Protocols for Regulated Pests, DP 6: *Xanthomonas citri* subsp. *citri*." August 2014.

German Social Accident Insurance. *GESTIS Biological Agents Database*. June 12, 2020 [http://gestis.itrust.de/nxt/gateway.dll/bioen/000000.xml?f=templates&fn=default.htm&vid=gestisbioeng:biosdbeng]. December 31, 2020.

Harkins, Deanna, Rose Overturf, Veronique Hauschild and Scott Goodison. *Safety and Health Guidance for Mortuary Affairs Operations: Infectious Materials and CBRN Handling, Technical Guide 195*. Washington, D.C.: Government Printing Office, May 2009.

Henderson, Donald A., Thomas V. Inglesby and Tara O'Toole, ed. *Bioterrorism: Guidelines for Medial and Public Health Management*. Chicago, Illinois: AMA Press, 2002.

Herenda, D. *Manual on Meat Inspection for Developing Countries*. Reprint 2000. Rome, Italy: Food and Agriculture Organization of the United Nations, 1994.

Heymann, David, ed. *Control of Communicable Diseases Manual*. 20th Edition. Washington, D.C.: American Public Health Association, December 2014.

Holty, Jon-Erik C., Dena M. Bravata, Hau Liu, Richard A. Olshen, Kathryn M. McDonald and Douglas K. Owens. "Systematic Review: A Century of Inhalational Anthrax Cases from 1900 to 2005." *Annals of Internal Medicine* 144 (February 21, 2006): 270–280.

Inglesby, Thomas V., David T. Dennis, Donald A. Henderson, John G. Bartlett, Michael S. Ascher, Edward Eitzen, Anne D. Fine, Arthur M. Friedlander, Jerome Hauer, John F. Koerner, Marcelle Layton, Joseph McDade, Michael T. Osterholm, Tara O'Toole, Gerald Parker, Trish M. Perl, Philip K. Russell, Monica Schoch-Spana and Kevin Tonat. "Plague as a Biological Weapon: Medical and Public Health Management." *Journal of the American Medical Association* 283 (2000): 2281–2290.

Inglesby, Thomas V., Donald A. Henderson, John G. Bartlett, Michael S. Ascher, Edward Eitzen, Arthur M. Friedlander, Jerome Hauer, Joseph McDade, Michael T. Osterholm, Tara O'Toole, Gerald Parker, Trish M. Perl, Philip K. Russell and Kevin Tonat. "Anthrax as a Biological Weapon: Medical and Public Health Management." *Journal of the American Medical Association* 281 (1999): 1735–1745.

Iowa State University. *The Center for Food Security & Public Health*. 2020 [http://www.cfsph.iastate.edu/]. December 31, 2020.

Leitenberg, Milton and Raymond A. Zilinskas. *The Soviet Biological Weapons Program: A History*. Cambridge, Massachusetts: Harvard University Press, 2012.

Lindler, Luther E., Frank J. Lebeda, and George W. Korch, ed. *Biological Weapons Defense: Infectious Diseases and Counterbioterrorism*. Totowa, New Jersey: Humana Press, Inc., 2005.

Melton-Celsa, Angela R. *Shiga Toxin (Stx) Classification, Structure, and Function*. Microbiol Spectr. 2014; 2(2): doi:10.1128/microbiolspec.EHEC-0024-2013.

Merck & Co. *Merck Manual Professional Version*. 2020 [https://www.merckmanuals.com/professional]. December 31, 2020.

Merck & Co. *Merck Veterinary Manual*. 2020 [https://www.merckvetmanual.com/]. December 31, 2020.

Pacific Community Land Resources Division. *AHP Disease Manual: Reference Guide for Animal Health Staff*. 2010 [http://lrd.spc.int/ext/Disease_Manual_Final/index.html]. December 31, 2020.

Pan American Health Organization. *Emergency Vector Control After Natural Disaster*. Scientific Publication No. 419. Washington, D.C.: Pan American Health Organization. 1982.

Plant Diseases.org. *A Comprehensive Resource for Bacterial Plant Disease Information*. 2019 [https://www.plantdiseases.org/]. December 31, 2020.

Public Health Agency of Canada. *Canadian Biosafety Handbook, 2nd Edition*. Ottawa, Canada, March 2016.

Rutala, William and David J. Weber. *Guideline for Disinfection and Sterilization in Healthcare Facilities, 2008*. Centers for Disease Control and Prevention, Updated May 2019.

Schmitt, Clare K., Karen C. Meysick and Alison D. O'Brien. "Bacterial Toxins: Friends or Foes?", *Emerging Infectious Diseases* 5, (April/June 1999): 224–234.

Sidell, Fredrick R., Ernest T. Takafuji and David R. Franz, ed. *Medical Aspects of Chemical and Biological Warfare, Textbook of Military Medicine Series, Part 1, Warfare, Weaponry, and the Casualty*. Washington, D.C.: Office of the Surgeon General, Department of the Army, 1997.

Siegel, Jane D., Emily Rhinehart, Marguerite Jackson and Linda Chiarello. *2007 Guideline for Isolation Precautions: Preventing Transmission of Infectious Agents in Healthcare Settings*. Centers for Disease Control and Prevention, June 2007.

Sifton, David W. ed. *PDR Guide to Biological and Chemical Warfare Response*. Montvale, New Jersey: Thompson/Physicians' Desk Reference, 2002.

United States Army Headquarters. *Potential Military Chemical/Biological Agents and Compounds, Field Manual No. 3-11.9*. Washington, D.C.: Government Printing Office, January 10, 2005.

United States Army Headquarters. *Technical Aspects of Biological Defense, Technical Manual No. 3–216*. Washington, D.C.: Government Printing Office, January 12, 1971.

United States Army Headquarters. *Treatment of Biological Warfare Agent Casualties, Field Manual No. 8–284*. Washington, D.C.: Government Printing Office, July 17, 2000.

United States Department of Agriculture Agricultural Research Service. *Recovery Plan for Rathayibacter Poisoning caused by Rathayibacter toxicus (syn. Clavibacter toxicus)*. Updated March 2015. [https://www.ars.usda.gov/ARSUserFiles/opmp/RathayibacterPoisoning_March2015.pdf]. December 31, 2020.

United States Department of Agriculture. Animal and Plant Health Inspection Service. 2020 [https://www.aphis.usda.gov/aphis/home]. December 31, 2020.

United States Department of Agriculture. Animal and Plant Health Inspection Service. *National Veterinary Accreditation Program Reference Guide*. June 2, 2020 [https://www.aphis.usda.gov/aphis/ourfocus/animalhealth/nvap/NVAP-Reference-Guide]. December 31, 2020.

United States Department of Health and Human Services, and Department of Agriculture. *Select Agents and Toxins List*. 2017 [https://www.selectagents.gov/SelectAgentsandToxinsList.html]. December 31, 2020.

United States Environmental Protection Agency. *Drinking Water Treatability Database: Find a Contaminant*. 2020 [https://tdb.epa.gov/tdb/findcontaminant]. December 31, 2020.

United States Food & Drug Administration, Center for Food Safety & Applied Nutrition. *Bad Bug Book, Foodborne Pathogenic Microorganisms and Natural Toxins Handbook*. 2nd Edition. October 24, 20217 [https://www.fda.gov/food/foodborne-pathogens/bad-bug-book-second-edition]. December 31, 2020.

University of Hawaii, Manoa. *EXTension ENTOmology & UH-CTAHR Integrated Pest Management Program*. August 30, 2011 [http://www.extento.hawaii.edu/kbase/crop/Type/Croppest.htm]. December 31, 2020.

United States Military Joint Chiefs of Staff. *Joint Tactics, Techniques, and Procedures for Mortuary Affairs in Joint Operations, Joint Publication No. 4-06*. Washington, D.C.: Government Printing Office, August 28, 1996.

University of Minnesota Center for Infectious Disease Research and Policy. *Infectious Disease Topics*. 2020 [http://www.cidrap.umn.edu/infectious-disease-topics]. December 31, 2020.

WebMD. *Medscape*. 2020 [https://emedicine.medscape.com/]. December 31, 2020.

Withers, Mark, ed. *Medical Management of Biological Casualties Handbook*. 8th Edition. Fort Detrick, Maryland: United States Army Medical Research Institute of Infectious Diseases, September 2014.

World Health Organization. *Health Aspects of Chemical and Biological Weapons: Report of A WHO Group of Consultants*. Geneva, Switzerland: World Health Organization, 1970.

World Health Organization. *Fact Sheets*. 2020 [https://www.who.int/news-room/fact-sheets]. December 31, 2020.

World Health Organization. *International Classification of Diseases, 11th Revision*. Geneva: World Health Organization. April 2019 [https://icd.who.int/browse11/l-m/en]. December 31, 2020.

World Health Organization. *Laboratory Biosafety Manual, 3rd Edition*. Geneva, Switzerland, 2004.

World Health Organization. *Public Health Response to Biological and Chemical Weapons: WHO Guidance*. Geneva, Switzerland: World Health Organization, 2004.

World Organization for Animal Health (OIE). *Animal Health of the World: Information on Aquatic and Terrestrial Animal Diseases*. 2020 [https://www.oie.int/en/animal-health-in-the-world/information-on-aquatic-and-terrestrial-animal-diseases/]. December 31, 2020.

World Organization for Animal Health (OIE) *Manual of Diagnostic Tests and Vaccines for Terrestrial Animals 2019*. [https://www.oie.int/en/standard-setting/terrestrial-manual/access-online/]. December 31, 2020.

World Organization for Animal Health (OIE) *OIE-Listed Diseases, Infections and Infestations in Force in 2020*. [https://www.oie.int/en/animal-health-in-the-world/oie-listed-diseases-2020/]. December 31, 2020.

World Organization for Animal Health (OIE) *Technical Disease Cards*. 2020 [https://www.oie.int/en/animal-health-in-the-world/technical-disease-cards/]. December 31, 2020.

18 Viral Pathogens

GENERAL INFORMATION

Viruses are the simplest type of microorganism consisting of a protein outer coat that contains genetic material (i.e., RNA, DNA). They are much smaller than bacteria and vary in size from 0.01 microns to 0.27 microns. They can only reproduce within a host's cells (obligate intracellular parasites) and lack a system for their own metabolism. They replicate by taking over metabolic processes of the invaded host cell and redirecting them toward virus production. Approximately 60% of infectious disease in humans are caused by a virus. Most treatment for viral infections is supportive because of limited anti-viral medications.

Viruses are very costly, time consuming, and difficult to grow in large quantities. They only replicate in living cells and cannot be grown in a growth media like bacteria (C17). In some cases, viruses can be grown on the chorioallantoic membrane of fertilized eggs.

Pathogens employed as biological weapons can be used for both lethal and incapacitating purposes. They cause disease by invading tissue or by producing toxins (C16) that are detrimental to the infected individual. Pathogens can be selected to target a specific host (e.g., humans, cows, fowl) or they may pose a broad threat to both animals and to people.

Pathogens deployed to target animals are generally used to produce lethal effects in an agriculturally significant species such as cows, pigs, or chickens. Although these pathogens are selected to target a specific animal species, there is the possibility that the disease may migrate to humans. The diseases produced by these crossover pathogens may be difficult for medical personnel not trained in exotic pathology to diagnose.

Other pathogens are selected to produce lethal effects in an agriculturally significant crop species such as wheat, corn, or rice. Symptoms in plants vary greatly and include mottling of leaves, stunting, lesions, leaf rolling, and death. Natural transmission from plant to plant is often by insects. There is little potential for migration of these pathogens to humans or animals.

A final group of biological warfare pathogens are those used as simulants to model the release of other, more hazardous agents. Pathogens employed as biological warfare simulants do not generally pose a significant risk to healthy people, animals, or plants. However, individuals with respiratory illness or suppressed immune systems may be at risk should they be exposed to an infectious dose of the agent.

Some viruses can be crystallized, similar to chemical compounds, while others can be stored as freeze-dried powders. In these forms, viruses are easy to disperse. However, because they are living organisms and can be killed during the dispersal process there are limitations to the methods that can be used. Unlike chemical or toxin agents, pathogens can also be stored and dispersed via infected vectors (e.g., mosquitoes, fleas). For more information on methods of dispersal, see appendix 1. In most cases, large-scale attacks will be clandestine and only detected through epidemiological analysis of resulting disease patterns. Even in the case of small-scale incidents or attacks directed at specific individuals (e.g., white-powder letters), without the inclusion of a threat the attack may go unrecognized until the disease appears in exposed individuals (e.g., the initial 2001 anthrax attack at American Media Inc., which claimed the life of Robert Stevens).

In general, unless a local reservoir (i.e., intermediate host that may or may not be affected by the virus) is established, pathogens are easily killed by unfavorable environmental factors such as fluctuations in temperature, humidity, food sources, or ultraviolet light. For this reason, their persistency is generally limited to days. However, pathogens can be freeze-dried (i.e., lyophilized) and remain in a preserved state almost indefinitely. Freeze-dried pathogens are reactivated when exposed to moisture.

DOI: 10.4324/9781003230564-9

Incubation times for diseases resulting from infection vary depending on the specific pathogen, but are generally on the order of days to weeks. Exposures to extremely high doses of some pathogens may reduce the incubation period to as short as several hours. The pathway of exposure (e.g., inhalation, ingestion) can also cause a significant change in the incubation time required as well as the clinical presentation of the disease.

Some diseases caused by viruses are communicable and easily transferred from an infected individual to anyone in close proximity. Typically, this occurs when the infected individual coughs or sneezes creating an infectious aerosol. These aerosols enter the body of a new host through inhalation and/or contact of the aerosol with the mucous membranes of the eyes, nose, or mouth. In addition, although intact skin is an effective barrier against most pathogens, abrasions, or lacerations circumvents this protective barrier and allows entry of the pathogen into the body.

Along with direct aerosol exposure, some viruses are potentially associated with exposure through ingestion of contaminated food and/or water. Pathogens that normally infect individuals through an ingestion pathway also often pose a significant risk of secondary infections from the fecal/oral cycle. Mechanical vectors, (e.g., flies, roaches) can also transport these types of pathogens to remote food items.

RESPONSE

PERSONAL PROTECTIVE REQUIREMENTS

Responding to the Scene of a Release

A number of conditions must be considered when selecting protective equipment for individuals at the scene of a release. For instances such as white-powder letters when the mechanism of release is known and it does not involve an aerosol generating device, then responders can use Level C with high-efficiency particulate air (HEPA) filters.

If an aerosol generating device is employed (e.g., sprayer), or the dissemination method is unknown and the release is ongoing, then responders should wear a Level A protective ensemble. Once the device has stopped generating the aerosol or has been rendered inoperable, and the aerosol has settled, then responders can downgrade to Level B.

In all cases, there is a significant hazard posed by contact of contaminated material with skin that has been cut or lacerated, or through injection of pathogens by contact with debris. Appropriate protection to avoid any potential abrasion, laceration, or puncture of the skin is essential. Individuals with damaged or open skin should not be allowed to enter the contaminated area.

Working with Infected Individuals

For most viruses, use infection control guidelines standard precautions. If appropriate, or the identity of the virus is unknown, use additional droplet and airborne precautions. Avoid direct contact with wounds or wound drainage.

Standard Precautions

Include hand hygiene; the use of gloves and eye protection; wearing a gown and mask as appropriate; and using care when handling sharps. If appropriate, have patients employ respiratory hygiene/cough etiquette practices (i.e., patients covering the mouth/nose with a tissue when coughing and prompt disposal of used tissues; using surgical masks on the coughing person when tolerated and appropriate; hand hygiene after contact with respiratory secretions; and spatial separation – ideally greater than 3 feet).

Contact Precautions

Isolate infected individuals in single-patient rooms. If this is not possible, ensure greater than 3 feet spatial separation between beds to reduce the opportunities for inadvertent sharing of items between patients.

Employ standard precaution and wear gowns. Protective garments are donned upon room entry and discarded before exiting the patient room.

Droplet Precautions

Isolate infected individuals in single-patient rooms. If this is not possible, ensure greater than 3 feet spatial separation with drawing curtain between patient beds.

Employ standard precaution and wear gowns. Wear a mask for close contact with infectious patient (i.e., within 3 feet of the patient). Protective garments are donned upon room entry and discarded before exiting the patient room.

Patients who must be transported outside of the room should wear a mask if tolerated and follow respiratory hygiene/cough etiquette practices (i.e., patients covering the mouth/nose with a tissue when coughing and prompt disposal of used tissues; hand hygiene after contact with respiratory secretions; and spatial separation – ideally greater than 3 feet).

Airborne Precautions

Isolate infected individuals in in an airborne infection isolation room with the door closed except during entry and exit. If such an isolation room is unavailable, mask the patient and place them in a private room with the door closed. Cohort patients who are presumed to have the same infection in areas of the facility that are away from other patients. Use portable exhaust fans to create a negative pressure environment in the converted area of the facility. Discharge air directly to the outside, away from people and air intakes, or direct exhaust through HEPA filters before it is introduced to other air spaces.

Employ standard precaution and wear gowns. Wear an N95 or higher-level mask or respirator. Masks and respirators must be fit-tested to each individual and users must perform a seal check each time they don the facepiece. Respiratory protection is donned prior to room entry. It is removed only after exiting the room and performing hand hygiene. Protective garments are donned upon room entry and discarded before exiting the patient room.

Patients who must be transported outside of the isolation area should wear a mask if tolerated and follow respiratory hygiene/cough etiquette practices (i.e., patients covering the mouth/nose with a tissue when coughing and prompt disposal of used tissues; hand hygiene after contact with respiratory secretions; and spatial separation – ideally greater than 3 feet).

ADDITIONAL HAZARDS

It is possible that local insects can become both a reservoir and a vector for the pathogen. Under these circumstances, the pathogen can survive well after the initial release and can rapidly spread beyond the immediately affected area. A host-vector cycle (i.e., an infected individual/animal/plant is fed on by multiple parasitic insects that then move on to infect other individuals/animals/plants) may be established that rapidly leads to vector amplification and an expediential expansion of the outbreak. In many cases, once a vector is infected, it is capable of transmitting the disease throughout its life span. Some pathogens are also transmitted directly to the young of the vector so that the next generation is born infected. Response activities must also include efforts to contain and eliminate these vectors (e.g., barriers to prevent exposures, application of pesticides).

DECONTAMINATION

Casualties/Personnel

Infected individuals

Unless the individual is reporting directly from the scene of an attack (e.g., white-powder letter, aerosol release, etc.), then decontamination is not necessary. Use standard protocols for individuals that may be infected with a communicable disease transmissible via an aerosol.

Direct Exposure

In the event that an individual is at the scene of a known or suspected attack (e.g., white-powder letter, aerosol release, etc.), have them wash their hands and face thoroughly with antimicrobial soap and water as soon as possible. If antimicrobial soap is not available, use any available soap or shampoo. They should also blow their nose to remove any agent particles that may have been captured by nasal mucous. If the release involved a powdered agent and it is practicable, dampen the agent with a water mist to help prevent aerosolization. Remove all clothing and seal in a plastic bag. To avoid further exposure of the head, neck, and face to the agent, cut off potentially contaminated clothing that must be pulled over the head. Shower using copious amounts of antimicrobial soap (if available) and water. Ideally, showers will be high volume with low pressure. Ensure that the hair has been washed and rinsed to remove potentially trapped agent. The CDC does not recommend that individuals use bleach or other disinfectants directly on their skin.

Animals

Unless the animals are at the scene of an attack then decontamination is not necessary.

Apply universal decontamination procedures using antimicrobial soap and water. If antimicrobial soap is not available, use any available soap, shampoo, or detergent. In some cases, severe infection may require euthanasia of animals and/or herds. Consult local/state veterinary assistance office. If the pathogen has not been identified, then wear a fitted N95 protective mask, eye protection, disposable protective coverall, disposable boot covers, and disposable gloves when dealing with infected animals.

Plants

May require removal and destruction of infected plants. Incineration of impacted fields may be required. Consult local/state agricultural assistance office. Some viruses are easily spread by mechanical vectors (e.g., farm implements, track out, running water) and extreme care must be taken to avoid further contamination. Wear disposable protective coveralls, disposable boot covers and disposable gloves to prevent spread of contamination. Many plant pathogens are spread via insects and in these situations response activities must also include efforts to contain and eliminate these vectors.

Food

All foodstuffs in the area of a release should be considered contaminated. Unopened items may be used after decontamination of the container. Opened or unpackaged items should be destroyed. Fruits and vegetables should be washed thoroughly with antimicrobial soap and water. Many pathogens can survive in food containers for extended periods.

Property

Surface Disinfectants

Compounds containing phenolics, chlorhexidine, potassium peroxymonosulfate, hypochlorites such as household bleach, and iodine/iodophors. Other disinfectants like alcohols and quaternary ammonium salts will work but may be less effective.

Although there is only limited data in the unclassified literature, preliminary studies indicate Sandia decontamination foam formulations containing high pH aqueous solutions of hydrogen peroxide, Canadian Aqueous System for Chemical-Biological Agent Decontamination (CASCAD) containing aqueous solutions of chloroisocyanurates and complexing agents, as well as Dahlgren decontamination formulations containing aqueous solutions of peracetyl borate and surfactants may be effective against viral agents, although with varying contact times.

The military also identifies the following nonstandard decontaminants: Detrochlorite (thickened bleach mixture of diatomaceous earth, anionic wetting agent, calcium hypochlorite, and water), 3%

aqueous peracetic acid solution, 1% aqueous hyamine solution, and 10% aqueous sodium or potassium hydroxide solution.

Large Area Fumigants

Gases, including formaldehyde, ozone, ethylene oxide, or chlorine dioxide are effective against many bacteria. However, these materials are highly toxic to humans and animals, and fumigation operations must be adequately controlled to prevent unnecessary exposure. Additional methods include vaporized hydrogen peroxide and an ionized hydrogen peroxide aerosol.

Fomites

Some pathogens may be absorbed into clothing or bedding causing these items to become infectious and capable of transmitting the disease. Others may contain vectors (e.g., lice, ticks) that pose a transmission hazard. Deposit items in an appropriate biological waste container and send to a medical waste disposal facility.

Alternatively, cotton or wool articles can be boiled in water for 30 minutes, autoclaved at 253°F for 45 minutes, or immersed in a 2% household-bleach solution (i.e., 1 liter of bleach in 2 liters of water) for 30 minutes followed by rinsing.

FATALITY MANAGEMENT

Unless the cadaver is coming directly from the scene of an attack (e.g., white-powder letter, aerosol release), then process the body according to established procedures for handling potentially infectious remains. Cadavers resulting from exposure to viral hemorrhagic fevers should be process with minimal handling and only by trained, essential personnel due to the risk of post-mortem transmission of the disease. These corpses should receive prompt burial or cremation.

Because of the nature of biological warfare agents, it is highly unlikely that a contaminated cadaver will be recovered from the scene of an attack unless it is from an individual who died from trauma or other complications while the attack was ongoing. If a fatality is grossly contaminated with a biological agent, wear disposable protective clothing with integral hood and booties, disposable gloves, eye protection, and an N95 respirator or powered air-purifying respirator (PAPR) equipped with N95 or high-efficiency particulate air (HEPA) filters.

Remove all clothing and personal effects. Items that will be retained for further processing should be double sealed in impermeable containers, ensuring that the inner container is decontaminated before placing it in the outer one. Otherwise, dispose of contaminated articles at an appropriate medical waste disposal facility.

Thoroughly wash the remains with antimicrobial soap and water. Pay particular attention to areas where agent may get trapped, such as hair, scalp, pubic areas, fingernails, folds of skin, and wounds. If deemed appropriate, the body can be washed with a 2% sodium hypochlorite bleach solution (i.e., 2 liters of water for every liter of household bleach), ensuring the solution is introduced into the ears, nostrils, mouth, and any wounds. This concentration of bleach will not affect remains but will disinfect the offending agent. Higher concentrations of bleach can harm remains. The bleach solution should remain on the cadaver for a minimum of 15 minutes. Wash with soap and water. If the body is to be embalmed (not recommended by the CDC), ensure that all the bleach solution is removed as it will react with embalming fluid.

If there is a potential that vectors may be involved, care must be taken to kill any vectors (e.g., lice, fleas) remaining either on the cadaver or residing in fomites. Remove all potentially infested clothing depositing it in a container that will trap and eliminate vectors. Dispose of contaminated articles at an appropriate medical waste disposal facility.

Once the remains have been thoroughly decontaminated, process the body according to established procedures for handling potentially infectious bodies. Use appropriate burial procedures.

The CDC has determined that the risks of occupational exposure to biological terrorism agents outweighs the advantages of embalming fatalities and recommends that bodies should be buried without embalming, unless death was due to a highly infectious virus (e.g., smallpox, hemorrhagic fever viruses). Fatalities due to a highly infectious virus should be cremated without embalming.

C18-A   AGENTS

C18-A001

African Horse Sickness
Reoviridae
ICD-11: —
UNII: —
EPPO: —

Endemic to Africa, although outbreaks have occurred in the Middle East. The natural reservoir is thought to be zebras, which are often asymptomatic. It can survive for months in urine, dried blood, feces, and serum at room temperature. It is resistant to common disinfectants and boiling for up to 15 minutes. Dogs can become ill after eating infected blood or meat; predominantly displaying the symptoms of the pulmonary form of the disease. This is a biosafety level 3 agent with special considerations for work involving infected arthropods.

It is on the Australia Group Core list, the U.S. Select Agents and Toxins list, and the OIE list of Notifiable Diseases for 2020.

In People:
Does not occur in humans.

In Domestic Animals:
Target species: Horses and other equids.

Communicability: Direct transmission does not occur.

Normal routes of exposure: Vectors (midges, possibly ticks and mosquitoes).

Secondary Hazards: Blood and body fluids; body tissue.

Incubation: 3–14 days.

Signs: Pulmonary form: high fever up to 106°F, followed by very marked and rapidly progressive respiratory failure including dilated nostrils and frothing from the nostrils and mouth, rapid breathing (tachypnea) exceeding 50 breaths per minute, difficulty breathing (dyspnea), spasmodic coughing. Expiration is frequently forced with the presence of abdominal heave lines. Affected animals tend to stand with forelegs spread apart, head extended, and nostrils dilated. There is profuse sweating and the eyes are red. Death is rapid.

Cardiac form: high fever up to 106°F followed by characteristic swelling of the depression above the eye (supraorbital fossae). There may be generalized severe swelling of the head and neck that can extend down to the chest. Small hemorrhages (petechia) may occur in the eyes and under the tongue. Animals become restless and may show signs of colic. Usually fatal within a week.

Post-mortem findings in the pulmonary form: trachea and bronchi filled with froth, edema of the lungs (distended and heavy), hydropericardium, pleural effusion, edema of thoracic lymph nodes, petechial hemorrhages in the pericardium. In the cardiac form: lungs are usually flaccid or slightly edematous, epicardial and endocardial ecchymoses, myocarditis, petechiae on the tongue and peritoneum, hemorrhagic gastritis, and subcutaneous and intramuscular yellow, gelatinous edema.

Suggested Alternatives for Differential Diagnosis: Anthrax, colic, encephalosis, infectious anemia, influenza, piroplasmosis, purpura hemorrhagica, rhinopneumonitis, trypanosomosis, viral arteritis. Field diagnosis may be virtually impossible.

Mortality Rate (untreated): ≤95% (horses), ≤50% (mules); ≤10% (donkeys).

C18-A002

African Swine Fever
Asfarviridae
ICD-11: —
UNII: —
EPPO: —

Endemic to Sub-Saharan Africa, but has spread to western and southern Russia, China, and Indochina. The natural reservoirs are African wild suidae (e.g., warthogs, bush pigs, giant forest hogs), which are often asymptomatic, and ticks (*Ornithodoros* spp.). Ticks remain infected for life. Infection is also transferred directly to tick eggs. It can survive for up to 6 months in uncooked pork products as well as extended periods in body fluids, blood, and feces. It is highly resistant to freeze/thaw cycles. This is a biosafety level 3-Ag agent with special considerations for work involving infected arthropods.

It is on the Australia Group Core list, the U.S. Select Agents and Toxins list, and the OIE list of Notifiable Diseases for 2020.

In People:
Does not occur in humans.

In Domestic Animals:
Target species: Pigs.

Communicability: Direct transmission is possible.

Normal routes of exposure: Ingestion; mucous membranes; vectors (ticks).

Secondary Hazards: Blood and body fluids; body tissue; fecal matter; fomites.

Incubation: 3–19 days.

Signs: In the peracute form there is sudden death of the infected pig. In the acute form, signs include high fever (up to 108°F), increased pulse and respiratory rate with vomiting, diarrhea (sometimes bloody), and eye discharges. On lighter-colored pigs, reddening of the skin may be visible, especially the tips of ears, tail, and distal extremities, as well as the ventral aspects of the chest and abdomen. Progresses to loss of appetite (anorexia), listlessness, blue coloration (cyanosis) and incoordination. Sows may abort. Infected animals that recover can become chronic carriers.

Post-mortem findings vary widely depending on the virus isolate and the species of pig, but are indistinguishable from death due to hog cholera. Findings include hemorrhage and blotchy cyanosis of the skin; fluid in the chest cavity (hydrothorax), in the sac surrounding the heart (hydropericardium), and in the abdomen (ascites); hemorrhage in the heart and of the serous membranes; enlarged and hemorrhagic gastrohepatic and renal lymph nodes; congestive enlarged spleen (splenomegaly); and small hemorrhages (petechia) on the kidneys.

Suggested Alternatives for Differential Diagnosis: Actinobacillosis, classical swine fever (hog cholera), eperythrozoonosis, erysipelas, *Haemophilus suis* infections, parvovirus, pasteurellosis, porcine dermatitis and nephropathy syndrome, porcine reproductive and respiratory syndrome, postweaning multisystemic wasting syndrome, pseudorabies, ruminant pestiviruse, salmonellosis, *Streptococcus suis* infection, thrombocytopenic purpura, poisonings from anticoagulants (e.g., warfarin) or salt.

Mortality Rate (untreated): ≤100%.

C18-A003

Andean Potato Latent Virus
Tymoviridae
ICD-11: —
UNII: —
EPPO: APLV00

Endemic to the Andean region of South America.

It is on the Australia Group Core list.

In Plants:

Target species: Potatoes.

Normal routes of transmission: Contact; vectors (beetles).

Secondary Hazards: Mechanical vectors (grafting); crop debris (tubers.

Signs: Initially, local yellow (chlorotic) lesions develop on leaves that progress to a systemic severe mosaic with necrotic flecking, curling and leaf-tip necrosis.

Crop Losses: Unknown

C18-A004

Andes Virus
Bunyaviridae
Hantavirus Pulmonary Syndrome
ICD-11: XN4AP
UNII: —
EPPO: —

Endemic to Chile and Argentina. The natural reservoir is the long-tailed pygmy rice rat (*Oligoryzomys longicaudatus*). Rodents can shed hantaviruses in saliva, feces, and urine. It can survive in the environment for up to several weeks at room temperature in shaded areas with high humidity. Dried virus is inactivated within 24 hours at room temperature. Andes virus is the only hantavirus that is directly transmitted from person to person. Virus has been found in respiratory secretions and urine of infected individuals. This is a biosafety level 3 agent.
It is on the Australia Group Core list.

In People:

CDC Case Definition for hantavirus pulmonary syndrome: Clinical criteria is an acute febrile illness (i.e., temperature greater than 101°F) with a prodrome consisting of fever, chills, myalgia, headache, and gastrointestinal symptoms, and one or more of the following clinical features:
1) bilateral diffuse interstitial edema; OR 2) clinical diagnosis of acute respiratory distress syndrome

(ARDS), or 3) radiographic evidence of noncardiogenic pulmonary edema; OR 4) an unexplained respiratory illness resulting in death, and includes an autopsy examination demonstrating noncardiogenic pulmonary edema without an identifiable cause; OR 5) healthcare record with a diagnosis of hantavirus pulmonary syndrome; OR 6) death certificate lists hantavirus pulmonary syndrome as a cause of death or a significant condition contributing to death.

Laboratory criteria for diagnosis is one or more of the following laboratory findings: 1) detection of hantavirus-specific immunoglobulin M or rising titers of hantavirus-specific immunoglobulin G; OR 2) detection of hantavirus-specific ribonucleic acid in clinical specimens; OR detection of hantavirus antigen by immunohistochemistry in lung biopsy or autopsy tissues.

Communicability: Direct person-to-person transmission is possible. When dealing with infected individuals, use contact and droplet precautions.

Normal Routes of Exposure: Inhalation; mucous membranes; v skin.

Infectious Dose: Unknown.

Secondary Hazards: Aerosols; fecal matter (rat); blood and body fluids (rat); fomites (from rat habitation).

Incubation: 7–49 days.

Signs & Symptoms: Flu-like with fever, headache, muscle aches (myalgia), nausea, vomiting, and diarrhea with sudden progression to cough, rapid breathing (tachypnea), rapid heart rate (tachycardia), followed by acute pulmonary edema with difficulty breathing (dyspnea). As the disease progresses, there may be a decrease in blood pressure (hypotension) and blood volume (hypovolemia), followed by cardiogenic shock. Andes virus also causes more incidents of renal failure than other New World hantaviruses.

Suggested Alternatives for Differential Diagnosis: Acute respiratory distress syndrome (ARDS), anthrax, chlamydophilial infections, community acquired pneumonia, congestive heart failure and pulmonary edema, cytomegalovirus, dengue fever, ehrlichiosis, Epstein-Barr virus, hepatitis, influenza, legionnaires disease, Lyme disease, malaria, mycoplasmas infections, pneumonia, pneumonic plague, pulmonary embolism, Q fever, shock, silent myocardial infarction, tularemia, poisonings from phosphine, phosgene, and salicylate toxicity with pulmonary edema.
Mortality Rate (untreated): ≤50%.

C18-A005

Aujeszky's Disease
Herpesviridae
ICD-11: —
UNII: —
EPPO: —

Endemic worldwide. Several countries have eliminated the disease from domestic animals but it remains present in feral pigs and wild boar. The natural reservoir is pigs. It can infect a wide range of other animals including cattle, sheep, goats, cats, dogs, and most mammals. It can persist for up to 7 hours as an aerosol with a relative humidity of greater than 55% and be spread by the wind for up to 2 kilometers. It can survive for up to 7 hours in non-chlorinated well water; 2 days in anaerobic lagoon effluent; for 2 days in green grass, soil, feces, and shelled corn; for 3 days in nasal washings on plastic and pelleted hog feed; and for 4 days in straw bedding. It can survive up to 5 weeks in pig tissue. This is a biosafety level 2 agent.

It is on the Australia Group Core list and the OIE list of Notifiable Diseases for 2020.

In People:

No documented infections in humans.

In Domestic Animals:

Target species: Pigs, cattle, sheep, goats.

Communicability: Direct transmission is possible.

Normal routes of exposure: Mucous membranes; inhalation; ingestion.

Secondary Hazards: Aerosols (respiratory secretions); body fluids (including milk and semen); body tissue; fecal matter; fomites.

Incubation: 2–6 days pigs; ≤ 9 cattle; 2–10 days dogs and cats.

Signs: In some cases, sudden death may occur with no apparent signs of illness. Piglets, fever, loss of appetite (anorexia), trembling, incoordination (ataxia), rapid involuntary movements of the eyes (nystagmus), backward bridging position (opisthotonos), seizures, paralysis.

Older pigs, fever, loss of appetite (anorexia), weight loss, eye infection (conjunctivitis), coughing, sneezing, nasal infection (rhinitis), nasal discharge, and difficulty breathing (dyspnea). May progress to pneumonia, mild muscle tremors, convulsions.

Other animals, suffer an intense itch (pruritus) concentrated in a patch of skin and will lick, rub or chew at that spot. Fever, incoordination (ataxia), convulsions, excessive and abnormal bellowing or howling, difficulty breathing (dyspnea), panting (tachypnea), difficulty swallowing, excessive salivation or drooling (ptyalism), spasms, paralysis. Usually die within a few days of exposure.

Post-mortem often does not present any characteristic gross findings. In pigs, there may be serous or fibrinonecrotic rhinitis, necrotic tonsillitis, hemorrhagic pulmonary lymph nodes, pulmonary edema, with lesions in the cerebrum and cerebellum. In cattle, there may be areas of edema, congestion and hemorrhage in the spinal cord. Gross lesions are often minimal or absent in dogs.

Suggested Alternatives for Differential Diagnosis: African swine fever, classical swine fever, congenital tremor, erysipela, hemagglutination encephalomyelitis virus, Nipah virus, Porcine polioencephalomyelitis, rabies, streptococcal meningoencephalitis, swine influenza, hypoglycemia, poisonings from organic arsenic, organic mercury or salt.

Mortality Rate (untreated): ≤100% (piglets); ≤50% (nursery pigs); ≤2% (adult pigs); ≤100% (cattle, sheep, dog, cats).

C18-A006

Highly Pathogenic Avian Influenza (Agent OE)

Orthomyxoviridae

ICD-11: XN4TT

UNII: R9HH0NDE2E

EPPO: —

1918 Influenza (H1N1)

Orthomyxoviridae

ICD-11: XN297

UNII: R9HH0NDE2E

EPPO: —

Normal variations of influenza virus are endemic worldwide. The natural reservoir is wild birds (avian) – particularly gulls, terns, shorebirds, and waterfowl – and pigs (swine). The avian disease is highly lethal to domestic poultry with a mortality rate ≤60%. Pigs show signs of infection, but it is rarely fatal. Pigs can be infected by the avian strains to form chimera variants that pose a significantly greater risk to people. Infected birds can shed the virus in their saliva, nasal secretions, and feces. At room temperature, avian virus can remain viable for up to 5 days in dry feces and up to 40 days in wet feces. The virus can survive for on hard surfaces for up to 9 hours and up to 4 hours on porous surfaces. This is a biosafety level 3-Ag agent, with the addition of HEPA filtration attached to the laboratory exhaust system.

It is on the Australia Group Core list, the U.S. Select Agents and Toxins list, and the OIE list of Notifiable Diseases for 2020.

In People:

CDC Case Definition for novel influenza A virus: Clinical criteria is an illness compatible with influenza virus infection (fever >100 degrees Fahrenheit, with cough and/or sore throat).

Laboratory criteria for diagnosis is a human case of infection with an influenza A virus subtype that is different from currently circulating human influenza H1 and H3 viruses. Novel subtypes include, but are not limited to, H2, H5, H7, and H9 subtypes. Influenza H1 and H3 subtypes originating from a non-human species or from genetic reassortment between animal and human viruses are also novel subtypes. Novel subtypes will be detected with methods available for detection of currently circulating human influenza viruses at state public health laboratories (e.g., real-time reverse transcriptase polymerase chain reaction [RT-PCR]). Confirmation that an influenza A virus represents a novel virus will be performed by CDC's influenza laboratory. Once a novel virus has been identified by CDC, confirmation may be made by public health laboratories following CDC-approved protocols for that specific virus, or by laboratories using an FDA-authorized test specific for detection of that novel influenza virus.

Communicability: Direct person-to-person transmission is possible. Aerosol transmission may occur a day before the appearance of symptoms. When dealing with infected individuals, use contact and droplet precautions. Consider use of a negative pressure, HEPA-filtered respirator or positive air-purifying respirator (PAPR).

Normal Routes of Exposure: Inhalation; mucous membranes.

Infectious Dose: 3 viral units (inhalation).

Secondary Hazards: Aerosols (respiratory secretions, dust); fomites (human and animal).

Incubation: 1 to 4 days.

Signs & Symptoms: Cannot be diagnosed by clinical evaluation. Signs and symptoms included fever, nonproductive cough, sore throat, muscle pain (myalgia), headache, eye infection (pink eye – conjunctivitis), and nasal discharge (rhinorrhea); sometimes accompanied by nausea, abdominal pain, vomiting, diarrhea (watery, non-bloody), shortness of breath (tachypnea), difficulty breathing (dyspnea), pneumonia, acute respiratory distress, and respiratory failure. May progress to altered mental status, incoordination (ataxia), coma, seizure.

Suggested Alternatives for Differential Diagnosis: Adenoviruses, arenaviruses, bacterial sepsis, California encephalitis, community-acquired pneumonia, coxsackieviruses, cytomegalovirus, dengue fever, eastern equine encephalitis, echoviruses, hantavirus pulmonary syndrome, infectious mononucleosis, influenza, Japanese encephalitis, Lyme disease, meningitis, Middle East respiratory syndrome (MERS), parainfluenza virus, pneumococcal infections, rhinoviruses, severe acute respiratory syndrome (SARS), St. Louis encephalitis, upper respiratory infection, Venezuelan encephalitis, West Nile encephalitis.

Mortality Rate (untreated): ≤60%.

In Domestic Animals:

Susceptible species: Poultry.

Communicability: Direct transmission is possible.

Normal routes of exposure: Ingestion; inhalation; mucous membranes.

Secondary Hazards: Blood and body fluids; fecal matter; fomites; mechanical vectors (track out, flies). *Incubation:* hours–7 days.

Signs: Varies based on species. In some cases, sudden death may occur with no apparent signs of illness. Otherwise, ruffled feathers, depression, lack of energy (lethargy), loss of appetite (anorexia), watery greenish diarrhea, sinus infection (sinusitis), coughing, sneezing, blood-tinged oral and nasal discharges, tearing (lacrimation), cloudiness (opacity) of the cornea, swelling (edema) and blue coloring (cyanosis) of the unfeathered skin on the head, comb and wattle/snood, swelling (edema) and red discoloration (ecchymoses) on the shanks and feet, sideways twitching of the neck (torticollis), backward bridging position (opisthotonos), incoordination (ataxia), paralysis, and drooping wings. Egg production decreases or stops, and depigmented, deformed or shell-less eggs may be produced.

Post-mortem findings vary depending the bird species but include edema, hemorrhage, congestion and cyanosis of the head, wattle and comb; congestion, swelling, and hemorrhages of the conjunctivae; edema and diffuse subcutaneous hemorrhages on the feet and shanks; fluid (possibly blood-stained) in the nares and oral cavity; mucoid exudate with possible hemorrhage in the trachea; petechiae on the viscera and in the muscles; kidneys can be severely congested and may be plugged with urate deposits; and splenomegaly with parenchymal mottling.

Suggested Alternatives for Differential Diagnosis: Aspergillosis, avian pneumovirus, chlamydiosis, duck viral enteritis, *E. coli* infections, fowl cholera, infectious bronchitis, infectious coryza, infectious laryngotracheitis, mycoplasmosis, Newcastle disease, ornithobacteriosis, turkey coryza, heat exhaustion, severe water deprivation.

Mortality Rate (untreated): ≤100%, varies with species.

C18-A007

Banana Bunchy Top Virus
Nanoviridae
ICD-11: —
UNII: —
EPPO: BBTV00

Endemic to Africa, Asia, Australia, and South Pacific islands. The virus is transmitted by the banana aphid, Pentalonia nigronervosa, which has worldwide distribution. Once infected, aphids remain infected through their adult life.

It is on the Australia Group Awareness list.

In Plants:

Target species: Bananas.

Normal routes of transmission: Vectors (aphids).

Secondary Hazards: Mechanical vectors (planting, harvesting); crop debris.

Signs: Signs are distinctive and include darker green streaks in a pattern of dots and dashes (Morse code streaking) that form hooks where they enter the edge of the leaf midrib. Post infection leaves are smaller, both in length and in width and often have yellowed (chlorotic), upturned margins. Leaves become dry and brittle and stand more erect than normal giving the plant a rosetted and bunchy-top appearance. Plants rarely produce a fruit bunch after infection.

Crop Losses: ≤100%

C18-A008

Bluetongue

Reoviridae

ICD-11: —

UNII: —

EPPO: —

Endemic to Africa, Europe, the Americas, Australia, and parts of Asia. The natural reservoirs are cattle and midges, which infect each other in a cycle. Midges remain infective for life. Affect cattle and goats are typically asymptomatic and rarely diagnosed. However, calves may be aborted or born with fatal brain damage. Can also infect deer, where mortality rates can reach 90%. It can survive for years in blood stored at 68°F. This is a biosafety level 3 agent with special considerations for work involving infected arthropods.

It is on the Australia Group Core list and the OIE list of Notifiable Diseases for 2020.

In People:

Does not occur in humans.

In Domestic Animals:

Target species: Sheep.

Communicability: Direct transmission does not occur.

Normal routes of exposure: Vectors (midges).

Secondary Hazards: Blood; vector amplification.

Incubation: 4–14 days.

Signs: High fever (up to 108°F) followed by swelling (edema) of lips, tongue, and jaw. There is excessive salivation and nasal discharge, becoming a combination of mucus and pus (mucopurulent) or even bloody. Animals have difficulty breathing (dyspnea) and may die from asphyxiation. Ulceration develops where the teeth come in contact with lips and tongue. Small hemorrhages (petechia) can be seen on the mucous membranes of the nose and mouth. Sheep eat less because of oral soreness and will hold food in their mouths to soften before chewing. Some affected sheep may develop a severely swollen tongue, which may infrequently become blue (cyanotic) and even protrude from the mouth. There is inflammation in the band just above the hoof (coronary band) that may develop a purple-red color. Animals become lame.

Post-mortem findings include congestion, edema, hemorrhages, and ulcerations of the mouth, esophagus, stomach, and intestine. Petechia, ecchymoses, or hemorrhages in the wall of the base of the pulmonary artery and focal necrosis of the papillary muscle of the left ventricle. Subcutaneous and intermuscular edema and hemorrhages, skeletal myonecrosis, myocardial and intestinal hemorrhages, hydrothorax, hydropericardium, pericarditis, splenomegaly, and pneumonia.

Suggested Alternatives for Differential Diagnosis: Akabane disease, bovine viral diarrhea, coe-nurosis, contagious ecthyma, epizootic hemorrhagic disease of deer, foot abscesses, foot-and-mouth disease, footrot, infectious bovine rhinotracheitis, laminitis, lungworm infestation and pneumonia, malignant catarrhal fever, muscular dystrophy in lambs, nasal botfly infestation, parainfluenza-3 infection, peste des petits ruminants, photosensitization, pneumonia, polyarthritis, sheep pox, vesicular stomatitis, white muscle disease, poisonings from toxic plants.

Mortality Rate (untreated): ≤70% (sheep), ≤1% (cattle).

C18-A009

Chapare Hemorrhagic Fever
Arenaviridae
ICD-11: XN2WG
UNII: —
EPPO: —

Endemic to Bolivia. The natural reservoir is unknown but suspected to be rodents. It is likely that the virus is shed in their urine. This is a biosafety level 4 agent.

It is on the Australia Group Core list and the U.S. Select Agents and Toxins list.

In People:

CDC Case Definition: None established.

Communicability: Direct person-to-person transmission is unknown, but may be possible. When dealing with infected individuals, use contact, and droplet precautions.

Normal Routes of Exposure: Unknown, but believed to be inhalation; abraded skin; mucous membranes.

Infectious Dose: Unknown.

Secondary Hazards: Unknown, but believed to be aerosols; blood and body fluids; body tissue; fecal matter (rodent); fomites (from rodent habitation).

Incubation: Unknown.

Signs & Symptoms: Based on limited cases, fever, headache, joint pain (arthralgia), muscle pain (myalgia) and vomiting with rapid progression to shock, multiple hemorrhagic signs.

Suggested Alternatives for Differential Diagnosis: Acute anemia, acute leukemias, anaphylactoid purpura, chikungunya, Colorado tick fever, dengue fever, disseminated intravascular coagulation, encephalitis, epidemic typhus, falciparum malaria, hemolytic uremic syndrome, Henoch-Schonlein purpura, hepatitis, herpes, infectious mononucleosis, influenza, leptospirosis, malaria, measles, meningitis, meningococ-cemia, plague, pleural effusion, Q fever, Rocky Mountain spotted fever, rubella, sepsis, shigellosis, Sindbis fever, thrombocytopenic purpura, tick bite fever, trypanosomiasis, typhoid fever, yellow fever, other New World viral hemorrhagic fevers, drug eruptions, allergies, food poisoning.

Rate (untreated): ≤30%.

C18-A010

Chikungunya
Togaviridae
ICD-11: 1D40
UNII: NGU16C4NV2
EPPO: —

Endemic to African, Southern Asia, Southeast Asia, parts of the Mediterranean, and the Caribbean. The natural reservoirs are monkeys, rodents, bats, and birds. Humans are the primary host during epidemic periods. Mosquitoes remains infective for life. This is a biosafety level 3 agent, with the addition of HEPA filtration attached to the laboratory exhaust system.

It is on the Australia Group Core list.

In People:

CDC Case Definition: Clinical criteria for cases of *Neuroinvasive disease* is defined as follows: 1) meningitis, encephalitis, acute flaccid paralysis, or other acute signs of central or peripheral neurologic dysfunction; AND 2) absence of a more likely clinical explanation.

Clinical criteria for cases of *non-neuroinvasive disease* is defined as follows: 1) fever (chills) as reported by the patient or a health-care provider; AND 2) absence of neuroinvasive disease; AND 3) absence of a more likely clinical explanation. Other clinically compatible symptoms include: headache, myalgia, rash, arthralgia, vertigo, vomiting, paresis and/ or nuchal rigidity.

Laboratory criteria for diagnosis is isolation of virus from, or demonstration of specific viral antigen or nucleic acid in, tissue, blood, CSF, or other body fluid, OR 1) fourfold or greater change in virus-specific quantitative antibody titers in paired sera, OR 2) virus-specific IgM antibodies in serum with confirmatory virus-specific neutralizing antibodies in the same or a later specimen, OR 3) virus-specific IgM antibodies in CSF or serum.

Criteria for case confirmation for *Neuroinvasive disease* consists of meeting the clinical criteria and one or more of the following laboratory criteria: 1) isolation of virus from, or demonstration of specific viral antigen or nucleic acid in, tissue, blood, CSF, or other body fluid, OR 2) fourfold or greater change in virus-specific quantitative antibody titers in paired sera, OR 3) virus-specific IgM antibodies in serum with confirmatory virus-specific neutralizing antibodies in the same or a later specimen, OR 4) virus-specific IgM antibodies in CSF, with or without a reported pleocytosis, and a negative result for other IgM antibodies in CSF for arboviruses endemic to the region where exposure occurred.

Criteria for case confirmation for *non-neuroinvasive disease* consists of meeting the clinical criteria and one or more of the following laboratory criteria: 1) isolation of virus from, or demonstration of specific viral antigen or nucleic acid in, tissue, blood, or other body fluid, excluding CSF, OR 2) fourfold or greater change in virus-specific quantitative antibody titers in paired sera, OR 3) virus-specific IgM antibodies in serum with confirmatory virus-specific neutralizing antibodies in the same or a later specimen.

Communicability: Direct person-to-person transmission does not normally occur, although rare in utero transmission has been reported. When dealing with infected individuals, use standard precautions and protect from vectors.

Normal Routes of Exposure: Vectors (mosquitoes).

Infectious Dose: Unknown.

Secondary Hazards: Vector amplification; blood; aerosols (blood)

Incubation: 1–12 days.

Signs & Symptoms: Sudden high fever with chills, eye infection (pink eye – conjunctivitis), sensitivity to light (photophobia), headache, nausea, fatigue and joint (arthralgia) and muscle (myalgia) pain. Joint symptoms are typically severe and can be debilitating. Skin may become red (erythema) or develop a rash of mixed flat and raised (maculopapular) spots. There may be bleeding of the gums and nasal mucous membranes. The fever subsides but then returns with the joint and muscle pain worsening. After the second round of fever has abated, the pain may still persist for weeks or months.

Suggested Alternatives for Differential Diagnosis: Adenovirus, Barmah Forest fever, Bussuquara fever, Crimean-Congo fever, dengue fever, Ebola, enteroviruses, epidemic typhus, gonococcemia, group A Streptococcus, hantavirus infection, hepatitis, herpes, infectious mononucleosis, influenza, Kyasanur forest disease, Lassa fever, leptospirosis, malaria, Mayaro fever, measles, mumps, O'nyong-nyong fever, parvovirus, postinfectious arthritis, Q fever, rheumatologic conditions, Rocky Mountain spotted fever, Ross River fever, rubella, Sindbis fever, tick bite fever, trypanosomiasis, typhoid fever, West Nile fever, Zika. *Mortality Rate (untreated):* <1%.

C18-A011
Choclo Virus
Bunyaviridae
Hantavirus Pulmonary Syndrome
ICD-11: —
UNII: —
EPPO: —

Endemic to Panama. The natural reservoir is the pygmy rice rat (*Oligoryzomys fulvescens*). Does not produce disease In Domestic Animals. Rodents can shed hantaviruses in saliva, feces and urine. This is a biosafety level 3 agent.

It is on the Australia Group Core list.

In People:
CDC Case Definition for hantavirus pulmonary syndrome: Clinical criteria is an acute febrile illness (i.e., temperature greater than 101°F) with a prodrome consisting of fever, chills, myalgia, headache, and gastrointestinal symptoms, and one or more of the following clinical features: 1) bilateral diffuse interstitial edema; OR 2) clinical diagnosis of acute respiratory distress syndrome (ARDS), or 3) radiographic evidence of noncardiogenic pulmonary edema; OR 4) an unexplained respiratory illness resulting in death, and includes an autopsy examination demonstrating noncardiogenic pulmonary edema without an identifiable cause; OR 5) healthcare record with a diagnosis of hantavirus pulmonary syndrome; OR 6) death certificate lists hantavirus pulmonary syndrome as a cause of death or a significant condition contributing to death.

Laboratory criteria for diagnosis is one or more of the following laboratory findings: 1) detection of hantavirus-specific immunoglobulin M or rising titers of hantavirus-specific immunoglobulin G; OR 2) detection of hantavirus-specific ribonucleic acid in clinical specimens; OR detection of hantavirus antigen by immunohistochemistry in lung biopsy or autopsy tissues.

Communicability: Direct person-to-person transmission does not occur. When dealing with infected individuals, use standard precautions.

Normal Routes of Exposure: Inhalation; mucous membranes; abraded skin.

Infectious Dose: Unknown.

Secondary Hazards: Aerosols; fecal matter (mouse); blood and body fluids; fomites (from mouse habitation).

Incubation: Unknown.

Signs & Symptoms: Flu-like with fever, headache, muscle aches (myalgia), nausea, vomiting, and diarrhea with sudden progression to cough, rapid breathing (tachypnea), rapid heart rate (tachycardia) followed by acute pulmonary edema with difficulty breathing (dyspnea). As the disease progresses, there may be a decrease in blood pressure (hypotension) and blood volume (hypovolemia), followed by cardiogenic shock.

Suggested Alternatives for Differential Diagnosis: Acute respiratory distress syndrome (ARDS), anthrax, chlamydophilial infections, community acquired pneumonia, congestive heart failure and pulmonary edema, cytomegalovirus, dengue fever, ehrlichiosis, Epstein-Barr virus, hepatitis, influenza, legionnaires disease, Lyme disease, malaria, mycoplasmas infections, pneumonia, pneumonic plague, pulmonary embolism, Q fever, shock, silent myocardial infarction, tularemia, poisonings from phosphine, phosgene, and salicylate toxicity with pulmonary edema.

Rate (untreated): ≤30%.

C18-A012

Classic Swine Fever

Flaviviridae

ICD-11: —

UNII: —

EPPO: —

Endemic to Central and South America, the Caribbean, Europe, and Asia. The natural reservoir is pigs and wild boar. It can survive for months in in refrigerated meat and for years in frozen meat. While it is sensitive to drying (desiccation), it is not affected by mild forms of curing. This is a biosafety level 3-Ag agent. Laboratory workers should have no contact with susceptible hosts for 5 days after working with the agent.

It is on the Australia Group Core list, the U.S. Select Agents and Toxins list, and the OIE list of Notifiable Diseases for 2020.

In People:

Does not occur in humans.

In Domestic Animals:

Target species: Pigs.

Communicability: Direct transmission is possible.

Normal routes of exposure: Ingestion; mucous membranes; abraded skin.

Secondary Hazards: Blood and body fluids (including semen); body tissue; fomites; fecal matter; mechanical Vectors (track out, insects).

Incubation: 2–14 days.

Signs: High fever, vomiting, constipation, severe depression and loss of appetite (anorexia). There are multiple superficial hemorrhages, particularly on the ears, lower abdomen, and legs, that result in a purple discoloration (cyanosis) of the skin. Eyes become infected (pink eye – conjunctivitis) producing a discharge that may ultimately result in encrustation of the eyelids to the point where they are completely adhered. Initial constipation is followed by yellowish diarrhea and vomit. Animals exhibit incoordination (ataxia) with a staggering gait and goose stepping, which may progress to posterior paralysis.

Post-mortem findings include widespread petechial and ecchymotic hemorrhages, especially in lymph nodes, tonsils, larynx, kidneys, spleen, urinary bladder, and ileum. Infarction may be seen, particularly in the periphery of the spleen. There are ulcers and necrosis in the intestines. There is an accumulation of straw-colored fluid in the thoracic cavity, which may also be present in the peritoneal cavity and pericardial sac. Encephalitis is common; pneumonia may also be seen in some cases.

Suggested Alternatives for Differential Diagnosis: Actinobacillosis, African swine fever, Aujeszky disease, border disease, eperythrozoonosis, erysipelas, hemolytic disease of the newborn, *Hemophilus suis,* parvovirus, pasteurellosis, pneumonia, porcine dermatitis and nephropathy syndrome, porcine reproductive and respiratory syndrome, postweaning multisystemic wasting syndrome, pseudorabies, salmonellosis, thrombocytopenic purpura, poisonings from anticoagulants (e.g., warfarin) or salts.

Mortality Rate (untreated): ≤100%.

C18-A013

Crimean-Congo Hemorrhagic Fever
Bunyaviridae
ICD-11: 1D49
UNII: —
EPPO: —

Endemic to the Middle East, Turkey, Asia, Africa, and southern Europe. The natural reservoir is ticks. Herbivores, which show no clinical symptoms, act as amplifying hosts. This is a biosafety level 4 agent.

It is on the Australia Group Core list, the U.S. Select Agents and Toxins list, and the OIE list of Notifiable Diseases for 2020.

In People:

CDC Case Definition for viral hemorrhagic fever: Clinical criteria is an illness with acute onset with ALL of the following clinical findings: 1) a fever >104°F; and 2) one or more of the following clinical findings: a) severe headache; OR b) muscle pain; OR c) erythematous maculopapular rash on the trunk with fine desquamation 3–4 days after rash onset; OR d) vomiting; OR e) diarrhea; OR f) abdominal pain; OR g) bleeding not related to injury; OR h) thrombocytopenia.

Laboratory criteria for diagnosis is one or more of the following laboratory findings: 1) detection of viral hemorrhagic fever (VHF) viral antigens in blood by enzyme-linked Immunosorbent Assay (ELISA) antigen detection; 2) VHF viral isolation in cell culture for blood or tissues; 3) detection of VHF-specific genetic sequence by Reverse Transcription-Polymerase Chain Reaction (RT-PCR) from blood or tissues; 4) detection of VHF viral antigens in tissues by immunohistochemistry.

Communicability: Direct person-to-person transmission is possible. It is highly communicable in a hospital setting. Post-mortem transmission is a serious risk. When dealing with infected individuals, use contact and droplet precautions. For procedures that could generate infectious aerosols, use special air handling and ventilation systems, and N95 or better respirators.

Normal Routes of Exposure: Inhalation; mucous membranes; abraded skin; ingestion; vectors (ticks).

Infectious Dose: Unknown.

Secondary Hazards: Aerosols (blood and body fluids; rodent fecal matter); blood and body fluids; body tissue; fecal matter (rodent); fomites (with vectors).

Incubation: 1–13 days.

Signs & Symptoms: Sudden onset of fever, fatigue, chills, headache, pain in the muscles (myalgia) and joints (arthralgias), back and abdominal pain, nausea, vomiting, and diarrhea. Appearance of red and sore throat, small hemorrhages (petechia) on the palate, red eyes (conjunctivitis), sensitivity to light (photophobia), and flushed face (erythema) with swelling (edema). Symptoms may also include jaundice, and in severe cases, changes in mood and sensory perception. A rash from small hemorrhages (petechia) appears on the chest and stomach, spreading to the rest of the body. May progress to hemorrhagic symptoms including large areas of severe bruising, bleeding from all body openings, bloody diarrhea and urine, and uncontrolled bleeding where the skin is broken.

Suggested Alternatives for Differential Diagnosis: Acute anemia, acute leukemias, anaphylactoid purpura, chikungunya, Colorado tick fever, dengue fever, disseminated intravascular coagulation, encephalitis, epidemic typhus, falciparum malaria, hemolytic uremic syndrome, Henoch-Schonlein purpura, hepatitis, herpes, infectious mononucleosis, influenza, leptospirosis, malaria, measles, meningitis, meningococcemia, plague, pleural effusion, Q fever, Rocky Mountain spotted fever, rubella, sepsis, shigellosis, Sindbis fever, thrombocytopenic purpura, tick bite fever, trypanoso-miasis, typhoid fever, yellow fever, other New World viral hemorrhagic fevers, drug eruptions, allergies, food poisoning.

Mortality Rate (untreated): ≤50%.

C18-A014

Dengue Fever
Flaviviridae
ICD-11: 1D2Z
UNII: —
EPPO: —

Endemic throughout the tropics and subtropics worldwide. There are multiple strains of the virus, each able to cause disease in humans. Subsequent infection with a different strain increases the risk of more serious symptoms during the follow-on course of the disease. The natural reservoir is humans, primates, and mosquitoes (*Aedes* spp). In some cases, infection is also transferred directly to mosquito eggs. Human-to-mosquito transmission can occur up to 2 days before someone shows symptoms of the illness, up to 2 days after the fever has resolved. The virus is stable in dried blood for up to 9 weeks at room temperature. This is a biosafety level 2 agent.

In People:

CDC Case Definition: Clinical criteria *for dengue* is an illness with fever as reported by the patient or healthcare provider and the presence of one or more of the following signs and symptoms: 1) nausea/vomiting; OR 2) rash; OR 3) aches and pains (e.g., headache, retro-orbital pain, joint pain, myalgia, arthralgia); OR 4) tourniquet test positive; OR 5) leukopenia (a total white blood cell count of <5,000/mm^3); OR 6) any warning sign for severe dengue including abdominal pain or tenderness, persistent vomiting, extravascular fluid accumulation (e.g., pleural or pericardial effusion, ascites), mucosal bleeding at any site, liver enlargement >2 cm, or increasing hematocrit concurrent with rapid decrease in platelet count.

Clinical criteria *for severe dengue* is dengue with any one or more of the following scenarios: 1) severe plasma leakage evidenced by hypovolemic shock and/or extravascular fluid accumulation (e.g., pleural or pericardial effusion, ascites) with respiratory distress. A high hematocrit value for patient age and sex offers further evidence of plasma leakage; OR 2) severe bleeding from the gastrointestinal tract (e.g., hematemesis, melena) or vagina (menorrhagia) as defined by requirement for medical intervention including intravenous fluid resuscitation or blood transfusion; OR 3) severe organ involvement, including any of the following: elevated liver transaminases: aspartate aminotransferase (AST) or alanine aminotransferase (ALT) ≥1,000 per liter (U/L), or impaired level of consciousness and/or diagnosis of encephalitis, encephalopathy, or meningitis, or heart or other organ involvement including myocarditis, cholecystitis, and pancreatitis.

Laboratory criteria for diagnosis is one or more of the following laboratory findings: 1) detection of DENV nucleic acid in serum, plasma, blood, cerebrospinal fluid (CSF), other body fluid or tissue by validated reverse transcriptase-polymerase chain reaction (PCR), OR 2) detection of DENV antigens in tissue by a validated immunofluorescence or immunohistochemistry assay, OR 3) detection in serum or plasma of DENV NS1 antigen by a validated immunoassay; OR 4) cell culture isolation of DENV from a serum, plasma, or CSF specimen; OR 5) detection of IgM anti-DENV by validated immunoassay in a serum specimen or CSF in a person living in a dengue endemic or non-endemic area of the United States without evidence of other flavivirus transmission (e.g., WNV, SLEV, or recent vaccination against a flavivirus (e.g., YFV, JEV)); OR 6) detection of IgM anti-DENV in a serum specimen or CSF by validated immunoassay in a traveler returning from a dengue endemic area without ongoing transmission of another flavivirus (e.g., WNV, JEV, YFV), clinical evidence of co-infection with one of these flaviviruses, or recent vaccination against a flavivirus (e.g., YFV, JEV); OR 7) IgM anti-DENV seroconversion by validated immunoassay in acute (i.e., collected <5 days of illness onset) and convalescent (i.e., collected >5 days after illness onset) serum specimens; OR 8) IgG anti-DENV seroconversion or ≥fourfold rise in titer by a validated immunoassay in serum specimens collected >2 weeks apart, and confirmed by a neutralization test (e.g., plaque reduction neutralization test) with a >fourfold higher end point titer as compared to other flaviviruses tested.

Communicability: Direct person-to-person transmission does not occur. When dealing with infected individuals, use standard precautions. Humans can infect new mosquitoes during the fever phase, protect from vectors.

Normal Routes of Exposure: Abraded Skin; vectors (mosquitoes).

Infectious Dose: ≤10 organisms.

Secondary Hazards: Blood and body fluids; vector amplification.

Incubation: 2–15 days.

Signs & Symptoms: Abrupt onset of high fever, chills, facial flushing, sore throat, red eyes with retro-orbital pain, extreme headache, severe joint (arthralgia) and muscle pain (myalgia), abdominal pain, nausea, vomiting, minor bleeding, and a rash of mixed flat and raised (maculopapular) spots that appears on the upper body then spreads to the arms, legs, and face. Fever may be biphasic; subsiding after 3 to 4 days then returning 1 to 3 days later. May progress to the hemorrhagic form with small red or purple spots on the surface of the skin (petechia), dermal hemorrhages (purpura), livid spots (ecchymosis), nose bleeds (epistaxis), vomiting blood (hematemesis), blood in the urine (hematuria), bloody stool, enlargement of the liver (hepatomegaly), and low blood pressure (hypotension).

Suggested Alternatives for Differential Diagnosis: Chikungunya virus, hanta viruses, hemorrhagic fever viruses, hepatitis, immune thrombocytopenia, influenza, leptospirosis, malaria, Mayaro fever, measles, meningitis, orbivirus, rickettsial infection, Rift Valley fever, roseola infantum, Ross River fever, rubella, scarlet fever, scrub typhus, Sindbis fever, St. Louis encephalitis, typhoid fever, typhus, West Nile virus, yellow fever, Zika virus.

Mortality Rate (untreated): ≤1% (classic dengue); ≤20% (hemorrhagic).

C18-A015

Dobrava-Belgrade Hemorrhagic Fever
Bunyaviridae
Hemorrhagic Fever with Renal Syndrome
ICD-11: XN16H
UNII: —
EPPO: —

Endemic to Eastern Europe and Russia. The natural reservoir is *Apodemus* spp. mice. Does not produce disease In Domestic Animals. Rodents can shed hantaviruses in saliva, feces, and urine. This is a biosafety level 3 agent.

It is on the Australia Group Core list.

In People:

CDC Case Definition: None established.

Communicability: Direct person-to-person transmission does not occur. When dealing with infected individuals, use standard precautions.

Normal Routes of Exposure: Inhalation; mucous membranes; abraded skin.

Infectious Dose: Unknown.

Secondary Hazards: Aerosols; fecal matter (mouse); blood and body fluids; fomites (from mouse habitation).

Incubation: 5–60 days.

Signs & Symptoms: Abrupt onset of fever with headache, chills, abdominal pain, vague feeling of bodily discomfort (malaise), blurred vision, lower back pain, and a slow heartbeat (bradycardia). There may be flushing of the face, neck, and chest with bleeding around the eyes (subconjunctival hemorrhage) and small hemorrhages (petechia) in the arm pits (axilla) and mouth (soft palate). When the fever ends (defervescence), individuals may develop rapid heartbeat (tachycardia) with acute abdominal pain and convulsions. Kidney failure is indicted by limited urine production (oliguria), high blood pressure (hypertension), bleeding (hemorrhage), and edema. Can rapidly progress to dehydration and severe shock. May also develop acute respiratory distress.

Suggested Alternatives for Differential Diagnosis: Acute anemia, acute leukemias, anaphylactoid purpura, chikungunya, Colorado tick fever, dengue fever, disseminated intravascular coagulation, encephalitis, epidemic typhus, falciparum malaria, hemolytic uremic syndrome, Henoch-Schonlein purpura, hepatitis, herpes, infectious mononucleosis, influenza, leptospirosis, malaria, measles, meningitis, meningococcemia, plague, pleural effusion, Q fever, Rocky Mountain spotted fever, rubella, sepsis, shigellosis, Sindbis fever, thrombocytopenic purpura, tick bite fever, trypanosomiasis, typhoid fever, yellow fever, other New World viral hemorrhagic fevers, drug eruptions, allergies, food poisoning.

Mortality Rate (untreated): ≤12%.

C18-A016

Eastern Equine Encephalitis
Togaviridae
ICD-11: XN78T
UNII: —
EPPO: —

Endemic to Canada, the United States east of the Mississippi River, some Caribbean Islands, Central America, and parts of South America. The natural reservoir is birds and mosquitoes. Mosquitoes remains infective for life. Infection is also transferred directly to mosquito eggs. Equidae (e.g., horses, donkeys) may develop sufficient viremia to serve as amplifying hosts. Horse-to-human transmission does not occur. It does not survive outside of a host. This is a biosafety level 3 agent.

It is on the Australia Group Core list, the U.S. Select Agents and Toxins list and the OIE list of Notifiable Diseases for 2020.

In People:

CDC Case Definition: Clinical criteria for cases of *Neuroinvasive disease* is defined as follows: 1) meningitis, encephalitis, acute flaccid paralysis, or other acute signs of central or peripheral neurologic dysfunction; AND 2) absence of a more likely clinical explanation.

Clinical criteria for cases of *Non-neuroinvasive disease* is defined as follows: 1) fever (chills) as reported by the patient or a health-care provider; AND 2) absence of neuroinvasive disease; AND 3) absence of a more likely clinical explanation. Other clinically compatible symptoms include: headache, myalgia, rash, arthralgia, vertigo, vomiting, paresis, and/ or nuchal rigidity.

Laboratory criteria for diagnosis is isolation of virus from, or demonstration of specific viral antigen or nucleic acid in, tissue, blood, CSF, or other body fluid, OR 1) fourfold or greater change in virus-specific quantitative antibody titers in paired sera, OR 2) virus-specific IgM antibodies in serum with confirmatory virus-specific neutralizing antibodies in the same or a later specimen, OR 3) virus-specific IgM antibodies in CSF or serum.

Criteria for case confirmation for *Neuroinvasive disease* consists of meeting the clinical criteria and one or more of the following laboratory criteria: 1) isolation of virus from, or demonstration of specific viral antigen or nucleic acid in, tissue, blood, CSF, or other body fluid, OR 2) fourfold or greater change in virus-specific quantitative antibody titers in paired sera, OR 3) virus-specific IgM antibodies in serum with confirmatory virus-specific neutralizing antibodies in the same or a later specimen, OR 4) virus-specific IgM antibodies in CSF, with or without a reported pleocytosis, and a negative result for other IgM antibodies in CSF for arboviruses endemic to the region where exposure occurred.

Criteria for case confirmation for *non-neuroinvasive disease* consists of meeting the clinical criteria and one or more of the following laboratory criteria: 1) isolation of virus from, or demonstration of specific viral antigen or nucleic acid in, tissue, blood, or other body fluid, excluding CSF, OR 2) fourfold or greater change in virus-specific quantitative antibody titers in paired sera, OR 3) virus-specific IgM antibodies in serum with confirmatory virus-specific neutralizing antibodies in the same or a later specimen.

Communicability: Direct person-to-person transmission does not occur. When dealing with infected individuals, use standard precautions.

Normal Routes of Exposure: Vectors (mosquitoes); abraded skin.

Infectious Dose: Unknown.

Secondary Hazards: None.

Incubation: 4–10 days.

Signs & Symptoms: Abrupt onset of fever, headache, chills, joint (arthralgia) and muscle (myalgia) pain, rapid heart rate (tachycardia), a vague feeling of bodily discomfort (malaise), nausea, and vomiting. May develop abdominal pain, diarrhea and sore throat along with neurologic signs including neck stiffness (nuchal rigidity), confusion, irritability, restlessness, semiconsciousness (somnolence), stupor, disorientation, tremors, seizures, paralysis, and coma. Many of who survive will have mild to severe permanent neurologic damage including convulsions, paralysis, and mental retardation.

Suggested Alternatives for Differential Diagnosis: Bartonellosis, brucellosis, California encephalitis, coxsackieviruses, cryptococcosis, cysticercosis, cytomegalovirus, echinococcosis hydatid cyst, histoplasmosis, Japanese encephalitis, Legionnaires disease, leptospirosis, listeriosis, Lyme disease, malaria, Naegleria infection, rabies, tuberculosis, western equine encephalitis.

Mortality Rate (untreated): ≤75%.

In Domestic Animals:

Susceptible species: Horses and other equids.

Communicability: Direct transmission does not occur.

Normal routes of exposure: Vectors (mosquitoes – *Culex* spp.).

Secondary Hazards: None.

Incubation: 5–14 days.

Signs: Fever, loss of appetite (anorexia), difficulty swallowing (dysphagia), severe depression, blindness, increased sensitivity to sensory stimuli such as noise, light, or touch (hyperesthesia), periods of excitement or intense itching (pruritus), facial swelling (edema), aimless wandering, head pressing, circling, incoordination (ataxia), paralysis, recumbency, seizures, convulsions.

Post-mortem does not present any characteristic gross findings. There is severe inflammation of the gray matter with possible neuronal degeneration, infiltration by inflammatory cells, gliosis, perivascular cuffing, and hemorrhages.

Suggested Alternatives for Differential Diagnosis: Botulism, equine herpesvirus 1, Everglades virus, hepatoencephalopathy, highlands J virus, Japanese encephalitis, Kunjin virus, leukoencephalo-malacia, louping ill, meningitis, Murray Valley encephalitis virus, protozoal encephalomyelitis, rabies, Ross River virus, Semliki Forest virus, St. Louis encephalitis, tickborne encephalitis, Una virus, Usutu, Venezuelan equine encephalitis virus, verminous meningoencephalomyelitis, West Nile encephalitis, western equine encephalitis, cranial trauma.

Mortality Rate (untreated): ≤90%.

C18-A017

Ebolavirus Fever
Filoviridae
ICD-11: XN1EN
UNII: 04N5737KTQ (Bundibugyo); L28X8242OW (Sudan); 0F7ZNZ0O50 (Tai Forest); 3R8G5WNW24 (Zaire)
EPPO: —

Endemic to sub-Saharan Africa. The natural reservoir is unknown but suspected to be fruit bats. It can survive at room temperature on dry surfaces for several hours, in dried blood for up to 10 days, and in body fluids up to several days. It can remain viable cotton clothing and goggles for up to 24 hours, on medical gloves for up to 1 hour, and other medical personal protective equipment (e.g., respirators, suits, hoods) for up to 72 hours. Also produces disease in primates with symptoms similar to those seen in humans. Ebola Reston, the only species known to be communicable through inhalation, only infects primates other than man. This is a biosafety level 4 agent.

It is on the Australia Group Core list, a Tier-1 agent on the U.S. Select Agents and Toxins list and the OIE list of Notifiable Diseases for 2020.

In People:

CDC Case Definition for viral hemorrhagic fever: Clinical criteria is an illness with acute onset with ALL of the following clinical findings: 1) a fever >104°F; and 2) one or more of the following clinical findings: a) severe headache; OR b) muscle pain; OR c) erythematous maculopapular rash on the trunk with fine desquamation 3–4 days after rash onset; OR d) vomiting; OR e) diarrhea; OR f) abdominal pain; OR g) bleeding not related to injury; OR h) thrombocytopenia.

Laboratory criteria for diagnosis is one or more of the following laboratory findings: 1) detection of viral hemorrhagic fever (VHF) viral antigens in blood by enzyme-linked Immunosorbent Assay (ELISA) antigen detection; 2) VHF viral isolation in cell culture for blood or tissues; 3) detection of VHF-specific genetic sequence by Reverse Transcription-Polymerase Chain Reaction (RT-PCR) from blood or tissues; 4) detection of VHF viral antigens in tissues by immunohistochemistry.

Communicability: Direct person-to-person transmission is possible. When dealing with infected individuals, use contact and droplet precautions. For procedures that could generate infectious aerosols, use special air handling and ventilation systems, and N95 or better respirators.

Normal Routes of Exposure: Inhalation (Reston only); ingestion; abraded skin; mucous membranes.

Infectious Dose: 1–10 viral units (inhalation).

Secondary Hazards: Aerosols; blood and body fluids (including semen); body tissue; fecal matter; fomites.

Incubation: 2–21 days.

Signs & Symptoms: Sudden onset of fever, headache, pain in the joints (arthralgias) and muscles (myalgia), sore throat, and weakness, followed by diarrhea, vomiting, and abdominal pain progressing in some cases to rash of mixed flat and raised (maculopapular) spots, bloodshot eyes (conjunctival injection), and sore throat (pharyngitis). Hemorrhagic signs may include bleeding from the eyes (hemorrhagic conjunctivitis), the gums, and the nose (epistaxis); vomiting blood (hematemesis), blood in the urine (hematuria), and black tar-like stools (melena). Terminally ill patients often have a normal temperature, but appear to be dull (obtunded), with rapid breathing (tachypnea), experience hiccups, are unable to urinate (anuria), then progress into shock.

Suggested Alternatives for Differential Diagnosis: Acute leukemia, cholera, Crimean-Congo hemorrhagic fever, dengue fever, disseminated intravascular coagulation, hemolytic uremic syndrome, influenza, Lassa, leptospirosis, malaria, Marburg hemorrhagic fever, meningococcemia, norovirus, plague, rickettsial infections, salmonella infection, shigellosis, thrombocytopenic purpura, typhoid fever.

Mortality Rate (untreated): ≤100%.

C18-A018

Far Eastern Tick-Borne Encephalitis Virus
Flaviviridae
ICD-11: 1C89
UNII: —
EPPO: —

Endemic to far-eastern Russia, China and Japan. The natural reservoir is ticks (*Ixodidae* spp.). Ticks remain infected for life. Infection is also transferred directly to tick eggs. This is a biosafety level 4 agent.

It is on the Australia Group Core list and the U.S. Select Agents and Toxins list.

In People:
CDC Case Definition: None established.
European Centre for Disease Prevention and Control Case Definition: Clinical criteria is any person with symptoms of inflammation of the CNS (for example, meningitis, meningo-encephalitis, encephalomyelitis, encephaloradiculitis).

Laboratory criteria for diagnosis is one or more of the following laboratory findings: 1) TBE specific IgM AND IgG antibodies in blood; OR 2) TBE specific IgM antibodies in CSF; OR 3) seroconversion or fourfold increase of TBE-specific antibodies in paired serum samples; OR 4) detection of TBE viral nucleic acid in a clinical specimen; OR 5) isolation of TBE virus from clinical specimen.

Communicability: Direct person-to-person transmission does not occur. When dealing with infected individuals, use standard precautions.

Normal Routes of Exposure: Vectors (ticks).

Infectious Dose: Unknown.

Secondary Hazards: Aerosols; abraded skin; bites (infected rodents); ingestion (unpasteurized products); fomites (with vectors).

Incubation: 2–28 days.

Signs & Symptoms: Fever, headache, sensitivity to light (photophobia), fatigue, muscle pain (myalgia), loss of appetite (anorexia), nausea and vomiting, progressing to neck stiffness (nuchal rigidity), sensorial changes, visual disturbances, rapid involuntary movements of the eyes (nystagmus), incoordination (ataxia), tremors, seizures, convulsions, or flaccid paralysis from inflammation of the membranes that surround the brain and spinal cord (meningitis), inflammation of the brain and membranes covering it (meningoencephalitis), or inflammation of the spinal cord (myelitis).

Suggested Alternatives for Differential Diagnosis: Aseptic meningitis, babesiosis, basilar artery blood clots, cardioembolic stroke, cavernous sinus syndromes, cerebral venous blood clots, encephalitis, epileptic and epileptiform encephalopathies, febrile seizures, granulocytic anaplasmosis, hemophilus meningitis, intracranial hemorrhage, leptomeningeal carcinomatosis, Lyme disease, meningitis, subdural empyema or hematoma, tick-transmitted rickettsioses, tularemia, confusional states and acute memory disorders.

Mortality Rate (untreated): ≤35%.

In Domestic Animals:

Susceptible species: Sheep, goats, dogs.

Communicability: Direct transmission does not occur.

Normal routes of exposure: Vectors (ticks).

Secondary Hazards: Fomites (with vectors).

Incubation: Unknown. Believed to be 7–14 days.

Signs: Usually asymptomatic. May present with fever, lethargy, muscle pain (myalgia), loss of appetite (anorexia), eyes have unequally sized pupils (anisocoria), rapid involuntary movements of the eyes (nystagmus), incoordination (ataxia), tremors, seizures, convulsions, and paralysis.

Suggested Alternatives for Differential Diagnosis: Other causes of meningitis, encephalitis, or meningoencephalitis.

Mortality Rate (untreated): ≤100% (species dependent).

C18-A019

Foot-and-Mouth Disease Virus (Agent OO)
Picornaviridae
ICD-11: 1F05.3
UNII: —
EPPO: FMDV00

Endemic to parts of South America, Asia, Africa, and the Middle East. The natural reservoir is domestic cattle and wild African buffalo. It can survive for up to 14 days in dry fecal material and 6 months in a fecal slurry, 39 days in urine, 5 months in hay or straw bedding, and in soil for up to 3 days during the summer and 28 days during the winter. The virus survives in milk and milk products during regular pasteurization, but is inactivated by ultra-high temperature pasteurization. Viral aerosols can travel for over 160 kilometers over water and over 8 kilometers over land. Humans, unaffected by the virus, can still harbor the virus in their respiratory tract for up to 2 days; passing it on to animals through aerosols of respiratory secretions. Recovered animals can become carriers and, depending on the species, spread the virus to other animals for months or years. This is a biosafety level 3-Ag agent. Laboratory workers should have no contact with susceptible hosts for 5 days after working with the agent.

It is on the Australia Group Core list, a Tier-1 agent on the U.S. Select Agents, and Toxins list, and the OIE list of Notifiable Diseases for 2020.

In People:
Rarely infects humans. When it occurs, it produces a short-lived mild illness with flu-like symptoms and blisters (vesicles) in the mouth and mucous membranes. Blisters can sometimes appear on the hands and feet.

In Domestic Animals:
Target species: Cattle, pigs, sheep, goats, water buffalo, yaks.

Communicability: Direct transmission is possible.

Normal routes of exposure: Inhalation; ingestion; mucous membranes; open lesions.

Secondary Hazards: Aerosols (respiratory secretions); blood and body fluids (including milk and semen); body tissue; fecal matter; fomites; mechanical vectors (track out).

Incubation: 2–14 days.

Signs: Characterized by fever, sudden appearance of blisters (vesicles) on the mouth, nose, feet and teats. Ruptured oral blisters can coalesce and form erosions. Severely affected animals may smack their lips and suffer excessive salivation with drooling. There is a loss of appetite (anorexia). Pain from blisters on the feet and the band just above the hoof (coronary band) cause animals to shake their feet and become lame. There is a drastic drop in milk production from lactating animals. Young animals usually die from heart failure (myocarditis).

Post-mortem findings include ulcerative lesions on the nostrils, lips, tongue, palate, gums, pillars of the rumen, coronary bands, teats, udders, and feet. In young animals, gray or yellow streaking in the heart (tiger heart) from degeneration and necrosis of the myocardium.

Suggested Alternatives for Differential Diagnosis: Indistinguishable from vesicular stomatitis and vesicular exanthema. Other considerations include allergic stomatitis, bluetongue, bovine mammillitis, bovine papular stomatitis, bovine viral diarrhea, contagious ecthyma, epizootic hemorrhagic disease, feedlot glossitis, foot rot, infectious bovine rhinotracheitis, malignant catarrhal fever, mycotoxicosis, ovine pox, peste des petits ruminants, photosensitization, pseudocowpox, rinderpest, and foreign bodies or trauma.

Mortality Rate (untreated): ≤5%; ≤50% (young animals).

C18-A020

Goatpox
Poxviridae
ICD-11: —
UNII: —
EPPO: —

Endemic to Africa north of the equator, the Middle East, Central Asia, and India. The natural reservoir is goats, although chronically infected carriers do not occur. Infection results in solid and enduring immunity. The virus can persist for up to 6 months in shaded animal pens and in dried scabs, and for up to 3 months in on the fleece, skin and hair from infected animals. It can survive freeze–thaw cycles, although the infectivity may be reduced. This is a biosafety level 3 agent.

It is on the Australia Group Core list, the U.S. Select Agents and Toxins list, and the OIE list of Notifiable Diseases for 2020.

In People:
Does not occur in humans.

In Domestic Animals:
Target species: Goats, sheep (minor).

Communicability: Direct transmission is possible.

Normal routes of exposure: Inhalation; abraded skin; mucous membranes.

Secondary Hazards: Aerosols (respiratory secretions); blood and body fluids (including milk); fecal matter; fomites; mechanical vectors (track out, biting insects).

Incubation: 4–14 days.

Signs: Rapid onset of fever with excessive salivation or drooling (ptyalism), nasal discharge and eye infection (pink eye – conjunctivitis). Eyelids become swollen, and a discharge of mucus and pus (mucopurulent) crusts the nostrils. Animals may have difficulty breathing (dyspnea). Red, flat (macular) spots areas appear on the skin that evolve into pimples (papules) and possibly blisters (vesicles or pustules). Ultimately, skin lesions become dark-colored scabs. Scabs may take up to 6 weeks to heal. The skin is sensitive in the affected area. Scabs leave small star-shaped scars that remain free of hair. Pustules may alternatively develop into nodules that die and slough off the animal leaving hairless patches.

Post-mortem skin lesions are often less obvious than on a live animal. Other findings include reddish to whitish firm nodules in the mucosa of the pharynx and trachea, as well as hard, reddish to whitish nodules in the lungs. There are rarely signs of pneumonia. Lymph nodes in the thorax are enlarged and edematous.

Suggested Alternatives for Differential Diagnosis: Bluetongue, caseous lymphadenitis, contagious pustular dermatitis, dermatophilosis, mange, mycotic dermatitis, orf, parasitic pneumonia, peste des petits ruminants, photosensitization, and urticaria from multiple insect bites.

Mortality Rate (untreated): ≤100%.

C18-A021

Guanarito Virus
Arenaviridae
Venezuelan Hemorrhagic Fever
ICD-11: XN56K
UNII: —
EPPO: —

Endemic to Venezuela. The natural reservoir is the short-tailed cane mouse (*Zygodontomys brevicauda*). The virus is shed in their urine. Does not produce disease in domestic animals. This is a biosafety level 4 agent.

It is on the Australia Group Core list and the U.S. Select Agents and Toxins list.

In People:

CDC Case Definition for viral hemorrhagic fever: Clinical criteria is an illness with acute onset with ALL of the following clinical findings: 1) a fever >104°F; and 2) one or more of the following clinical findings: a) severe headache; OR b) muscle pain; OR c) erythematous maculopapular rash on the trunk with fine desquamation 3–4 days after rash onset; OR d) vomiting; OR e) diarrhea; OR f) abdominal pain; OR g) bleeding not related to injury; OR h) thrombocytopenia; OR i) retrosternal chest pain; OR j) proteinuria; OR k). pharyngitis.

Laboratory criteria for diagnosis is one or more of the following laboratory findings: 1) detection of viral hemorrhagic fever (VHF) viral antigens in blood by enzyme-linked Immunosorbent Assay (ELISA) antigen detection; 2) VHF viral isolation in cell culture for blood or tissues; 3) detection of VHF-specific genetic sequence by Reverse Transcription-Polymerase Chain Reaction (RT-PCR) from blood or tissues; 4) detection of VHF viral antigens in tissues by immunohistochemistry.

Communicability: Direct person-to-person transmission is unlikely. When dealing with infected individuals, use standard precautions.

Normal Routes of Exposure: Inhalation; abraded skin; mucous membranes.

Infectious Dose: Unknown.

Secondary Hazards: Aerosols; blood and body fluids (mouse); body tissue (mouse); fecal matter (mouse); fomites (from mouse habitation).

Incubation: 7–16 days.

Signs & Symptoms: Fever, headache, muscle pain (myalgia), sore throat, weakness, loss of appetite (anorexia), nausea, vomiting, and dehydration. Hemorrhagic manifestations include nose bleeds (epistaxis), bleeding gums, vomiting blood (hematemesis), black tar-like stools (melena). Manifestation of convulsions carries a pore prognosis.

Suggested Alternatives for Differential Diagnosis: Acute anemia, acute leukemias, anaphylactoid purpura, chikungunya, Colorado tick fever, dengue fever, disseminated intravascular coagulation, encephalitis, epidemic typhus, falciparum malaria, hemolytic uremic syndrome, Henoch-Schonlein purpura, hepatitis, herpes, infectious mononucleosis, influenza, leptospirosis, malaria, measles, meningitis, meningococcemia, plague, pleural effusion, Q fever, Rocky Mountain spotted fever, rubella, sepsis, shigellosis, Sindbis fever, thrombocytopenic purpura, tick bite fever, trypanosomiasis, typhoid fever, yellow fever, other New World viral hemorrhagic fevers, drug eruptions, allergies, food poisoning.

Mortality Rate (untreated): ≤33%.

C18-A022

Hantaan Hemorrhagic Fever
Bunyaviridae
Hemorrhagic Fever with Renal Syndrome
ICD-11: XN3GW
UNII: —
EPPO: —

Endemic to Central and Eastern Asia. The natural reservoir is the striped field mouse (*Apodemus agrarius*). Does not produce disease in domestic animals. Rodents can shed hantaviruses in saliva, feces, and urine. This is a biosafety level 3 agent.

It is on the Australia Group Core list.

In People:
CDC Case Definition: None established.

Communicability: Direct person-to-person transmission does not occur. When dealing with infected individuals, use standard precautions.

Normal Routes of Exposure: Inhalation; mucous membranes; abraded skin.
Infectious Dose: Unknown.

Secondary Hazards: Aerosols; fecal matter (mouse); blood and body fluids; fomites (from mouse habitation).

Incubation: 5–60 days.

Signs & Symptoms: Abrupt onset of fever with headache, chills, abdominal pain, vague feeling of bodily discomfort (malaise), blurred vision, lower back pain and a slow heartbeat (bradycardia). There may be flushing of the face, neck, and chest with bleeding around the eyes (subconjunctival hemorrhage) and small hemorrhages (petechia) in the arm pits (axilla) and mouth (soft palate). When the fever ends (defervescence), individuals may develop rapid heartbeat (tachycardia) with acute abdominal pain and convulsions. Kidney failure is indicted by limited urine production (oliguria), high blood pressure (hypertension), severe bleeding (hemorrhage), and edema. Can rapidly progress to dehydration and severe shock.

Suggested Alternatives for Differential Diagnosis: Acute abdominal diseases, acute kidney injury, acute poststreptococcal glomerulonephritis, Colorado tick fever, disseminated intravascular coagulation, heat stroke, hemolytic uremic syndrome, leptospirosis, malaria, murine typhus, non-a, non-b hepatitis, scrub typhus, septicemia, spotted fevers.

Mortality Rate (untreated): <15%

C18-A023

Hendra Virus
Paramyxoviridae
ICD-11: XN53N
UNII: —
EPPO: —

Endemic to Australia. The natural reservoir is fruit bats (*Pteropid* spp.). There is no evidence of direct transmission from bats to humans. Horses are the only other significant host and special precautions should be taken when examining a horse suspected of having the disease or performing a necropsy. Horse may be asymptomatic during prodromal period but are able to shed the virus. Dogs may also suffer subclinical infections. It can survive for more than four days in bat urine and up to a few days on fruit or in fruit juice. This is a biosafety level 4 agent.

It is on the Australia Group Core list, the U.S. Select Agents and Toxins list and the OIE list of Notifiable Diseases for 2020.

In People:

CDC Case Definition: None established.

Communicability: Direct person-to-person transmission does not occur. When dealing with infected individuals, use standard precautions.

Normal Routes of Exposure: Not established, possibly mucous membranes.

Infectious Dose: Unknown.

Secondary Hazards: Blood and body fluids (horse); body tissue (horse); fecal matter (horse).
Incubation: 5–16 days.

Signs & Symptoms: Initially a flu-like illness with fever, dry cough, sore throat, swollen and painful lymph nodes (lymphadenopathy) in the neck, fatigue, and muscle pain (myalgia). May progress to pneumonitis, multiorgan failure, arterial blood clots (thrombosis) respiratory failure and death. Alternatively, patients may develop inflammation of the brain and membranes covering it (meningoencephalitis) with drowsiness, headache, vomiting, neck stiffness (nuchal rigidity), incoordination (ataxia), vertigo, mild confusion, bilateral drooping upper eyelids (ptosis), unable to speak normally (dysarthria), progressing to seizures and coma. Individuals may recover from either of these syndromes only to have fatal inflammation of the brain (encephalitis) reappear after a delay of up to 1 year.

Suggested Alternatives for Differential Diagnosis: Anthrax, eastern equine encephalitis, hantavirus pulmonary syndrome, herpes encephalitis, Japanese encephalitis, measles encephalitis, Murray Valley encephalitis, pneumonia, Venezuelan equine encephalitis, West Nile virus, western equine encephalitis, poisonings from botulism or fluoroacetate.

Mortality Rate (untreated): ≤60%.

In Domestic Animals:

Susceptible species: Horses.

Communicability: Direct transmission does occur, although it requires close contact.

Normal routes of exposure: Not established, possibly ingestion; mucous membranes.

Secondary Hazards: Blood and body fluids; body tissue; fomites.

Incubation: 3–16 days.

Signs: In some cases, sudden death may occur with no apparent signs of illness. Otherwise, fever, loss of appetite (anorexia), rapid heart rate (tachycardia), depression, sweating and uneasiness, muscle tremors, colic and straining to defecate, foul odor to the breath, delayed blood clotting times, difficult or painful urination (dysuria), excessively warm hooves. Animals may shift their weight constantly from leg to leg, or alternate weight shifting with a rigid stance. Respiratory signs include difficulty breathing (dyspnea), congested and jaundiced mucus membranes with a cyanotic border, incoordination (ataxia), facial swelling (edema) or swelling of the lips, and accumulation of fluid in the lungs (pulmonary edema). Just before death, animals may develop a copious nasal discharge, which becomes frothy and may be bloodstained. Neurological signs include an altered or wobbly gait that progresses to incoordination (ataxia), altered consciousness, head pressing, blindness in one or both eyes, head tilt, circling, muscle twitches or tremors, locked jaw, spasms of the jaw or involuntary chomping, or facial paralysis.

Post-mortem findings include the airway is filled with thick, often blood-tinged, froth, severe pulmonary edema, dilation of the pulmonary lymphatics, congestion and ventral consolidation of the lungs. May also see increased pleural and pericardial fluids, congestion of lymph nodes, hemorrhages in various organs, and slight jaundice.

Suggested Alternatives for Differential Diagnosis: African horse sickness, anthrax, colic, eastern equine encephalitis, equine herpes virus, equine influenza, hantavirus pulmonary syndrome, inhalation pneumonia, Japanese encephalitis, Murray Valley encephalitis, pasteurellosis, peracute equine herpesvirus 1 infection, purpura hemorrhagica, purulent bronchopneumonia, Venezuelan equine encephalitis, West Nile virus, western equine encephalitis, poisonings from botulism, chemicals (e.g., agricultural materials, lead, fluoroacetate, ionophores), or toxic plants (e.g., Crofton weed, avocado), snake bite.

Mortality Rate (untreated): ≤90%.

C18-A024

Japanese Encephalitis (Agent AN)
Flaviviridae
ICD-11: XN9ZK
UNII: P07E7XWU9D
EPPO: —

Endemic to Southeast Asia, the Indian subcontinent, and parts of Oceania. The natural reservoir is water birds and mosquitoes. Mosquitoes remains infective for life. Infection is also transferred directly to mosquito eggs. Pigs act as amplifying hosts. The disease in pigs usually manifests through reproductive disorders. Other than piglets, the disease in swine is essentially nonfatal. This is a biosafety level 3 agent.

It is on the Australia Group Core list and the OIE list of Notifiable Diseases for 2020.

In People:
CDC Case Definition: None established.

Communicability: Direct person-to-person transmission does not occur. When dealing with infected individuals, use standard precautions.

Normal routes of exposure: Vectors (mosquitoes); abraded skin.

Infectious Dose: Unknown.

Secondary Hazards: Aerosol; blood and body fluids; body tissue; vector amplification (pigs).

Incubation: 4–15 days.

Signs & Symptoms: Abrupt onset of high fever, headache, chills, nausea, vomiting, diarrhea, muscle pain (myalgia), sensitivity to light (photophobia), neck stiffness (nuchal rigidity), stupor, disorientation, seizures, spastic or flaccid paralysis, coma. Up to 50% of survivors continue to have neurologic, cognitive, or psychiatric consequence of the disease (sequela).

Suggested Alternatives for Differential Diagnosis: Amebic meningoencephalitis, Behcet disease, California encephalitis, central nervous system lupus, central nervous system tumors, cerebrovascular diseases / accident, dengue fever, eastern equine encephalitis, ehrlichiosis, enterovirus infection, fungal infections, herpes simplex, leptospirosis, malaria, meningitis, Murray Valley encephalitis, neurocysticercosis, Nipah virus, pyogenic focal brain abscess, Rocky Mountain spotted fever, St. Louis encephalitis, tuberculous meningitis, typhoid fever, West Nile virus, western equine encephalitis.

Mortality Rate (untreated): ≤30%.

In Domestic Animals:

Susceptible species: Horses and other equids.

Communicability: Direct transmission does not occur.

Normal routes of exposure: Vectors (mosquitoes).

Secondary Hazards: Pigs are an amplifying host.

Incubation: 4–14 days (horses).

Signs: Horses: Fever, loss of appetite (anorexia), lethargy, incoordination (ataxia), yellow discoloration (jaundice) or small hemorrhages (petechia) of the mucous membranes, stupor, grinding or chewing motions, difficulty swallowing (dysphagia), transient neck stiffness (nuchal rigidity), impaired vision, profuse sweating, muscle tremors, aimless wandering, aggression, loss of vision, collapse, paralysis, coma.

Pigs: Reproductive disease with stillbirths or mummified fetuses. Infected piglets born alive often have tremors and convulsions, then die soon after birth.

Post-mortem of horses does not present any characteristic gross findings. There is diffuse non-suppurative encephalomyelitis with apparent perivascular cuffing; phagocytic destruction of nerve cells, perivascular cuffing and focal gliosis. Blood vessels appear dilated with numerous mono-nuclear cells.

Suggested Alternatives for Differential Diagnosis: Horses: Acute babesiosis, African horse sickness, bacterial or toxic encephalitis, Borna disease, cerebral nematodiasis or protozoodiasis, eastern equine encephalitis, equine herpes myelencephalopathy, equine infectious anaemia, equine protozoal myelencephalitis, hepatic encephalopathy, leucoencephalomalacia, Murray Valley encephalitis, rabies, Venezuelan equine encephalitis, viral equine rhinopneumonitis, West Nile encephalitis, western equine encephalitis, tetanus, botulism, and poisoning from lead.

Pigs: Aujeszky's disease, brucellosis, classical swine fever, coronavirus, encephalomyocarditis virus, hemagglutinating encephalomyelitis, La Piedad Michoacan paramyxovirus, Menangle virus, Nipah virus, parvovirus, porcine reproductive and respiratory syndrome, Teschen disease, water deprivation, and poisoning from salt.

Mortality Rate (untreated): ≤40% (horses); 0% (adult pigs).

C18-A025

Junin Virus

Arenaviridae

Argentine Hemorrhagic Fever

ICD-11: XN2ZL

UNII: —

EPPO: —

Endemic to Argentina. The natural reservoir is vesper mice (*Calomys* spp.). The virus is shed in their urine. Does not produce disease in domestic animals. This is a biosafety level 3 agent, with the addition of HEPA filtration attached to the laboratory exhaust system.

It is on the Australia Group Core list and the U.S. Select Agents and Toxins list.

In People:

CDC Case Definition for viral hemorrhagic fever: Clinical criteria is an illness with acute onset with ALL of the following clinical findings: 1) a fever >104°F; and 2) one or more of the following clinical findings: a) severe headache; OR b) muscle pain; OR c) erythematous maculopapular rash on the trunk with fine desquamation 3–4 days after rash onset; OR d) vomiting; OR e) diarrhea; OR f) abdominal pain; OR g) bleeding not related to injury; OR h) thrombocytopenia; OR i) retrosternal chest pain; OR j) proteinuria; OR k). pharyngitis.

Laboratory criteria for diagnosis is one or more of the following laboratory findings: 1) detection of viral hemorrhagic fever (VHF) viral antigens in blood by enzyme-linked Immunosorbent Assay (ELISA) antigen detection; 2) VHF viral isolation in cell culture for blood or tissues; 3) detection of VHF-specific genetic sequence by Reverse Transcription-Polymerase Chain Reaction (RT-PCR) from blood or tissues; 4) detection of VHF viral antigens in tissues by immunohistochemistry.

Communicability: Direct person-to-person transmission is rare. When dealing with infected individuals, use standard precautions.

Normal Routes of Exposure: Inhalation; abraded skin; mucous membranes.

Infectious Dose: Unknown.

Secondary Hazards: Aerosols; blood and body fluids (mouse); body tissue (mouse); fecal matter (mouse); fomites (from mouse habitation).

Incubation: 5–21 days.

Signs & Symptoms: Initial onset is insidious with mild fever, chills, vague feeling of bodily discomfort (malaise), loss of appetite (anorexia), headache, bloodshot eyes (conjunctival injection), swelling around the eyes (periorbital edema), flushing (erythroderma) of the face, neck, and upper chest, and lower back pain. May also experience pain behind the eyes (retro-orbital), nausea, vomiting, abdominal pain, sensitivity to light (photophobia), dizziness, and either constipation or mild diarrhea. May progress to hemorrhagic signs including blood in the lungs, vomiting blood (hematemesis), black tar-like stools (melena), blood in the urine (hematuria), nose bleeds (epistaxis), and bruising (hematoma); with neurological manifestations including mental confusion, marked incoordination (ataxia), increased irritability and tremors that are followed by delirium, generalized convulsions, and coma.

Suggested Alternatives for Differential Diagnosis: Acute anemia, acute leukemias, anaphylactoid purpura, chikungunya, Colorado tick fever, dengue fever, disseminated intravascular coagulation, encephalitis, epidemic typhus, falciparum malaria, hemolytic uremic syndrome, Henoch-Schonlein purpura, hepatitis, herpes, infectious mononucleosis, influenza, leptospirosis, malaria, measles, meningitis, meningococcemia, plague, pleural effusion, Q fever, Rocky Mountain spotted fever, rubella, sepsis, shigellosis, Sindbis fever, thrombocytopenic purpura, tick bite fever, trypanoso-miasis, typhoid fever, yellow fever, other New World viral hemorrhagic fevers, drug eruptions, allergies, food poisoning.

Mortality Rate (untreated): ≤30%.

C18-A026

Kyasanur Forest Disease
Flaviviridae
ICD-11: 1D4B
UNII: —
EPPO: —

Endemic to southern India. The natural reservoir is ticks (*Haemaphysalis spinigera*). Ticks remains infective for life. Monkeys, rodents, and other small vertebrates act as amplifying hosts for the disease. Causes severe febrile illness in most monkeys, often resulting in death. This is a biosafety level 4 agent.

It is on the Australia Group Core list and the U.S. Select Agents and Toxins list.

In People:

CDC Case Definition: None established.

Communicability: Direct person-to-person transmission does not occur. When dealing with infected individuals, use standard precautions.

Normal Routes of Exposure: Inhalation; abraded skin; mucous membranes; vectors (ticks). *Infectious Dose:* Unknown.

Secondary Hazards: Aerosols; blood and body fluids (monkey); body tissue (monkey); fomites (with vectors).

Incubation: 3–8 days.

Signs & Symptoms: Often biphasic, with sudden onset of fever, chills, frontal headache, sensitivity to light (photophobia), eye infection (pink eye – conjunctivitis), coughing, sore throat (pharyngitis), blisters (papulovesicular eruption) on the upper and inner mouth, muscle pain (myalgia), insomnia, loss of appetite (anorexia), abdominal pain, diarrhea, vomiting, intermittent nose bleeds (epistaxis), vomiting blood (hematemesis), bloody or black tar-like stools (melena), slow heart rate (brady-cardia), low blood pressure (hypotension), enlarged liver (hepatomegaly), and in extreme cases hemorrhagic pneumonia. After 1 or 2 weeks, symptoms seem to disappear then return. Signs and symptoms resemble the first phase with the addition of neurological manifestations such as mental confusion, tremors, stiff neck, abnormal reflexes, and vision deficits.

Suggested Alternatives for Differential Diagnosis: Acute leukemia, African trypanosomiasis, bacterial sepsis, borreliosis, *Chlamydia* infection, collagen-vascular diseases, Crimean-Congo hemorrhagic fever, dengue fever, ebola, Gram-negative bacterial septicemia, hantavirus pulmonary syndrome, hemolytic-uremic syndrome, hemorrhagic fever with renal syndrome, influenza, leptospirosis, malaria, Marburg virus, measles, meningococcemia, Omsk hemorrhagic fever, plague, psittacosis, rickettsialpox, Rift Valley fever, Rocky Mountain spotted fever, rubella, salmonellosis, secondary syphilis, shigellosis, smallpox, thrombocytopenic purpura, toxic shock syndrome, trypanosomiasis, typhoid fever, typhus, varicella, viral hepatitis, aflatoxicosis.

Mortality Rate (untreated): ≤10%.

C18-A027

Laguna Negra Virus
Bunyaviridae
Hantavirus Pulmonary Syndrome
ICD-11: XN6P0
UNII: —
EPPO: —

Endemic to South America. The natural reservoir is the small and large vesper mice (*Calomys laucha* and *Calomys callosus*). This is a biosafety level 3 agent.

It is on the Australia Group Core list.

In People:

CDC Case Definition for hantavirus pulmonary syndrome: Clinical criteria is an acute febrile illness (i.e., temperature greater than 101°F) with a prodrome consisting of fever, chills, myalgia, headache, and gastrointestinal symptoms, and one or more of the following clinical features: 1) bilateral diffuse interstitial edema; OR 2) clinical diagnosis of acute respiratory distress syndrome (ARDS), or 3) radiographic evidence of noncardiogenic pulmonary edema; OR 4) an unexplained respiratory illness resulting in death, and includes an autopsy examination demonstrating noncardiogenic pulmonary edema without an identifiable cause; OR 5) healthcare record with a diagnosis of hantavirus pulmonary syndrome; OR 6) death certificate lists hantavirus pulmonary syndrome as a cause of death or a significant condition contributing to death.

Laboratory criteria for diagnosis is one or more of the following laboratory findings: 1) detection of hantavirus-specific immunoglobulin M or rising titers of hantavirus-specific immunoglobulin G; OR 2) detection of hantavirus-specific ribonucleic acid in clinical specimens; OR detection of hantavirus antigen by immunohistochemistry in lung biopsy or autopsy tissues.

Communicability: Direct person-to-person transmission does not occur. When dealing with infected individuals, use standard contact precautions.

Normal Routes of Exposure: Inhalation; mucous membranes; abraded skin.

Infectious Dose: Unknown.

Secondary Hazards: Aerosols; fecal matter (mouse); blood and body fluids (mouse); fomites (from mouse habitation).

Incubation: Unknown.

Signs & Symptoms: Flu-like with fever, headache, fatigue, muscle pain (myalgia), dizziness, chills, vomiting, diarrhea, and nausea followed by an abrupt onset of diffuse accumulation of fluid in the lungs (pulmonary edema), low oxygen levels in the blood (hypoxia), and low blood pressure (hypotension).

Suggested Alternatives for Differential Diagnosis: Acute Respiratory Distress Syndrome, anthrax, congestive heart failure and pulmonary edema, HIV infection, influenza, plague, pneumonia, shock, silent myocardial infarction, tularemia, poisonings from phosgene, phosphine, or salicylate.

Mortality Rate (untreated): ≤30%

C18-A028

Lassa Fever
Arenaviridae
ICD-11: XN0CU
UNII: —
EPPO: —

Endemic to west Africa. The natural reservoir is the multimammate rat (*Mastomys natalensis*). Does not produce disease in domestic animals. Rats shed the virus in their feces and urine. This is a biosafety level 4 agent.

It is on the Australia Group Core list and the U.S. Select Agents and Toxins list.

In People:

CDC Case Definition for viral hemorrhagic fever: Clinical criteria is an illness with acute onset with ALL of the following clinical findings: 1) a fever >104°F; and 2) one or more of the following clinical findings: a) severe headache; OR b) muscle pain; OR c) erythematous maculopapular rash on the trunk with fine desquamation 3–4 days after rash onset; OR d) vomiting; OR e) diarrhea; OR f) abdominal pain; OR g) bleeding not related to injury; OR h) thrombocytopenia; OR i) retrosternal chest pain; OR j) proteinuria; OR k). pharyngitis.

Laboratory criteria for diagnosis is one or more of the following laboratory findings: 1) detection of viral hemorrhagic fever (VHF) viral antigens in blood by enzyme-linked Immunosorbent Assay (ELISA) antigen detection; 2) VHF viral isolation in cell culture for blood or tissues; 3) detection of VHF-specific genetic sequence by Reverse Transcription-Polymerase Chain Reaction (RT-PCR) from blood or tissues; 4) detection of VHF viral antigens in tissues by immunohistochemistry.

Communicability: Direct person-to-person transmission is possible. The virus is excreted in urine for 3 to 9 weeks and in semen for up to 3 months. When dealing with infected individuals, use contact and droplet precautions. For procedures that could generate infectious aerosols, use special air handling and ventilation systems, and N95 or better respirators.

Normal Routes of Exposure: Inhalation; ingestion; abraded skin; mucous membranes.

Infectious Dose: 1 to 10 viral units (inhalation).

Secondary Hazards: Aerosols; blood and body fluids (including semen); body tissue; fecal matter (rat); fomites (from rat habitation).

Incubation: 5–21 days.

Signs & Symptoms: Gradual onset of fever, chills, weakness, and a vague feeling of bodily discomfort (malaise) progressing to headache, sore throat, muscle pain (myalgia), chest pain, cough, nausea, vomiting, diarrhea, and abdominal pain. Severe cases may progress to facial swelling (edema), fluid in the lung cavity, bleeding from gums and nose. Further complications include low blood pressure (hypotension), limited urine production (oliguria), shock, seizures, tremor, disorientation, and coma. Hearing loss is a frequent consequence of the disease (sequela).

Suggested Alternatives for Differential Diagnosis: Acute anemia, acute leukemias, anaphylactoid purpura, chikungunya, Colorado tick fever, dengue fever, disseminated intravascular coagulation, encephalitis, epidemic typhus, falciparum malaria, hemolytic uremic syndrome, Henoch-Schonlein purpura, hepatitis, herpes, infectious mononucleosis, influenza, leptospirosis, malaria, measles, meningitis, meningococcemia, plague, pleural effusion, Q fever, Rocky Mountain spotted fever, rubella, sepsis, shigellosis, Sindbis fever, thrombocytopenic purpura, tick bite fever, trypanosomiasis, typhoid fever, yellow fever, other New World viral hemorrhagic fevers, drug eruptions, allergies, food poisoning.

Mortality Rate (untreated): ≤50%.

C18-A029

Lujo Virus

Arenaviridae

ICD-11: XN77P

UNII: —

EPPO: —

Endemic to southern Africa. The natural reservoir is unknown but suspected to be rodents. It is likely that the virus is shed in their urine. This is a biosafety level 4 agent.

It is on the Australia Group Core list and the U.S. Select Agents and Toxins list.

In People:

CDC Case Definition for viral hemorrhagic fever: Clinical criteria is an illness with acute onset with ALL of the following clinical findings: 1) a fever >104°F; AND 2) one or more of the following clinical findings: a) severe headache; OR b) muscle pain; OR c) erythematous maculopapular rash on the trunk with fine desquamation 3–4 days after rash onset; OR d) vomiting; OR e) diarrhea; OR f) abdominal pain; OR g) bleeding not related to injury; OR h) thrombocytopenia; OR i) retrosternal chest pain; OR j) proteinuria; OR k) pharyngitis.

Laboratory criteria for diagnosis is one or more of the following laboratory findings: 1) detection of viral hemorrhagic fever (VHF) viral antigens in blood by enzyme-linked Immunosorbent Assay (ELISA) antigen detection; 2) VHF viral isolation in cell culture for blood or tissues; 3) detection of VHF-specific genetic sequence by Reverse Transcription-Polymerase Chain Reaction (RT-PCR) from blood or tissues; 4) detection of VHF viral antigens in tissues by immunohistochemistry.

Communicability: Direct person-to-person transmission is possible. When dealing with infected individuals, use contact and droplet precautions.

Normal Routes of Exposure: Inhalation; abraded skin; mucous membranes.

Infectious Dose: Unknown.

Secondary Hazards: Aerosols; blood and body fluids (possibly semen); body tissue; fecal matter (rodent); fomites (from rodent habitation).

Incubation: 7–13 days.

Signs & Symptoms: Fever, headache, and muscle pain (myalgia), followed by diarrhea and sore throat (pharyngitis). May develop a red rash of mixed flat and raised (maculopapular) spots on the face and trunk along with neck and facial swelling. While bleeding is not a prominent feature, but there may be a rash of small hemorrhages (petechia) or bleeding gums. Individuals may experience a brief, transient period of improvement that is followed by rapid deterioration with respiratory distress, neurologic signs, and circulatory collapse.

Suggested Alternatives for Differential Diagnosis: Acute anemia, acute leukemias, anaphylactoid purpura, chikungunya, Colorado tick fever, dengue fever, disseminated intravascular coagulation, encephalitis, epidemic typhus, falciparum malaria, hemolytic uremic syndrome, Henoch-Schonlein purpura, hepatitis, herpes, infectious mononucleosis, influenza, leptospirosis, malaria, measles, meningitis, meningococcemia, plague, pleural effusion, Q fever, Rocky Mountain spotted fever, rubella, sepsis, shigellosis, Sindbis fever, thrombocytopenic purpura, tick bite fever, trypanosomiasis, typhoid fever, yellow fever, other New World viral hemorrhagic fevers, drug eruptions, allergies, food poisoning.

Rate (untreated): ≤80%.

C18-A030

Louping III
Flaviviridae
ICD-11: —
UNII: —
EPPO: —

Endemic to the British Isles. Viral agents producing disease indistinguishable from louping ill have been reported in Norway, Spain, Turkey, and Bulgaria. The natural reservoir is sheep and ticks (*Ixodes ricinus*). Ticks remains infective for life. Sheep and red grouse act as amplifying hosts. This is a biosafety level 3 agent, with the addition of HEPA filtration attached to the laboratory exhaust system.

It is on the Australia Group Core list and the OIE list of Notifiable Diseases for 2020.

In People:
CDC Case Definition: None established.

Communicability: Direct person-to-person transmission does not occur. When dealing with infected individuals, use standard precautions.

Normal Routes of Exposure: Vectors (ticks); abraded skin; ingestion.

Infectious Dose: Unknown.

Secondary Hazards: Aerosols; blood and body fluids (animals); body tissue (animals); fomites (with vectors).

Incubation: 2–8 days.

Signs & Symptoms: Disease is often biphasic. Initially, fever, headache, joint pain (arthralgia) and a vague feeling of bodily discomfort (malaise). After a period of remission, may progress to fatigue, stiff neck, muscle pain (myalgia), muscle weakness, confusion, agitation, hallucinations, paralysis, and seizures.

Suggested Alternatives for Differential Diagnosis: Influenza, other tickborne encephalitis, poliomyelitis.

Rate (untreated): ~0%.

In Domestic Animals:

Susceptible species: Sheep, goats, cattle.

Communicability: Direct transmission does not occur.

Normal routes of exposure: Vectors (ticks).

Secondary Hazards: Body fluids (goat milk); body tissue; fomites (with vectors); vector amplification (in sheep).

Incubation: 6–18 days.

Signs: In some cases, sudden death may occur with no apparent signs of illness. Otherwise, disease is usually biphasic. Fever, depression, and loss of appetite (anorexia) progressing to depression, panting, increased sensitivity to noise or touch (hyperesthesia), muscle tremors, incoordination (ataxia), circling, sideways twitching of the neck (torticollis), progressive paralysis, convulsions and coma. Sheep develop a characteristic hopping gait with both front legs being moved forward simultaneously followed by both back ones.

Post-mortem does not present any characteristic gross findings. There is nonsuppurative polioen-cephalomyelitis with lesions predominantly in the brain stem.

Suggested Alternatives for Differential Diagnosis: Bovine spongiform encephalopathy, coenurosis, hydatid disease, listeriosis, maedi-visna, malignant catarrhal fever, pregnancy toxemia, pseudorabies, rabies, scrapie, tetanus, tick pyemia, hypocalcemia, hypocuprosis, hypomagnesemia, poisonings from lead or toxic plants.

Mortality Rate (untreated): ≤60% (sheep).

C18-A031

Lumpy Skin Disease
Poxviridae
ICD-11: —
UNII: YBZ5T12O30
EPPO: —

Endemic to Africa, the Middle East, and Turkey. The natural reservoir is cattle and African buffalo. It is very resistant to physical and chemical agents. It can persist in necrotic skin for forty days and remains viable in air-dried hides for over 18 days at ambient temperature. This is a biosafety level 3-Ag agent.

It is on the Australia Group Core list, the U.S. Select Agents and Toxins list, and the OIE list of Notifiable Diseases for 2020.

In People:
Does not occur in humans.

In Domestic Animals:
Target species: Cattle, Asian buffalo, water buffalo.

Communicability: Direct transmission is possible.

Normal routes of exposure: Vectors (mosquitoes, biting flies); ingestion; abraded skin; mucous membranes.

Secondary Hazards: Blood and body fluids (including semen); fecal matter; body tissue; fomites (hides).
Incubation: 6–9 days.

Signs: Fluctuating fever, eye infection (pink eye – conjunctivitis), tearing (lacrimation), nasal discharge, and excessive salivation or drooling (ptyalism). The discharge from the eyes and nose becomes a mixture of mucus and pus (mucopurulent), and the cornea may become inflamed (keratitis). Animals develop characteristic eruptions on the skin, mucous membranes, and internal organs. Theses nodules are well circumscribed, round, slightly raised, firm, and painful; containing a firm, creamy-gray or yellow mass of tissue. Nodules become swollen (edema) with inflamed blood vessels (vasculitis) and hemorrhage. They may become necrotic plugs that penetrate the full thickness of the hide; which the animal ultimately sheds, leaving deep holes. Regional lymph nodes are swollen, congested and hemorrhage. Swelling (edema) develops in the udders, brisket, and legs. Animals become emaciated and subject to secondary bacterial infections.

Post-mortem findings include nodules penetrating the thickness of the skin and involving the subcutaneous tissue which appears edematous and infiltrated with blood tinged fluid. Lymph nodes may be grossly enlarged due to lymphoid hyperplasia and edema. There are ulcerative lesions in the mucous membranes of the respiratory and digestive tract, with interlobular edema and nodules in the lungs.

Suggested Alternatives for Differential Diagnosis: Pseudo-lumpy skin disease, *Dermatophilus congolensis*, besnoitiosis, bovine ephemeral fever, bovine farcy, bovine herpes dermophatic infection, bovine papular stomatitis, cattle grubs, cutaneous tuberculosis, demodicosis, dermatophilosis (streptothricosis), *Hypoderma bovis* infection, onchocercosis, photosensitization, pseudo-cowpox, rinderpest, ringworm, screw-worm myiasis, sporadic bovine lymphomatosis, sweating weakness of calves, urticaria, vesicular disease, allergies.

Mortality Rate (untreated): ≤10%.

C18-A032

Lymphocytic Choriomeningitis
Arenaviridae
ICD-11: XN4ZL
UNII: —
EPPO: —

Endemic worldwide. The natural reservoir is the common house mouse (*Mus musculus*). The virus is shed in their saliva, urine, and fecal matter. Disease can be passed to rodent pets (e.g., mice, hamsters, guinea pigs). The virus is quickly inactivated outside a host. This is a biosafety level 2 agent unless there is a high risk of aerosol production or dealing with high concentrations of infectious materials, then it should be treated as a biosafety level 3 agent.

It is on the Australia Group Core.

In People:

CDC Case Definition: Clinical criteria is an illness with acute onset with ALL of the following clinical findings: 1) a fever >104°F; AND 2) one or more of the following clinical findings: a) severe headache; OR b) muscle pain; OR c) erythematous maculopapular rash on the trunk with fine desquamation 3–4 days after rash onset; OR d) vomiting; OR e) diarrhea; OR f) abdominal pain; OR g) bleeding not related to injury; OR h) thrombocytopenia; OR i) retrosternal chest pain; OR j) proteinuria; OR k) pharyngitis.

Laboratory criteria for diagnosis is one or more of the following laboratory findings: 1) detection of viral hemorrhagic fever (VHF) viral antigens in blood by enzyme-linked Immunosorbent Assay (ELISA) antigen detection; 2) VHF viral isolation in cell culture for blood or tissues; 3) detection of VHF-specific genetic sequence by Reverse Transcription-Polymerase Chain Reaction (RT-PCR) from blood or tissues; 4) detection of VHF viral antigens in tissues by immunohistochemistry.

Communicability: Direct person-to-person transmission does not occur. Can be transmitted through solid organ transplants. When dealing with infected individuals, use standard precautions.

Normal Routes of Exposure: Inhalation; abraded skin; mucous membranes; ingestion.

Infectious Dose: Unknown.

Secondary Hazards: Aerosols; blood and body fluids (rodent); fecal matter (rodent); fomites (from rodent habitation).

Incubation: 5–13 days.

Signs & Symptoms: Often a biphasic febrile illness. During the first phase fever, vague feeling of bodily discomfort (malaise), muscle pain (myalgia), weakness, nausea, vomiting, retro-orbital headache, sensitivity to light (photophobia), loss of appetite (anorexia), light-headedness. Additional symptoms may include sore throat, cough, and pain in the joints (arthralgias), chest, testicles, and salivary gland. Symptoms may subside for several days and then reoccur with fever, increased headache, rashes, neck stiffness (nuchal rigidity), inflammation of the testicles (orchitis), inflammation of the salivary glands (parotitis), hair loss (alopecia), arthritis of the hand, drowsiness, confusion, sensory disturbances, coma. Individuals may also experience inflammation of the membranes that surround the brain and spinal cord (meningitis), inflammation of the brain (encephalitis), inflammation of the spinal cord (myelitis), inflammation of the heart muscle (myocarditis), inflammation of lung tissue (pneumonitis), inflammation of the sac surrounding the heart (pericarditis), water on the brain (hydrocephalus), Guillain-Barre-type syndrome, cranial nerve palsies, paralysis. Recovery is generally complete but may be prolonged.

Suggested Alternatives for Differential Diagnosis: Chikungunya virus, coccidioidomycosis, crypto-coccosis, eastern equine encephalitis, enteroviruses, herpes simplex, herpes simplex virus, influenza, Japanese encephalitis, leptospirosis, Malignancy, mumps, St. Louis encephalitis, syphilis, tuberculous, typhoid fever, varicella-zoster virus, West Nile virus, western equine encephalitis, yellow fever, zika virus, poisonings from nonsteroidal anti-inflammatory drugs (NSAIDs) (e.g., aspirin, ibuprofen, naproxen).

Mortality Rate (untreated): <1%.

C18-A033

Machupo Virus
Arenaviridae
Bolivian Hemorrhagic Fever
ICD-11: XN45B
UNII: —
EPPO: —

Endemic to Bolivia. The natural reservoir is the large vesper mouse (*Calomys callosus*). Rodents can shed the virus in saliva, feces, and urine. Does not produce disease in domestic animals. It can survive in blood specimens outside of a host for up to 2 weeks. This is a biosafety level 4 agent. It is on the Australia Group Core list and the U.S. Select Agents and Toxins list.

In People:
CDC Case Definition for viral hemorrhagic fever: Clinical criteria is an illness with acute onset with ALL of the following clinical findings: 1) a fever >104°F; and 2) one or more of the following clinical findings: a) severe headache; OR b) muscle pain; OR c) erythematous maculopapular rash on the trunk with fine desquamation 3–4 days after rash onset; OR d) vomiting; OR e) diarrhea; OR f) abdominal pain; OR g) bleeding not related to injury; OR h) thrombocytopenia; OR i) retrosternal chest pain; OR j) proteinuria; OR k) pharyngitis.

Laboratory criteria for diagnosis is one or more of the following laboratory findings: 1) detection of viral hemorrhagic fever (VHF) viral antigens in blood by enzyme-linked Immunosorbent Assay (ELISA) antigen detection; 2) VHF viral isolation in cell culture for blood or tissues; 3) detection of VHF-specific genetic sequence by Reverse Transcription-Polymerase Chain Reaction (RT-PCR) from blood or tissues; 4) detection of VHF viral antigens in tissues by immunohistochemistry.

Communicability: Direct person-to-person transmission is rare. When dealing with infected individuals, use contact and droplet precautions. For procedures that could generate infectious aerosols, use special air handling and ventilation systems, and N95 or better respirators.
Normal Routes of Exposure: Inhalation; ingestion; abraded skin; mucous membranes.
Infectious Dose: 1–10 viral units (inhalation).

Secondary Hazards: Aerosols; blood and body fluids (mouse); body tissue (mouse); fecal matter (mouse); fomites (from mouse habitation).

Incubation: 4–21 days.

Signs & Symptoms: Fever, headache, fatigue, muscle pain (myalgia), and joint pain (arthralgia). May progress to shock, hemorrhage from mucous membrane, nose bleeds (epistaxis), vomiting blood (hematemesis), passing black tar-like stools (melena), and blood in the urine (hematuria), accompanied by neurological damage such as tremor, seizures, and coma.

Suggested Alternatives for Differential Diagnosis: Acute anemia, acute leukemias, anaphylactoid purpura, chikungunya, Colorado tick fever, dengue fever, disseminated intravascular coagulation, encephalitis, epidemic typhus, falciparum malaria, hemolytic uremic syndrome, Henoch-Schonlein purpura, hepatitis, herpes, infectious mononucleosis, influenza, leptospirosis, malaria, measles, meningitis, meningococcemia, plague, pleural effusion, Q fever, Rocky Mountain spotted fever, rubella, sepsis, shigellosis, Sindbis fever, thrombocytopenic purpura, tick bite fever, trypanoso-miasis, typhoid fever, yellow fever, other New World viral hemorrhagic fevers, drug eruptions, allergies, food poisoning.

Mortality Rate (untreated): ≤35%.

C18-A034

Marburgvirus Disease
Filoviridae
ICD-11: XN3F2
UNII: UJP06Q687J
EPPO: —

Endemic to Africa. The natural reservoir is the African fruit bat (Rousettus aegyptiacus). It is unknown how the virus crosses from bats to people, however, contact with bat feces or inhalation with viral aerosols has been postulated. Other than humans, the disease has only been documented in primates, with a similar clinical picture. Can survive up to 5 days on surfaces and can survive in liquid or dried material for a number of days. This is a biosafety level 4 agent.

It is on the Australia Group Core list and a Tier-1 agent on the U.S. Select Agents and Toxins list.

In People:
CDC Case Definition for viral hemorrhagic fever: Clinical criteria is an illness with acute onset with ALL of the following clinical findings: 1) a fever >104°F; and 2) one or more of the following clinical findings: a) severe headache; OR b) muscle pain; OR c) erythematous maculopapular rash on the trunk with fine desquamation 3–4 days after rash onset; OR d) vomiting; OR e) diarrhea; OR f) abdominal pain; OR g) bleeding not related to injury; OR h) thrombocytopenia.

Laboratory criteria for diagnosis is one or more of the following laboratory findings: 1) detection of viral hemorrhagic fever (VHF) viral antigens in blood by enzyme-linked Immunosorbent Assay (ELISA) antigen detection; 2) VHF viral isolation in cell culture for blood or tissues; 3) detection of VHF-specific genetic sequence by Reverse Transcription-Polymerase Chain Reaction (RT-PCR) from blood or tissues; 4) detection of VHF viral antigens in tissues by immunohistochemistry.

Communicability: Direct person-to-person transmission is possible. When dealing with infected individuals, use contact and droplet precautions. For procedures that could generate infectious aerosols, use special air handling and ventilation systems, and N95 or better respirators.

Normal Routes of Exposure: mucous membranes; abraded skin; ingestion.

Infectious Dose: 1–10 viral units (inhalation).

Secondary Hazards: Aerosols; blood and body fluids (including semen); body tissue; fecal matter; fomites.

Incubation: 2–14 days.

Signs & Symptoms: Sudden onset of nonspecific symptoms including fever, vague feeling of bodily discomfort (malaise), muscle pain (myalgia), joint pain (arthralgia), and headache. After approximately five days, a centrally located rash of mixed flat and raised (maculopapular) spots appears. The rash is most prominent on the chest, back, and stomach. Progresses to abdominal pain, nausea, vomiting, diarrhea, and sore throat (pharyngitis). Hemorrhagic features may include small red or purple spots on the surface of the skin (petechia) and bleeding from mucous membranes. Severe bleeding typically occurs in cases that are fatal and tends come from the stomach and intestines (gastrointestinal tract). Death is due to a combination of hemorrhage, capillary leakage, shock, and end-organ failure.

Suggested Alternatives for Differential Diagnosis: Acute leukemia, cholera, Crimean-Congo hemorrhagic fever, dengue fever, disseminated intravascular coagulation, Ebola, hemolytic uremic syndrome, influenza, Lassa, leptospirosis, lupus, malaria, meningococcemia, norovirus, plague, rickettsial infections, salmonella infection, shigellosis, thrombocytopenic purpura, typhoid fever.

Mortality Rate (untreated): ≤90%.

C18-A035

Monkeypox
Poxviridae
ICD-11: XN2GM
UNII: —
EPPO: —

Endemic to Central and West Africa. The natural reservoir is unknown but suspected to be an African rodent. It is highly stable and may remain infectious for weeks in the desiccated form. This is a biosafety level 3 agent.

It is on the Australia Group Core list and the U.S. Select Agents and Toxins list

In People:
CDC Case Definition: None established.

Communicability: Direct person-to-person transmission is possible. When dealing with infected individuals, use contact and airborne precautions.

Normal Routes of Exposure: Inhalation; mucous membranes; abraded skin.

Infectious Dose: Unknown.

Secondary Hazards: Aerosols (respiratory secretions); blood and body fluids; body tissue (skin lesions, scabs); fomites.

Incubation: 4–21 days.

Signs & Symptoms: Fever, chills, cough, severe headache, muscle pain (myalgia), vague feeling of bodily discomfort (malaise), loss of appetite (anorexia), sore throat (pharyngitis), difficulty breathing (dyspnea), fatigue, with swollen lymph nodes (lymphadenopathy). There may be ulcers on the mucous membranes of the oral cavity, eye infection (pink eye – conjunctivitis) with swollen eyelids and painful lesions in the genital regions. A rash of flat spots (macules) appears on the face then spreads to other part of the body. The rash is centrifugal, with greatest concentration of lesions on the face and distal extremities rather than the truck. Rash progress to bumps, then hard blisters (pustules) and then scabs. Rash may be accompanied by severe itching (pruritus).

In the epidemic of 2022, a number of cases presented atypical signs including as little as only 1 or 2 pox lesions that only appeared on the genitals or around the anus. Some lesions did not follow the classic progression of bumps to pustules to scabs, but merely appeared as ulcers or craters.

Suggested Alternatives for Differential Diagnosis: Acne, bullous pemphigoid, chickenpox, cowpox, coxsackievirus, cytomegalovirus, eczema herpeticum, eczema vaccinatum, ehrlichiosis, general-ized vaccinia, herpes simplex, herpes zoster, impetigo, Kawasaki disease, malaria, measles, meningococcemia, molluscum contagiosum, parvovirus B19, plague, pseudocowpox, rickett-sialpox, Rocky Mountain spotted fever, rubella, scarlet fever, smallpox, syphilis, contact dermatitis, drug eruptions, scabies and other insect bites.

Mortality Rate (untreated): ≤33%.

In Domestic Animals:
Susceptible species: Rodent and Lagomorph pets.

Communicability: Direct transmission is possible.

Normal routes of exposure: Inhalation; abraded skin; mucous membranes.

Secondary Hazards: Aerosols (respiratory secretions); blood and body fluids; body tissue; fomites.
Incubation: 6–7 days.

Signs: Varies on virus strain and animal species. Fever, conjunctivitis, cough, lethargy, loss of appetite (anorexia), rash, profuse discharge from the nose and eyes, difficulty breathing (dyspnea), swollen lymph nodes (lymphadenopathy).

Suggested Alternatives for Differential Diagnosis: Ectromelia virus, mites, myxomatosis, other poxviruses.

Mortality Rate (untreated): Varies greatly, species dependent.

C18-A036

Murray Valley Encephalitis
Flaviviridae
ICD-11: 1C88
UNII: —
EPPO: —

Endemic to Australia and New Guinea. The natural reservoir is water birds and mosquitoes. Mosquitoes remains infective for life. This is a biosafety level 3 agent.

It is on the Australia Group Core list.

In People:

CDC Case Definition: None established.

Communicability: Direct person-to-person transmission does not occur. When dealing with infected individuals, use standard precautions.

Normal Routes of Exposure: Vectors (mosquitoes).

Infectious Dose: Unknown.

Secondary Hazards: Aerosols; blood and body fluids.

Incubation: 5–28 days.

Signs & Symptoms: Rapid onset of high fever, headache, muscle pain (myalgia), vague feeling of bodily discomfort (malaise), loss of appetite (anorexia), nausea, vomiting, and diarrhea. May progress to neurological symptoms including sensitivity to light (photophobia), numbness, neck stiffness (nuchal rigidity), trouble speaking, incoordination (ataxia), drowsiness, stupor, tremors, spastic paralysis, seizures, and coma. Up to 50% of survivors continue to have neurologic consequence of the disease (sequela) including paralysis of the lower half of the body (paraplegia), impaired gait and motor control, and decreased intellect.

Suggested Alternatives for Differential Diagnosis: Bartonellosis, basilar artery blood clots, California encephalitis, cardioembolic stroke, cavernous sinus syndromes, cerebral venous blood clots, coxsackieviruses, cytomegalovirus, eastern equine encephalitis, epileptic and epileptiform encephalopathies, Epstein-Barr virus, febrile seizures, *Haemophilus* meningitis, Hendra virus, herpes simplex, histoplasmosis, intracranial hemorrhage, Japanese encephalitis, leptomeningeal carcinomatosis, leptospirosis, Lyme disease, malaria, meningitis, Naegleria infection, rabies, Rocio encephalitis, St. Louis encephalitis, subdural empyema or hematoma, tuberculosis, Venezuelan equine encephalitis, West Nile fever, western equine encephalitis, confusional states, and acute memory disorders.

Mortality Rate (untreated): ≤20%.

C18-A037

Newcastle Disease (Agent OE)
Paramyxoviridae
ICD-11: 1D84.Y
UNII: —
EPPO: —

Endemic throughout the world. The natural reservoir is birds. It can survive in poultry houses for up to 7 days in summer, 14 days in spring, and 30 days in winter; and at least 3 months in liquid manure. The virus can also survive freezing. It has been recovered from soil after 22 days and lake water after 19 days. This is a biosafety level 3-Ag agent. Laboratory workers should have no contact with susceptible hosts for 5 days after working with the agent.

It is on the Australia Group Core list, the U.S. Select Agents and Toxins list, and the OIE list of Notifiable Diseases for 2020.

In People:

CDC Case Definition: None established.

Communicability: Direct person-to-person transmission does not occur. When dealing with infected individuals, use standard precautions.

Normal Routes of Exposure: Inhalation; mucous membranes.

Infectious Dose: Unknown.

Secondary Hazards: Aerosols; blood and body fluids (bird); fecal matter (bird); fomites (from bird habitation).

Incubation: 1–4 days.

Signs & Symptoms: Unilateral or bilateral eye infections with reddening, excessive tearing (lachrymation), swollen eyelids, and hemorrhage under the whites of the eyes (sub-conjunctival). Infections are usually transient and the cornea is not affected. Rarely produces a mild, self-limiting influenza-like disease with fever, headache, and vague feeling of bodily discomfort (malaise).

Suggested Alternatives for Differential Diagnosis: None.

Mortality Rate (untreated): 0%.

In Domestic Animals:

Target species: Poultry.

Communicability: Direct transmission is possible.

Normal routes of exposure: Inhalation; ingestion; mucous membranes.

Secondary Hazards: Aerosols (respiratory secretions, fomites); blood and body fluids; fecal matter; fomites; mechanical vectors (track out).

Incubation: 2–15 days.

Signs: In some cases, sudden death may occur without any or only few signs. Otherwise, swelling of the neck and head with red eyes and bluish (cyanotic) discoloration; loss of appetite (anorexia), greenish or white watery diarrhea; difficulty breathing (dyspnea) with gasping, coughing, sneezing, noncontinuous clicking or rattling sounds when breathing (rales); tremors, spasms, drooping wings, dragging legs, twisted neck, circling, paralysis. Hens have a partial or complete drop in egg production may occur. Eggs may be abnormal in color, shape, or surface. Surviving birds may have permanent neurological damage and/or a permanent decrease in egg production.

There are no specific diagnostic postmortem lesions seen with Newcastle disease. However, findings may include swelling of periorbital area or entire head with edema of the interstitial or peritracheal tissue of the neck, especially at the thoracic inlet; congestion and sometimes hemorrhages in the caudal pharynx and tracheal mucosa; diphtheritic membranes may be evident in the oropharynx, trachea, and esophagus; petechiae and small ecchymoses on the mucosa of the proventriculus, concentrated around the orifices of the mucous glands; edema, hemorrhages, necrosis or ulcerations of respiratory/digestive lymphoid tissue, including cecal tonsils and Peyer's patches; the spleen may be enlarged, friable, and dark red or mottled.

Suggested Alternatives for Differential Diagnosis: Adenovirus, aspergillosis, avian chlamydiosis, avian encephalomyelitis, avian influenza, bronchitis, coryza, fowl cholera, fowl pox, infectious bronchitis, laryngotracheitis, Marek's disease, mycoplasmosis, Pacheco's disease, psittacosis, salmonellosis, toxicosis, botulism, vitamin E deficiency, water or feed deprivation, poor ventilation.

Mortality Rate (untreated): ≤100%.

C18-A038

Nipah Virus
Paramyxoviridae
ICD-11: XN931
UNII: —
EPPO: —

Endemic to Malaysia, Singapore, Bangladesh, and India. The natural reservoir is fruit bats (*Pteropid* spp.). Pigs are the amplifying host and special precautions should be taken when examining a pig suspected of having the disease or performing a necropsy. The virus can survive for up to 3 days on some fruits and in some fruit juices. The half-life is reported to be 18 hours in the urine of fruit bats. This is a biosafety level 4 agent.

It is on the Australia Group Core list, the U.S. Select Agents and Toxins list, and the OIE list of Notifiable Diseases for 2020.

In People:
CDC Case Definition: None established.

Communicability: Direct person-to-person transmission is possible. When dealing with infected individuals, use contact and droplet precautions. For procedures that could generate infectious aerosols, use special air handling and ventilation systems, and N95 or better respirators.

Normal Routes of Exposure: Inhalation; mucous membranes; ingestion.

Infectious Dose: Unknown.

Secondary Hazards: Aerosol (respiratory secretions); blood and body fluids; body tissue; fomites.
Incubation: 2–30 days.

Signs & Symptoms: Fever, severe headache, sore throat, muscle pain (myalgia), nonproductive cough, nausea, vomiting, followed by inflammation of the brain (encephalitis) with drowsiness, disorientation and mental confusion progressing to segmental irregular muscular twitching (myoclonus), convulsions, seizures, and coma. Some individuals experience atypical pneumonia or acute respiratory distress syndrome (ARDS) with or without developing neurological signs. Survivors of encephalitis may have residual neurological deficits or remain in a vegetative state. Others can develop relapsed encephalitis months or years later.

Suggested Alternatives for Differential Diagnosis: Basilar artery blood clots, cardioembolic stroke, cavernous sinus syndromes, cerebral venous blood clots, epileptic and epileptiform encephalopathies, febrile seizures, hemophilus meningitis, intracranial hemorrhage, leptomeningeal carcinomatosis, meningitis, subdural empyema or hematoma, confusional states, and acute memory disorders.

Mortality Rate (untreated): ≤75%.

In Domestic Animals:

Susceptible species: Pigs. May also affect horses, cats, and dogs.

Communicability: Direct transmission is possible.

Normal routes of exposure: Inhalation; ingestion; mucous membranes.

Secondary Hazards: Aerosols (respiratory secretions); blood and body fluids; body tissues; fomites.

Incubation: 4–14 days.

Signs: Presentation varies with the age of the pig. Piglets may present with labored breathing and muscle tremors with limb weakness. Young swine may present with acute fever, labored open-mouthed breathing, nasal discharge, loud nonproductive cough (i.e., barking pig syndrome), and coughing up blood (hemoptysis). Neurologic signs can include involuntary muscular contractions or twitching (fasciculations), limb weakness, spastic paresis, and tetanic spasms. Mature pigs may suffer sudden death without signs. Otherwise, may present acute fever, open-mouthed breathing, nasal discharge and excessive salivation or drooling (ptyalism), rapid involuntary movements of the eyes (nystagmus), teeth grinding (bruxism), head pressing, aggressive behavior, trembling, twitching, spastic paresis, tetanic spasms, and seizures. Sows may abort.

Post-mortem findings include the trachea and bronchi may be filled with clear or blood-tinged frothy exudate, mild to severe pulmonary consolidation with petechial or ecchymotic hemorrhages, and distended interlobular septa. There may be meningeal edema with congestion of the cerebral blood vessels.

Suggested Alternatives for Differential Diagnosis: Classical swine fever, porcine pleuropneumonia caused by *Actinobacillus pleuropneumonia* or *Pasteurellosis* species, porcine reproductive and respiratory syndrome, pseudorabies, swine enzootic pneumonia caused by *Mycoplasma hyopneumoniae.*

Mortality Rate (untreated): ≤5%; ≤40% (piglets).

C18-A039

Omsk Hemorrhagic Fever
Flaviviridae
ICD 11: 1D4A
UNII: —
EPPO: —

Endemic to western Siberia. The natural reservoir is ticks. Ticks remains infective for life. Rodents are potent amplifying hosts of virus, although they become ill and die when infected. It can survive for up to 2 weeks in warm water, and up to 3.5 months in cold water. This is a biosafety level 4 agent.

It is on the Australia Group Core list and the U.S. Select Agents and Toxins list.

In People:

CDC Case Definition: None established.

Communicability: Direct person-to-person transmission does not occur. When dealing with infected individuals, use standard precautions.

Normal Routes of Exposure: Inhalation; ingestion; abraded skin; mucous membranes; vectors (ticks).

Infectious Dose: Unknown.

Secondary Hazards: Aerosols; body tissue (rodent); blood and body fluids (rodent, milk from goats or sheep); fecal matter (rodent); fomites (with vectors).

Incubation: 1–10 days.

Signs & Symptoms: Often biphasic, with sudden onset of fever, chills, headache, cough, muscle pain (myalgia), swollen lymph nodes, diarrhea, slow heart beat (bradycardia), and low blood pressure (hypotension). The liver becomes enlarged (hepatomegaly). The face, eyes, and throat are swollen and red, becoming more pronounced and intense as the disease progresses. Hemorrhagic manifestations including small red or purple spots on the surface of the skin (petechia), nose bleeds (epistaxis), bleeding gums, coughing up blood (hemoptysis), blood in the urine (hematuria), and vomiting blood (hematemesis). After 1 or 2 weeks, symptoms seem to disappear then return. Presentation with fever, pneumonia, kidney inflammation (nephritis) and inflammation of the brain (encephalitis).

Suggested Alternatives for Differential Diagnosis: Acute anemia, acute leukemias, anaphylactoid purpura, chikungunya, Colorado tick fever, dengue fever, disseminated intravascular coagulation, encephalitis, epidemic typhus, falciparum malaria, hemolytic uremic syndrome, Henoch-Schonlein purpura, hepatitis, herpes, infectious mononucleosis, influenza, leptospirosis, malaria, measles, meningitis, meningococcemia, plague, pleural effusion, Q fever, Rocky Mountain spotted fever, rubella, sepsis, shigellosis, Sindbis fever, thrombocytopenic purpura, tick bite fever, trypanosomiasis, typhoid fever, yellow fever, other New World viral hemorrhagic fevers, drug eruptions, allergies, food poisoning.

Mortality Rate (untreated): ≤3%.

C18-A040

Oropouche Virus
Bunyaviridae
ICD-11: 1D43
UNII: —
EPPO: —

Endemic to Central and South America. The natural reservoir is unknown but it has been isolated from three-toed sloths and black-tufted marmosets. Antibodies, but not the actual virus, have been identified in rodents and birds. It is transmitted to humans by midges (Culicoides paraenesis). This is a biosafety level 3 agent, with the addition of HEPA filtration attached to the laboratory exhaust system.

It is on the Australia Group Core list.

In People:

CDC Case Definition: None established.

Communicability: Direct person-to-person transmission does not occur. When dealing with infected individuals, use standard precautions. Humans can infect new midges during the fever phase, protect from vectors.

Normal Routes of Exposure: Vectors (midges).

Infectious Dose: Unknown.

Secondary Hazards: Blood and body fluids (animal); vector amplification.

Incubation: 3–12 days.

Signs & Symptoms: High fever, chills, vague feeling of bodily discomfort (malaise), severe headache, severe retro-orbital pain, joint pain (arthralgia), muscle pain (myalgia), loss of appetite (anorexia), weakness, dizziness, sensitivity to light (photophobia), abdominal pain, nausea, vomiting, and diarrhea. May also develop a rash of mixed flat and raised (maculopapular) spots on the trunk and arms and sometimes the legs. Other possible complications include hemorrhagic phenomena (e.g., small hemorrhages (petechia), nose bleeds (epistaxis), and gingival bleeding) and neurological issues such as vertigo, lethargy, double vision (diplopia), rapid involuntary movements of the eyes (nystagmus), and neck stiffness (nuchal rigidity).

Suggested Alternatives for Differential Diagnosis: Chikungunya, dengue fever, Mayaro virus disease, zika.

Mortality Rate (untreated): 0%.

C18-A041

Peste des Petits Ruminants
Paramyxoviridae
ICD-11: —
UNII: —
EPPO: —

Endemic to Africa, the Middle East, and Asia. The natural reservoir is sheep and goats, although wild ruminants also pose a source of infection. Infection results in long-term immunity. Aerosols can travel up to approximately 10 meters. It has a long survival time in chilled and frozen tissues. This is a biosafety level 3-Ag agent.

It is on the Australia Group Core list, the U.S. Select Agents and Toxins list, and the OIE list of Notifiable Diseases for 2020.

In People:
Does not occur in humans.

In Domestic Animals:
Target species: Goats, sheep.

Communicability: Direct transmission is possible.

Normal routes of exposure: Inhalation; mucous membranes.

Secondary Hazards: Aerosols (respiratory secretions); body fluids (including milk and semen); fecal matter; fomites.

Incubation: 2–10 days.

Signs: Fever followed by severe depression and loss of appetite (anorexia). There is a watery nasal discharge that progress to a profuse discharge of mucus and pus (mucopurulent), which eventually occludes the nostrils and creates respiratory distress. Eyes become infected (conjunctivitis), congested, and may produce copious mucus discharge. There is painful, necrotic inflammation of the mouth producing an unpleasant, fetid odor on the animal's breath – and of the intestinal tract (enteritis) with severe diarrhea, dehydration, and weight loss. Animals may develop bronchop-neumonia. Death is usually from dehydration or pneumonia.

Post-mortem findings include an overall appearance of emaciation and dehydration with sunken eyes, dried nasal and ocular discharges, and fecal soiling. There are necrotic lesions in the mouth, nose and throughout the gastrointestinal tract. The small intestines are congested with small streaks of hemorrhages and some erosions. Peyer's patches are the site of extensive necrosis, which may result in severe ulceration. The large intestine is usually more severely affected with congestion around the ileocecal valve, at the cecocolic junction, and in the rectum. The mucosal folds in the cecum, colon and rectum have hemorrhaged or darkened tissue resulting in 'tiger striping'. There is congestion and enlargement of spleen and liver (hepatosplenomegaly). Lymph nodes, particularly those associated with the respiratory and gastrointestinal tracts, are generally congested, enlarged and edematous. Respiratory lesions may include congestion of the lungs; small erosions and petechiae in the nasal mucosa, turbinates, larynx and trachea; and bronchopneumonia.

Suggested Alternatives for Differential Diagnosis: Bluetongue, coccidiosis, contagious caprine pleuropneumonia, contagious ecthyma, foot and mouth disease, gastrointestinal helminth infestations, heartwater, Nairobi sheep disease, orf, pasteurellosis, rinderpest, salmonellosis, sheep pox, poisonings from minerals (e.g., arsenic).

Mortality Rate (untreated): ≤100%.

C18-A042

Plum Pox
Potyviridae
ICD-11: —
UNII: —
EPPO: PPV000

Endemic to Europe, it has spread through the Middle East and Chile; and has limited distribution in North America, India, and China. It is transmitted by many species of aphids. Aphids do not remain infected; they lose the virus particles at the next feeding.

In Plants:

Target species: Peaches, apricots, plums, nectarines, cherries, almonds.

Normal routes of transmission: Vectors (aphids).

Secondary Hazards: Crop debris (nursery stock, budwood); mechanical vectors (grafting).

Signs: Presentation of signs is variable and depends on the host, age of the tree, and the environment. They may be observed on only a few leaves or fruit, along one limb, or they may be expressed throughout the entire tree. They may not appear until several months post-infection. Infected trees rarely die but do become less productive as the disease progresses. Crop loss up to 30% has been reported

Plums: Diffuse yellowed (chlorotic) or olive-green rings or mottling on the leaves. Fruit has dark-colored rings or depressions on the skin, brown or reddish necrotic flesh with gumming, and brown spots on the stones. Fruit drops prematurely.

Peaches and nectarines: Yellowed (chlorotic) spotting and distortion of the leaves. Fruit has yellowish rings or diffuse bands on the skin. Fruit may drop prematurely.

Apricots: Yellowed (chlorotic) or pale-green rings and lines on the leaves. Fruit is deformed and has light colored, depressed, necrotic rings and yellow rings on the stones. Fruit may drop prematurely.

Cherries: Pale green patterns and rings appear on the leaves. Fruit is slightly deformed, with yellowed (chlorotic) and necrotic rings and notched marks. Fruit drops prematurely.

Almonds: Infection is often symptomless.

Crop Losses: ≤100%

C18-A043

Porcine Teschovirus
Picornaviridae
ICD-11: —
UNII: —
EPPO: —

Endemic to eastern Europe and Madagascar. The natural reservoir is pigs and wild boar. It can survive in the environment for months and longer in liquid manure. This is a biosafety level 3 agent.

It is on the Australia Group Core list and the OIE list of Notifiable Diseases for 2020.

In People:
Does not occur in humans.

In Domestic Animals:
Target species: Pigs.

Communicability: Direct transmission is possible.

Normal routes of exposure: Ingestion; mucous membranes (nasal).

Secondary Hazards: Body fluids (oral, urine); fecal matter; fomites.

Incubation: 5–28 days.

Signs: Incoordination (ataxia), fever, loss of appetite (anorexia), depression, tremors, incoordination, stiffness of the limbs, rapid involuntary movements of the eyes (nystagmus), seizures, backward bridging (opisthotonos), sustained rhythmical jerking (clonic) convulsions of the legs, progressive paralysis beginning in the hind legs and traveling cranially, and coma. Once paralysis affects the respiratory muscles, the animal dies of suffocation.

Post-mortem does not present any characteristic gross findings. Meningitis is common over the cerebellum.

Suggested Alternatives for Differential Diagnosis: African swine fever, bacterial meningoencephalitis, classical swine fever, edema disease, hemagglutinating encephalomyelitis, hypoglycemia, Japanese encephalitis, porcine reproductive and respiratory syndrome, pseudorabies, rabies, *Streptococcus suis*, water deprivation, and poisonings from salt, lead, or insecticides.

Mortality Rate (untreated): ≤90%.

C18-A044

Potato Spindle Tuber Viroid
Pospiviroidae
ICD-11: —
UNII: —
EPPO: PSTVD0

Endemic to Europe, Southern Asia, Africa, Central and northern South America, New Zealand, and parts of Australia.

It is on the Australia Group Core list.

In Plants:

Target species: Potatoes, tomatoes, eggplant.

Normal routes of transmission: Contact.

Secondary Hazards: Crop debris (tubers, seeds, pollen); mechanical vectors (cultivation, track out); vectors (aphids – source plant must also be infected with potato leafroll virus for this to occur). *Signs:* May be mild at initial infection but progressively worsen in following generations.

Potatoes: impact to foliage is often difficult to recognize as symptoms resemble those caused by nutrient deficiency, root disease or spray drift damage. Vines may be smaller, more upright, and produce smaller leaves with fluted margins. Leaflets may be twisted and leaf surfaces may be wrinkled. Infected tubers may be small, cracked, elongated, and misshapen with pointed ends (spindle shaped). Eyes may be deeper and more prominent.

Tomatoes: initially leaves at the top of the plant become yellowed (chlorotic). Plants become stunted growth with a 'bunchy top' caused by shortened internodes. Leaves turn red to purple, often becoming curled, twisted and brittle. Flowers may abort and fruit can be dark green, fail to ripen normally, and have thicker outer walls.

Crop Losses: ≤65% (potatoes); ≤50% (tomatoes).

C18-A045

Powassan Encephalitis
Flaviviridae
ICD-11: —
UNII: —
EPPO: —

Endemic to Canada, the northern United States, and parts of Russia. The natural reservoir is ticks (*Ixodes* spp.) and small mammals (e.g., groundhogs, woodchucks, squirrels, mice). Ticks remains infective for life. The disease can be transmitted from a tick to a person within 15 minutes of the initial bite. Does not produce disease In Domestic Animals. It does not survive outside a host. This is a biosafety level 3 agent.

It is on the Australia Group Core list.

In People:

CDC Case Definition: Clinical criteria for cases of *Neuroinvasive disease* is defined as follows: 1) meningitis, encephalitis, acute flaccid paralysis, or other acute signs of central or peripheral neurologic dysfunction; AND 2) absence of a more likely clinical explanation.

Clinical criteria for cases of *non-neuroinvasive disease* is defined as follows: 1) fever (chills) as reported by the patient or a health-care provider; AND 2) absence of neuroinvasive disease; AND 3) absence of a more likely clinical explanation. Other clinically compatible symptoms include: headache, myalgia, rash, arthralgia, vertigo, vomiting, paresis, and/ or nuchal rigidity.

Laboratory criteria for diagnosis is isolation of virus from, or demonstration of specific viral antigen or nucleic acid in, tissue, blood, CSF, or other body fluid, OR 1) fourfold or greater change in virus-specific quantitative antibody titers in paired sera, OR 2) virus-specific IgM antibodies in serum with confirmatory virus-specific neutralizing antibodies in the same or a later specimen, OR 3) virus-specific IgM antibodies in CSF or serum.

Criteria for case confirmation for *Neuroinvasive disease* consists of meeting the clinical criteria and one or more of the following laboratory criteria: 1) isolation of virus from, or demonstration of specific viral antigen or nucleic acid in, tissue, blood, CSF, or other body fluid, OR 2) fourfold or greater change in virus-specific quantitative antibody titers in paired sera, OR 3) virus-specific IgM antibodies in serum with confirmatory virus-specific neutralizing antibodies in the same or a later specimen, OR 4) virus-specific IgM antibodies in CSF, with or without a reported pleocytosis, and a negative result for other IgM antibodies in CSF for arboviruses endemic to the region where exposure occurred.

Criteria for case confirmation for *Non-neuroinvasive disease* consists of meeting the clinical criteria and one or more of the following laboratory criteria: 1) isolation of virus from, or demonstration of specific viral antigen or nucleic acid in, tissue, blood, or other body fluid, excluding CSF, OR 2) fourfold or greater change in virus-specific quantitative antibody titers in paired sera, OR 3) virus-specific IgM antibodies in serum with confirmatory virus-specific neutralizing antibodies in the same or a later specimen.

Communicability: Direct person-to-person transmission does not occur. When dealing with infected individuals, use standard precautions.

Normal Routes of Exposure: Vectors (ticks); abraded skin.

Infectious Dose: Unknown.

Secondary Hazards: Aerosols; blood and body fluids; fomites (with vectors).

Incubation: 7–34 days.

Signs & Symptoms: Fever, headache, nausea, vomiting, generalized weakness, progressing to neck stiffness (nuchal rigidity), sensitivity to light (photophobia), confusion, incoordination (ataxia), unable to speak normally (dysarthria), memory loss, seizures, paralysis, and coma.

Suggested Alternatives for Differential Diagnosis: Acute disseminated encephalomyelitis, basilar artery blood clots, cardioembolic stroke, cavernous sinus syndromes, cerebral venous blood clots, eastern equine encephalitis virus, enteroviruses, epileptic and epileptiform encephalopathies, febrile seizures, hemophilus meningitis, herpes simplex virus, intracranial hemorrhage, Jamestown Canyon virus, La Crosse virus, leptomeningeal carcinomatosis, meningitis, St. Louis encephalitis virus, subdural empyema or hematoma, West Nile virus, confusional states, and acute memory disorders.

Mortality Rate (untreated): ≤10%.

C18-A046

Rabies
Rabies-related Lyssaviruses
Rhabdoviridae
ICD-11: XN796
UNII: I0V66KI1LD
EPPO: —

With few exceptions, it is endemic worldwide. It is a highly lethal disease that can affect all mammals, but only a limited number can act as a reservoir; *Canidae* spp. (e.g., dogs, coyotes, foxes), *Mustelidae* spp. (e.g., skunks, ferrets), *Herpestidae* spp. (e.g., mongooses), and *Procyonidae* (raccoons), and *Chiroptera* spp. (bats). The two main reservoirs of concern are dogs and bats. In saliva on exterior surfaces, the virus is subject to sunlight and desiccation; surviving no more than a few hours at room temperature. Can persist for weeks in brain tissue in carcasses and for years in frozen tissues. This is a biosafety level 2 agent unless there is a high risk of aerosol production then it should be treated as a biosafety level 3 agent.

It is on the Australia Group Core list and the OIE list of Notifiable Diseases for 2020.

In People:

CDC Case Definition: Laboratory criteria for diagnosis is one or more of the following laboratory findings: 1) detection of Lyssavirus antigens in a clinical specimen (preferably the brain or the nerves surrounding hair follicles in the nape of the neck) by direct fluorescent antibody test; OR 2) isolation (in cell culture or in a laboratory animal) of a Lyssavirus from saliva or central nervous system tissue; OR 3) identification of Lyssavirus specific antibody (i.e. by indirect fluorescent antibody (IFA) test or complete rabies virus neutralization at 1:5 dilution) in the cerebrospinal fluid (CSF); OR 4) identification of Lyssavirus specific antibody (i.e. by indirect fluorescent antibody (IFA) test or complete rabies virus neutralization at 1:5 dilution) in the serum of an unvaccinated person; OR 5) detection of Lyssavirus viral RNA (using reverse transcriptase-polymerase chain reaction [RT-PCR]) in saliva, CSF, or tissue.

Communicability: Direct person-to-person transmission is possible but unlikely. Can be transmitted through corneal or solid organ transplants. When dealing with infected individuals, use standard precautions.

Normal Routes of Exposure: Abraded skin; mucous membranes; inhalation.

Infectious Dose: Unknown.

Secondary Hazards: Aerosols; body fluids (saliva, cerebrospinal fluid).

Incubation: 9–90 days; although may last up to 1 year. The nearer the site of infection is to the central nervous system (i.e., brain, spine), the shorter the incubation period.

Signs & Symptoms: By the time symptoms are apparent, rabies is almost invariably fatal. Initially fever, headache, malaise, rapid heart rate (tachycardia), high blood pressure (hypertension), hyperventilation, excessive dilation of the pupil (mydriasis), eyes have unequally sized pupils (anisocoria), facial paralysis, tearing (lacrimation), excessive salivation, perspiration, postural low blood pressure (hypotension).

In furious rabies, (approximately 80% of cases), loss of appetite (anorexia), excessive salivation, restlessness, confusion, agitation, increased sensitivity to sensory stimuli such as noise or light (hyperesthesia), abnormal behavior, hallucinations, insomnia. Attempting to drink or having air blown in the face produces severe spasms in the throat and diaphragm producing a sensation of asphyxia. This instills a fear of water (hydrophobia) and of drafts or of fresh air (aerophobia); which are symptom that are decisively characteristic (pathognomonic) of rabies.

In paralytic rabies (approximately 20% of cases), there are wide variations in blood pressure, cardiac arrhythmias, low body temperature (hypothermia), slow heart rate (bradycardia), progressing to ascending paralysis with quadriplegia, delirium, stupor, and coma.

Suggested Alternatives for Differential Diagnosis: Autonomic instability, cerebrovascular accident, Creutzfeldt-Jacob disease, dysphagia, encephalitis, epilepsy, Guillain-Barre syndrome, herpes simplex, hydrophobia, intracranial mass, paresthesia, poliomyelitis, pseudohydrophobia, psychosis, tetanus, transverse myelitis, poisonings from atropine or botulism.

Mortality Rate (untreated): >99%.

In Domestic Animals:

Susceptible species: All mammals.

CDC Case Definition: Laboratory criteria for diagnosis is 1) positive direct fluorescent antibody test (preferably performed on central nervous system tissue); OR 2) isolation of rabies virus (in cell culture or in a laboratory animal).

Communicability: Direct transmission is possible.

Normal routes of exposure: Injection; mucous membranes; ingestion.

Secondary Hazards: Body fluids; body tissue.

Incubation: 2 weeks–6 months.

Signs: Fever, vomiting, loss of appetite (anorexia), vomiting, diarrhea, restlessness, excessive and abnormal bellowing or howling, difficulty breathing, difficulty swallowing, excessive salivation, frequent urination, abnormal behavior, overreaction to stimuli such as noises or lights, aggression or uncharacteristic affection, self-mutilation, incoordination (ataxia), sagging and swaying of the hind quarters, weakness, paralysis, seizures.

Post-mortem does not present any characteristic gross findings. There is polioencephalomyelitis; craniospinal ganglionitis with mononuclear perivascular infiltrates, diffuse glial proliferation, and regressive changes in neuronal cells; and glial nodules. Negri bodies in neurons may also be seen. *Suggested Alternatives for Differential Diagnosis:* Acetonemia, African swine fever, bacterial meningoencephalitis, Borna disease, canine distemper, canine hepatitis, coenurosis, cryptococcosis traumatic injuries, encephalomyelitis, enterotoxemia, erysipela, indigestion, lactation tetany, listeriosis, milk fever, polioencephalomalacia, pregnancy ketosis, pseudorabies, Teschen's disease, foreign body in the mouth, vitamin A or B1 deficiency, poisonings from, chlorinated hydrocarbons, lead, organophosphate pesticides, or sodium fluoroacetate.

Mortality Rate (untreated): >95%.

C18-A047

Rift Valley Fever
Bunyaviridae
ICD-11: 1D44
UNII: —
EPPO: —

Endemic to Africa and the Middle East. The natural reservoir is mosquitoes. Mosquitoes remains infective for life. Infection is also transferred directly to mosquito eggs. It can survive in unhatched eggs for several years. Other arthropods (e.g., ticks, sandflies, blackflies) may become infected and transmit the disease. Under optimal conditions, aerosols remain viable for more than an hour. This is a biosafety level 3-Ag agent, with the addition of HEPA filtration attached to the laboratory exhaust system.

It is on the Australia Group Core list, the U.S. Select Agents and Toxins list, and the OIE list of Notifiable Diseases for 2020.

In People:
CDC Case Definition: None established.

Communicability: Direct person-to-person transmission does not occur. When dealing with infected individuals, use standard precautions. In some instances, humans can infect new mosquitoes during the fever phase, protect from vectors.

Normal Routes of Exposure: Inhalation; ingestion; abraded skin; mucous membranes; vectors (mosquitoes).

Infectious Dose: Unknown.

Secondary Hazards: Aerosols (blood, body fluids); blood and body fluids (including unpasteurized milk); body tissue; Vector amplification.

Incubation: 2–6 days.

Signs & Symptoms: Fever, headache, fatigue, dizziness, loss of appetite (anorexia), muscle pain (myalgia), back pain as well as possibly neck stiffness (nuchal rigidity), sensitivity to light (photophobia), nausea, vomiting, and diarrhea. May progress to pain (arthralgia) in the large joints (i.e., elbows, knees shoulders), enlargement of the liver (hepatomegaly), jaundice, and delirium. Possible hemorrhagic symptoms include nose bleeds (epistaxis), vomiting blood (hematemesis), black tar-like stools (melena), small red or purple spots on the surface of the skin (petechia), and bleeding from the gums. Additional complications may include inflammation of the brain and membranes covering it (meningoencephalitis) with memory loss, vertigo, hallucinations, confusion, disorientation, coma, and seizures; or inflammation of the retinas (retinitis) with blurred vision and loss of acuity. If lesions appear on the retina, permanent vision loss is possible.

Suggested Alternatives for Differential Diagnosis: Acute anemia, acute leukemias, anaphylactoid purpura, chikungunya, Colorado tick fever, dengue fever, disseminated intravascular coagulation, encephalitis, hemolytic uremic syndrome, Henoch-Schönlein purpura, hepatitis, herpes, influenza, leptospirosis, malaria, measles, meningitis, meningococcemia, mononucleosis, other New World viral hemorrhagic fevers, plague, pleural effusion, Q fever, Rocky Mountain spotted fever, rubella, shigellosis, Sindbis fever, thrombocytopenic purpura, tick bite fever, trypanosomiasis, typhoid fever, typhus, yellow fever, drug eruptions, allergies, food poisoning.

Mortality Rate (untreated): ≤50% (hemorrhagic).

In Domestic Animals:

Target species: Sheep, goats, cattle.

Communicability: Direct transmission does not occur.

Normal routes of exposure: Vectors (mosquitoes).

Secondary Hazards: Vector amplification; mechanical vectors (flies, ticks).

Incubation: 1–3 days.

Signs: In some cases, sudden death may occur with no apparent signs of illness. New born animals are particularly susceptible to disease. Fever, loss of appetite (anorexia), weakness, listlessness, excessive salivation, tearing (lacrimation), sensitivity to light (photophobia), rapid breathing (tachypnea), bloodstained nasal discharge of mucus and pus (mucopurulent), yellow discoloration of the skin and eyes (jaundice), swollen lymph nodes (lymphadenopathy) abdominal pain, vomiting, bloody or foul-smelling diarrhea, and black tar-like stools (melena). Abortions, unrelated to the gestation period, may reach 100%.

Post-mortem findings include hepatic necrosis; edematous and hemorrhagic gall bladder; enlarged and edematous spleen and peripheral lymph nodes with possible petechia; hemorrhage of the gastrointestinal tract with the contents of the small intestine possibly a dark chocolate-brown color; disseminated intravascular coagulation; mild to moderate effusion of fluid in body cavities; and congestion and edema of the lungs.

Suggested Alternatives for Differential Diagnosis: Anthrax, bluetongue, bovine ephemeral fever, brucellosis, campylobacteriosis, chlamydiosis, *Coxiella burnetii*, East Coast fever, enterotoxemia, ephemeral fever, fungal conditions, heartwater, leptospirosis, Nairobi sheep disease, ovine enzootic abortion, pasteurellosis, peste des petits ruminants, rinderpest, salmonellosis, trichomoniasis, vibriosis, Wesselbron disease, defects in porphyrin metabolism, poisonings from hepatotoxic plant or algae.

Mortality Rate (untreated): ≤30% (sheep); ≤10% (cattle); higher in lambs and calves.

C18-A048

Rinderpest (Agent R)
Paramyxoviridae
ICD-11: —
UNII: —
EPPO: —

Rinderpest has been eradicated worldwide. The natural reservoir was cattle. Sheep and goats are susceptible but epidemiologically unimportant. It is quickly inactivated (within 12 hours) in the environment by drying or sunlight, but can remain viable for long periods in chilled or frozen tissues. Recovered animals acquired lifelong immunity. This is a biosafety level 3-Ag agent.

It is on the Australia Group Core list, a Tier-1 agent on the U.S. Select Agents and Toxins list, and the OIE list of Notifiable Diseases for 2020.

In People:
Does not occur in humans.

In Domestic Animals:
Target species: Cattle, buffalo.

Communicability: Direct transmission is possible.

Normal routes of exposure: mucous membranes; inhalation (limited).

Secondary Hazards: Blood and body fluids; body tissue; fecal matter; mechanical vectors (track out); fomites (minor).

Incubation: 3–15 days.

Signs: Fever, loss of appetite (anorexia), depression, and watery discharge from the eyes and nose develops, progressing to necrotic lesions on the gums, cheeks, and tongue. The discharge from the eyes and nose becomes a mixture of mucus and pus (mucopurulent), leading to difficulty breathing (dyspnea). The muzzle appears dry and cracked. Initial constipation is followed by diarrhea, which may be watery and bloody, leading to dehydration, emaciation, and prostration.

Post-mortem findings include an overall appearance of emaciation and dehydration, with mucopurulent nasal exudate and fecal soiling. There are erosions of the mucosa in the mouth, pharynx and esophagus with edema or emphysema of the lungs. Hemorrhagic or ulcerative lesions are found in the omasum as well as congestion, edema and erosion of the abomasal mucosa. There is severe congestion and hemorrhage in the intestine and enlarged and necrotic Peyer's patches. The last portion of the large intestine and rectum are hemorrhagic showing 'tiger stripping' of longitudinal folds. Lymph nodes are generally swollen and edematous, and there is evidence of hemorrhage in the spleen, gallbladder, and urinary bladder.

Suggested Alternatives for Differential Diagnosis: Bluetongue, bovine papular stomatitis, bovine viral diarrhea, *Brachyspira hyodyesntereiae*, *Campylobacter* spp., coccidiosis, contagious pleuropneumonia, East Coast fever, foot and mouth disease, infectious bovine rhinotracheitis, Jembrana disease, malignant catarrhal fever, mucosal disease, Nairobi sheep disease, necrobacillosis, necrotic stomatitis, papular stomatitis, paratuberculosis, pasteurellosis, peste des petits ruminants, salmonellosis, vesicular stomatitis, poisonings from arsenic.

Mortality Rate (untreated): ≤100%.

C18-A049

Rocio Encephalitis
Flaviviridae
ICD-11: 1C87
UNII: —
EPPO: —

Endemic to Brazil. The natural reservoir is unknown but suspected to be mosquitoes and birds. This is a biosafety level 3 agent, with the addition of HEPA filtration attached to the laboratory exhaust system.

It is on the Australia Group Core list.

In People:

CDC Case Definition: None established.

Communicability: Direct person-to-person transmission does not occur. When dealing with infected individuals, use standard precautions.

Normal Routes of Exposure: Vectors (mosquitoes).

Infectious Dose: Unknown.

Secondary Hazards: Aerosol.

Incubation: 5–15 days.

Signs & Symptoms: Sudden onset of fever, severe headache, vague feeling of bodily discomfort (malaise), muscle pain (myalgia), loss of appetite (anorexia), nausea, vomiting, unable to urinate (anuria), tearing (lacrimation), sensitivity to light (photophobia), feeling of weariness or diminished energy (lassitude), lack of energy (lethargy), and stupor; progressing to neck stiffness (nuchal rigidity), blindness, deafness, unable to speak normally (dysarthria), impaired gait and motor control, impaired equilibrium, mental confusion, convulsions, and coma. Up to 20% of survivors continue to have neuropsychiatric consequences of the disease (sequela) including burning or prickling skin sensations (paresthesia), inability to speak normally (dysarthria), inability of the eyes to point in the same direction (strabismus), difficulty swallowing (dysphagia), memory disturbances, impaired gait and motor control, and impaired equilibrium, and variable degrees of paralysis.

Suggested Alternatives for Differential Diagnosis: Bacterial meningitis, brain abscess, California encephalitis, carcinomatous meningitis, cerebrovascular disease, Chikungunya virus, CNS vasculitis, dengue fever, eastern equine encephalitis, fungal meningitis, hanta viruses, hepatitis, herpes simplex encephalitis, immune thrombocytopenia, influenza, leptospirosis, malaria, Mayaro fever, measles, meningitis, orbivirus, rickettsial infection, Rift Valley fever, roseola infantum, Ross River fever, rubella, scarlet fever, scrub typhus, Sindbis fever, St. Louis encephalitis, typhoid fever, typhus, West Nile virus, West Nile virus, western equine encephalitis, yellow fever, Zika virus.

Mortality Rate (untreated): ≤13%.

C18-A050

Sabia Virus
Arenaviridae
Brazilian Hemorrhagic Fever
ICD-11: XN55S
UNII: —
EPPO: —

Endemic to Brazil. The natural reservoir is unknown but suspected to be rodents. It is likely that the virus is shed in their urine. This is a biosafety level 4 agent.

It is on the Australia Group Core list and the U.S. Select Agents and Toxins list.

In People:

CDC Case Definition for viral hemorrhagic fever: Clinical criteria is an illness with acute onset with ALL of the following clinical findings: 1) a fever >104°F; and 2) one or more of the following clinical findings: a) severe headache; OR b) muscle pain; OR c) erythematous maculopapular rash on the trunk with fine desquamation 3–4 days after rash onset; OR d) vomiting; OR e) diarrhea; OR f) abdominal pain; OR g) bleeding not related to injury; OR h) thrombocytopenia; OR i) retrosternal chest pain; OR j) proteinuria; OR k) pharyngitis.

Laboratory criteria for diagnosis is one or more of the following laboratory findings: 1) detection of viral hemorrhagic fever (VHF) viral antigens in blood by enzyme-linked Immunosorbent Assay (ELISA) antigen detection; 2) VHF viral isolation in cell culture for blood or tissues; 3) detection of VHF-specific genetic sequence by Reverse Transcription-Polymerase Chain Reaction (RT-PCR) from blood or tissues; 4) detection of VHF viral antigens in tissues by immunohistochemistry.

Communicability: Direct person-to-person transmission is unknown, but may be possible. When dealing with infected individuals, use contact and droplet precautions.

Normal Routes of Exposure: Unknown, but believed to be inhalation; abraded skin; mucous membranes.

Infectious Dose: Unknown.

Secondary Hazards: Unknown, but believed to be aerosols; blood and body fluids (rodent); fecal matter (rodent); fomites (from rodent habitation).

Incubation: Unknown.

Signs & Symptoms: Based on limited cases, fever, chills, vague feeling of bodily discomfort (malaise), headache, muscle pain (myalgia), sore throat, eye infection (pink eye – conjunctivitis), nausea, vomiting, diarrhea, abdominal pain, bleeding gums, and possibly gastrointestinal hemorrhage.

Suggested Alternatives for Differential Diagnosis: Acute anemia, acute leukemias, anaphylactoid purpura, chikungunya, Colorado tick fever, dengue fever, disseminated intravascular coagulation, encephalitis, epidemic typhus, falciparum malaria, hemolytic uremic syndrome, Henoch-Schonlein purpura, hepatitis, herpes, infectious mononucleosis, influenza, leptospirosis, malaria, measles, meningitis, meningococcemia, plague, pleural effusion, Q fever, Rocky Mountain spotted fever, rubella, sepsis, shigellosis, Sindbis fever, thrombocytopenic purpura, tick bite fever, trypanoso-miasis, typhoid fever, yellow fever, other New World viral hemorrhagic fevers, drug eruptions, allergies, food poisoning.

Mortality Rate (untreated): <30%.

C18-A051

SARS Associated Coronavirus
Coronaviridae
ICD-11: XN1V8
UNII: R7D54R07N5
EPPO: —

Middle East Respiratory Syndrome Coronavirus (MERS)
Coronaviridae
ICD-11: XN3BD
UNII: ZS35O9OIY0
EPPO: —

SARS is endemic to Asia; MERS is endemic to the Middle East. The natural reservoir for SARS is the Chinese horseshoe bat (*Rhinolophus sinicus*); for MERS is unknown but thought to be camels. It can survive for more than a day on hard surfaces such as glass and metal, less than 3 days in soil/water, for 4 days in diarrhea, more than 4 days in urine, more than 7 days in respiratory secretions at room temperature, and 6 days in a dried state. Manipulation of untreated specimens can be performed in biosafety 2 following biosafety 3 practices, but otherwise this is a biosafety 3 agent.

It is on the Australia Group Core list and the U.S. Select Agents and Toxins list. Middle East Respiratory Syndrome is on the OIE list of Notifiable Diseases for 2020.

In People:

CDC Case Definition: Clinical criteria is an *Early illness* presenting two or more of the following features: fever (might be subjective), chills, rigors, myalgia, headache, diarrhea, sore throat, or rhinorrhea. *Mild-to-moderate respiratory illness* with 1) temperature of >100.4°F; AND 2) one or more clinical findings of lower respiratory illness (e.g., cough, shortness of breath, or difficulty breathing). *Severe respiratory illness* meeting the clinical criteria of a mild-to-moderate respiratory illness, and one or more of the following findings: 1) radiographic evidence of pneumonia; OR 2) acute respiratory distress syndrome; OR 3) autopsy findings consistent with pneumonia or acute respiratory distress syndrome without an identifiable cause.

Laboratory criteria for diagnosis is one or more of the following general criteria for laboratory findings: 1) detection of serum antibody to SARS-CoV by a test validated by CDC (e.g., enzyme immunoassay); OR 2) isolation in cell culture of SARS-CoV from a clinical specimen; OR 3) detection of SARS-CoV RNA by a reverse transcription polymerase chain reaction test validated by CDC and with subsequent confirmation in a reference laboratory.

Communicability: Direct person-to-person transmission is possible. When dealing with infected individuals, use contact, droplet and use contact, droplet and airborne precautions.

Normal Routes of Exposure: Inhalation; mucous membranes; ingestion; abraded skin.

Infectious Dose: Unknown.

Secondary Hazards: Aerosols (respiratory secretions); blood and body fluids; fecal matter; fomites.

Incubation: 1–14 days.

Signs & Symptoms: Initially indistinguishable from other common respiratory viruses. High fever, headache, chills, nasal discharge (rhinorrhea), fatigue, vague feeling of bodily discomfort (malaise), muscle pain (myalgia), cough, shortness of breath, difficulty breathing (dyspnea), rapid heart rate (tachycardia), low blood pressure (hypotension), dizziness, with possible loss of appetite (anorexia), sore throat, abdominal pain, nausea, vomiting, diarrhea. May progress to pneumonia, inflammation of the liver (hepatitis), kidney failure, and epileptic fits.

Suggested Alternatives for Differential Diagnosis: Adenovirus, arenaviruses, aspiration pneumonia, atelectasis, bronchiectasis, bronchiolitis, bronchitis, brucellosis, chronic obstructive pulmonary disease, common cold, community-acquired pneumonia, coxsackieviruses, cytomegalovirus, echovirus infection, emphysema, foreign body aspiration, influenza, *Legionella*, metapneumovirus infection, mycobacterium avium-intracellulare and other atypical mycobacterial diseases, myco-plasma infections, parainfluenza virus, pleural effusion, pneumonia, psittacosis, Q fever, respiratory syncytial virus, rhinoviruses, rickettsialpox, upper respiratory infections.

Rate (untreated): ≤18% (SARS), ≤35% (MERS).

C18-A052

Seoul Hemorrhagic Fever
Bunyaviridae
Hemorrhagic Fever with Renal Syndrome
ICD-11: AN3PV
UNII: —
EPPO: —

Endemic worldwide. The natural reservoir is Norway rats (*Rattus norvegicus*) and the black rat (Rattus rattus). Does not produce disease in domestic animals. Rodents can shed hantaviruses in saliva, feces and urine. This is a biosafety level 3 agent.

It is on the Australia Group Core.

In People:

CDC Case Definition: None established.

Communicability: Direct person-to-person transmission does not occur. When dealing with infected individuals, use standard precautions.

Normal Routes of Exposure: Inhalation; mucous membranes; abraded skin.

Infectious Dose: Unknown.

Secondary Hazards: Aerosols; fecal matter (mouse); blood and body fluids; fomites.

Incubation: 7–56 days.

Signs & Symptoms: Fever, headache, back and abdominal pain, chills, nausea, blurred vision, flushing of the face, inflammation or redness of eyes, and rash. In rare cases, may progress to low blood pressure (hypotension), bleeding (hemorrhage), acute shock, and kidney failure.

Suggested Alternatives for Differential Diagnosis: Acute abdominal diseases, acute kidney injury, acute poststreptococcal glomerulonephritis, Colorado tick fever, disseminated intravascular coagulation, heat stroke, hemolytic uremic syndrome, leptospirosis, malaria, murine typhus, non-a, non-b hepatitis, scrub typhus, septicemia, spotted fevers.

Mortality Rate (untreated): ≤3%.

C18-A053

Sheeppox
Poxviridae
ICD-11: —
UNII: —
EPPO: —

Endemic to Africa north of the equator, the Middle East, Central Asia, and India. The natural reservoir is sheep, although chronically infected carriers do not occur. Infection results in solid and enduring immunity. The virus can persist for up to 6 months in shaded animal pens and in dried scabs, and for up to 3 months in on the fleece, skin and hair from infected animals. It can survive freeze–thaw cycles, although the infectivity may be reduced. This is a biosafety level 3 agent.

It is on the Australia Group Core list, the U.S. Select Agents and Toxins list, and the OIE list of Notifiable Diseases for 2020.

In People:

Does not occur in humans.

In Domestic Animals:

Target species: Sheep, goats (minor).

Communicability: Direct transmission is possible.

Normal routes of exposure: Inhalation; abraded skin; mucous membranes.

Secondary Hazards: Aerosols (respiratory secretions); blood and body fluids (including milk); fecal matter; fomites; mechanical vectors (track out, insects).

Incubation: 4–8 days.

Signs: Rapid onset of fever with excessive salivation or drooling (ptyalism), nasal discharge, and eye infection (pink eye – conjunctivitis). Eyelids become swollen, and a discharge of mucus and pus (mucopurulent) crusts the nostrils. Animals may have difficulty breathing (dyspnea). Red, flat (macular) spots areas appear on the skin that evolve into pimples (papules) and possibly blisters (vesicles or pustules). Ultimately, skin lesions become dark-colored scabs. Scabs may take up to 6 weeks to heal. The skin is sensitive in the affected area. Scabs leave small star-shaped scars that remain free of hair. Pustules may alternatively develop into nodules that die and slough off the animal leaving hairless patches.

Post-mortem skin lesions are often less obvious than on a live animal. Other findings include reddish to whitish firm nodules in the mucosa of the pharynx and trachea, as well as hard, reddish to whitish nodules in the lungs. There are rarely signs of pneumonia. Lymph nodes in the thorax are enlarged and edematous.

Suggested Alternatives for Differential Diagnosis: Bluetongue, caseous lymphadenitis, contagious pustular dermatitis, dermatophilosis, mange, mycotic dermatitis, orf, parasitic pneumonia, peste des petits ruminants, photosensitization, and urticaria from multiple insect bites.

Mortality Rate (untreated): ≤100%.

C18-A054

Siberian Tick-Borne Encephalitis Virus
Flavivirus
ICD-11: —
UNII: —
EPPO: —

Endemic to Russia and northeastern Europe. The natural reservoir is ticks (*Ixodidae* spp.). Ticks remain infected for life. Infection is also transferred directly to tick eggs. This is a biosafety level 4 agent.

It is on the U.S. Select Agents and Toxins list.

In People:

CDC Case Definition: None established.

European Centre for Disease Prevention and Control Case Definition: Clinical criteria is any person with symptoms of inflammation of the CNS (for example, meningitis, meningo-encephalitis, encephalomyelitis, encephaloradiculitis).

Laboratory criteria for diagnosis is one or more of the following laboratory findings: 1) TBE specific IgM AND IgG antibodies in blood; OR 2) TBE specific IgM antibodies in CSF; OR 3) seroconversion or fourfold increase of TBE-specific antibodies in paired serum samples; OR 4) detection of TBE viral nucleic acid in a clinical specimen; OR 5) isolation of TBE virus from clinical specimen.

Communicability: Direct person-to-person transmission does not occur. When dealing with infected individuals, use standard precautions.

Normal Routes of Exposure: Vectors (ticks).

Infectious Dose: Unknown.

Secondary Hazards: Aerosols; abraded skin; bites (infected rodents); ingestion (unpasteurized products); fomites (with vectors).

Incubation: 2–28 days.

Signs & Symptoms: Fever, headache, sensitivity to light (photophobia), fatigue, muscle pain (myalgia), loss of appetite (anorexia), nausea and vomiting, progressing to neck stiffness (nuchal rigidity), sensorial changes, visual disturbances, rapid involuntary movements of the eyes (nystagmus), incoordination (ataxia), tremors, seizures, convulsions, or flaccid paralysis from inflammation of the membranes that surround the brain and spinal cord (meningitis), inflammation of the brain and membranes covering it (meningoencephalitis), or inflammation of the spinal cord (myelitis).

Suggested Alternatives for Differential Diagnosis: Aseptic meningitis, babesiosis, basilar artery blood clots, cardioembolic stroke, cavernous sinus syndromes, cerebral venous blood clots, encephalitis, epileptic and epileptiform encephalopathies, febrile seizures, granulocytic anaplasmosis, hemophilus meningitis, intracranial hemorrhage, leptomeningeal carcinomatosis, Lyme disease, meningitis, subdural empyema or hematoma, tick-transmitted rickettsioses, tularemia, confusional states, and acute memory disorders.

Rate (untreated): ≤3%.

In Domestic Animals:

Susceptible species: Sheep, goats, dogs.

Communicability: Direct transmission does not occur.

Normal routes of exposure: Vectors (ticks).

Secondary Hazards: Fomites (with vectors).

Incubation: Unknown. Believed to be 7–14 days.

Signs: Usually asymptomatic. May present with fever, lethargy, muscle pain (myalgia), loss of appetite (anorexia), eyes have unequally sized pupils (anisocoria), rapid involuntary movements of the eyes (nystagmus), incoordination (ataxia), tremors, seizures, convulsions, and paralysis.

Suggested Alternatives for Differential Diagnosis: Other causes of meningitis, encephalitis, or meningoencephalitis.

Mortality Rate (untreated): ≤100% (species dependent).

C18-A055

Sin Nombre
Bunyaviridae
Hantavirus Pulmonary Syndrome
ICD-11: XN7R1
UNII: —
EPPO: —

Endemic to western United States. The natural reservoir is the deer mouse (*Peromyscus maniculatus*). Does not produce disease in domestic animals. Rodents can shed hantaviruses in saliva, feces and urine. It can survive in the environment for up to 3 days. This is a biosafety level 3 agent.

It is on the Australia Group Core list.

In People:

CDC Case Definition for hantavirus pulmonary syndrome: Clinical criteria is an acute febrile illness (i.e., temperature greater than 101°F) with a prodrome consisting of fever, chills, myalgia, headache, and gastrointestinal symptoms, and one or more of the following clinical features: 1) bilateral diffuse interstitial edema; OR 2) clinical diagnosis of acute respiratory distress syndrome (ARDS), or 3) radiographic evidence of noncardiogenic pulmonary edema; OR 4) an unexplained respiratory illness resulting in death, and includes an autopsy examination demonstrating noncardiogenic pulmonary edema without an identifiable cause; OR 5) healthcare record with a diagnosis of hantavirus pulmonary syndrome; OR 6) death certificate lists hantavirus pulmonary syndrome as a cause of death or a significant condition contributing to death.

Laboratory criteria for diagnosis is one or more of the following laboratory findings: 1) detection of hantavirus-specific immunoglobulin M or rising titers of hantavirus-specific immunoglobulin G; OR 2) detection of hantavirus-specific ribonucleic acid in clinical specimens; OR detection of hantavirus antigen by immunohistochemistry in lung biopsy or autopsy tissues.

Communicability: Direct person-to-person transmission does not occur. When dealing with infected individuals, use standard precautions.

Normal Routes of Exposure: Inhalation; mucous membranes; abraded skin.

Infectious Dose: Unknown.

Secondary Hazards: Aerosols; fecal matter (mouse); blood and body fluids (mouse); fomites (from mouse habitation).

Incubation: 2–51 days.

Signs & Symptoms: Flu-like with fever, headache, muscle aches (myalgia), nausea, vomiting, and diarrhea with sudden progression to cough, rapid breathing (tachypnea), rapid heart rate (tachycardia) followed by acute pulmonary edema with difficulty breathing (dyspnea). As the disease progresses, there may be a decrease in blood pressure (hypotension) and blood volume (hypovolemia), followed by cardiogenic shock.

Suggested Alternatives for Differential Diagnosis: Acute respiratory distress syndrome (ARDS), anthrax, chlamydophilial infections, community acquired pneumonia, congestive heart failure and pulmonary edema, cytomegalovirus, dengue fever, ehrlichiosis, Epstein-Barr virus, hepatitis, influenza, legionnaires disease, Lyme disease, malaria, mycoplasmas infections, pneumonia, pneumonic plague, pulmonary embolism, Q fever, shock, silent myocardial infarction, tularemia, poisonings from phosphine, phosgene and salicylate toxicity with pulmonary edema.

Mortality Rate (untreated): ≤50%.

C18-A056

Smallpox (Agent N1)

Poxviridae

ICD-11: XN4Q0

UNII: —

EPPO: —

Smallpox has been eradicated worldwide. There are only two remaining known repositories of the virus – one in the United States and one in Russia. However, other countries may still retain unacknowledged sample caches. The natural reservoir was humans. Aerosolized virus can survive for up to 24 hours under optimal conditions (i.e., temperatures near 50°F, humidity ≤ 20%, and no UV light). Dried body fluids, crusts and scrapings from skin lesions containing virus remain infectious for up to 1 year at room temperature. Samples of freeze-dried virus survive for at least 20 years. This is a biosafety level 4 agent.

It is on the Australia Group Core list and a Tier-1 agent on the U.S. Select Agents and Toxins list.

In People:

CDC Case Definition: CDC Case Definition: Clinical criteria is an illness with acute onset of fever ≥101°F followed by a rash characterized by firm, deep seated vesicles or pustules in the same stage of development without other apparent cause. Clinically consistent cases are those presentations of smallpox that do not meet this classical clinical case definition: a) hemorrhagic type, b) flat type, and c) *variola sine eruptione*.

Laboratory criteria for diagnosis is one or more of the following laboratory findings: 1) polymerase chain reaction (PCR) identification of variola DNA in a clinical specimen, OR 2) isolation of smallpox (variola) virus from a clinical specimen (Level D laboratory only; confirmed by variola PCR).

Communicability: Direct person-to-person transmission is possible. It is communicable from the onset of the rash until the last scab falls off. Infectivity declines once crusts form on the skin lesions. When dealing with infected individuals, use contact and airborne precautions.

Normal Routes of Exposure: Inhalation; abraded skin; mucous membranes.

Infectious Dose: 10–100 viral units (inhalation).

Secondary Hazards: Aerosols (respiratory secretions); blood and body fluids; body tissue (skin lesions, scabs); fomites.

Incubation: 7–19 days.

Signs & Symptoms: High fever, vague feeling of bodily discomfort (malaise), headache, muscle pain (myalgia) particularly in the back, and sometimes vomiting. A rash appears on the tongue and in the mouth that progress into open sores. A rash of flat spots (macules) appears on the face, spreads to the arms and legs, and then to the hands and feet. The rash is centrifugal, with greatest concentration of lesions on the face and distal extremities rather than the truck. Skin lesions are all at the same stage of development on a given body part. These spots progress to bumps filled with a thick, opaque fluid that often have a depression in the center that looks like a bellybutton. The bumps become hard blisters (pustules) and then scabs. Blisters and crusts are accompanied by severe itching (pruritus). Some individuals may develop generalized reddening of the skin (erythema) with cutaneous and mucosal hemorrhage. When the scabs fall off, they leave a mark on the skin that will become a pitted scar.

Suggested Alternatives for Differential Diagnosis: Acne, acute leukemia, bullous pemphigoid, chickenpox, cowpox, coxsackievirus, cytomegalovirus, eczema vaccinatum, ehrlichiosis, enteroviruses, erythema multiforme, generalized vaccinia, herpes simplex, herpes zoster, impetigo, infectious mononucleosis, influenza, Kawasaki disease, malaria, measles, meningitis, meningococcemia, molluscum contagiosum, monkeypox, parvovirus B19, rat-bite fever, rickettsialpox, Rocky Mountain spotted fever, rubella, scarlet fever, syphilis, viral hemorrhagic fevers, contact dermatitis, drug eruptions, scabies, and other insect bites.

Mortality Rate (untreated): ≤30%.

C18-A057

St. Louis Encephalitis
Flaviviridae
ICD-11: 1C86
UNII: —
EPPO: —

It is endogenous to the United States. The natural reservoir is birds and mosquitoes. Mosquitoes remains infective for life. It is stable for ≤6 hours at room temperature and almost indefinitely when freeze-dried. This is a biosafety level 3 agent.

It is on the Australia Group Core list.

In People:
CDC Case Definition: Clinical criteria for cases of *Neuroinvasive disease* is defined as follows: 1) meningitis, encephalitis, acute flaccid paralysis, or other acute signs of central or peripheral neurologic dysfunction; AND 2) absence of a more likely clinical explanation.

Clinical criteria for cases of *non-neuroinvasive disease* is defined as follows: 1) fever (chills) as reported by the patient or a health-care provider; AND 2) absence of neuroinvasive disease; AND 3) absence of a more likely clinical explanation. Other clinically compatible symptoms include: headache, myalgia, rash, arthralgia, vertigo, vomiting, paresis, and/ or nuchal rigidity.

Laboratory criteria for diagnosis is isolation of virus from, or demonstration of specific viral antigen or nucleic acid in, tissue, blood, CSF, or other body fluid, OR 1) fourfold or greater change in virus-specific quantitative antibody titers in paired sera, OR 2) virus-specific IgM antibodies in serum with confirmatory virus-specific neutralizing antibodies in the same or a later specimen, OR 3) virus-specific IgM antibodies in CSF or serum.

Criteria for case confirmation for *Neuroinvasive disease* consists of meeting the clinical criteria and one or more of the following laboratory criteria: 1) isolation of virus from, or demonstration of specific viral antigen or nucleic acid in, tissue, blood, CSF, or other body fluid, OR 2) fourfold or greater change in virus-specific quantitative antibody titers in paired sera, OR 3) virus-specific IgM antibodies in serum with confirmatory virus-specific neutralizing antibodies in the same or a later specimen, OR 4) virus-specific IgM antibodies in CSF, with or without a reported pleocytosis, and a negative result for other IgM antibodies in CSF for arboviruses endemic to the region where exposure occurred.

Criteria for case confirmation for *non-neuroinvasive disease* consists of meeting the clinical criteria and one or more of the following laboratory criteria: 1) isolation of virus from, or demonstration of specific viral antigen or nucleic acid in, tissue, blood, or other body fluid, excluding CSF, OR 2) fourfold or greater change in virus-specific quantitative antibody titers in paired sera, OR 3) virus-specific IgM antibodies in serum with confirmatory virus-specific neutralizing antibodies in the same or a later specimen.

Communicability: Direct person-to-person transmission does not occur. When dealing with infected individuals, use standard precautions.

Normal Routes of Exposure: Vectors (mosquitoes); abraded skin.

Infectious Dose: Unknown.

Secondary Hazards: Aerosols; blood and body fluids; body tissue; fomites (from bird habitation).

Incubation: 4–21 days.

Signs & Symptoms: Abrupt onset of fever, headache, cough, sore throat, joint pain (arthralgia), dizziness, nausea, vomiting, and a vague feeling of bodily discomfort (malaise). May develop neurologic signs including neck stiffness (nuchal rigidity), confusion, disorientation, irritability, tremors, incoordination (ataxia), convulsions, and coma.

Suggested Alternatives for Differential Diagnosis: Bacterial meningitis, brain abscess, California encephalitis, carcinomatous meningitis, cerebrovascular disease, CNS vasculitis, eastern equine encephalitis, fungal meningitis, herpes simplex encephalitis, West Nile virus, western equine encephalitis.

Mortality Rate (untreated): ≤20%.

C18-A058

Swine Vesicular Disease
Picornaviridae
ICD-11: —
UNII: —
EPPO: —

Endemic to Italy. The natural reservoir is pigs. Virus may continue to be shed in the feces of pigs for up to 3 months after full recovery. Resistant to fermentation and smoking processes. May remain in hams for 180 days, dried sausages for more than a year, and in processed intestinal casings for more than 2 years. Contaminated farm equipment can remain infectious for up to 4 months. The virus is not inactivated by normal pH change associated with rigor mortis. This is a biosafety level 3 agent. It is on the Australia Group Core list, the U.S. Select Agents and Toxins list and the OIE list of Notifiable Diseases for 2020.

In People:
Rarely infects humans; only occurring in a laboratory setting. Infected individuals generally report mild flu-like symptoms, with weakness, and abdominal pain. There are no reports of infections occurring in farmers or veterinarians.

In Domestic Animals:
Target species: Pigs.

Communicability: Direct transmission is possible.

Normal routes of exposure: Inhalation; ingestion; abraded skin; mucous membranes.
Secondary Hazards: Blood and body fluids; body tissue; fecal matter; mechanical vectors (track out).

Incubation: 2–7 days.

Signs: Fever, loss of appetite (anorexia), blisters (vesicles) on the feet, especially the coronary band, and less often on the mouth, lips, teats, and snout. Animals may become lame. Vesicles may rupture leaving shallow ulcers. May develop neurologic symptoms including shivering, unsteady walking, and rhythmic jerking. Affected pigs usually do not lose condition and usually recover in less than 3 weeks.

Suggested Alternatives for Differential Diagnosis: Foot and mouth disease, foot rot, swine pox, vesicular exanthema, vesicular stomatitis, phototoxic dermatitis, chemical and thermal burns, traumatic injuries.

Mortality Rate (untreated): 0%.

C18-A059

T3 Coliphage
Podoviridae
ICD-11: —
UNII: —
EPPO: —

Biological warfare simulant. A bacteriophage that infects the *E. coli* bacterium.

No significant harmful health effects to humans or animals are expected from exposure to this pathogen.

C18-A060

Variola Minor
Poxviridae
ICD-11: XN4Q0
UNII: —
EPPO: —

Milder form of smallpox caused by a less virulent form of the virus. The toxemia is less, lesions are more superficial, and healing time was more rapid. The natural reservoir is humans. This is a biosafety level 4 agent.

It is on the Australia Group Core list and a Tier-1 agent on the U.S. Select Agents and Toxins list.

In People:
See Smallpox (C18-A056).

Mortality Rate (untreated): ≤1%.

C18-A061

Venezuelan Equine Encephalitis (Agent NU)
Togaviridae
ICD-11: XN445
UNII: —
EPPO: —

Endemic to Central and South America. The natural reservoir is rodents and mosquitoes. Mosquitoes remains infective for life. Infection is also transferred directly to mosquito eggs. Equidae (e.g., horses, donkeys) serve as amplifying hosts. Horse-to-human transmission does not occur. It is stable in dried blood and body fluids. This is a biosafety level 3 agent, with the addition of HEPA filtration attached to the laboratory exhaust system.

It is on the Australia Group Core list, the U.S. Select Agents and Toxins list, and the OIE list of Notifiable Diseases for 2020.

In People:
CDC Case Definition: None established.

Communicability: Direct person-to-person transmission does not occur. When dealing with infected individuals, use standard precautions. Humans can infect new mosquitoes during the fever phase, protect from vectors.

Normal Routes of Exposure: Vectors (mosquitoes); abraded skin.

Infectious Dose: 1 viral unit (injection); 10–100 viral units (inhalation).

Secondary Hazards: Aerosols; vector amplification; fomites (from rodent habitation).

Incubation: 1–6 days.

Signs & Symptoms: Abrupt onset of fever, severe frontal headache, sore throat, chills, muscle (myalgia) pain, rapid heart rate (tachycardia), a vague feeling of bodily discomfort (malaise), cough, sore throat, nausea, vomiting, and diarrhea. May develop neurologic signs including sensitivity to light (photophobia), neck stiffness (nuchal rigidity), confusion, semiconsciousness (somnolence), incoordination (ataxia), seizures, paralysis, and coma.

Suggested Alternatives for Differential Diagnosis: Acute HIV infection, arenaviruses, Colorado tick fever, coxsackieviruses, cytomegalovirus, dengue fever, eastern equine encephalitis, echoviruses, Epstein-Barr virus, herpes simplex, infectious mononucleosis, influenza, Japanese encephalitis, leptospirosis, listeria monocytogenes, Lyme disease, malaria, measles, meningitis, meningococcal infections, meningococcemia, Naegleria infection, Norwalk virus, picornavirus, poliomyelitis, Q fever, St. Louis encephalitis, viral hepatitis, West Nile virus, western equine encephalitis, yellow fever.

Mortality Rate (untreated): ≤1%; ≤25% (with encephalitis).

In Domestic Animals:

Susceptible species: Horses and other equids.

Communicability: Direct transmission does not occur.

Normal routes of exposure: Vectors (mosquitoes); mucous membranes.

Secondary Hazards: Aerosols; body fluids; vector amplification; mechanical vectors (blackflies, mites).

Incubation: 1–5 days.

Signs: In some cases, sudden death may occur with no apparent signs of illness. Otherwise, fever, loss of appetite (anorexia), difficulty swallowing (dysphagia), severe depression, blindness, increased sensitivity to sensory stimuli such as noise, light, or touch (hyperesthesia), periods of excitement or intense itching (pruritus), facial swelling (edema), aimless wandering, head pressing, circling, incoordination (ataxia), paralysis, recumbency, seizures, convulsions.

Post-mortem does not present any characteristic gross findings. There is severe inflammation of the gray matter with possible neuronal degeneration, infiltration by inflammatory cells, gliosis, perivascular cuffing and hemorrhages.

Suggested Alternatives for Differential Diagnosis: Botulism, eastern equine encephalitis, equine herpesvirus 1, Everglades virus, hepatoencephalopathy, highlands J virus, Japanese encephalitis, Kunjin virus, leukoencephalomalacia, louping ill, meningitis, Murray Valley encephalitis virus, protozoal encephalomyelitis, rabies, Ross River virus, Semliki Forest virus, St. Louis encephalitis, tickborne encephalitis, Una virus, Usutu, verminous meningoencephalomyelitis, West Nile encephalitis, western equine encephalitis, cranial trauma.

Mortality Rate (untreated): ≤90%

C18-A062

Vesicular Stomatitis Fever
Rhabdoviridae
ICD-11: —
UNII: —
EPPO: —

Endemic to the Americas. The natural reservoir is sand flies (Phlebotomus spp.). Black flies and sandflies remain infected for life. Infection is also transferred directly to their eggs. Fluid from blisters on animals contains large concentrations of infective virus. It can survive for up to 4 days in saliva on hard surfaces in the environment, on grass for several weeks and extended periods in soil at moderate temperatures. This is a biosafety level 2 agent.

It is on the Australia Group Core list and the OIE list of Notifiable Diseases for 2020.

In People:

CDC Case Definition: None established.

Communicability: Direct person-to-person transmission does not occur. When dealing with infected individuals, use standard precautions.

Normal Routes of Exposure: Inhalation; abraded skin; mucous membranes; vectors (sandflies, black flies, midges, mosquitoes).

Infectious Dose: Unknown.

Secondary Hazards: Aerosols; body fluids (animal).

Incubation: 1–6 days.

Signs & Symptoms: Produces flu-like symptoms including fever, headache, enlarged lymph nodes, eye infection (pink eye – conjunctivitis), muscle pain (myalgia), joint pain (arthralgia), eye pain, a vague feeling of bodily discomfort (malaise), nausea. Although oral blisters similar to cold sores are possible, they rarely occur. Fever may be biphasic.

Suggested Alternatives for Differential Diagnosis: Influenza.

Mortality Rate (untreated): ≤1%.

In Domestic Animals:

Target species: Horses and other equids, cattle, pigs, llamas.

Communicability: Direct transmission is possible.

Normal routes of exposure: Unclear. Vectors (black flies, sandflies); mucous membranes; abraded skin.

Secondary Hazards: Aerosols; body fluids; fomites; vector amplification; mechanical vectors (track out, feeding and milking equipment).

Incubation: 1–21 days.

Signs: Easily confused with foot-and-mouth disease (C18-A019). Fever, excessive salivation, and chewing movements; blanched flat spots (macules) appear on the oral or nasal mucosa, gums, mammary glands, external genitalia, or the band just above the hoof (coronary band). These spots become vesicular lesions that progress to ulcers and erosions. These areas coalesce resulting in extensive necrosis and sloughing. Raw tissue is so painful that infected animals often refuse to eat. If the hooves are involved, animals often become lame.

Animals rarely die, postmortem findings include epithelial vesicles, ulcers, and erosions of the mouth, nostrils, teats and feet; crusting of the muzzle and lips.

Suggested Alternatives for Differential Diagnosis: Bluetongue, bovine herpes mammillitis, bovine papular stomatitis, bovine viral diarrhea, cowpox, epizootic hemorrhagic disease, foot and mouth disease, foot rot, infectious bovine rhinotracheitis, malignant catarrhal fever, mycotic stomatitis, Potomac valley fever, pseudo-cowpox, pseudo-lumpy skin disease, rinderpest, swine vesicular disease, swine vesicular exanthema, photosensitization, chemical or thermal burns.

Mortality Rate (untreated): ≤1%.

C18-A063

West Nile Fever
Flaviviridae
ICD-11: 1D46
UNII: J067L6I0TY
EPPO: —

Endemic to Africa, Europe, the Middle East, West Asia, and North America. The natural reservoir is birds and mosquitoes. Mosquitoes remains infective for life. Infection is also transferred directly to mosquito eggs. Mortality in birds in the Old World is rare, in contrast birds in to the New World, especially crows, where the virus causes significant death. Horses are also affected by the virus but do not spread the infection. This is a biosafety level 3 agent.

It is on the OIE list of Notifiable Diseases for 2020.

In People:

CDC Case Definition: Clinical criteria for cases of *Neuroinvasive disease* is defined as follows: 1) meningitis, encephalitis, acute flaccid paralysis, or other acute signs of central or peripheral neurologic dysfunction; AND 2) absence of a more likely clinical explanation.

Clinical criteria for cases of *non-neuroinvasive disease* is defined as follows: 1) fever (chills) as reported by the patient or a health-care provider; AND 2) absence of neuroinvasive disease; AND 3) absence of a more likely clinical explanation. Other clinically compatible symptoms include: headache, myalgia, rash, arthralgia, vertigo, vomiting, paresis and/ or nuchal rigidity.

Laboratory criteria for diagnosis is isolation of virus from, or demonstration of specific viral antigen or nucleic acid in, tissue, blood, CSF, or other body fluid, OR 1) fourfold or greater change in virus-specific quantitative antibody titers in paired sera, OR 2) virus-specific IgM antibodies in serum with confirmatory virus-specific neutralizing antibodies in the same or a later specimen, OR 3) virus-specific IgM antibodies in CSF or serum.

Criteria for case confirmation for *Neuroinvasive disease* consists of meeting the clinical criteria and one or more of the following laboratory criteria: 1) isolation of virus from, or demonstration of specific viral antigen or nucleic acid in, tissue, blood, CSF, or other body fluid, OR 2) fourfold or greater change in virus-specific quantitative antibody titers in paired sera, OR 3) virus-specific IgM antibodies in serum with confirmatory virus-specific neutralizing antibodies in the same or a later specimen, OR 4) virus-specific IgM antibodies in CSF, with or without a reported pleocytosis, and a negative result for other IgM antibodies in CSF for arboviruses endemic to the region where exposure occurred.

Criteria for case confirmation for *Non-neuroinvasive disease* consists of meeting the clinical criteria and one or more of the following laboratory criteria: 1) isolation of virus from, or demonstration of specific viral antigen or nucleic acid in, tissue, blood, or other body fluid, excluding CSF, OR 2) fourfold or greater change in virus-specific quantitative antibody titers in paired sera, OR 3) virus-specific IgM antibodies in serum with confirmatory virus-specific neutralizing antibodies in the same or a later specimen.

Communicability: Direct person-to-person transmission does not occur. When dealing with infected individuals, use standard precautions.

Normal Routes of Exposure: Vectors (mosquitoes); abraded skin.

Infectious Dose: 1 viral unit (injection).

Secondary Hazards: Aerosols; blood and body fluids; body tissue; fecal matter (bird).

Incubation: 2–14 days.

Signs & Symptoms: Most infections are asymptomatic or produce a nonspecific mild flu-like illness. Otherwise, symptoms include fever, headache, fatigue, body aches, joint (arthralgia) and muscle (myalgia) pain, sore throat, swollen lymph glands, cough, nasal discharge (rhinorrhea), nausea, vomiting, diarrhea, and a transient rash of mixed flat and raised (maculopapular) spots on the trunk of the body. May progress to inflammation of the brain (encephalitis) or of the membranes that surround the brain and spinal cord (meningitis) with symptoms of neck stiffness (nuchal rigidity), muscle weakness, stupor, disorientation, sensitivity to light (photophobia), vision loss, tremors, convulsions, numbness, acute flaccid paralysis, coma.

Suggested Alternatives for Differential Diagnosis: Acute poliomyelitis, brain abscess, brain tumor, cat-scratch disease, eastern equine encephalitis, enterovirus, Guillain-Barre syndrome, herpes simplex, hypoglycemia, La Crosse virus, leptospirosis, Lyme diseases, meningitis, multiple sclerosis, myasthenia gravis, postpolio syndrome, Powassan virus, Rocky Mountain spotted fever, St. Louis encephalitis, stroke, subarachnoid hemorrhage, toxoplasmosis, tuberculosis, western equine encephalitis, zika virus.

Mortality Rate (untreated): ≤10% (encephalitis).

In Domestic Animals:

Susceptible species: Horses and other equids, geese.

Communicability: Direct transmission does not occur.

Normal routes of exposure: Vectors (mosquitoes).

Secondary Hazards: Blood and body tissue (dead birds); vector amplification (birds).
Incubation: 3–15 days (horses), approximately 5 days (birds).

Signs: Horses suffer loss of appetite (anorexia), depression, muscles twitching in the face or neck, teeth grinding, difficulty swallowing (dysphagia), impaired vision, aimless wandering, head pressing, circling, incoordination (ataxia), irregular gait, trembling, weakness, muscle contractions, semiconsciousness (somnolence), paralysis, convulsions. Horses may become apprehensive or have an increased sensitivity to sensory stimuli such as noise, light, or touch (hyperesthesia).

In geese it primarily affects young birds. In some cases, sudden death may occur with no apparent signs of illness. Otherwise weight loss, diarrhea, ruffled feathers, decreased activity, depression, sideways twitching of the neck (torticollis), rhythmic side-to-side head movements, backward bridging position (opisthotonos), paralysis. Signs of incoordination (ataxia) are pronounced; some birds flip over while attempting to stand. There may be copious oral and nasal secretions with labored breathing.

Post-mortem in horses does not present any characteristic gross findings. There may be small multifocal areas of discoloration and hemorrhage in the spinal cord, brain stem and midbrain; the meninges may be congested or hemorrhagic. There may also be mild nonsuppurative myocarditis, scattered hemorrhages in the renal medulla, and lymphoid depletion of the spleen.

In birds, post-mortem findings include emaciation, dehydration, splenomegaly, hepatomegaly, myocardial pallor, pale mottling of the liver, spleen or kidney, with multiorgan hemorrhages, petechiae and congestion.

Suggested Alternatives for Differential Diagnosis: Horses: Borna disease, eastern equine encephalitis, equine herpesvirus-1, equine protozoal myelitis, Japanese encephalitis, rabies, Venezuelan equine encephalitis, and western equine encephalitis.

Birds: Aspergillosis, avian influenza, erysipelothricosis, listeriosis, Newcastle disease, riemerellosis, salmonellosis, streptococcal infection, and poisonings from ionophores (e.g., antibiotics).

Mortality Rate (untreated): ≤60% (horses); ≤75% (geese)

C18-A064

Western Equine Encephalitis
Togaviridae
ICD-11: 1C83
UNII: —
EPPO: —

The natural reservoir is birds and mosquitoes. Mosquitoes remains infective for life. Infection is also transferred directly to mosquito eggs. The virus can also be transmitted by the tick *Dermacentor andersoni*. Horse-to-human transmission does not occur. It does not survive outside of a host. This is a biosafety level 3 agent.

It is on the Australia Group Core list and the OIE list of Notifiable Diseases for 2020.

In People:

CDC Case Definition: Clinical criteria for cases of *Neuroinvasive disease* is defined as follows: 1) meningitis, encephalitis, acute flaccid paralysis, or other acute signs of central or peripheral neurologic dysfunction; AND 2) absence of a more likely clinical explanation.

Clinical criteria for cases of *non-neuroinvasive disease* is defined as follows: 1) fever (chills) as reported by the patient or a health-care provider; AND 2) absence of neuroinvasive disease; AND 3) absence of a more likely clinical explanation. Other clinically compatible symptoms include: headache, myalgia, rash, arthralgia, vertigo, vomiting, paresis and/ or nuchal rigidity.

Laboratory criteria for diagnosis is isolation of virus from, or demonstration of specific viral antigen or nucleic acid in, tissue, blood, CSF, or other body fluid, OR 1) fourfold or greater change in virus-specific quantitative antibody titers in paired sera, OR 2) virus-specific IgM antibodies in serum with confirmatory virus-specific neutralizing antibodies in the same or a later specimen, OR 3) virus-specific IgM antibodies in CSF or serum.

Criteria for case confirmation for *Neuroinvasive disease* consists of meeting the clinical criteria and one or more of the following laboratory criteria: 1) isolation of virus from, or demonstration of specific viral antigen or nucleic acid in, tissue, blood, CSF, or other body fluid, OR 2) fourfold or greater change in virus-specific quantitative antibody titers in paired sera, OR 3) virus-specific IgM antibodies in serum with confirmatory virus-specific neutralizing antibodies in the same or a later specimen, OR 4) virus-specific IgM antibodies in CSF, with or without a reported pleocytosis, and a negative result for other IgM antibodies in CSF for arboviruses endemic to the region where exposure occurred.

Criteria for case confirmation for *non-neuroinvasive disease* consists of meeting the clinical criteria and one or more of the following laboratory criteria: 1) isolation of virus from, or demonstration of specific viral antigen or nucleic acid in, tissue, blood, or other body fluid, excluding CSF, OR 2) fourfold or greater change in virus-specific quantitative antibody titers in paired sera, OR 3) virus-specific IgM antibodies in serum with confirmatory virus-specific neutralizing antibodies in the same or a later specimen.

Communicability: Direct person-to-person transmission does not occur. When dealing with infected individuals, use standard precautions.

Normal Routes of Exposure: Vectors (mosquitoes); abraded skin.

Infectious Dose: Unknown.

Secondary Hazards: None.

Incubation: 2–15 days.

Signs & Symptoms: In severe cases, abrupt onset of fever, headache, chills, joint (arthralgia) and muscle (myalgia) pain, rapid heart rate (tachycardia), a vague feeling of bodily discomfort (malaise), nausea and vomiting. May develop abdominal pain, diarrhea, and sore throat along with neurologic signs including neck stiffness (nuchal rigidity), confusion, irritability, restlessness, semiconsciousness (somnolence), stupor, disorientation, tremors, seizures, paralysis and coma.

Suggested Alternatives for Differential Diagnosis: Bartonellosis, cytomegalovirus, Epstein-Barr virus, herpes simplex, histoplasmosis, infective endocarditis, leptospirosis, Lyme disease, malaria, metabolic encephalopathy, mumps, *Mycoplasma pneumoniae,* Naegleria infection, rabies, Reye syndrome, rheumatoid arthritis, spinal cord abscess, St. Louis encephalitis, stroke, subarachnoid hemorrhage surgery, superficial thrombophlebitis, systemic lupus erythematosus, toxoplasmosis, tuberculosis, Venezuelan equine encephalitis, West Nile virus.

Mortality Rate (untreated): ≤7%.

In Domestic Animals:

Susceptible species: Horses and other equids.

Communicability: Direct transmission does not occur.

Normal routes of exposure: Vectors (mosquitoes).

Secondary Hazards: None.

Incubation: 5–14 days.

Signs: Fever, loss of appetite (anorexia), difficulty swallowing (dysphagia), severe depression, blindness, increased sensitivity to sensory stimuli such as noise, light, or touch (hyperesthesia), periods of excitement or intense itching (pruritus), facial swelling (edema), aimless wandering, head pressing, circling, incoordination (ataxia), paralysis, recumbency, seizures, convulsions.

Post-mortem does not present any characteristic gross findings. There is severe inflammation of the gray matter with possible neuronal degeneration, infiltration by inflammatory cells, gliosis, perivascular cuffing, and hemorrhages.

Suggested Alternatives for Differential Diagnosis: Botulism, eastern equine encephalitis, equine herpesvirus 1, Everglades virus, hepatoencephalopathy, highlands J virus, Japanese encephalitis, Kunjin virus, leukoencephalomalacia, louping ill, meningitis, Murray Valley encephalitis virus, protozoal encephalomyelitis, rabies, Ross River virus, Semliki Forest virus, St. Louis encephalitis, tickborne encephalitis, Una virus, Usutu, Venezuelan equine encephalitis virus, verminous meningoencephalomyelitis, West Nile encephalitis, cranial trauma.

Mortality Rate (untreated): ≤30%.

C18-A065

Yellow Fever (Agent UT)
Flaviviridae
ICD-11: XN9S3
UNII: 4G601DAM77
EPPO: —

Endemic to Africa and South America. The natural reservoir is monkeys and mosquitoes. Mosquitoes remains infective for life. Infection is also transferred directly to mosquito eggs. It does not survive outside of a host. This is a biosafety level 3 agent, with the addition of HEPA filtration attached to the laboratory exhaust system.

It is on the Australia Group Core list and the OIE list of Notifiable Diseases for 2020.

In People:

CDC Case Definition: Clinical criteria is an acute illness with at least one of the following: fever, jaundice, or elevated total bilirubin ≥ 3 mg/dl and absence of a more likely clinical explanation.

Laboratory criteria for diagnosis is one of the following laboratory findings: 1) isolation of yellow fever virus from, or demonstration of yellow fever viral antigen or nucleic acid in, tissue, blood, CSF, or other body fluid; OR 2) fourfold or greater rise or fall in yellow fever virus-specific neutralizing antibody titers in paired sera; OR 3) yellow fever virus-specific IgM antibodies in CSF or serum with confirmatory virus-specific neutralizing antibodies in the same or a later specimen.

Communicability: Direct person-to-person transmission does not occur. When dealing with infected individuals, use standard precautions. Humans can infect new mosquitoes during the fever phase, protect from vectors.

Normal Routes of Exposure: Vectors (mosquitoes); abraded skin.

Infectious Dose: Unknown.

Secondary Hazards: Aerosols; blood and body fluids; body tissue; vector amplification.

Incubation: 3–6 days.

Signs & Symptoms: Abrupt onset of fever, chill, severe headache, fatigue, vague feeling of bodily discomfort (malaise), lower back pain, nausea, loss of appetite (anorexia), and dizziness. The heart rate is slower than expected for a fever (pulse-fever dissociation). The face is flushed and the eyes are bloodshot (conjunctival injection).

A period of remission (up to 24 hours) is followed by a return of fever and more severe symptoms including vomiting, nausea, abdominal pain, enlargement of the liver (hepatomegaly), limited urine production (oliguria), yellow discoloration of the skin and eyes (jaundice), and slow heart rate (bradycardia); with hemorrhagic manifestations including bloody diarrhea or black tar-like stools (melena), blood in the urine (hematuria), non-menstrual uterine bleeding, small red or purple spots on the surface of the skin (petechia), bruises (ecchymoses), nose bleads (epistaxis), and blood oozing from the gums. Death is typically preceded by profound low blood pressure (hypotension) and shock.

Suggested Alternatives for Differential Diagnosis: Acute liver failure, arenavirus, Argentine hemorrhagic fever, Bolivian hemorrhagic fever, dengue fever, disseminated intravascular coagulation, ebola hemorrhagic fever, flavivirus, hantavirus, hepatitis, Japanese encephalitis, Lassa fever, leptospirosis, louse-borne relapsing fever, malaria, Marburg hemorrhagic fever, multiple organ dysfunction syndrome in sepsis, Rift Valley Fever, typhoid fever, typhus, Venezuelan hemorrhagic fever, viral hepatitis, West Nile virus, food poisoning.

Mortality Rate (untreated): ≤90%.

C18-A066

Akabane Virus

Bunyaviridae

ICD-11: —

UNII: —

EPPO: —

Endemic to Africa, the Middle East, Southeast Asia, and Australia. The natural reservoir is midges (*Culicoides* spp.). Insects remain infected for life. It affects the fetus in utero but generally causes no clinical signs in adult animals, although some cases of inflammation of the brain and spinal cord (encephalomyelitis) in adult cattle have been reported. This is a biosafety level 3-Ag agent with special considerations for work involving infected arthropods.

Removed from the Australia Group Core list and also the U.S. Select Agents and Toxins list in 2010.

In People:

Does not occur in humans.

In Domestic Animals:

Target species: Cattle, sheep, goats.

Communicability: Direct transmission does not occur.

Normal routes of exposure: Vectors (midges).

Secondary Hazards: None.

Incubation: 1–6 days.

Signs: Depend on the stage of the pregnancy. Causes abortions, stillbirths and fetal deformities including blindness, rapid involuntary movements of the eyes (nystagmus), deafness, dullness, incoordination, multiple joints and limbs bent and locked in position (arthrogryposis), and paralysis. *Suggested Alternatives for Differential Diagnosis:* Aino virus, bovine viral diarrhea virus, Cache Valley virus, Chuzan virus, Schmallenberg virus.

Mortality Rate (untreated): Very high. Most offspring animals die soon after birth or must be euthanized.

C18-A067

Camelpox Virus
Poxviridae
ICD-11: —
UNII: —
EPPO: —

Endemic to northern Africa, the Middle East, and southwestern Asia. The natural reservoir is camels. It can remain virulent outside of a host for up to four months. This is a biosafety level 2 agent. Removed from the Australia Group Core list and also the U.S. Select Agents and Toxins list in 2010. It is on the OIE list of Notifiable Diseases for 2020.

In People:

Genetically closely related to the smallpox virus, it was investigated by Iraq as a potential antipersonnel weapon. However, it infrequently caused disease in humans; rarely appearing as lesions on the hands of individuals working with camels or as ulcers on the lips and mouth of individual who consumed camel milk from infected animals. It has never caused serious disease in humans.

In Domestic Animals:

Target species: Camels.

Communicability: Direct transmission is possible.

Normal routes of exposure: Inhalation; abraded skin; mucous membranes.

Secondary Hazards: Aerosols (respiratory secretions); blood and body fluids; body tissue (skin lesions, scabs); fomites; vectors (ticks).

Incubation: 3–15 days.

Signs: Fever and enlarged lymph nodes. Skin lesions first appear on the head, eyelids, nostrils and the margins of the ears one to three days after the onset of fever. Lesions begin as red flat spots (macules) that progress to pimples (papules), then to blisters (vesicles) that eventually harden (pustules), and finally turn to scabs. Lesions may eventually cover the entire body and can take up to six weeks to heal. Animals experience salivation, tearing (lacrimation), a nasal discharge of mucus and pus (mucopurulent), diarrhea and loss of appetite (anorexia).

Suggested Alternatives for Differential Diagnosis: Contagious ecthyma, orthopoxvirus, papillomatosis, parapoxvirus, sarcoptic mange.

Mortality Rate (untreated): ≤28% in adults; ≤100% in young animals.

C18-A068

Flexal Fever Virus
Arenaviridae
ICD-11: —
UNII: —
EPPO: —

A member of the New World Hemorrhagic Fever viruses normally found in Brazil. The natural reservoir is the rice rat (*Oryzomys* spp.). The virus is shed in their urine. Infection occurs after inhalation of dust contaminated with excreta from infected rats or from aerosol of animal blood or fluids. Does not produce disease in animals. This is a biosafety level 3 agent.

Removed from the Australia Group Core list and also the U.S. Select Agents and Toxins list in 2010.

In People:
CDC Case Definition: None established.

Communicability: Unknown. When dealing with infected individuals, use contact, and droplet precautions.

Normal Routes of Exposure: Inhalation; ingestion; abraded skin; mucous membranes.

Infectious Dose: Unknown.

Secondary Hazards: Aerosols; blood and body fluids (rat); body tissue (rat); fecal matter (rat); fomites (from rat habitation).

Incubation: 7–16 days.

Signs & Symptoms: Gradual onset of sustained fever accompanied by a vague feeling of bodily discomfort (malaise), headache, severe muscle pain (myalgia) and joint pain (arthralgias), dizziness, sensitivity to light (photophobia), sore throat, nausea, vomiting, and diarrhea progressing to hemorrhagic symptoms including nosebleed (epistaxis), bloody vomit and blood in stool or urine. Patients may deteriorate due to vascular or neurologic complications. The illness lasts 2 to 14 days after the onset of symptoms. Recovery may be prolonged.

Suggested Alternatives for Differential Diagnosis: Acute leukemia, disseminated intravascular coagulation, hemolytic uremic syndrome, idiopathic thrombocytopenic purpura, leptospirosis, lupus, malaria, meningococcemia, rickettsial infections, salmonella infection, shigellosis, thrombocytopenic purpura, thrombotic thrombocytopenic purpura, typhoid fever.

Mortality Rate (untreated): Unknown.

C18-A069

Malignant Catarrhal Fever Virus
Herpesviridae
ICD-11: —
UNII: —
EPPO: —

Overall endemic throughout the world; various strains found with their natural reservoir. The natural reservoir is sheep, goats and wild ruminants (e.g., white tailed deer, African wildebeest, ibex) where the virus produces an inapparent infection. Virus is typically shed by young animals, but also intermittently by adults. Viral aerosols can travel up to 3 miles. Morbidity is typically low but mortality is high. This is a biosafety level 2 agent.

Removed from the Australia Group Core list and also the U.S. Select Agents and Toxins list in 2010.

In People:
Does not occur in humans.

In Domestic Animals:
Target species: Cattle, bison, water buffalo, pigs.

Communicability: Direct transmission is only between the carrier species (i.e., sheep, wild ruminants) and susceptible domestic species. Affected animals of susceptible domestic species do not transmit the disease to their cohorts.

Normal routes of exposure: Inhalation; ingestion.

Secondary Hazards: Aerosols (respiratory secretions); body fluids (nasal and ocular secretions); fomites.

Incubation: 11 days–9 months.

Signs: Highly variable. In some cases, sudden death preceded by no clinical signs or only by depression with diarrhea. More typically, high fever, nasal and ocular discharge becoming a mixture of mucus and pus (mucopurulent), with a loss of appetite (anorexia). Animals have difficulty breathing (dyspnea) due to nasal cavity obstruction. There is involuntary blinking (blepharospasm) and both corneas progressively become opaque. Erosions or ulcers may develop on the tongue, hard palate, and gums. Superficial lymph nodes may be enlarged and limb joints may be swollen. There may be reddening of the skin (erythema) or skin ulcers with hardened scabs. Animals may have diarrhea, inflammation and hemorrhage in the stomach and intestines (hemorrhagic gastroenteritis), or blood in their urine (hematuria). Nervous disorders may include hyperexcitability and aggression, rapid involuntary movements of the eyes (nystagmus), muscle tremor, incoordination, or paralysis.

Post-mortem findings include diffuse or focal bilateral corneal opacity. There are small focal erosions or ulcers on the nasal mucosa and oral cavity with petechiae or ecchymoses on the tongue, buccal mucosa, in the gastrointestinal and respiratory tracts, urinary bladder, and on the serosa of internal organs. Erosions, ulcers, and hemorrhages can occur throughout the gastrointestinal tract; in severe cases, the contents of the intestines can be bloody. Lymph nodes are enlarged. Catarrhal exudate, erosions and diphtheritic membranes are often observed in the respiratory tract. There are raised pale foci on the surfaces of the kidneys that may extend into the cortex. There is necrotizing vasculitis and obliterative arteriopathy.

Suggested Alternatives for Differential Diagnosis: Bluetongue, bovine viral diarrhea mucosal disease, epizootic hemorrhagic disease, East Coast fever, foot and mouth disease, infectious bovine rhinotracheitis, photosensitive dermatitis, pneumonic pasteurellosis, rabies, rinderpest, theileriosis, tickborne encephalitides, vesicular stomatitis, poisonings from caustic materials or toxic plants.

Mortality Rate (untreated): ≤100%.

C18-A070

Puumala Hemorrhagic Fever
Bunyaviridae
ICD-11: XN28L
UNII: —
EPPO: —

It is a hantavirus that is normally found in Europe, Russia, and Scandinavia. The natural reservoir is the bank vole (Myodes glaerolus). The virus is shed in their urine. Infection occurs after inhalation of dust contaminated with excreta from infected voles or from aerosol of animal blood or fluids. This is a biosafety level 3 agent.

Removed from the Australia Group Core list and also the U.S. Select Agents and Toxins list in 2010.

In People:
CDC Case Definition: None established.

Communicability: Direct person-to-person transmission does not occur. When dealing with infected individuals, use standard precautions.

Normal Routes of Exposure: Inhalation; ingestion; abraded skin; mucous membranes.

Infectious Dose: Unknown.

Secondary Hazards: Aerosols; blood and body fluids (vole); fecal matter (vole); fomites (from vole habitation).

Incubation: 2–4 weeks.

Signs & Symptoms: Produces a relatively mild form of Hemorrhagic Fever with Renal Syndrome (HFRS). Characterized by sudden onset of high fever, headache, a vague feeling of bodily discomfort (malaise), dizziness, anorexia, abdominal and/or lower back pain, nausea and vomiting. Progresses to rapid heart rate (tachycardia), hypoxemia, renal failure and protein in the urine (proteinuria). This is followed by excessive urination (diuresis) which may progress to dehydration and severe shock. Normal blood pressure returns and there is a dramatic drop in urine production. Recovery may be prolonged.

Suggested Alternatives for Differential Diagnosis: Acute poststreptococcal glomerulonephritis, Colorado tick fever, disseminated intravascular coagulation, heat stroke, hemolytic uremic syndrome., hepatitis, leptospirosis, malaria, septicemia, spotted fevers, typhus.

Mortality Rate (untreated): ≤1%

BIBLIOGRAPHY

Acha, Pedro N. and Boris Szyfres. *Zoonoses and Communicable Diseases Common to Man and Animals, Scientific and Technical Publication No. 580.* 3rd Edition. Volume 2, *Chlamydioses, Rickettsioses, and Viroses.* Washington, D.C.: Pan American Health Organization, 2003.

American Committee of Medical Entomology; American Society of Tropical Medicine and Hygiene. *Arthropod Containment Guidelines, Version 3.2.* Vector Borne and Zoonotic Diseases 19, #3, (2019): 152–173.

American Society for Microbiology. *Sentinel Level Clinical Laboratory Guidelines for Suspected Agents of Bioterrorism and Emerging Infectious Diseases: Smallpox.* Revised June 2013.

Argonne National Laboratory. *Australia Group Common Control List Handbook Volume II: Biological Weapons-Related Common Control Lists*, Revision 4. Washington, D.C.: U.S. Government Printing Office, February 2018.

The Australia Group. *Australia Group Common Control Lists.* February 28, 2020 [https://www.dfat.gov.au/publications/minisite/theaustraliagroupnet/site/en/human_animal_pathogens.html]. December 31, 2020.

Bayard, Vicente, Paul T. Kitsutani, Eduardo O. Barria, Luis A. Ruedas, David S. Tinnin, Carlos Muñoz, Itza B. de Mosca, Gladys Guerrero, Rudick Kant, Arsenio Garcia, Lorenzo Caceres, Fernando G. Gracia, Evelia Quiroz, Zoila de Castillo, Blas Armien, Marlo Libel, James N. Mills, Ali S. Khan, Stuart T. Nichol, Pierre E. Rollin, Thomas G. Ksiazek and Clarence J. Peters. "Outbreak of Hantavirus Pulmonary Syndrome, Los Santos, Panama, 1999–2000." *Emerging Infectious Diseases* 10 (September 2004): 1635–1642.

Borio, Luciana, Thomas Inglesby, C. J. Peters, Alan L. Schmaljohn, James M. Huges, Peter B. Jahrling, Thomas Ksiazek, Karl M. Johnson, Andrea Meyerhoff, Tara O'Toole, Michael S. Ascher, John Bartlett, Joel G. Breman, Edward. M. Eitzen, Jr., Margaret Hamburg, Jerry Hauer, D. A. Henderson, Richard T. Johnson, Gigi Kwik, Marci Layton, Scott Lillibridge, Gar J. Nabel, Michael T. Osterholm, Trish M. Perl, Philip Russell and Kevin Tonat. "Hemorrhagic Fever Viruses as Biological Weapons: Medical and Public Health Management." *Journal of the American Medical Association* 287 (2002): 2391–2405.

Bozue, Joel, Christopher K. Cote and Pamela J. Glass, ed. *Medical Aspects of Biological Warfare, Textbooks of Military Medicine Series.* Washington, D.C.: Office of the Surgeon General, Department of the Army, 2018.

Brunette, Gary W. and Jeffrey B. Nemhauser, ed. *CDC Yellow Book 2020: Health Information for International Travel.* July 18, 2019 [https://wwwnc.cdc.gov/travel/page/yellowbook-home-2014]. December 31, 2020.

Burrows, W. Dickinson and Sara E. Renner. "Biological Warfare Agents as Threats to Potable Water." *Environmental Health Perspectives* 107 (1999): 975–984.

California Department of Food and Agriculture. Animal Health and Food Safety Services. Animal Health Branch. *Biosecurity: Selection and Use of Surface Disinfectants*, Revision June 2002.

Canada Centre for Biosecurity. *Pathogen Safety Data Sheets.* December 6, 2019 [https://www.canada.ca/en/public-health/services/laboratory-biosafety-biosecurity/pathogen-safety-data-sheets-risk-assessment.html]. December 31, 2020.

Canada Centre for Biosecurity. *Canadian Biosafety Standard*, 2nd Edition, Ottawa, Canada, March 2015.

Centers for Disease Control and Prevention. *National Notifiable Diseases Surveillance System: Surveillance Case Definitions for Current and Historical Conditions.* August 2, 2017 [https://wwwn.cdc.gov/nndss/conditions/]. December 31, 2020.

Centers for Disease Control and Prevention. "Biological and Chemical Terrorism: Strategic Plan for Preparedness and Response. Recommendations of the CDC Strategic Planning Workgroup." *Morbidity and Mortality Weekly Report* 49 (RR-4) (2000): 1–14.

Centre for Agriculture and Bioscience International. *Invasive Species Compendium.* 2020 [https://www.cabi.org/isc]. December 31, 2020.

Centre for Agriculture and Bioscience International. *Plantwise Knowledge Bank.* 2020 [https://www.plantwise.org/KnowledgeBank]. December 31, 2020.

Chosewood, L. Casey and Deborah E. Wilson, ed. *Biosafety in Microbiological and Biomedical Laboratories.* 5th Edition. Washington, D.C.: U.S. Government Printing Office, 2009.

Committee on Foreign Animal Diseases of the United States Animal Health Association. *Foreign Animal Diseases.* 7th Edition. Revised 2008. Boca Raton, Florida: Boca Publishing Group, Inc., 2008.

Compton, James A.F. *Military Chemical and Biological Agents: Chemical and Toxicological Properties.* Caldwell, New Jersey: The Telford Press, 1987.

de Manzione, Nuris, Rosa Alba Salas, Hector Paredes, Oswaldo Godoy, Luis Rojas, Francisco Araoz, Charles Fulhorst, Thomas Ksiazek, James Mills, Barbara Ellis, Clarence Peters and Robert Tesh. "Venezuelan Hemorrhagic Fever: Clinical and Epidemiological Studies of 165 Cases." *Clinical Infectious Diseases* 26 (1998): 308–313.

Department of Agriculture. 9 CFR Part 121 – "Possession, Use, and Transfer of Biological Agents and Toxins," 2005: 817–832.

European and Mediterranean Plant Protection Organization. *EPPO Global Database.* 2020 [https://gd.eppo. int/]. December 31, 2020.

European Association of Zoo and Wildlife Veterinarians. *Transmissible Disease Handbook, 5th Edition.* 2020 [https://www.eazwv.org/page/inf_handbook]. December 31, 2020.

European Centre for Disease Prevention and Control. *All Topics.* December 31, 2020 [https://www.ecdc. europa.eu/en/all-topics]. 2020.

German Social Accident Insurance. *GESTIS Biological Agents Database.* June 12, 2020 [http://gestis.itrust. de/nxt/gateway.dll/bioen/000000.xml?f=templates&fn=default.htm&vid=gestisbioeng:biosdbeng]. December 31, 2020.

Hardcastle, K., D. Scott, D. Safronetz, D. L. Brining, H. Ebihara, H. Feldmann and R. A. LaCasse. "Laguna Negra Virus Infection Causes Hantavirus Pulmonary Syndrome in Turkish Hamsters (*Mesocricetus brandti*)." *Veterinary Pathology* 53 (January 01, 2016) 182–189.

Harkins, Deanna, Rose Overturf, Veronique Hauschild and Scott Goodison. *Safety and Health Guidance for Mortuary Affairs Operations: Infectious Materials and CBRN Handling, Technical Guide 195.* Washington, D.C.: Government Printing Office, May 2009.

Henderson, Donald A., Thomas V. Inglesby, John G. Bartlett, Michael S. Ascher, Edward Eitzen, Peter B. Jahrling, Jerome Hauer, Marcelle Layton, Joseph McDade, Michael T. Osterholm, Tara O'Toole, Gerald Parker, Trish M. Perl, Philip K. Russel and Kevin Tonat. "Smallpox as a Biological Weapon: Medical and Public Health Management." *Journal of the American Medical Association* 281 (1999): 2127–2137.

Henderson, Donald A., Thomas V. Inglesby and Tara O'Toole, ed. *Bioterrorism: Guidelines for Medial and Public Health Management.* Chicago, Illinois: AMA Press, 2002.

Herenda, D. *Manual on Meat Inspection for Developing Countries.* Reprint 2000. Rome, Italy: Food and Agriculture Organization of the United Nations, 1994.

Heymann, David, ed. *Control of Communicable Diseases Manual.* 20th Edition. Washington, D.C.: American Public Health Association, December 2014.

Iowa State University. *The Center for Food Security & Public Health.* 2020 [http://www.cfsph.iastate.edu/]. December 31, 2020.

Leitenberg, Milton and Raymond A. Zilinskas. *The Soviet Biological Weapons Program: A History.* Cambridge, Massachusetts: Harvard University Press, 2012.

Levis, Silvana, Jorge Garcia, Noemi' Pini, Gladys Caldero' N, Josefina Rami'Rez, Daniel Bravo, Stephen St. Jeor, Carlos Ripoll, Mariana Bego, Elena Lozano, Rube' N Barquez, Thomas G. Ksiazek and Delia Enria. "Hantavirus Pulmonary Syndrome in Northwestern Argentina: Circulation of Laguna Negra Virus Associated with *Calomys Callosus.*" *American Journal of Tropical Medicine and Hygiene* 71 (November 2004) 658–663.

Lindler, Luther E., Frank J. Lebeda and George W. Korch, ed. *Biological Weapons Defense: Infectious Diseases and Counterbioterrorism.* Totowa, New Jersey. Humana Press, Inc., 2005.

Martinez, Valeria P., Carla Bellomo, Jorge San Juan, Diego Pinna, Raul Forlenza, Malco Elder and Paula J. Padula. "Person-to-Person Transmission of Andes Virus." *Emerging Infectious Diseases* 11 (December 2005): 1848–1853.

Merck & Co. *Merck Manual Professional Version.* 2020 [https://www.merckmanuals.com/professional]. December 31, 2020.

Merck & Co. *Merck Veterinary Manual.* 2020 [https://www.merckvetmanual.com/]. December 31, 2020.

Nelson, Randin, Raul Cañate, Juan Miguel Pascale, Jerry W. Dragoo, Blas Armien, Anibal G. Armien and Frederick Koster. "Confirmation of Choclo Virus as the Cause of Hantavirus Cardiopulmonary Syndrome and High Serum Antibody Prevalence in Panama." *Journal of Medical Virology* 82 (September 2010): 1586–1593.

Nikitin, Nikolai, Ekaterina Petrova, Ekaterina Trifonova and Olga Karpova. "Review Article: Influenza Virus Aerosols in the Air and Their Infectiousness." *Advances in Virology* (2014), Article ID 859090, 6 pages, 10.1155/2014/859090.

Pacific Community Land Resources Division. *AHP Disease Manual: Reference Guide for Animal Health Staff.* 2010 [http://lrd.spc.int/ext/Disease_Manual_Final/index.html]. December 31, 2020.

Pan American Health Organization. *Emergency Vector Control After Natural Disaster*. Scientific Publication No. 419. Washington, D.C.: Pan American Health Organization, 1982.

Paweska, Janusz, Nivesh Sewlall, Thomas Ksiazek, Lucille Blumberg, Martin Hale, Ian Lipkin, Jacqueline Weyer, Stuart Nichol, Pierre Rollin, Laura McMullan, Christopher Paddock, Thomas Briese, Joy Mnyaluza, Thu-Ha Dinh, Victor Mukonka, Pamela Ching, Adriano Duse, Guy Richards, Gillian de Jong, Cheryl Cohen, Bridget Ikalafeng, Charles Mugero, Chika Asomugha, Mirriam Malotle, Dorothy Nteo, Eunice Misiani, Robert Swanepoel and Sherif Zaki, "Nosocomial Outbreak of Novel Arenavirus Infection, Southern Africa." *Emerging Infectious Diseases* 15, (October 2009): 1598–1602.

Public Health Agency of Canada. *Canadian Biosafety Handbook, 2nd Edition*. Ottawa, Canada, March 2016.

Rutala, William and David J. Weber. *Guideline for Disinfection and Sterilization in Healthcare Facilities, 2008*. Centers for Disease Control and Prevention, Updated May 2019.

Sidell, Fredrick R., Ernest T. Takafuji and David R. Franz, ed. *Medical Aspects of Chemical and Biological Warfare, Textbook of Military Medicine Series, Part 1, Warfare, Weaponry, and the Casualty*. Washington, D.C.: Office of the Surgeon General, Department of the Army, 1997.

Siegel, Jane D., Emily Rhinehart, Marguerite Jackson and Linda Chiarello. *2007 Guideline for Isolation Precautions: Preventing Transmission of Infectious Agents in Healthcare Settings*. Centers for Disease Control and Prevention, June 2007.

Sifton, David W. ed. *PDR Guide to Biological and Chemical Warfare Response*. Montvale, New Jersey: Thompson/Physicians' Desk Reference, 2002.

United States Army Headquarters. *Potential Military Chemical/Biological Agents and Compounds, Field Manual No. 3-11.9*. Washington, D.C.: Government Printing Office, January 10, 2005.

United States Army Headquarters. *Technical Aspects of Biological Defense, Technical Manual No. 3–216*. Washington, D.C.: Government Printing Office, January 12, 1971.

United States Army Headquarters. *Treatment of Biological Warfare Agent Casualties, Field Manual No. 8–284*. Washington, D.C.: Government Printing Office, July 17, 2000.

United States Department of Agriculture. Animal and Plant Health Inspection Service. 2020 [https://www.aphis.usda.gov/aphis/home]. December 31, 2020.

United States Department of Agriculture. Animal and Plant Health Inspection Service. *National Veterinary Accreditation Program Reference Guide*. June 2, 2020 [https://www.aphis.usda.gov/aphis/ourfocus/animalhealth/nvap/NVAP-Reference-Guide]. December 31, 2020.

United States Department of Health and Human Services, and Department of Agriculture. *Select Agents and Toxins List*. 2017 [https://www.selectagents.gov/SelectAgentsandToxinsList.html]. December 31, 2020.

United States Environmental Protection Agency. *Drinking Water Treatability Database: Find a Contaminant*. 2020 [https://tdb.epa.gov/tdb/findcontaminant]. December 31, 2020.

United States Food & Drug Administration, Center for Food Safety & Applied Nutrition. *Bad Bug Book, Foodborne Pathogenic Microorganisms and Natural Toxins Handbook*. 2nd Edition. October 24, 20217 [https://www.fda.gov/food/foodborne-pathogens/bad-bug-book-second-edition]. December 31, 2020.

University of Hawaii, Manoa. *EXTension ENTOmology & UH-CTAHR Integrated Pest Management Program*. August 30, 2011 [http://www.extento.hawaii.edu/kbase/crop/Type/Croppest.htm]. December 31, 2020.

United States Military Joint Chiefs of Staff. *Joint Tactics, Techniques, and Procedures for Mortuary Affairs in Joint Operations, Joint Publication No. 4-06*. Washington, D.C.: Government Printing Office, August 28, 1996.

University of Minnesota Center for Infectious Disease Research and Policy. *Infectious Disease Topics*. 2020 [http://www.cidrap.umn.edu/infectious-disease-topics]. December 31, 2020.

WebMD. *Medscape*. 2020 [https://emedicine.medscape.com/]. December 31, 2020.

Withers, Mark, ed. *Medical Management of Biological Casualties Handbook*. 8th Edition. Fort Detrick, Maryland: United States Army Medical Research Institute of Infectious Diseases, September 2014.

World Health Organization. *Health Aspects of Chemical and Biological Weapons: Report of A WHO Group of Consultants*. Geneva, Switzerland: World Health Organization, 1970.

World Health Organization. *Fact Sheets*. 2020 [https://www.who.int/news-room/fact-sheets]. 2020.

World Health Organization. *International Classification of Diseases, 11th Revision*. Geneva, Switzerland: World Health Organization. April 2019 [https://icd.who.int/browse11/l-m/en]. 2020.

World Health Organization. *Laboratory Biosafety Manual, 3rd Edition*. Geneva, Switzerland, 2004.

World Health Organization. *Public Health Response to Biological and Chemical Weapons: WHO Guidance*. Geneva, Switzerland: World Health Organization, 2004.

World Organization for Animal Health (OIE). *Animal Health of the World: Information on Aquatic and Terrestrial Animal Diseases*. 2020 [https://www.oie.int/en/animal-health-in-the-world/information-on-aquatic-and-terrestrial-animal-diseases/]. December 31, 2020.

World Organization for Animal Health (OIE). *Manual of Diagnostic Tests and Vaccines for Terrestrial Animals 2019.* [https://www.oie.int/en/standard-setting/terrestrial-manual/access-online/]. December 31, 2020.

World Organization for Animal Health (OIE). *OIE-Listed Diseases, Infections and Infestations in Force in 2020.* [https://www.oie.int/en/animal-health-in-the-world/oie-listed-diseases-2020/]. December 31, 2020.

World Organization for Animal Health (OIE). *Technical Disease Cards.* 2020 [https://www.oie.int/en/animal-health-in-the-world/technical-disease-cards/]. December 31, 2020.

Yahnke, Christopher J., Peter L. Meserve, Thomas G. Ksiazek and James N. Mills. "Patterns of Infection with Laguna Negra Virus in Wild Populations of *Calomys Laucha* in the Central Paraguayan Chaco." *American Journal of Tropical Medicine and Hygiene* 65 (December 2001) 768–776.

19 Fungal Pathogens

GENERAL INFORMATION

Fungi are unicellular or multicellular organisms that are more highly evolved than bacteria (C17). They are members of the plant kingdom and include molds, mildew, smuts, rusts, and yeasts. They range in size from 3 to 50 microns. With the exception of yeasts, they are usually rod shaped and arranged end to end in strands or filaments. Yeasts are usually oval.

Fungi reproduce by forming spores. Spores are part of the reproductive cycle and are not a protective mechanism as used by some bacteria (C17) when they are subjected to adverse conditions. Fungal spores can lie dormant, sometimes for decades, waiting for conditions that allow germination. Infections occur when a spore germinates on or in a host. Fungi are relatively easy to grow. Production, isolation, harvesting, and storage of spores is also relatively uncomplicated.

Although several notable antipersonnel fungal agents have been investigated as biological warfare agents, fungi have primarily been selected because of their ability to attack agriculturally significant crop species such as wheat, corn, or rice. Most anti-plant agents are host specific, and some are even specific to individual varieties of the host species. There is little potential for anti-plant fungi to attack humans or animals.

A final group of biological warfare pathogens are those used as simulants to model the release of other, more hazardous agents. Pathogens employed as biological warfare simulants do not generally pose a significant risk to healthy people, animals, or plants. However, individuals with respiratory illness or suppressed immune systems may be at risk should they be exposed to an infectious dose of the agent.

Fungi can be stored as active cultures or isolated as spores. Spores are relatively easy to disperse. However, because they are living organisms and can be killed during the dispersal process there are limitations to the methods that can be used. For more information on methods of dispersal, see appendix 1. In most cases, large-scale attacks will be clandestine and only detected through epidemiological analysis of resulting disease patterns. Even in the case of small-scale incidents or attacks directed at specific individuals (e.g., white-powder letters), without the inclusion of a threat the attack may go unrecognized until the disease appears in exposed individuals (e.g., the initial 2001 anthrax attack at American Media Inc., which claimed the life of Robert Stevens).

Incubation times for diseases resulting from infection vary depending on the specific pathogen, but are generally on the order of days to weeks. Exposures to extremely high doses of some pathogens may reduce the incubation period to as short as several hours. The pathway of exposure (e.g., inhalation, lacerations) can also cause a significant change in the incubation time required as well as the clinical presentation of the disease. Diseases caused by fungi are not communicable and cannot be transferred directly from an infected individual to anyone else.

RESPONSE

PERSONAL PROTECTIVE REQUIREMENTS

Responding to the Scene of a Release

A number of conditions must be considered when selecting protective equipment for individuals at the scene of a release. For instances such as white-powder letters when the mechanism of release is known and it does not involve an aerosol generating device, then responders can use Level C with N95 or higher-level filters.

DOI: 10.4324/9781003230564-10

If an aerosol-generating device is employed (e.g., sprayer), or the dissemination method is unknown and the release is ongoing, then responders should wear a Level A protective ensemble. Once the device has stopped generating the aerosol or has been rendered inoperable, and the aerosol has settled, then responders can downgrade to Level B.

In all cases, there is a significant hazard posed by contact of contaminated material with skin that has been cut or lacerated, or through injection of pathogens by contact with debris. Appropriate protection to avoid any potential abrasion, laceration, or puncture of the skin is essential. Individuals with damaged or open skin should not be allowed to enter the contaminated area.

Working with Infected Individuals

Use infection control guidelines standard precautions. Avoid direct contact with wounds or wound drainage. Standard precautions include hand hygiene; the use of gloves and eye protection; wearing a gown and mask as appropriate; and using care when handling sharps.

DECONTAMINATION

Casualties/Personnel

Infected individuals

Unless the individual is reporting directly from the scene of an attack (e.g., white-powder letter, aerosol release, etc.), then decontamination is not necessary.

Direct Exposure

In the event that an individual is at the scene of a known or suspected attack (e.g., white-powder letter, aerosol release), have them wash their hands and face thoroughly with soap and water as soon as possible. They should also blow their nose to remove any agent particles that may have been captured by nasal mucous. If the release involved a powdered agent and it is practicable, dampen the agent with a water mist to help prevent aerosolization. Remove all clothing and seal in a plastic bag. To avoid further exposure of the head, neck, and face to the agent, cut off potentially contaminated clothing that must be pulled over the head. Shower using copious amounts of soap and water. Ideally, showers will be high volume with low pressure. Ensure that the hair has been washed and rinsed to remove potentially trapped agent. The CDC does not recommend that individuals use bleach or other disinfectants directly on their skin.

Animals

Unless the animals are at the scene of an attack, then decontamination is not necessary.

Apply universal decontamination procedures using soap and water. Consult local/state veterinary assistance office. If the pathogen has not been identified, then wear a fitted N95 protective mask, eye protection, disposable protective coveralls, disposable boot covers, and disposable gloves when dealing with infected animals.

Plants

May require removal and destruction of infected plants. Incineration of impacted fields may be required. Consult local/state agricultural assistance office. Many fungi are easily spread by mechanical vectors (e.g., farm implements, track out, running water, insects) and extreme care must be taken to avoid further contamination. Wear disposable protective coveralls, disposable boot covers, and disposable gloves to prevent spread of contamination.

Food

All foodstuffs in the area of a release should be considered contaminated. Unopened items may be used after decontamination of the container. Opened or unpackaged items should be destroyed. Fruits and vegetables should be washed thoroughly with soap and water.

Property

Surface Disinfectants

Compounds containing phenolics, chlorhexidine, quaternary ammonium salts (additional activity if bis-n-tributyltin oxide present), alcohols such as 70 to 90% ethanol and isopropyl, potassium peroxymonosulfate, and iodine/iodophors.

The military also identifies the following nonstandard decontaminants: 3% aqueous peracetic acid solution, 1% aqueous hyamine solution, and 10% aqueous sodium or potassium hydroxide solution.

Large Area Fumigants

Gases including formaldehyde, ozone, ethylene oxide, or chlorine dioxide are effective against many fungi. These materials are highly toxic to humans and animals, and fumigation operations must be adequately controlled to prevent unnecessary exposure. Additional methods include vaporized hydrogen peroxide and an ionized hydrogen peroxide aerosol.

Fomites

Clothing or bedding may become contaminated with spores. Deposit items in an appropriate biological waste container and send to a medical waste disposal facility.

Alternatively, cotton or wool articles can be boiled in water for 30 minutes, autoclaved at 253°F for 45 minutes, or immersed in a 2% household-bleach solution (i.e., 1 liter of bleach in 2 liters of water) for 30 minutes followed by rinsing.

FATALITY MANAGEMENT

Unless the cadaver is coming directly from the scene of an attack (e.g., white-powder letter, aerosol release), then process the body according to established procedures for handling potentially infectious remains.

Because of the nature of biological warfare agents, it is highly unlikely that a contaminated cadaver will be recovered from the scene of an attack unless it is from an individual who died from trauma or other complications while the attack was ongoing. If a fatality is grossly contaminated with a biological agent, wear disposable protective clothing with integral hood and booties, disposable gloves, eye protection, and an N95 respirator or powered air-purifying respirator (PAPR) equipped with N95 or high-efficiency particulate air (HEPA) filters.

Remove all clothing and personal effects. Items that will be retained for further processing should be double sealed in impermeable containers, ensuring that the inner container is decontaminated before placing it in the outer one. Otherwise, dispose of contaminated articles at an appropriate medical waste disposal facility.

Thoroughly wash the remains with soap and water. Pay particular attention to areas where the agent may get trapped, such as hair, scalp, pubic areas, fingernails, folds of skin, and wounds. If deemed appropriate, the body can be washed with a 2% sodium hypochlorite bleach solution (i.e., 2 liters of water for every liter of household bleach), ensuring the solution is introduced into the ears, nostrils, mouth, and any wounds. This concentration of bleach will not affect remains but will disinfect the offending agent. Higher concentrations of bleach can harm remains. The bleach solution should remain on the cadaver for a minimum of 15 minutes. Wash with soap and water. If the body is to be embalmed, ensure that all the bleach solution is removed as it will react with embalming fluid.

Once the remains have been thoroughly decontaminated, process the body according to established procedures for handling potentially infectious bodies. Use appropriate burial procedures.

C19-A   AGENTS

C19-A001

Aspergillus fumigatus
BW Simulant
ICD-11: XN5Z7
UNII: X88DF51T48
EPPO: ASPEFU

It is endemic throughout the world in decaying organic matter and soil (saprophyte). It has a pronounced tolerance to heat (up to 158°F) and exploits this to survive in compost pile. Spores are commonly inhaled pollutants that do not cause any significant effects in healthy humans or animals.

May cause disease (i.e., allergic bronchopulmonary aspergillosis, chronic necrotizing *Aspergillus pneumonia*, aspergilloma, or invasive aspergillosis) in sensitive populations or individuals with compromised immune system.

C19-A002

Blumeria graminis
Powdery Mildew of Grasses and Cereals
ICD-11: —
UNII: —
EPPO: ERYSGR

It is endemic throughout the world on numerous species comprising more than 100 genera of grasses. Epidemics tend to occur during conditions of alternating wet and dry weather. Growth is encouraged by excessive use of nitrogen fertilizer and can be particularly severe in dense crops grown in a sheltered, humid environment.

In Plants:

Target species: Wheat, barley, oats, rye.

Normal routes of transmission: Airborne.

Secondary Hazards: Spores; mechanical vectors (cultivation, track out); crop debris.

Signs: Initial symptoms are yellowed (chlorotic) flecks that quickly become white patches with a powdery appearance on leaves, stems and ears. If the plant is shaken even gently, clouds of spores are released. In later stages, the colonies turn a gray-tan color with black bodies imbedded throughout.

Crop Losses: ≤60%

C19-A003

Coccidioides immitis (Agent OC)
Coccidioidomycosis
ICD-11: XN53F
UNII: JNF66W9FFW
EPPO: CCDIIM

It is endemic to soils in the southwest United States, northern Mexico, and certain areas in Central and South America. A highly virulent soil-fungus that reproduces asexually to form spores (arthroconidia) <10 μm in diameter that are highly resistant to desiccation. Does not cause clinical disease in cattle, sheep, and pigs. However, can cause disease in dogs and cats. This is a biosafety level 2 agent for clinical samples. Biosafety level 3 should be used for activities with a high potential for aerosol production or high concentrations of agent.

It is on the Australia Group Core list.

In People:

CDC Case Definition: Clinical criteria may be asymptomatic or may produce an acute or chronic disease. Although the disease initially resembles an influenza-like or pneumonia-like febrile illness primarily involving the bronchopulmonary system, dissemination can occur to multiple organ systems. An illness is typically characterized by one or more of the following: influenza-like signs and symptoms (e.g., fever, chest pain, cough, myalgia, arthralgia, and headache); pneumonia or other pulmonary lesion, diagnosed by chest radiograph; erythema nodosum or erythema multiforme rash; and/or involvement of bones, joints, or skin by dissemination; meningitis; involvement of viscera and lymph nodes.

Confirmatory laboratory criteria is 1) Cultural, histopathologic, or molecular evidence of presence of *Coccidioides* species; OR 2) Positive serologic test for coccidioidal antibodies in serum, cerebrospinal fluid, or other body fluids by detection of coccidioidal immunoglobulin M by immunodiffusion, enzyme immunoassay, latex agglutination, or tube precipitin; OR 3) Positive serologic test for coccidioidal antibodies in serum, cerebrospinal fluid, or other body fluids by detection of coccidioidal immunoglobulin G by immunodiffusion, EIA, or complement fixation; OR 4) Positive serologic test for coccidioidal antibodies in serum, cerebrospinal fluid, or other body fluids by coccidioidal skin-test conversion from negative to positive after onset of clinical signs and symptoms.

Communicability: Direct person-to-person transmission does not occur. When dealing with infected individuals, use standard precautions.

Normal Routes of Exposure: Inhalation; abraded skin; mucous membranes.

Infectious Dose: 1–10 spores (inhalation).

Secondary Hazards: Spores; aerosols (dust, body tissue); fomites (spores).

Incubation: 1–3 weeks.

Signs & Symptoms: Fatigue, cough, chest pain, difficulty breathing (dyspnea), coughing up blood (hemoptysis), night sweats, muscle pain (myalgia), joint pain (arthralgia), headache, nausea, vomiting. Pneumonia is usually one-sided. A diffuse rash of mixed flat and raised (maculopapular) spots that does not itch (nonpruritic); reddish, painful, tender lumps (erythema nodosum); or lesions with a pink-red center, pale border ring, and an outer pink-red ring (erythema multiforme) on the upper body or legs that may be transient and seemingly inconsequential. Most people develop lifelong immunity post-infection.

Suggested Alternatives for Differential Diagnosis: Acute respiratory distress syndrome, babesiosis, blastomycosis, blebs, bullae, congenital cystic lesions, diffuse bronchiectasis, enteropathic arthropathies, eosinophilic pneumonia, histoplasmosis, honeycomb lung associated with advanced fibrosis, hydatid disease, localized bronchiectasis, lung abscess, lung cancer, lymphocytic interstitial pneumonia, lymphoma, metastatic disease, myelophthisic anemia, neoplasms, old granuloma, otolaryngologic manifestations of granulomatosis with polyangiitis, paracoccidioidomycosis, pericarditis, pneumatoceles, pneumonia, pulmonary eosinophilia, pulmonary infarct, pulmonary langerhans cell histiocytosis, pulmonary lymphangioleiomyomatosis, rheumatoid nodule, sarcoidosis, septic embolism, solitary pulmonary nodule, traumatic lesions, tuberculosis, viral meningitis.

Mortality Rate (untreated): ≤65% (disseminated form).

C19-A004

Coccidioides posadasii
Coccidioidomycosis
ICD-11: XN5TT
UNII: DV0VFN5F4T
EPPO: —

Until 2002, *Coccidioides posadasii* was believed to be a non-California variant of *Coccidioides immitis*. The two species can only be distinguished by genetic analysis and by the fact that *C. posadasii* grows more slowly in the presence of high salt concentrations. There is no apparent difference in pathogenicity between the two species. For more information on coccidioidomycosis, the disease caused by these two fungi, see *Coccidioides immitis* (C19-A003).

It is on the Australia Group Core list.

C19-A005

Cochliobolus miyabeanus (Agent E)
Brown Spot
ICD-11: —
UNII: —
EPPO: COCHMI

It is endemic throughout the world. In addition to rice, it can survive on wild rice, corn, and cutgrasses. Favorable conditions for growth include nutrient-deficient soils, temperatures between 77°F and 86°F, and a relative humidity above 80%. Rice leaves must be wet for 8 to 24 hours for infection to occur. The fungus can survive in the seed for more than 4 years.

It is on the Australia Group Core list.

In Plants:

Target species: Rice.

Normal routes of transmission: Airborne.

Secondary Hazards: Spores; cropdebris; infected seeds.

Signs: Young lesions appear as small, dark-brown spots. More developed infection presents as evenly distributed oval-shaped brown lesions with grayish centers. Spots may coalesce causing the leaf to dry up. Seeds and bottom scales around the seeds (glumes) may develop black or dark lesions with a velvety aspect. Infected seeds may not germinate, result in seedling mortality, or reduce the gain quality and weight if the seedling survives.

Crop Losses: ≤90%

C19-A006

Colletotrichum coffeanum var. virulans
Brown Blight of Coffee
ICD-11: —
UNII: —
EPPO: COLLCO

It is endemic to Africa. The fungus lives in the bark of the coffee tree and produces spores that attack the immature or green coffee berries.

It is on the Australia Group Core list.

In Plants:

Target species: Coffee.

Normal routes of transmission: Airborne.

Secondary Hazards: Spores; mechanical vectors (cultivation, birds, rain); infected seeds.

Signs: Appears as small dark sunken spots which spread rapidly over the berry. Spots may have a pale pink crust on their surface. Berry may become mummified prior to full development. Otherwise, the berry ripens and the bean can become infected.

Crop Losses: ≤80%

C19-A007
Coniothyrium glycines
Red Leaf Blotch
ICD-11: —
UNII: —
EPPO: DACHGY

It is endemic to central and southern Africa. The fungus does not produce significant airborne spores and is primarily spread through movement of infested plant debris and contaminated soil. Plants are infected when rain splashes the fungus onto plants leaves. It overseasons in infected plant debris and the soil as tiny balls of cells (sclerotia) that grow when conditions are right. These hibernating cells can survive wide temperature extremes (41°F to 212°F). The disease is favored by wet, humid conditions.

It is on the U.S. Select Agents and Toxins list.

In Plants:

Target species: Soybeans.

Normal routes of transmission: Contact (rain-splashed).

Secondary Hazards: Mechanical vectors (cultivation, track out, rain); crop debris.

Signs: Initial signs include small dark red to brown spots that appear on the upper and lower surfaces of leaves. These spots coalesce to form larger, buff-colored blotches with dark centers and margins, that are eventually surrounded by yellowed (chlorotic) halos and can cover over half of the leaf. The dead centers of the blotches may disintegrate leaving a hole. Lesions on other parts of the plant are ovoid and mauve to reddish-purple. Plants suffer premature leaf drop (defoliation) returning the fungus back to the soil.

Crop Losses: ≤70%.

C19-A008
Fusarium langsethiae
Fusarium Head Blight
ICD-11: —
UNII: —
EPPO: FUSALA

It is endemic throughout the world. One of a cohort of *Fusarium* species that collectively cause Head Blight in grains.

Primary concern is the production of toxins, notably T2 toxin (C16-A045), HT-2 toxin (C16-A042), nivalenol (C16-A044), neosolaniol (C16-A043), and diacetoxyscirpenol (C16-A039). Fusarium mycotoxins are produced in the field as part of the fungal colonization of the ear. Levels of ear blight seen in the field do not always correlate with mycotoxin occurrence. Kernels that are infected later in development may show no outward signs, but may still be contaminated with the mycotoxin. In properly stored grains – reduced to less than 15% moisture – the pathogen stops growing and ceases production of mycotoxins. This will not reduce the toxin levels already present in harvested grains.

It is on the Australia Group Warning list.

In Plants:

Target species: Oats, wheat, barley.

Normal routes of transmission: Airborne.

Secondary Hazards: Spores; mechanicalvectors (cultivation, rain); crop debris.

Signs: First symptoms appear shortly after flowering when the weather is warm and wet – temperatures between 59°F and 86°F with extended periods of rain or dew, or relative humidity at or above 90%. Fungus bleaches the spikelets as it grows and spreads within the head. While it may progress to the entire head, the fungus may only infect individual spikelets scattered throughout the head. Affected grains shrink and dry up, with a color ranging from pink or chalky-white to light-brown, commonly referred to as a tombstone appearance. Sporing can also give the seed a blue-black scabbed appearance. There is no obvious odor.

Crop Losses: > 45% (fusarium head blight)

C19-A009

Fusarium sporotrichioides
Fusarium Head Blight
ICD-11: —
UNII: Z0HNB8C468
EPPO: FUSASR

It is endemic throughout the world. One of a cohort of *Fusarium* species that collectively cause Head Blight in grains.

Primary concern is the production of toxins, notably T2 toxin (C16-A045), HT-2 toxin (C16-A042), nivalenol (C16-A044), neosolaniol (C16-A043), and diacetoxyscirpenol (C16-A039). Fusarium mycotoxins are produced in the field as part of the fungal colonization of the ear. Levels of ear blight seen in the field do not always correlate with mycotoxin occurrence. Kernels that are infected later in development may show no outward signs, but may still be contaminated with the mycotoxin. In properly stored grains – reduced to less than 15% moisture – the pathogen stops growing and ceases production of mycotoxins. This will not reduce the toxin levels already present in harvested grains.

It is on the Australia Group Warning list.

In Plants:

Target species: Wheat, barley oats, corn.

Normal routes of transmission: Airborne.

Secondary Hazards: Spores; mechanical vectors (cultivation, rain); crop debris.

Signs: First symptoms appear shortly after flowering when the weather is warm and wet – temperatures between 59°F and 86°F with extended periods of rain or dew, or relative humidity at or above 90%. Fungus bleaches the spikelets as it grows and spreads within the head. While it may progress to the entire head, the fungus may only infect individual spikelets scattered throughout the head. Affected grains shrink and dry up, with a color ranging from pink or chalky-white to light-brown, commonly referred to as a tombstone appearance. Sporing can also give the seed a blue-black scabbed appearance. There is no obvious odor.

Crop Losses: >45% (fusarium head blight)

C19-A010

Histoplasma capsulatum
Histoplasmosis
ICD-11: XN8VH
UNII: 6EF1RL8Z5O
EPPO: AJELCP

It is endemic to soils in central and eastern United States, and parts of Central and South America, Africa, Asia, and Australia. A dimorphic fungus, appearing as a mold in the environment and a yeast in the body. Spores are <5 μm in diameter and easily deposited in the lung. Normally found in soils and the dung of bats. Does not cause clinical disease in cattle, sheep, and pigs. However, can cause disease in dogs and cats. This is a biosafety level 2 agent for clinical samples. Biosafety level 3 should be used for activities with a high potential for aerosol production or high concentrations of agent.

In People:
CDC Case Definition: Clinical criteria is either 1) at least two of the following – fever, chest pain, cough, myalgia, shortness of breath, headache, erythema nodosum/erythema multiforme rash; OR 2) at least one of the following – abnormal chest imaging (e.g., pulmonary infiltrates, cavitation, enlarged hilar or mediastinal lymph nodes, pleural effusion), clinical evidence of disseminated disease (gastrointestinal ulcerations or masses; skin or mucosal lesions; peripheral lymphadenopathy; pancytopenia, as evidence of bone marrow involvement; enlargement of the liver, spleen, or abdominal lymph nodes; meningitis, encephalitis, or focal brain lesion).

Confirmatory laboratory criteria is 1) culture of *H. capsulatum* from a clinical specimen, OR 2) identification of characteristic *H. capsulatum* yeast in tissue or sterile body fluid by histopathology, OR 3) ≥fourfold rise in *H. capsulatum* serum complement fixation antibody titers taken at least 2 weeks apart, OR 4) detection in serum of H band by *H. capsulatum* immunodiffusion antibody test, OR 5) detection in serum of M band by *H. capsulatum* immunodiffusion antibody test after a documented lack of M band on a previous test (i.e., seroconversion), OR 6) demonstration of *H. capsulatum*-specific nucleic acid in a clinical specimen using a validated assay (i.e., polymerase chain reaction).

Communicability: Direct person-to-person transmission does not occur. When dealing with infected individuals, use standard precautions.

Normal Routes of Exposure: Inhalation; abraded skin; mucous membranes.

Infectious Dose: Unknown.

Secondary Hazards: Spores; aerosols(contaminated dust).

Incubation: 3–17 days.

Signs & Symptoms: Symptoms vary, depending on the exposure, the host immune status, and any underlying disease. Generally, symptoms include a mild, flu-like respiratory illness with a vague feeling of bodily discomfort (malaise), fever, chest pain, dry or nonproductive cough, headache, loss of appetite (anorexia), difficulty breathing (dyspnea), joint (arthralgia) and muscle (myalgia) pain, chills, and hoarseness. May progress to low oxygen levels in the blood (hypoxemia), an adult respiratory distress syndrome (ARDS)-like illness, inflammation around the brain and spinal cord (meningitis), seizures, and infection of the heart (endocarditis). Cutaneous signs may include a red rash of mixed flat and raised (maculopapular) spots, ulcerations, dermal bleeding (purpura), lesions with a pink-red center, pale border ring, and an outer pink-red ring (erythema multiforme), and reddish, painful, tender lumps (erythema nodosum).

Suggested Alternatives for Differential Diagnosis: Aspergillosis, blastomycosis, carcinoid lung tumors, coccidioidomycosis, fungal pneumonia, Legionnaire's disease, lung abscess, mediastinal cysts, mediastinal lymphoma, mycoplasma infections, Pancoast syndrome, pneumococcal infections, pneumonia, pneumonitis, respiratory distress syndrome, sarcoidosis, sarcoidosis, small cell lung cancer, tuberculosis.

Mortality Rate (untreated): ≤8%.

C19-A011

Magnaporthe oryzae (Agent IE)
Rice Blast
ICD-11: —
UNII: —
EPPO: PYRIOR

It is endemic throughout the world. Primarily a disease of rice, but also infects a number of wild grasses. Favorable conditions include frequent and prolonged showers with temperatures between 75°F and 82°F. Symptoms can begin to appear in just 4 days after spores infect a young plant.

It is on the Australia Group Core list.

In Plants:

Target species: Rice.

Normal routes of transmission: Airborne.

Secondary Hazards: Spores; mechanical vectors (irrigation); crop debris.

Signs: Oval or diamond-shaped spots with dark borders appear on the leaves; often with yellow haloes. Spots become longer with age; the centers turn whitish-gray and the borders become wider and red-brown. Ultimately, the spots join together and the leaves die. Severely infected fields have a scorched appearance. Infection spreads to the rest of the plant and causes rot on the stems at the nodes (slightly swollen parts of the stem where the leaves and tillers develop), collars (junction of the leaf base and leaf sheath), neck (stem below the flower heads), and flower heads.

Crop Losses: ≤100% Rice blast is said to be responsible for loss of approximately 30% of the worlds annual rice crop.

C19-A012

Microcyclus ulei
South American Leaf Blight of Rubber
ICD-11: —
UNII: —
EPPO: MICCUL

It is endemic to Central and South America.

It is on the Australia Group Core list.

In Plants:

Target species: Rubber plants.

Normal routes of transmission: Airborne.

Secondary Hazards: Spores; mechanical vectors (rain); crop debris (leaves).

Signs: Only infects young leaflets that are less than 15 days old. Mature leaves are completely resistant. Young rubber trees, up to 4 years, produce new leaves throughout the year. Older trees normally change leaves once a year with the onset of the dry season. Infections occur when leaves have been wet for several hours. Symptoms vary with the age of the leaves. In leaves 4 to 10 days old, it causes discolored green masses. These spots grow together, consuming the leaf and causing it to die. In older leaves, the infection may cause holes in the leaves that are susceptible to further damage or infection. This stage persists on mature diseased leaves and provides the survival stage of the fungus. Infection and rapid re-infection of young leaflets can cause successive defoliations which lead to dieback of terminal twigs and branches and ultimately to death of young trees.

Crop Losses: ≤33%

C19-A013

Moniliophthora roreri
Frosty Pod Rot
ICD-11: —
UNII: —
EPPO: MONPRO

It is endemic to Central America and northwest South America. Only grows on the pods of the cocoa tree. Spores on pods left on trees can remain viable for up to 9 months; fungi on pods that have fallen to the ground are rapidly killed by microbes.

It is on the Australia Group Awareness list.

In Plants:

Target species: Cocoa trees.

Normal routes of transmission: Airborne.

Secondary Hazards: Spores; mechanical vectors (rain); crop debris.

Signs: Depends on the age of the pod. Less than 3 months old, infection may initially appear as slightly yellowed (chlorotic) swellings or distortion on the surface. Irregular brown, oily, necrotic spots form that eventually cover the fruit surface and then progress to a thick, white, or cream-colored velvety fungal growth. Pods infected after 3 months of age may show no external symptoms or only limited necrosis. They are often slightly sunken and surrounded by areas of premature ripening. Infected pods are noticeably heavier than healthy ones. While the pod surface remains firm, necrosis spreads internally and the seed mass may become soft and watery, with a reddish-brown tint. In all cases, infected pods remain attached to the tree as they shrink, dry, and become mummified.

Crop Losses: ≤90%

C19-A014

Peronosclerospora philippinensis
Philippine Downy Mildew
ICD-11: —
UNII: —
EPPO: PRSCPH

It is endemic to South and Southeast Asia as well as parts of Africa. In addition to agriculturally significant crops, it infects many weedy grass species. Favorable conditions for spore production include darkness, temperatures greater than 61°F, high humidity, and a film of water on the leaf for 4 to 5 hours. Spores can germinate in less than an hour on a new host. Seedborne transmission is possible in corn; there are no visible signs of infection on seeds. However, once a seed or grain has been dried to below 14% moisture, it will not produce an infected plant.

It is on the Australia Group Core list and on the U.S. Select Agents and Toxins list.

In Plants:

Target species: Corn, oats, sugarcane, sorghum.

Normal routes of transmission: Airborne.

Secondary Hazards: Spores; mechanical vectors (rain); crop debris; infected seeds.

Signs: Yellowed (chlorotic)stripes or overall yellowing of the leaves. Stripes may have a downy appearance. Spore formation is more abundant on the lower surface. There are no external symptoms on stems but early-infected plants may be stunted. As the plant ages, leaves may narrow, become abnormally erect, and appear dried out. Post-infection, new leaves are narrow and rigid; shoots and roots are stunted. Tassels may be malformed and produce less pollen. Ears may be aborted.

Crop Losses: ≤100% (sweet corn), ≤60% (other corn), ≤25% (sugarcane)

C19-A015

Phoma tracheiphila
Mal Secco
ICD-11: —
UNII: —
EPPO: DEUTTR

It is endemic to the Mediterranean and Black Sea basins. The fungal spores enter through wounds in leaves, branches and roots when the temperature is between 57°F and 82°F. Optimum temperature for growth is between 68°F and 77°F. *P. tracheiphila* produces the phytotoxin malseccin that mimics the symptoms of veinal chlorosis, necrosis and wilt.

It is on the Australia Group Awareness list.

In Plants:

Target species: Lemons, other citrus trees.

Normal routes of transmission: Airborne; contact (through wounds).

Secondary Hazards: Spores; mechanical vectors (cultivation, rain); crop debris.

Signs: Initial symptoms are yellowed (chlorotic) shoots and leaves followed by dieback of twigs and branches. Raised black points within gray areas appear on withered twigs. Sprouts grow from the base of the affected branches and sucker from the rootstock. If the wood of infected trees is cut or stripped of bark, there is a characteristic salmon-pink or orange-red discoloration. Ultimately the entire tree is affected and it dies.

Crop Losses: ≤60%

C19-A016

Phytophthora infestans (Agent LO)
Phytophthora Blight
ICD-11: —
UNII: 8606X4NXMR
EPPO: PHYTIN

It is endemic throughout the world. Although it's filamentous structure and metabolic strategies have historically caused *P. infestans* to be classified as a fungus, it is currently recognized as an oomycete or water mold. It has two reproductive cycles. Asexually, it produces a somewhat fragile spore (sporangia) that can survive days or weeks in soil, and hours in the air when protected from solar radiation. These spores cannot survive outside a host over the winter. Sexual reproduction produces a hardier spore (oogonium) that can withstand much harsher conditions and survive winter in northern temperate zones. Oospores germinate and produce sporangia.

Epidemics of late blight occur when night temperatures are cool, followed by warm days with mists and rains. Under those conditions, *P. infestans* spreads rapidly and fields of potato are destroyed in a less than 2 weeks.

In Plants:

Target species: Potatoes; tomatoes.

Normal routes of transmission: Airborne.

Secondary Hazards: Spores; mechanical vectors (cultivation; rain/irrigation); crop debris (potato tubers; tomato seeds).

Signs: Symptoms on potato and tomato are similar. Small brown irregular-shaped spots appear on the leaves that expand rapidly. Older lesions are more circular and not usually delimited by the leaf veins. They are surrounded by a zone of non-necrotic collapsed tissue. There is a furry white growth on the underside and within days the leaves become yellowed (chlorotic), shrivel, and die. Black or brown spots occur on the stems. Spores wash off the leaves to infect potato tubers in the soil. Infected potato tubers exhibit wet and dry rots. On tomato fruit, lesions are firm, large, irregular, brownish-green blotches with a greasy, rough appearance. Patches of infected plants have a characteristic odor.

Crop Losses: ≤100%

C19-A017

Puccinia graminis (Agent IE)
Stem Rust of Cereals
ICD-11: —
UNII: O0HJ02QBWN
EPPO: PUCCGR

It is endemic throughout the world. Over 400 graminaceous species are known hosts.

It is on the Australia Group Core list.

In Plants:

Target species: Wheat, oats, barley, rye.

Normal routes of transmission: Airborne; contact.

Secondary Hazards: Spores; mechanical vectors (track out); crop debris.

Signs: First appears as minute flecks, progressing into roughened reddish-brown to black oval lesions on leaves, stems, leaf sheaths, and spikes that are easily recognized against the normally color of healthy tissue. The lesions coalesce to cover large areas of the host tissue in heavy infection.

Crop Losses: ≤80%

C19-A018

Puccinia striiformis (Agent TX)
Stripe Rust of Wheat
ICD-11: —
UNII: 9NLW29GJAX
EPPO: PUCCST

It is endemic throughout the world. Appears in early spring when there are cool temperatures and high humidity. Spores germinate when temperatures are between 37°F and 59°F. Optimum conditions for growth are temperatures between 50°F and 60°F with free moisture (i.e., rain, dew) on the plant. Growth of *P. striiformis* is inhibited when night-time temperatures get above 68°F or there are several consecutive days of temperatures in the mid-80s. The fungus cannot survive for extended periods outside a living host (e.g., on plant debris).

It is on the Australia Group Core list.

In Plants:

Target species: Wheat, barley, rye, triticale.

Normal routes of transmission: Airborne.

Secondary Hazards: Spores; mechanical vectors (track out).

Signs: Yellow-orange blister-like lesions appear in long stripes on the leaves; can also affect the bottom scales around the seeds (glumes) and base of the bristles of the seeds (awns) of some varieties. As the plant matures, the stripes turn from yellow to black. Severe infections affect yield by reducing kernel numbers, weight, and overall quality.

Crop Losses: ≤50%

C19-A019

Sclerophthora rayssiae var. zeae
Brown Stripe Downy Mildew of Maize
ICD-11: —
UNII: —
EPPO: SCPHRZ

It is endemic to India and parts of Southeast Asia. In addition to corn, also infects several species of crabgrass. Spores in air-dried leaf tissue can remain viable for 3 to 5 years, although infected seed dried to 14% moisture or less and stored for 4 or more weeks are not capable of transmitting the disease.

It is on the Australia Group Core list and on the U.S. Select Agents and Toxins list.

In Plants:

Target species: Corn.

Normal routes of transmission: Airborne; contact.

Secondary Hazards: Spores; mechanical vector (rain); crop debris.

Signs: Only leaves show outward signs of infection. Initially appears as vein-limited yellowed (chlorotic) flecks or blobs that enlarge lengthwise and coalesce into well-defined stripes. The stripes may extend to the full length of the leaf (lamina) and range in width from 3 mm to 7 mm. The under surfaces of the stripes develop a grayish-white, granular, downy growth. There are no other vegetative or floral malformations of any kind. As the disease progresses, stripes become yellowish-tan, reddish-brown, or purple and necrotic; and the downy growth disappears. Affected leaves remain intact and do not shred, even after severe storms. When disease occurs prior to flowering, it suppresses formation of the ear or reduces it to an embryotic form.

Crop Losses: ≤90%

C19-A020

Synchytrium endobioticum
Potato Wart Disease
ICD-11: —
UNII: —
EPPO: SYNCEN

It is endemic to the Andean zone of South America; now also reported in Europe, and parts of Southern Asia, Africa, Canada, and New Zealand. It is a soil borne fungus that becomes active in the spring when soil temperatures rise above 46°F. Optimal soil temperatures for growth are between 54°F and 75°F. Only a few spores are needed for infection to occur. Spores can persist in the soil for over 30 years and at depths of over 50 cm. This limits the use of the land not only for future production of potatoes but also for any plants intended for export.

It is on the Australia Group Core list and on the U.S. Select Agents and Toxins list.

In Plants:

Target species: Potatoes.

Normal routes of transmission: Contact (spores in soil).

Secondary Hazards: Spores; mechanical vectors (cultivation); crop debris; fomites (soil); fecal (animals that have eaten infected tubers).

Signs: Other than possible reduction in overall vigor, sighs of infection on the above-ground part of the plant are not usually apparent. Infection may not be evident until the crop is harvested. Underground, causes warty, cauliflower-like outgrowths on all tissue other than the roots. Warts are soft and pulpy, and easier to cut than a healthy tuber. They are white, tan, green, or brown, turning black as they decay. The whole tuber may be replaced by the warty proliferation. If galls are small and undetected during harvest, they may continue to develop in storage.

Crop Losses: ≤100%

C19-A021

Thecaphora solani
Potato Smut
ICD-11: —
UNII: —
EPPO: THPHSO

It is endemic to the Andean zone of South America and parts of Central America. It is a soil borne fungus that favors high humidity and saline soils. It is reported to survive for up to 7 years in gall fragments and spores are thought to be long-lived in soil. In addition to potatoes, it infects tomatoes and jimsonweed. In tomatoes, galls develop particularly at the junction of the stem and roots.

It is on the Australia Group Core list.

In Plants:

Target species: Potatoes.

Normal routes of transmission: Contact (spores in soil).

Secondary Hazards: Spores; mechanical vectors (cultivation); crop debris; fomites (soil).

Signs: There are no signs of infection on the above-ground part of the plant. Infection may not be evident until the crop is harvested. Although roots are not infected, galls, resembling deformed tubers, form on underground stems and runners. Infected tubers are hard and misshapen or have warty swellings on the surface. There are light-brown to brown-black specks filled with spores throughout the flesh. Ultimately, tubers become a dry, brown, powdery mass containing spores.

Crop Losses: ≤90%

C19-A022

Tilletia indica
Wheat Cover Smut
ICD-11: —
UNII: —
EPPO: NEOVIN

It is endemic to India and parts of the Middle East and Mexico. Plants are infected within the first 3 weeks of heading. Optimal conditions for infection are high humidity, light rain showers with cloud cover, and temperatures are between 46°F and 68°F. Dry weather, temperatures between 68°F and 77°F with bright sunshine are unfavorable. Spores can survive for up to 4 years in the soil.

It is on the Australia Group Core list.

In Plants:

Target species: Wheat, rye, triticale.

Normal routes of transmission: Airborne.

Secondary Hazards: Spores; mechanical vectors (cultivation; rain); infected seeds; crop debris; fomites (soil); fecal (animals that have eaten infected grains).

Signs: Not easily detected prior to harvest since only a few kernels per spike are typically affected and diseased heads do not look different than healthy ones. When the grain is threshed, diseased kernels can be easily identified. Visually, infections range from pinpoint sized spots to thick black spore masses running the length of the groove in the grain. Infected parts crush easily producing a greasy, black powder and an odor of decaying fish.

Crop Losses: ≤1% The main effect is on seed quality. Grain lots containing more than 3% infected kernels are considered unsatisfactory for human consumption.

BIBLIOGRAPHY

Acha, Pedro N. and Boris Szyfres. *Zoonoses and Communicable Diseases Common to Man and Animals, Scientific and Technical Publication No. 580.* 3rd Edition. Volume 1, *Bacterioses and Mycoses.* Washington, D.C.: Pan American Health Organization, 2003.

Argonne National Laboratory. *Australia Group Common Control List Handbook Volume II: Biological Weapons-Related Common Control Lists,* Revision 4. Washington, D.C.: US Government Printing Office, February 2018.

The Australia Group. *Australia Group Common Control Lists.* February 28, 2020 [https://www.dfat.gov.au/publications/minisite/theaustraliagroupnet/site/en/human_animal_pathogens.html]. December 31, 2020.

Bozue, Joel, Christopher K. Cote and Pamela J. Glass, ed. *Medical Aspects of Biological Warfare, Textbooks of Military Medicine Series.* Washington, D.C.: Office of the Surgeon General, Department of the Army, 2018.

Brunette, Gary W. and Jeffrey B. Nemhauser, ed. *CDC Yellow Book 2020: Health Information for International Travel.* July 18, 2019 [https://wwwnc.cdc.gov/travel/page/yellowbook-home-2014]. December 31, 2020.

Burrows, W. Dickinson and Sara E. Renner. "Biological Warfare Agents as Threats to Potable Water." *Environmental Health Perspectives* 107 (1999): 975–984.

California Department of Food and Agriculture. Animal Health and Food Safety Services. Animal Health Branch. *Biosecurity: Selection and Use of Surface Disinfectants,* Revision June 2002.

Canada Centre for Biosecurity. *Pathogen Safety Data Sheets.* December 6, 2019 [https://www.canada.ca/en/public-health/services/laboratory-biosafety-biosecurity/pathogen-safety-data-sheets-risk-assessment.html]. December 31, 2020.

Canada Centre for Biosecurity. Canadian Biosafety Standard, 2nd Edition, Ottawa, Canada, March 2015.

Centers for Disease Control and Prevention. *National Notifiable Diseases Surveillance System: Surveillance Case Definitions for Current and Historical Conditions.* August 2, 2017 [https://wwwn.cdc.gov/nndss/conditions/]. December 31, 2020.

Centers for Disease Control and Prevention. "Biological and Chemical Terrorism: Strategic Plan for Preparedness and Response. Recommendations of the CDC Strategic Planning Workgroup." *Morbidity and Mortality Weekly Report* 49 (RR-4) (2000): 1–14.

Centre for Agriculture and Bioscience International. *Invasive Species Compendium.* 2020 [https://www.cabi.org/isc]. December 31, 2020.

Centre for Agriculture and Bioscience International. *Plantwise Knowledge Bank.* 2020 [https://www.plantwise.org/KnowledgeBank]. December 31, 2020.

Chosewood, L. Casey and Deborah E. Wilson, ed. *Biosafety in Microbiological and Biomedical Laboratories.* 5th Edition. Washington, D.C.: US Government Printing Office, 2009.

Compton, James A.F. *Military Chemical and Biological Agents: Chemical and Toxicological Properties.* Caldwell, New Jersey: The Telford Press, 1987.

Department of Agriculture. 9 CFR Part 121 – "Possession, Use, and Transfer of Biological Agents and Toxins," 2005: 817–832.

European and Mediterranean Plant Protection Organization. *EPPO Global Database.* 2020 [https://gd.eppo.int/]. December 31, 2020.

German Social Accident Insurance. *GESTIS Biological Agents Database.* June 12, 2020 [http://gestis.itrust.de/nxt/gateway.dll/bioen/000000.xml?f=templates&fn=default.htm&vid=gestisbioeng:biosdbeng]. December 31, 2020.

Henderson, Donald A., Thomas V. Inglesby and Tara O'Toole, ed. *Bioterrorism: Guidelines for Medial and Public Health Management.* Chicago, Illinois: AMA Press, 2002.

Heymann, David, ed. *Control of Communicable Diseases Manual.* 20th Edition. Washington, D.C.: American Public Health Association, December 2014.

International Plant Protection Convention. "International Standard for Phytosanitary Measures 27, Diagnostic Protocols for Regulated Pests, DP 4: *Tilletia indica* Mitra." January 2014.

Lenhart, Steven, Millie Schafer, Mitchell Singal and Rana Hajjeh. *Histoplasmosis: Protecting Workers at Risk*, Revised Edition. Centers for Disease Control and Prevention, National Institute for Occupational Safety and Health, December 2004.

Lindler, Luther E., Frank J. Lebeda and George W. Korch, ed. *Biological Weapons Defense: Infectious Diseases and Counterbioterrorism.* Totowa, New Jersey: Humana Press, Inc., 2005.

Martin, Charlotte, Torsten Schöneberg, Susanne Vogelgsang, Carla Susana Mendes Ferreira, Romina Morisoli, Mario Bertossa, Thomas D. Bucheli, Brigitte Mauch-Mani, Fabio Mascher. "Responses of Oat Grains to *Fusarium poae* and *F. langsethiae* Infections and Mycotoxin Contaminations." *Toxins* 10 (2018): 47 (January 20, 2018).

Merck & Co. *Merck Manual Professional Version.* 2020 [https://www.merckmanuals.com/professional]. December 31, 2020.

Mycoses Study Group Education and Research Consortium. 2020. [https://drfungus.org/]. December 31, 2020.

Public Health Agency of Canada. *Canadian Biosafety Handbook*, 2nd Edition. Ottawa, Canada, March 2016.

Sidell, Fredrick R., Ernest T. Takafuji and David R. Franz, ed. *Medical Aspects of Chemical and Biological Warfare, Textbook of Military Medicine Series, Part 1, Warfare, Weaponry, and the Casualty.* Washington, D.C.: Office of the Surgeon General, Department of the Army, 1997.

Siegel, Jane D., Emily Rhinehart, Marguerite Jackson and Linda Chiarello. *2007 Guideline for Isolation Precautions: Preventing Transmission of Infectious Agents in Healthcare Settings.* Centers for Disease Control and Prevention, June 2007.

Sifton, David W. ed. *PDR Guide to Biological and Chemical Warfare Response.* Montvale, New Jersey: Thompson/Physicians' Desk Reference, 2002.

United States Army Headquarters. *Potential Military Chemical/Biological Agents and Compounds, Field Manual No. 3-11.9.* Washington, D.C.: Government Printing Office, January 10, 2005.

United States Army Headquarters. *Technical Aspects of Biological Defense, Technical Manual No. 3–216.* Washington, D.C.: Government Printing Office, January 12, 1971.

United States Army Headquarters. *Treatment of Biological Warfare Agent Casualties, Field Manual No. 8–284.* Washington, D.C.: Government Printing Office, July 17, 2000.

United States Department of Agriculture. Animal and Plant Health Inspection Service. 2020 [https://www.aphis.usda.gov/aphis/home]. December 31, 2020.

United States Department of Agriculture. Agricultural Research Service Fungal Diagnostic Fact Sheets. October 25, 2010 [https://nt.ars-grin.gov/sbmlweb/fungi/diagnosticfactsheets.cfm]. December 31, 2020.

United States Department of Health and Human Services, and Department of Agriculture. *Select Agents and Toxins List.* 2017 [https://www.selectagents.gov/SelectAgentsandToxinsList.html]. December 31, 2020.

University of Hawaii, Manoa. *EXTension ENTOmology & UH-CTAHR Integrated Pest Management Program.* August 30, 2011 [http://www.extento.hawaii.edu/kbase/crop/Type/Croppest.htm]. December 31, 2020.

Valley Fever Center for Excellence. "Valley Fever (Coccidioidomycosis): Tutorial for Primary Care Professionals." The University of Arizona, 2016.

WebMD. *Medscape.* 2020 [https://emedicine.medscape.com/]. December 31, 2020.

Wheat, Joseph, Alison Freifeld, Martin Kleiman, John Baddley, David McKinsey, James Loyd and Carol Kauffman. "Clinical Practice Guidelines for the Management of Patients with Histoplasmosis: 2007 Update by the Infectious Diseases Society of America." *Clinical Infectious Diseases* 45 (2007): 807–825.

World Health Organization. *Health Aspects of Chemical and Biological Weapons: Report of A WHO Group of Consultants.* Geneva, Switzerland: World Health Organization, 1970.

World Health Organization. *International Classification of Diseases, 11th Revision.* Geneva, Switzerland: World Health Organization. April 2019 [https://icd.who.int/browse11/l-m/en]. December 31, 2020.

World Health Organization. *Public Health Response to Biological and Chemical Weapons: WHO Guidance.* Geneva, Switzerland: World Health Organization, 2004.

20 Bioregulators

GENERAL INFORMATION

Bioregulators are naturally occurring compounds that are normally present in the body in minute quantities. They regulate and modulate a diverse set of key physiological and psychological processes. They are active in extremely low doses. Because of their complexity, many bioregulators are difficult to synthesize in large quantities by traditional chemical means. However, they can be harvested from cultures of genetically engineered species, or these species can be used to infect a host and deliver the agent directly to the target. Bioregulators are odorless, tasteless, and nonvolatile.

Bioregulators are often referred to as mid-spectrum agents; meaning they fall between classical manmade chemical agents (e.g., tabun (C01-A001, Volume 1)) and living biological agents (e.g., *Bacillus anthracis* (C17-A001)). They are covered under the general-purpose criteria of both the Chemical Weapons Convention and the Biological Weapons Convention.

Bioregulators have been investigated by numerous countries for both combat and riot control purposes. In the latter case, these agents can be used to trigger psychological changes including altering human cognition, perceptions, emotions, moods, and behavior similar to mind altering agents (C12).

TOXICOLOGY

EFFECTS

Bioregulators govern cellular processes such as blood pressure, heart rate, breathing, muscle contraction, temperature, mood control, consciousness, sleep, emotions, immune responses, and other critical functions. Bioregulators are fast acting with a short biological half-life. They cannot be traced by pathologists.

Although sometimes classified as biological weapons, bioregulators are chemicals. They are not alive and do not replicate themselves like pathogens (C17–C19). They are not communicable; to be affected an individual must come into direct contact with the bioregulator. If microorganisms are used as the delivery mechanism, however, the organisms themselves can replicate and may be contagious. This situation is essentially the same as with a bacterial (C17) or viral (C18) agent that produces a toxin during the course of its infection.

PATHWAYS AND ROUTES OF EXPOSURE

Bioregulators are primarily hazardous through inhalation and injection, including entry through broken, abraded, or lacerated skin (e.g. penetration of skin by debris). Some bioregulators pose an ingestional hazard. Bioregulators can also be delivered by employing genetically engineered microorganism that produce the bioregulator when introduced into the body of a susceptible host.

GENERAL EXPOSURE HAZARDS

In general, bioregulators do not have good warning properties. They are nonvolatile and do not have an odor.

DOI: 10.4324/9781003230564-11

LATENCY PERIOD

Unless modified, effects from exposure to bioregulators will appear within minutes. The route of exposure may have some effect on the latency period.

CHARACTERISTICS

PHYSICAL APPEARANCE/ODOR

Laboratory Grade

Pure bioregulators are typically colorless, white, tan, or yellow liquids or solids.

Modified Agents

Bioregulators can be dissolved in solvents to facilitate handling or stabilize them. Color and other properties of these solutions may vary from the pure agent. Odors will vary depending on the characteristics of the solvent(s) used.

Bioregulators can be microencapsulated to facilitate their dispersal and increase their persistency. Color and other physical properties may be affected by these modifications.

STABILITY

Very dependent on the specific bioregulator. Freeze-drying (lyophilization) or isolation as salts increases their stability and shelf life.

PERSISTENCY

Bioregulators are generally non-persistent and do not survive long in the environment. In cases where bioregulators have been microencapsulated or otherwise modified to facilitate their dispersal, reaerosolization by ground traffic, or strong winds may be a concern.

ENVIRONMENTAL FATE

All bioregulators are nonvolatile. Once the initial aerosol has settled, there is minimal inhalation hazard unless the bioregulator is released as an aerosolized powder that has been modified to increase the potential of reaerosolization. Solubility in water depends on the specific bioregulator, presence of solvents, and isolation as salts.

ADDITIONAL HAZARDS

EXPOSURE

All foodstuffs in the area of a release should be considered contaminated. Unopened items packaged in glass, metal or heavy plastic and exposed only to aerosols may be used after decontamination of the container. All unopened items exposed to bulk agents should be decontaminated within a few hours post-exposure or destroyed. Opened or unpackaged items, or those packaged only in paper or cardboard, should be destroyed.

LIVESTOCK/PETS

Animals can be decontaminated with shampoo/soap and water. If the animal's eyes have been exposed to agent, they should be irrigated with water or saline solution for a minimum of 30 minutes.

Unprotected feedstock (e.g., hay or grain) should be destroyed. Depending on the specific bioregulator released, the level of contamination and the weather conditions, leaves of forage vegetation could still retain sufficient agent to produce effects for several days post-release.

FIRE

Bioregulators are not volatile and the heat from a fire will destroy these agents. However, actions taken to extinguish the fire may spread the agent before it is destroyed. Runoff from firefighting efforts may pose a potential contact threat through exposure of broken, abraded, or lacerated skin, or though accidental ingestion. Smoke from a fire may contain acrid, irritating, and/or toxic decomposition products.

REACTIVITY

Varies depending on the specific bioregulator. Many react with strong acids, bases, or oxidizing agents.

PROTECTION

EVACUATION RECOMMENDATIONS

There are no published recommendations for isolation or protective action distances for bioregulators released in mass casualty situations.

PERSONAL PROTECTIVE REQUIREMENTS

Structural Firefighters' Gear

Structural firefighters' protective clothing is recommended for fire situations only; it is not effective in spill situations or release events. However, bioregulators have negligible vapor pressure and do not pose a vapor hazard. The primary risk of exposure is through contact with aerosolized agents, bulk agents (e.g., spilled liquids or solids) or solutions of agents. If chemical protective clothing is not available and it is necessary to rescue casualties from a contaminated area, then structural firefighters' gear will provide some skin protection against most bioregulator aerosols. Contact with bulk material and solutions should be avoided. However, any responder with pre-existing areas of cut, abraded, or lacerated skin should not make entry because this places the individual at extreme risk of subcutaneous exposure.

There is also a significant hazard posed by injection of bioregulators through contact with contaminated debris. Appropriate protection to avoid any potential laceration or puncture of the skin is essential.

Respiratory Protection

Self-contained breathing apparatuses (SCBAs) or air purifying respirators (APRs) should have a NIOSH CBRN certification. However, during emergency operations, other NIOSH-approved SCBAs or APRs that have been specifically tested by the manufacturer against chemical warfare agents may be used if deemed necessary by the incident commander. APRs should be equipped with a NIOSH-approved Chemical/Biological/Radiological/Nuclear (CBRN) filter or a combination organic vapor/acid gas/particulate cartridge.

Immediately dangerous to life or health (IDLH) levels are the ceiling limit for respirators other than SCBAs. However, IDLH levels have not been established for bioregulators. Therefore, any potential exposure to aerosols of these agents should be regarded with extreme caution and the use of SCBAs for respiratory protection should be considered.

Chemical Protective Clothing

Currently, there is no information available on performance testing of chemical protective clothing against bioregulators.

Because of the extreme hazard posed by an aerosolized agent to any area of cut or lacerated skin, responders should wear a Level A protective ensemble whenever there is any potential for exposure to an airborne agent.

Because there is a significant hazard posed by injection of bioregulators through contact with debris, appropriate protection to avoid any potential abrasion, laceration, or puncture of the skin is essential.

DECONTAMINATION

General

Apply universal decontamination procedures using soap and water.

Liquids, Solutions, or Liquid Aerosols

Casualties/Personnel

Cover all open wounds during the decontamination process. Remove all clothing immediately. To avoid further exposure of the head, neck, and face to the agent, cut off potentially contaminated clothing that must be pulled over the head. Use a sponge or cloth with liquid soap and copious amounts of water to wash the skin surface and hair at least three times. Ideally, showers will be high volume with low pressure. Do not delay decontamination to find warm or hot water if it is not readily available. Avoid rough scrubbing as this could abrade the skin and increase the potential for movement of any residual bioregulator through the skin barrier. Rinse with copious amounts of water. If there is a potential that the eyes have been exposed to the agent, irrigate with water or 0.9% saline solution for a minimum of 15 minutes.

Small areas

Ventilate to remove the aerosol. Puddles of liquid can be absorbed by covering with absorbent material such as vermiculite, diatomaceous earth, clay, sponges, or towels. Place the absorbed material into containers lined with high-density polyethylene. Wash the area with copious amounts of soap and water. Collect and containerize the rinseate in containers lined with high-density polyethylene.

Solids or Particulate Aerosols

Casualties/Personnel

Do not attempt to brush the agent off of the individual or their clothing as this can aerosolized the agent. Cover all open wounds during the decontamination process. If possible, dampen the agent with a water mist to help prevent aerosolization. Remove all clothing immediately. To avoid further exposure of the head, neck, and face to the agent, cut off potentially contaminated clothing that must be pulled over the head. Wash the skin surface and hair at least three times with copious amounts of soap and water. Ideally, showers will be high volume with low pressure. Do not delay decontamination to find warm or hot water if it is not readily available. Rinse with copious amounts of water. If there is a potential that the eyes have been exposed to bioregulators, irrigate with water or 0.9% saline solution for a minimum of 15 minutes.

Small areas

Extreme care must be exercised when dealing with dry or powdered agents as bioregulators may adhere to the skin or clothing then be spread to other areas. Because of the minute quantities needed to produce a response in an exposed individual, cross-contamination can pose a significant inhalation or puncture hazard later.

If indoors, close windows and doors in the area and turn off anything that could create air currents (e.g., fans, air conditioner, etc.). Allow aerosol to settle. Avoid actions that could aerosolize the agent such as sweeping or brushing. Collected the agent with a vacuum cleaner equipped with a high-efficiency particulate air (HEPA) filter. Do not use a standard home or industrial vacuum. Do not allow the vacuum exhaust to stir the air in the affected area. Vacuum all surfaces with extreme care in a very slow and controlled manner to minimize aerosolizing the agent. Place the collected material into containers lined with high-density polyethylene. Wash the area with copious amounts of soap and water. Collect and containerize the rinseate in containers lined with high-density polyethylene.

MEDICAL

CDC Case Definition

The CDC has not published a specific case definition for intoxication by bioregulators.

Differential Diagnosis

Varies greatly by individual bioregulator.

Signs and Symptoms

Highly variable depending on the specific bioregulator, route of exposure, and dose.

Mass-Casualty Triage Recommendations

There are no universal recommendations for triaging casualties exposed to bioregulators as a class. However, in general, anyone who has been exposed should be transported to a medical facility for evaluation. Individuals who are asymptomatic and have not been directly exposed to the agent can be discharged after their names, addresses, and telephone numbers have been recorded. They should be told to seek medical care immediately if symptoms develop.

Casualty Management

Decontaminate the casualty ensuring that all the bioregulator has been removed. Extreme care must be exercised when dealing with dry or powdered agents as bioregulators may adhere to the skin or clothing and present an inhalation hazard. If any agent has gotten into the eyes, irrigate the eyes with water or 0.9% saline solution for at least 15 minutes. Irrigate open wounds with water or 0.9% saline solution for at least 10 minutes.

Although these agents do not produce any significant vapor, aerosolization of residual dusts on casualties could cause impacts to medical responders. Once the casualty has been decontaminated, including the removal of foreign matter from wounds, medical personnel do not need to wear a chemical-protective mask.

Treatment primarily consists of supportive care. Ventilate patient if there is difficulty breathing and administer oxygen. Be prepared to treat for shock. Monitor and support cardiac and respiratory functions as necessary.

FATALITY MANAGEMENT

Remove all clothing and personal effects segregating them as either durable or non-durable items. While it may be possible to decontaminate durable items, it may be safer and more efficient to destroy non-durable items rather than attempt to decontaminate them. Items that will be retained for further processing should be double sealed in impermeable containers, ensuring that the inner container is decontaminated before placing it in the outer one.

Extreme care must be exercised when dealing with dry or powdered agents as bioregulators may adhere to the skin or clothing and present an inhalation hazard.

Bioregulators that have entered the body are metabolized, hydrolyzed, or bound to tissue and pose little threat. To remove agents on the outside of the body, wear appropriate respiratory and dermal protective clothing while washing the remains with copious amounts of soap and water, ensuring the solution is introduced into the ears, nostrils, mouth, and any wounds. Pay particular attention to areas where agent may get trapped, such as hair, scalp, pubic areas, fingernails, folds of skin, and wounds. All wash and rinse waste must be contained for proper disposal. Body fluids removed during the embalming process do not pose any additional risks and should be contained and handled according to established procedures.

Once the remains have been thoroughly decontaminated, no further protective action is necessary. Use standard burial procedures.

C20-A   AGENTS

C20-A001

Angiotensin II
CAS: 11128-99-7
RTECS: —
EC: —
UNII: M089EFU921

$C_{50}H_{71}N_{13}O_{12}$
Molecular Weight: 1,046

It is a potent vasoconstrictor and also stimulates production of other hormones.

C20-A002

Bombesin
CAS: 31362-50-2
RTECS: BD3480000
EC: —
UNII: PX9AZU7QPK

$C_{71}H_{110}N_{24}O_{18}S$
Molecular Weight: 1,620

Inhalation causes an inflammatory reaction within the airways and alveoli resulting in broncho-constriction. It also affects the central nervous system, influencing cardiac function.

C20-A003

Bradykinin
CAS: 58–82-2
RTECS: EE1530000

EC: 200-398-8
UNII: PX9AZU7QPK

$C_{50}H_{73}N_{15}O_{11}$
Molecular Weight: 1,060

It is a potent arteriolar dilator producing rapid hypotension and increased capillary permeability leading to edema. Inhalation causes bronchospasm and results in bronchial mucus hypersecretion. It is also a potent pain-inducing agent.

C20-A004

Cholecystokinin
CAS: 9011-97-6
RTECS: —
EC: —
UNII: —

$C_{166}H_{261}N_{51}O_{52}S_4$
Molecular Weight: 3,931

Signs and symptoms include nausea, diarrhea, vomiting, sneezing, numbness, excessive sweating (diaphoresis), either hypotension or hypertension, slow heart rate (bradycardia), dizziness, fainting, lightheadedness, shortness of breath, headache, and seizures. It can cause visual hallucinations as well as anxiety and panic attacks. In some individuals, it produces hypersensitivity, resulting in anaphylaxis and anaphylactic shock. It stimulates production of other hormones throughout the body.

Cytokines

Produce a systemic inflammatory response with multi-organ failure and death that is recognized as cytokine release syndrome (CRS), also referred to as a cytokine storm. Signs and symptoms include fever; fatigue; loss of appetite; muscle and joint pain (myalgia, arthralgia) on both sides of the body and above and below the waist; nausea, vomiting and diarrhea; rapid breathing (tachypnea), difficulty breathing (dyspnea), poorly oxygenated blood (hypoxemia), and lung inflammation with edema; liver and kidney failure; heart arrhythmias, rapid heartbeat (tachycardia), and low blood pressure (hypotension); headache, confusion, loss of coordination, delirium, hallucinations, tremor, and seizures.

C20-A005

Interleukin 1
CAS: —
RTECS: NM9730000
EC: —
UNII: —

C20-A006

Interleukin 6
CAS: 308067-66-5
RTECS: —
EC: —
UNII: 92QVL9080Y

C20-A007

Interleukin 13

CAS: 148157-34-0

RTECS: —

EC: —

UNII: —

C20-A008

Interleukin 18

CAS: —

RTECS: —

EC: —

UNII: —

C20-A009

Tumor Necrosis Factor alpha

CAS: 308079-78-9

RTECS: —

EC: —

UNII: 4TVX2I407K

C20-010

Interferon gamma

CAS: 82115-62-6

RTECS: NM9701150

EC: —

UNII: P050J5FWC5

C20-A011

Dopamine

CAS: 51–61-6; 62-31-7 (Hydrochloride salt); 645-31-8 (Hydrobromide salt)

RTECS: UX1088000

EC: 200-110-0; 200–527-8 (Hydrochloride salt); 211–436-8 (Hydrobromide salt)

UNII: VTD58H1Z2X; 7L3E358N9L (Hydrochloride salt)

$C_8H_{11}NO_2$
Molecular Weight: 153.2

Signs and symptoms include irregular heart rhythm (arrhythmia), extra or skipped heartbeats (ectopic beats), rapid heartbeat (tachycardia), chest pain (angina), palpitations, vasoconstriction, high blood pressure (hypertension) or low blood pressure (hypotension), low blood volume (hypovolemia), blue fingers or toes (peripheral cyanosis), difficulty breathing (dyspnea), nausea, vomiting, headache, dilated pupils, anxiety, and mood alterations.

Eicosanoids

These are inflammatory agents that cause tightening of airway muscles and the production of excess mucus and fluid. Signs and symptoms include bronchospasm, bronchial mucus hypersecretion, pulmonary eosinophilia (white blood cells in lungs), shortness of breath, and wheezing (asthmatic response).

C20-A012

Leukotriene B$_4$ ($C_{20}H_{32}O_4$)
CAS: 71160-24-2
RTECS: JX3852000
EC: —
UNII: 1HGW4DR56D
Molecular Weight: 336.5

C20-A013

Leukotriene C$_4$ ($C_{30}H_{47}N_3O_9S$)
CAS: 72025-60-6
RTECS: MB9143500
EC: —
UNII: 2CU6TT9V48
Molecular Weight: 625.8

C20-A014

Leukotriene D$_4$ ($C_{25}H_{40}N_2O_6S$)
CAS: 73836-78-9
RTECS: MB9143000
EC: —
UNII: 5FNY4416UE
Molecular Weight: 496.7

C20-A015

Leukotriene E$_4$ ($C_{23}H_{37}NO_5S$)
CAS: 75715-89-8
RTECS: —
EC: —
UNII: 8FYT8ATL7G
Molecular Weight: 439.6

C20-A016

Prostaglandin D$_2$ ($C_{20}H_{32}O_5$)
CAS: 41598-07-6
RTECS: UK7930000
EC: —
UNII: RXY07S6CZ2
Molecular Weight: 352.5

Endorphins

Endogenous opioids that inhibit pain, alter blood pressure, and depress respiration. These compounds are tens to hundreds of times more potent than morphine on a molar basis. Endorphins can further decompose to small fragments (oligomers) that are still active and will pass through the blood-brain barrier more readily. Beta-endorphin is the most potent of the three isomers.

C20-A017

α-Endorphin ($C_{77}H_{120}N_{18}O_{26}S$)

CAS: 61512-76-3

RTECS: —

EC: —

UNII: —

Molecular Weight: 1,746

C20-A018

β-Endorphin ($C_{158}H_{251}N_{39}O_{46}S$)

CAS: 60617-12-1

RTECS: —

EC: 262–330-3

UNII: 3S51P4W3XQ

Molecular Weight: 3,465

C20-A019

δ-Endorphin ($C_{83}H_{131}N_{19}O_{27}S$)

CAS: 61512-77-4

RTECS: —

EC: —

UNII: —

Molecular Weight: 1,859

Endothelins

They are powerful and long-lasting vasoconstrictors producing pulmonary and systemic hypertension. Signs and symptoms include acute renal failure, cerebral and coronary vasospasm, and cardiac failure. They also stimulate production of other hormones.

C20-A020

Endothelin-1 ($C_{109}H_{159}N_{25}O_{32}S_5$)

CAS: 117399-94-7

RTECS: KE9446500

EC: —

UNII: 2K62B8Z6XF

Molecular Weight: 2,492

C20-A021

Endothelin-2 ($C_{115}H_{160}N_{26}O_{32}S_4$)
CAS: 123562-20-9
RTECS: —
EC: —
UNII: —
Molecular Weight: 2,547

C20-A022

Endothelin-3 ($C_{121}H_{168}N_{26}O_{33}S_4$)
CAS: 117399-93-6
RTECS: —
EC: —
UNII: 0066WLJ02E
Molecular Weight: 2,643

Enkephalins

They are endogenous opioids that inhibit pain, increase the strength of contraction in heart muscle cells, affect memory, and emotional conditions. They also stimulate production of other hormones.

C20-A023

[Leu]Enkephalin ($C_{28}H_{37}N_5O_7$)
CAS: 58822-25-6
RTECS: QP1350000
EC: 261–457-1
UNII: RI01R707R6
Molecular Weight: 555.6

C20-A024

[Met]Enkephalin ($C_{27}H_{35}N_5O_7S$)
CAS: 58569-55-4
RTECS: —
EC: 261–335-8
UNII: 9JEZ9OD3AS
Molecular Weight: 573.7

C20-A025

Epinephrine
CAS: 51-43-4
RTECS: DO2625000
EC: 200-098-7
UNII: YKH834O4BH

$C_9H_{13}NO_3$

Molecular Weight: 183.2

While it may produce transient paradoxical hypotension and bradycardia, it is a potent vasoconstrictor producing hypertension with diffuse vasoconstriction and tachycardia with serious cardiac arrhythmias. Signs and symptoms include mood alterations (aggressive or assaultive behavior), hallucinations, psychosis, fear, agitation, anxiety, apprehension, restlessness, throbbing headache, tremor, dizziness, lightheadedness, nervousness, excitability, weakness, sleeplessness, and loss of consciousness. Inhalation can cause pulmonary edema with crackling (rales) or rattling sounds (rhonchi), difficulty breathing (dyspnea), and frothy or bloody sputum.

Exposure Hazards

$MEG_{(1hr)}$: Neg: —; Mar: —; Crit: 0.0025 mg/m^3; Cat: —

Neurokinins

Signs and symptoms include salivation, vasodilatation and hypotension, stimulation of smooth muscle including constriction of the bronchi, and plasma leakage into the extravascular tissue (extravasation).

C20-A026

Neurokinin A ($C_{50}H_{80}N_{14}O_{14}S$)

CAS: 86933-74-6

RTECS: —

EC: —

UNII: 94168F9W1D

Molecular Weight: 1,133

C20-A027

Neurokinin B ($C_{55}H_{79}N_{13}O_{14}S_2$)

CAS: 86933-75-7

RTECS: PC4360080

EC: —

UNII: —

Molecular Weight: 1,210

C20-A028

Substance P ($C_{63}H_{98}N_{18}O_{13}S$)

CAS: 33507-63-0; 137348-11-9 (Acetate salt)

RTECS: WM2660000

EC: 251–545-8

UNII: 675VGV5J1D

Molecular Weight: 1,348

C20-A029

Neuropeptide Y
CAS: 82785-45-3
RTECS: —
EC: —
UNII: BY7U39XXK0

$C_{190}H_{287}N_{55}O_{57}$
Molecular Weight: 4,254

It is a potent vasoconstrictor causing reduction of cerebral and coronary blood flow. Signs and symptoms include arrhythmias and heart failure. It also stimulates production of other hormones. This agent has a long biological half-life and duration of action.

C20-A030

Neurotensin
CAS: 39379-15-2; 55508-42-4 (Acetate salt)
RTECS: QQ4482000
EC: —
UNII: XHB61LG5QS

$C_{78}H_{121}N_{21}O_{20}$
Molecular Weight: 1,673

Signs and symptoms include hypotension, produces a spectrum of pharmacological effects resembling those of major tranquilizers, and can affect thermoregulation inducing hypothermia. It also stimulates production of other hormones.

C20-A031

Norepinephrine
CAS: 51-41-2; 108341-18-0 (Bitartrate salt)
RTECS: DN5950000
EC: 200-096-6
UNII: X4W3ENH1CV

$C_8H_{11}NO_3$
Molecular Weight: 169.2

It is a potent vasoconstrictor. Signs and symptoms include severe hypertension with bradycardia, respiratory difficulty, difficulty breathing (dyspnea), violent headache, anxiety, photophobia, stabbing retrosternal pain, pallor, intense sweating, vomiting, and convulsions. Tissue hypoxia may lead to blue lips or fingernails (cyanosis), and mottled skin.

C20-A032

Serotonin

CAS: 50–67-9; 153-98-0 (Hydrochloride salt); 18525-25-2 (Maleate salt)
RTECS: NM2450000
EC: 200-058-9; 242–399-6 (Maleate salt)
UNII: 333DO1RDJY; GKN429M9VS (Hydrochloride salt)

$C_{10}H_{12}N_2O$
Molecular Weight: 176.2

Signs and symptoms include irregular heartbeat with a rapid heart rate (tachycardia) and high blood pressure (hypertension). Signs and symptoms include muscle rigidity, tremors, twitching, loss of coordination, dilated pupils, high fever, shivering, heavy sweating (diaphoresis), nausea, vomiting, diarrhea, headache, panic attacks, agitation, restlessness, confusion, lethargy, hallucinations, unconsciousness, and seizures.

C20-A033

Vasopressin
CAS: 11000-17-2
RTECS: YW8200000
EC: 234–236-2
UNII: Y490706MFD

$C_{46}H_{65}N_{15}O_{12}S_2$
Molecular Weight: 1,084

An antidiuretic that regulates the osmotic pressure in body fluids. It is a potent vasoconstrictor. Signs and symptoms include hypertension with dizziness, tremors, sweating, chest pain or a sensation of tightness in the chest, slow heart rate, heart rhythm disturbances, and heart attacks. It can also cause convulsions and coma. Inhalation may cause airway constriction, producing a sensation of tightness in the chest with coughing and wheezing.

C20   CHEMICAL STRUCTURES

C20-A001
Angiotensin II

C20-A002
Bombesin

C20-A003
Bradykinin

C20-A004
Cholecystokinin

C20-A011
Dopamine

C20-A012
Leukotriene B$_4$

C20-A013
Leukotriene C$_4$

C20-A014
Leukotriene D$_4$

C20-A015
Leukotriene E$_4$

C20-A016
Prostaglandin D$_2$

C20-A017
α-Endorphin

C20-A018
β-Endorphin

C20-A019
δ-Endorphin

C20-A020
Endothelin-1

C20-A022
Endothelin-3

C20-A023
[Leu]Enkephalin

C20-A024
[Met]Enkephalin

C20-A025
Epinephrine

C20-A026

Neurokinin A

C20-A027
Neurokinin B

C20-A028
Substance P

C20-A029
Neuropeptide Y

C20-A030
Neurotensin

C20-A031
Norepinephrine

C20-A032
Serotonin

C20-A033
Vasopressin

BIBLIOGRAPHY

Alibek, Ken and Stephen Handelman. Biohazard: *The Chilling True Story of the Largest Covert Biological Weapons Program in the World-Told from Inside by the Man Who Ran It*. New York, New York: Random House, 1999, 154–155.

Antonov, Nikolai. "Khimicheskoye Oruzhiye na Rubezhe Dvukh Stoletiy." Moscow, Russia: *Progress* 1994: 1–175. Translated by the National Air Intelligence Center under the title "Chemical Weapons at the Turn of the Century." Wright-Patterson Air Force Base, Ohio. January 31, 1996, 103–108.

Bokan, Slavko. "The Toxicology of Bioregulators as Potential Agents of Bioterrorism." *Arhiv za higijenu rada i toksikologiju* 56 (2005): 205–211.

Bokan, Slavko and Zvonko Orahovec. "An Evaluation of Bioregulators/Modulators as Terrorism and Warfare Agents" in *Technology for Combating WMD Terrorism*. NATO Science Series (Series II: Mathematics, Physics and Chemistry), vol 174. Edited by P.J. Stopa and Z. Orahovec. Springer, Dordrecht, 2004, 29–39.

Drug Bank Database. 2020 [https://www.drugbank.ca/drugs]. December 31, 2020.

Drugs.com Database. 2020 [https://www.drugs.com/]. December 31, 2020.

International Union of Basic and Clinical Pharmacology, and the British Pharmacological Society. *IUPHAR/BPS Guide to Pharmacology*. 2020 [https://www.guidetopharmacology.org/]. December 31, 2020.

Kagan, Elliott. "Bioregulators as Prototypic Nontraditional Threat Agents," *Clinics in Laboratory Medicine* 26 (2006): 421–443.

Leitenberg, Milton and Raymond A. Zilinskas. *The Soviet Biological Weapons Program: A History*. Cambridge, Massachusetts: Harvard University Press, 2012, 178, 192-194, 236-238, 368.

Madsen, James. "Bio Warfare and Terrorism: Toxins and Other Mid-Spectrum Agents." in *Encyclopedia of Toxicology*. 2nd Edition. Amsterdam, Netherlands: Academic Press, 2005.

Trapp, Ralf. Bioregulators and Peptide Synthesis" in *Double-Edged Innovations: Preventing the Misuse of Emerging Biological/Chemical Technologies*, Edited by Jonathan B. Tucker. Defense Threat Reduction Agency, Ft. Belvoir, Virginia, July 2010.

Tucker, Jonathan. "The Body's Own Bioweapons." *Bulletin of the Atomic Scientists* 64 (March/April 2008) 16–22.

United States National Institute of Health, National Library of Medicine. *PubChem*. 2020 [https://pubchem.ncbi.nlm.nih.gov/]. December 31, 2020.

21 Non-Vector Entomological Agents

GENERAL INFORMATION

This class of agents includes arthropods, mollusks, etc. that directly pose a risk through their ability to bite, sting, eat, or to make an area or an object uninhabitable, unconsumable, or unusable. Entomological vectors – organisms that carry and transmits a disease – are not included in this chapter. For more information on vectors and their use as a means of delivering pathogens, see the chapters on pathogenic agents (C17–C18) as well as appendix 1.

Entomological warfare/terrorism involves the deliberate introduction of an organism to generate fear, cause economic losses, and/or undermine social stability. Vectors used to deliver pathogens are banned by article 1 of the Biological and Toxic Weapons Convention. However, the explicit use of arthropods (e.g., insects, arachnids) or mollusks (e.g., snails) as a weapon is not covered by the treaty. (Cuba did file a complaint in 1996 under the Biological Weapons Convention alleging that the United States had engaged in biological warfare by deliberately releasing *Thrips palmi* (C21-A008) over the island to destroy essential crops. The complaint was dismissed due for, among other reasons, the fact that insects, as weapons, were not covered by the treaty.)

Entomological agents are an ancient weapon of war. Historically, armies would hurl insect nests or pots filled with stinging insets (e.g., bees, wasps, scorpions) into enemy fortifications. In the Bible, three of the plagues brought down onto the Egyptians by Moses were insects (i.e., lice, flies and locus). The concept has carried over to modern times. Various countries have evaluated deploying the Colorado potato beetle – *Leptinotarsa decemlineata* (C21-A001) – as a means of destroying enemy crops. Governments have even considered using insects as a means of controlling illicit drug crops (e.g., caterpillars of the tussock moth – *Eloria noyesi* – that eat cocaine). Terrorists have also threatened or claimed the use of entomological agents (e.g., in 1989 a group called the Breeders claimed responsibility for the invasion of Mediterranean fruit flies – *Ceratitis capitata* (C21-007) – in California), and individuals or groups have used insects as a means of harassing or intimidating companies or groups (e.g., bedbugs – *Cimex lectularius* – released in a store in Washington Township, Pennsylvania, in January, 2020).

Non-vector entomological agents can generally be organized into two major categories: those used to attack people and those used to attack a country's agriculture in order to devastate its economy. While there are a large and diverse variety of arthropods that could be used to attack people or even to infest and temporarily deny access to key locations (e.g., wasps, bees, spiders, scorpions, ants, blister beetles, etc.), it is unlikely that sufficient numbers could be amassed to be more than a limited and momentary nuisance used for harassment and intimidation. This chapter focuses on pests that could prove to be a significant threat to a country's agriculture and damage its economy through lost production, cost of destroying infested products, cost of containment (e.g., diagnostics, pesticides), loss of foreign trade, and lost tourism.

In nature, entomological agents occur as invasive species accidentally introduced into a new habitat or as periodic swarms, historically often referenced as plagues. They can cause environmental changes (e.g., *Nipponaclerda biwakoensis*, an invasive scale, destroying the roseau cane that stabilizes the Louisiana coastline), strip the land of vegetation causing deforestation and famine (e.g., swarms of desert locust – *Schistocerca gregaria* – in Africa beginning in 2019), threaten indigenous animals or people (e.g., Africanized or killer bees, red imported fire ants – *Solenopsis invicta*), or even cause structural damage (e.g., Formosan termites – *Coptotermes formosanus*).

DOI: 10.4324/9781003230564-12

Many invasive insects could wreak havoc on a country's ecology but would not necessarily be effective as a potential weapon. Desirable characteristics of pests that could be used as weapons to attack agriculture include: abundance, ease of acquisition or breeding, ease of dissemination, mobility, highly adaptable to different environments, habitat generalists with a broad native range and invasive outside its native range, feeding on many different kinds of food (polyphagous) with a focus on key agricultural crops, resistance to pesticides or have protected life stages that are not exposed to insecticides or other suppressive tactics, asexual reproduction, high fecundity with multiple generations per year, fast growing and long lived, gregarious, highly likely to be transported accidentally on commodity items, and difficult to identify/detect as a commodity contaminant. The entomological agents identified in this chapter individually have many of these characteristics.

C21-A   AGENTS

C21-A001

Colorado Potato Beetle
Leptinotarsa decemlineata
EPPO: LPTNDE
UNII: 5G7820U646

Adults are a stout, oval, strongly convex and hard-backed beetle approximately 9 to 12 mm long. They are yellowish-orange with 5 black stripes down each wing casing (elytron). There are approximately 12 small, irregular black spots on the top of the head and thorax. The tips of the legs are dark-brown or black.

Eggs are yellow to bright orange and football-shaped. They are approximately 1.2 mm long and 0.8 mm wide. They are laid in clusters of 20 to 60 on the underside of leaves with the long axis of the egg almost perpendicular to the leaf. Eggs within a cluster hatch simultaneously within 4 to 14 days, depending in part on temperature and humidity.

Larvae are very plump with a large, convex abdomen. There are three thoracic segments, each with a pair of three-segmented legs ending in a claw. They are initially cherry-red with shiny black head and legs; changing to carrot-red, then pale-orange, with two rows of black spots down the sides marking the breathing holes (spiracles). Larvae are approximately 15 mm long when mature. Larval development ranges from 8 to 28 days, depending on temperature.

Mature larvae burrow 2 to 5 cm into the soil and form an orangish, oval pupa which develops into adults in approximately 6 days. The pupal stage typically lasts 8 to 18 days, depending on temperature. The overall length of time required to develop from egg to adult is usually 30 to 60 days.

Colorado potato beetles overwinter as adults by borrowing into the soil to a depth of approximately 8 to 12 cm. Beetles emerge from the ground over a period of several weeks in spring or early summer and usually feed for 5 to 10 days before mating and producing eggs. There are one to three generations per year, depending on latitude.

They are indigenous to Mexico and most of the United States (excluding Alaska, California, Hawaii, and Nevada). They are also found in southern Canada and Central America. They have been introduced into Europe and parts of Asia. In general, they are not found in tropical countries, nor in most of eastern Asia, Korea, Japan, India, northern Africa, or the temperate Southern Hemisphere.

While potatoes are the preferred food, Colorado potato beetles may feed and survive on tomato, eggplant, ground cherry, pepper (rarely), and tobacco. They also feed on a variety of nonagricultural plants in the nightshade family (e.g., belladonna, common nightshade, buffalo-bur, horse-nettle, thorn apple, henbane), which occur widely and can act as a reservoir for infestation.

Impact: The Colorado potato beetle is one of the most economically damaging insect pests of potatoes, and can result in crop loss of 50 to 100%. In the case of tomatoes, yields can be reduced by 67%. They have developed a resistance to many insecticides and rapidly develop additional resistance when new insecticides are used repeatedly.

Secondary Migration/Transport: Migration over large areas occurs through flying. Host plants are located largely by chance, through random searching. Adults falling into sea water can survive for several days, and be washed onto beaches by the tide.

Adults and larvae are also readily transported on infested plants and tubers. They can contaminate almost all forms of packaging material used to transport agricultural products. Adults may also be found on any vehicles that passes near an infested field.

Signs of Infestation: Because of their size and distinctive coloration, adults and larvae are not difficult to observe by visual inspection. Beetles have a tendency to release their hold when plants are shaken and this can be used to detect insects hidden among foliage.

Adults and larvae feed on the edges of leaves and may quickly defoliate young plants. They also eat tubers exposed at the soil surface. Eventually, they will strip all leaves from the stalks and stems. A characteristic black and sticky excrement is left on the affected plants.

C21-A002

Cotton Leafworm
Spodoptera littoralis
EPPO: SPODLI

Adults moths are 15 mm to 20 mm long with gray-brown bodies and a wingspan of 30 mm to 38 mm. Their simple eyes (ocellus) are marked by two or three oblique whitish stripes. Their forewings are brown with many pale cream streaks and gray-blue overtones. The hindwings are translucent white and with gray margins. Males may have bluish areas at the base of the wings and on their tips. Adult cotton leafworms are almost indistinguishable from taro caterpillars (see C21-A018).

Eggs are 0.6 mm in diameter. They are typically whitish-yellow with a flattened spherical shape. Eggs become black just prior to hatching. They are laid in clusters of 30 to 300 eggs. Within the clusters, eggs are laid in rows one to three layers deep. Each cluster is covered with hair scales from the female moth, which gives the mass a golden-brown color. Most egg clusters are placed on the lower parts of plants. Each female can up to 3,500 eggs, and there may be up to 10 generations per year.

Larvae are hairless with eight pairs of legs. Initially 1 mm, they can grow to a length of 45 mm. Young larvae are pale green, but become dark gray, dark green or reddish-brown with age. Their bellies remain a lighter color. Other markings are variable, but there are typically dark and light longitudinal bands on the sides of the body and two dark semilunar spots on the back of each body segment. Coloration and color patterns can change dramatically during development and vary greatly between individual specimens.

Mature larvae dig 3 to 5 cm into the soil close to the plant they were feeding on and form a pupa. Pupa are 15 mm to 20 mm long and red-brown. There are two small spines on the tip of pupal abdomen; but these are easily broken off and may not be present.

Cotton leafworms are found in Africa, Madagascar, Europe, and the Middle East. The species is often confused with the taro caterpillars (see C21-A018).

Impact: Cotton leafworms can infest over 87 different plant species (polyphagous) of economic importance including alfalfa, beans, cabbages, corns, cotton, eggplants, groundnuts, jute, peppers, potatoes, rice, soybeans, sweets potatoes, tea, tobacco, tomatoes, and many ornamental plants. Larvae are voracious defoliators.

Affected plants are also susceptible to damage from fungus and bacteria.

Migration/Secondary Transport: Adults migrate by flying, traveling up to 1.5 kilometers each night. Larvae can crawl short distances to infest an adjacent field in search of food. Migration also occurs through shipment/transportation of eggs, larvae, pupa or adults on infested plant material (e.g., flowers, vegetables, seedlings) and growing medium accompanying infested plants.

Signs of Infestation: Young larvae primarily feed on the lower surface of leaves while leaving the upper epidermis layer intact. Older larvae chew large areas of a leaf and can completely strip a plant. The stems may be mined and young grains may be damaged. There may be yellowish-green to dark-green excrement around the holes eaten in the leaves or fruit. Older larvae feed only at night and are usually found in the soil around the base of plants during the day.

C21-A003

False Codling Moth
Thaumatotibia leucotreta
EPPO: ARGPLE

The appearance of adult moths differs based on their sex (dimorphic). Females have bodies that are grayish brown to dark brown or black, and 19 mm to 20 mm long. Their wingspan is approximately 20 mm. Forewings are broad, elongated, fringed with hairs, and have a rounded apex. These wings have a mixture of gray, brown, black, and orange-brown markings. There is a small white dot near the center of the wing (end of the discal cell) and a distinct "question mark"-shaped band of dark scales near the outer edge. The hindwings are lighter grayish brown but darker towards the outer margins.

Males are smaller, 6 mm to 9 mm long with wingspans of approximately 16 mm. Their forewings have similar markings but are more triangular with an acute apex. Males have a distinct semicircular pocket of opalescent scales on the rear edge of the hindwing near the body. The hindlegs have long, dense, grayish-white scales; the upper hindlegs (tibia) are heavily tufted.

Adult are nocturnal and attracted to light.

Eggs are approximately 1 mm flattened oval shaped disks with a granulated surface. Initially translucent and white to cream colored, they become reddish prior to hatching. Eggs are deposited singularly or in small groups in depressions or on smooth (non-pubescent) surfaces of fruit or foliage. Each female can lay up to 800 eggs, and there may be as many as 10 non-discrete, overlapping generations per year.

Depending on temperature, eggs hatch in 2 to 22 days. They are extremely sensitive to cold temperatures or extended periods of low humidity. Temperatures below freezing for 2 to 3 days can kill the eggs.

Initially, larvae are approximately 1 mm long, creamy white with minute black spots and a brownish black head. They go through five stages of development (instars) and ultimately reach lengths of up to 20 mm. Fully developed larvae have a pink body tending to orange-yellow on the sides, top, and legs. Their heads become yellow-brown. Young larvae immediately burrow into the fruit, feeding near the surface. As they grow older, they bore toward the center. The larval period lasts 12 to 67 days depending on temperature.

Mature larvae leave the fruit, descend to the ground on a silken thread, and spin a silken cocoon in the soil, under leaf litter, or in bark crevices.

Pupa are 7 mm to 10 mm long, yellow to dark brown, with transverse row spines. Adults emerge in 13 to 60 days depending on temperature. Pupa are sensitive to cold temperatures and heavy rainfall.

The complete life cycle of false codling moths ranges from 30 days under optimal weather conditions to 174 days under poorer conditions. They do not hibernate (diapause) and adults are present year-round in their native habitat.

False codling moth are adapted for warm climate survival. Temperatures below 50°F impede development and reduce survival rates. Below 34°F, all life stages of the moth die.

False codling moths are believed to have originated in Sub-Saharan Africa. They have been reported in South Africa, Israel, and the islands of the Atlantic and Indian Oceans. They can survive in tropical, dry, or temperate climates.

Impact: False codling moths are extremely polyphagous, feeding on over 70 food plants including acorns, avocados, bananas, beans, cacao, castor beans, cherries, coffee, corn, cotton, eggplants, figs, grapes, grapefruit, grapes, guavas, hibiscus, lemons, lima beans, limes, litchis, macadamia nuts, mallows, mangos, okra, olives, oranges, peaches, pecans, peppers, persimmons, pineapples, plums, pomegranates, prunes, sorghum, stone fruit, tea, tomatoes, and walnuts.

Migration/Secondary Transport: Adult moths are poor fliers with a limited range typically less than 600 meters. They are active only at night, spending their days resting on shaded portions of host plants.

Migration primarily occurs through shipment/transportation of produce or commodities (e.g., flowers, buds, nursery stock with growing medium attached) containing eggs, larvae or pupas. Adults may also be present in cargo vessels such as trucks, railcars and aircraft. They may also be found on packing material and in vehicles (e.g., aircraft, railcars, trucks, etc.).

Signs of Infestation: Adult moths are large and easily seen. Young larvae may be visible, as they wander over fruit before burrowing into the fruit. Entry holes are approximately 1 mm in diameter and conspicuous due to the presence of excrement (frass pellets) and discoloration of the surrounding skin. However, discoloration can take days to become noticeable. Other than the initial scarring, internal feeding leaves few outward symptoms on infested fruit. Fruit may ripen prematurely and drop.

C21-A004

Flat Scarlet Mite
Cenopalpus pulcher
EPPO: BRVPOU

Adults are small, broadly rounded, and only noticeable because of their intense scarlet color. Females are approximately 0.32 mm long and 0.16 mm wide with eight legs. Males are shorter and paler with an almost transparent abdomen. They are relatively sedentary and normally live in groups on the undersurface of leaves along the midrib and leaf veins.

Eggs are oval, 0.11 mm by 0.07 mm, and bright red. The first seasonal eggs are deposited on the bark of the tree late in early spring, while subsequent egg cycles are deposited on the striations and natural indentations of leaves and fruits; primarily on the undersurface of leaves along the midrib, buried beneath the leaf hairs.

Larvae and nymphs resemble smaller adults.

Fertilized females and nymphs overwinter in bark crevices and under bud scales, usually in small groups. Overwintering mites can survive temperatures as low as −22°F. In colder climates, there is only one generation per year, while in warm temperate to subtropical climates there are as many as three generations.

Flat scarlet mites are found in Europe, Asia, and the Middle East. They have also been found in Oregon, USA. Populations occur in a range of climatic zones, from arid tropical to subarctic.

Impact: Flat scarlet mites can infest a wide variety of fruit and ornamental trees (polyphagous) including almond, apple, apricot, cherry, fig, grape, lemon, loquat, peach, pear, plum, pomegranate, quince, walnut, willow, lilac, dogwood, and Oriental sycamore.

Migration/Secondary Transport: Flat scarlet mites are relatively sedentary. While they cannot fly, young nymphs can be carried by air currents to new hosts in the same locality. Major migration occurs through shipment/transportation of eggs, larvae, nymphs, or adults on infested plants and plant material (e.g., nursery stock, budwood), as well as infested fruit.

Signs of Infestation: Feeding on leaves causes yellowing close to the veins that later develops into necrotic patches. Injured tissue may be stippled, leaves may shrivel and twigs may die-back. If infestation is severe, leaves and fruit drop prematurely. Flat scarlet mites also cause bronzing damage to lower leaf surfaces the same as other mites (e.g., red spider mite, apple rust mite). Signs of infestation may also be mistaken for nitrogen deficiency.

Mites are unlikely to be dislodged from fruit by harvesting and grading activities because of their small size.

C21-A005

Hessian Fly
Mayetiola destructor
EPPO: MAYEDE

Adults are brown or black and approximately 4 mm long. At times females appear reddish-brown due to the presence of the orange eggs developing inside the abdomen. In males the abdomen is an elongate cylindrical, while in females the abdomen is heavier and markedly tapered. They appear fragile and mosquito-like, with long antennae and legs as long as or longer than their bodies. Adults do not feed and live less than 3 days.

Eggs are found in the grooves on the upper side of the plant leaf parallel with the veins. They are small, elongated cylinders approximately 0.5 mm long with rounded ends. Eggs are laid end-to-end on the host leaves. They are initially a glossy orange but darken to red with age. They hatch in 3 to 12 days depending on the temperature.

Hessian fly larvae go through three stages (instars) of development. The first stage is 0.5 mm to 1.7 mm long, orange-red, and somewhat flattened. As they feed, they become cylindrical with a tapered posterior end and turn white with a translucent green stripe down the middle of the back. The second stage is 1.7 mm to 4.0 mm long and is almost uniformly covered with elongate spines.

After feeding for approximately 2 weeks, the outer skin loosens, turns brown and hardens, forming a protective case (puparium) that resembles a seed of flax. These cases are usually found at the base of old plant crowns or in straw, near the nodes, behind the leaf sheaths. Unseen inside this protective case, the larva develops into its third stage.

If it is a spring or summer generation, the larvae then further metamorphosizes into the pupa while still within the protective case. Adult flies typically emerge 2 to 5 days later. If it is an autumn generation, the third stage overwinters in a protective state of suspended development (diapause). In the spring, development resumes inside the casing and larvae transform into a pupa. Adult flies emerge within 2 to 5 days.

There are usually two generations per year but may be as many as six in favorable climates. Female lays between 250 to 300 eggs on plants, usually where the stems are covered by leaves. Eggs can withstand severe frost but are susceptible to drying (desiccation).

Because of the many potential unfavorable environmental factors that can affect the frequency and duration of suspended development for any given generation of flies, the total duration of the life-cycle is extremely variable with a minimum of approximately 20 days, and a maximum of approximately 49 months.

The Hessian fly originated in Asia, but is now present in the Middle East, Europe, North Africa, North and Central America, and New Zealand.

Impact: Wheat is the preferred host of the Hessian fly although they will feed on other cereal crops (e.g., barley, rye) and at least 16 wild grass species found around the world. Larvae mainly attack the plant stem, although they will eat any part of the plant if there is a shortage of food.

The Hessian fly is one of the most destructive pests of wheat. Larvae usually feed on the lower leaves and can cause heavy damage. A single larva feeding for 3 days, can stunt a young plant.

Migration/Secondary Transport: Adult fly migration occurs through flying, but the flight distance is typically limited to approximately 8 kilometers, and then only if the fly is aided by thermal air currents.

Major migration occurs through shipment/transportation of eggs, larvae, or pupa on infested plant material (e.g., seeds, leaves, culms, straw). Wooden packaging may also retain errant travelers.

Signs of Infestation: Infested plants become stunted and stiffly erect, and leaves are thickened with a bluish green color. Infested young plants are generally stunted and have leaves which are shorter, broader and more erect than healthy plants. In older plants, stems may be weakened by larval feeding and then collapse. Tillers may be prevented from heading or will only produce shriveled grain. As seed heads begin to fill, heavily infested plants will fall over (lodge), and yields can decline significantly. Pupa can be found at the bases of leaves, where they are attached to the stem.

C21-A006

Khapra Beetle
Trogoderma granarium
EPPO: TROGGA

Adults are oblong-oval shaped, densely hairy beetles that vary in size from 2 to 3 mm. They are brown to black with indistinct reddish-brown marking on the back plates covering the wings (elytra). The plate at the base of the head (pronotum) is a darker brown. They have six yellowish-brown legs and a simple eye (ocellus) located in the middle of the head between the two compound eyes. The adults possess wings, but do not fly. They rarely, if ever, eat or drink and live between 12 to 25 days.

Eggs are cylindrical, approximately 0.7 mm long by 0.25 mm diameter, with one rounded end and the other pointed with a number of spine-like projections. They are initially milky white but turn pale yellowish with age.

Larvae are typically very hairy. They are uniformly yellowish white except for the head and body hairs, which are brown. They are approximately 1.6 to 1.8 mm long at hatching with more than half of this length consisting of a tail made up of hairs on the last abdominal segment. They grow to approximately 6 mm long and 1.5 mm wide at maturity. As the larvae grow, their body color changes to a golden or reddish brown, more body hairs develop, and the tail becomes proportionally shorter. Larvae molt 4 to 15 times and development usually takes 4 to 6 weeks.

Under adverse conditions (e.g., temperature, humidity, overcrowding, etc.), larvae can enter a protective state of suspended development (diapause) that can last up to 6 years. In this state, larvae respiration rates drop to an extremely low level, but they continue to periodically feed and molt. Even in this state, they will also seek a new refuge if disturbed. During diapause, larvae are extremely resistant to the effects of contact insecticides or fumigants.

At the end of the larval phase, the beetle forms a pupa inside the skin of the larvae. After 2 to 5 days the metamorphosis is complete and the quiescent adult beetle pushes the pupal covering to the posterior end of the larval skin. It remains within the skin covering for an additional 1 to 2 days. Females only need to mate once and lay an average of 50 to 100 eggs. There can be as many as 10 generations per year.

Khapra beetles are generally believed to have originated from the Indian subcontinent. They have been reported in all continents where grain and grain products are stored except in Southeast Asia, South America, Australia, and New Zealand. They are especially prevalent in certain areas of the Middle East, Africa, and South Asia.

Khapra beetles prefer hot, dry conditions with temperatures above 68°F and less than 50% relative humidity. They live in close association with humans (synanthropic) and are found in grain stores, food stores, malthouses, seed processing plants, fodder production plants, dried milk factories, merchant stores, and a wide variety of expended packing materials (e.g., sacks, bags, crates). Larvae are most likely to be seen just before dusk.

Impact: Khapra beetles prefer grain and cereal products; particularly wheat, barley, oats, rye, maize, and rice. They feed on peanuts, walnuts, pecans, and almonds. They have been found in grocery commodities such as bread, dried coconuts, cornmeal, crackers, white and whole wheat flour, hominy grits, baby cereals, pearl barley, wheat germ, malt, and noodles. They will also feed on dried animal products as well as opportunistic bodies of dead mice and dried insects. If the beetle is left undisturbed in stored grain it can cause weight loss of 5 to 70%.

Large numbers of larval skins and insect hairs (seta) may cause dermatitis and/or allergic reactions in sensitive individuals.

Migration/Secondary Transport: Since khapra beetles do not fly, migration is through human movement (e.g., tourism) and shipment of bulk commodities such as grain and uncleaned containers.

Signs of Infestation: Khapra beetles may remain hidden deep in stored food for relatively long periods. Visual inspection of cracks and crevices, behind paneling on walls, and under timbers, tanks, shelves, etc. is critical. The obvious signs of a khapra beetle infestation are the larvae and molted skins. However, the larvae look very similar to those of other relatively unimportant insects. For true confirmation, larvae and adults are best identified by microscopic examination.

C21-A007

Mediterranean Fruit Fly
Ceratitis capitata
EPPO: CERTCA

The adult fly is 3 to 5 mm in length, or approximately two-thirds the size of a housefly. They are yellowish with brown tinge. The thorax is creamy white to yellow with a characteristic pattern of black blotches. The oval shaped abdomen has two narrow, transverse, light colored bands on the bottom. Wings, usually held in a drooping position when at rest, are broad and glassy with black, brown, and brownish-yellow markings. There is a wide brownish-yellow band across the middle of the wing. Eyes are reddish purple.

Eggs are slender, curved, 1 mm long, smooth and shiny white. They are deposited under the skin of fruit which is just beginning to ripen. Eggs hatch in 1.5 to 3 days in warm weather.

Larvae are initially 1 mm or less with a mostly transparent cylindrical and maggot-shaped body. They grow to 6 to 9 mm in length with a fully opaque cream-colored body.

Mature larvae typically leave the fruit near daybreak, borrow 2 to 5 cm below the surface of the soil, and form a 3 mm long, dark reddish-brown cylindrical pupa. During warm weather, metamorphosized adults emerge from the pupal cases in large numbers early in the day. During cooler weather, emergence is more sporadic. The overall length of time required to develop from egg to adult is usually 21 to 30 days under tropical conditions.

Mediterranean fruit flies can overwinter as adults, as eggs and larvae (in fruit), or as pupa in the ground.

Mediterranean fruit flies are indigenous to Sub-Saharan Africa but have spread to many other regions including the Mediterranean region, southern Europe, the Middle East, Western Australia, South and Central America, and Hawaii. Habitat includes cultivated and agricultural lands; forests, plantations, and orchards; urban and peri-urban areas. They prefer a warm temperate climate with temperatures above 26°F. They also tolerate either tropical monsoon or desert climates.

Impact: Mediterranean fruit flies attack over 200 species of fruits, vegetables and nuts (polyphagous). Susceptible species includes almonds, apples, apricots, avocados, bell peppers, cherries, citrus, coffee, eggplant, figs, grapes, grapefruit, kiwi, lemons, limes, mangos, nectarines, olives, papaya, peaches, pears, persimmons, plums, pomegranates, tangerines, tomatoes, and walnuts. Damage to fruit crops is frequently high and may reach 100%.

Migration/Secondary Transport: Adult fly migration occurs through flying, but the flight distance is typically limited to less than 800 meters. Major migration occurs through shipment/transportation of eggs and larvae in infested fresh produce, fruit, and vegetables; and pupas in transposed soil (e.g., growing medium accompanying plants).

Signs of Infestation: Affected fruit usually shows signs of egg implantation. Young fruits become distorted, usually drop from the tree, and show signs of accelerated decay. Mature fruit that has been infected may develop a water-soaked appearance on the surface around the points of entry.

C21-A008

Mellon Thrips
Thrips palmi
EPPO: THRIPL

Because of their small size, the various species of thrips cannot readily be identified in the field, even with a hand lens. Identification typically requires evaluation of adult specimens under high microscope magnification by an entomologist. Identification is restricted to adult specimens because there are no adequate keys for the identification of eggs, larvae or nymphs.

Adult melon thrips have clear, pale yellow to white bodies that are approximately 1 mm to 1.3 mm long, with numerous dark hairs (setae) on the body. Heads are wider than long. They have three simple eyes (ocelli) in a triangular formation on the top of their heads and antennae with seven segments. Antennae are gray except for the terminal segments, which are dark-brown. Coloration around the eyes is red. Males are smaller than females. Both sexes are fully winged, with the juncture of the wings presenting a black line that runs along the back of the body (thorax). The wings are pale, slender and fringed with hairs.

Females can lay eggs without mating (parthenogenesis).

Eggs are bean-shaped, colorless to pale white, and not visible without microscope magnification. They are deposited in a slit cut into plant tissue (e.g., leaves, flowers, fruit). One end of the egg protrudes from the surface of the tissue to allow the larvae to emerge. Depending on temperature, eggs hatch in 4 to 16 days. A single female will typically lay 50 eggs in her lifetime, but can lay as many as 200 eggs.

Larvae are 0.5 mm to 0.73 mm long with clear, pale yellow bodies that resemble adults, but without wings. Larvae feed in groups, particularly along the leaf midrib and veins, and usually on older leaves. Development is complete in 4 to 14 days, depending on temperature. Once mature, larvae fall to the ground and hide in the soil or plant litter.

Nymphs look like larvae except they have small, nonfunctional wing buds. They are relatively inactive and are non-feeding. They are primarily found in the soil around the infested plants. Nymphs are vulnerable to desiccation in drought conditions. Nymph development is complete in 3 to 12 days, depending on temperature.

Mellon thrips can overwinter in warmer climates. Adults and larvae overwinter in the soil or under plant litter on the ground, whereas the nymphs overwinter in the soil. Melon thrips are able to multiply during any season that crops are cultivated; however, when crops mature their growth rate will diminish. There may be as many as 26 generations per year under optimal conditions.

Mellon thrips originated in Southern Asia but are now present throughout the tropics and many subtropical areas. They have been found throughout Asia, the Pacific, and the Caribbean. They have also been identified in North, Central, and South America; Africa; Oceania; and Western Europe.

Impact: Melon thrips feed on many kinds of plants (polyphagous) including avocados, beans, cabbage, cantaloupe, carnations, chili, Chinese cabbage, chrysanthemum, citrus flowers, cotton, cowpeas, cucumbers, eggplants, hibiscus, lettuce, mangos melons, okra, onions, peaches, peas, peppers, plums, potatoes, pumpkins, soybeans, squash, tobacco, tomatoes, and watermelons. Both adults and larvae are gregarious feeders on leaves, stems, flowers and fruit.

Melon thrips can also act as vectors for capsicum chlorosis virus, groundnut bud necrosis virus, melon yellow spot virus, watermelon bud necrosis virus, and watermelon silver mottle virus.

Migration/Secondary Transport: Adults are good flyers, but their small size makes them susceptible to wind and weather. Their activity peaks during hot weather when updrafts can carry them great distances. Migration also occurs through shipment/transportation of eggs, larvae, nymphs or adults on infested plant material (e.g., flowers, fruits, leaves, seedlings, stems, shoots), and growing medium accompanying infested plants.

Signs of Infestation: Melon thrips can be found in pockets, cracks or crevices on host plants. They prefer to eat foliage, but will also attack flowers and fruit. Feeding causes a silvery or bronzed appearance on the plant surface – especially on the midrib and veins of leaves – and walls of fruit. Leaves and terminal shoots become stunted, fruit is scarred and deformed. Damaged leaves generally show a darkened, glossy, pearly appearance with the oldest tissue becoming thickened, warped, and crinkled. Heavy infestations may kill entire plants.

C21-A009

Mexican Fruit Fly
Anastrepha ludens
EPPO: ANSTLU

Adults are larger than a housefly, 7 mm to 11 mm long, with a yellowish-brown body and iridescent green eyes. There are longitudinal yellow stripes on the back of the first body section (dorsal thorax). Wings are hyaline and transparent, with distinctive yellow to light brown markings. On the outer half of the wing there are two inverted "V"-shaped markings, one fitting within the other. There is also a stripe along the forward edge of the wing, running from near the base of the wing to approximately half-way along the wing length. The medial vein in the wing curves forward at the wing tip. The forewings are 7 mm to 9 mm long.

Females have tubular telescopic egg-laying organs (ovipositors) that are relatively long – 3 mm to 5 mm – compared to their body size.

Eggs are laid either singly or in groups up to 23 beneath the skin of a fruit that is beginning to show color. They are approximately 1.5 mm long, white and spindle-shaped; approximately 0.2 mm at the widest end. They hatch in 6 to 12 days. A single female can lay up to 1,500 eggs.

Larvae are legless, elongated and cylindrical with a flattened tail. They go through three stages of development (instars). Newly hatched larvae take on the color of their food so they are easily overlooked. By the third stage, they have become white and are 6 mm to 12 mm long and 1 mm to 3 mm wide. They have two mouth hooks that are strongly developed and equal in size.

Depending on temperature, larvae feed for 11 to 35 days before emerging through a conspicuous exit hole and burrow into the soil to form a pupa.

Pupa are tan to dark brownish-yellow. Depending on temperature, adults emerge after 15 to 19 days.

Adult females can live up to 11 months and adult males can live as long as 16 months. Reproduction occurs year-round; there are up six generations per year. In cooler areas, they overwinter as pupas.

These flies are indigenous to Mexico but are also found in Central America, the Caribbean, South America, and in Texas in the United States. They have also been detected but have not become established in Arizona, California, and Florida. They prefer tropical climates or warm temperate climates with dry summer and winter.

Impact: Mexican fruit flies have a strong preference for laying their eggs in grapefruit, but attack a wide variety of fruits and nuts (polyphagous) including almond, apples, avocados, bully trees, cashews, cherimoyas, citrons, coffee, granadillas, guavas, lemons, limes, mangos, mombins, nectarines, oranges, papayas, passionfruit, peaches, pears, peppers, persimmons, pomegranates, pomelos, quinces, tangelos, tangerines, and sapotes.

Migration/Secondary Transport: Adult fly migration occurs through flying, reaching distances of up to 140 kilometers. Major migration occurs through shipment/transportation of eggs and larvae in infested fresh produce, fruit, and vegetables, or on wood containers and packaging; and pupas in transposed soil (e.g., growing medium accompanying plants).

Signs of Infestation: Adult flies are large and easily seen. Holes left in fruit from egg laying are visible but often difficult to detect in the early stages of infestation. Because of their coloration and size, young larvae are extremely difficult to find in affected fruits. However, their feeding often produces a network of tunnels. By the third stage of development larvae have become white and grown large enough to be easily identified.

C21-A010

Moroccan Locust
Dociostaurus maroccanus
EPPO: DICIMA

Adult females are 20 mm to 38 mm long with grayish to yellowish bodies covered with irregular dark patches or spots. There is a pale cross shape on the plate at the base of the head (pronotum). The hard forewings (elytra) are well developed, almost transparent, and stretch past the leg joint of the rear jumping legs. There may be sparse small-sized brownish or grayish spots on the forewings. The flight wings are colorless with blackish veins. The top half of the rear jumping legs (femora) is the color of the body and may or may not have dark bands on them. The bottom half (tibia) is usually red, but less frequently may also be yellow, pinkish, or even whitish.

Adult males are smaller and range from 16 mm to 28 mm long.

Adults presents a continuous and reversible phase polymorphism between two extremes known as the solitary phase and gregarious phase. The differences between the two phases – including changes in their size, coloration, and behavior – is to the extent that for many years they were considered two different species. Individuals of the solitary phase are smaller, have brighter general body color, with the spots on the forewings and jumping legs more marked than those of gregarious phase.

Eggs are elongated, whitish to pale yellow, and approximately 5 mm long. Eggs are usually laid in a cylindrical, curved egg pod (ootheca) 17 mm to 24 mm long, that is buried 3 cm to 4 cm in a pocket dug below the surface of firm soil. These locusts will only emplace egg pods in bare patches of ground that is not being used for agriculture (undisturbed, virgin land). A single female will produce two to four egg pod that contain 16 to 45 eggs each. In other instances, eggs may also be laid directly in plant clumps or in cracks of very dry soil.

During swarming outbreaks, Moroccan locusts tend to aggregate when laying eggs and the density of egg pods may reach as high as several thousand per square meter.

Nymphs look like small adults but are usually darker. They go through 5 stages (instars), each lasting 5 to 10 days. By the second stage, they can start marching together. At the third stage they have visible wing buds. The overall length of time required to reach adulthood is 25 to 40 days. Moroccan locusts have a single annual generation, overwintering as eggs. Hatching occurs from February to April, depending on latitude and altitude. Nymphs move to areas of vegetation to begin feeding.

They are found in northern Africa, southern and Eastern Europe, and western and central Asia. They characteristically inhabit semi-arid steppe or semi-desert areas with abundant spring vegetation.

Impact: Moroccan locusts have the ability to eat almost all forms of vegetation (polyphagous), devouring leaves, stalk and grain. Species of economic importance include alfalfa, beans, barley, cabbage, carrots, corn, cotton, grape vines, lentil, lettuce and other salads, millet, onions, peas, rice, sunflowers, tobacco, wheat, clover, citrus, date palms, figs, olives, and other fruit trees. Once developed, a locust plague is almost impossible to stop or control, with population density reported to reach up to more than 1,000 locusts per square meter.

Migration/Secondary Transport: Increase in locust numbers usually occurs when temperatures are above average and rainfall is below normal for two or three consecutive years. Under these conditions, nymphs show gregarious behavior, marching in long lines that may be kilometers long. At adulthood, they begin to swarm. Swarms typically travel at an altitude between 20 meters and 100 meters with a speed of 30–35 kilometers per hour. While typically limited to a distance of less than 80 kilometers during an adult's lifetime, travel distances of up to 250 kilometers have been reported.

Signs of Infestation: The Moroccan locust is a relatively large chewing insect and defoliation is visible. All green masses of the host plants can be destroyed when the population density of the pest is high.

C21-A011

Onion Thrips

Thrips tabaci

EPPO: THRITB

Because of their small size, the various species of thrips cannot readily be identified in the field, even with a hand lens. Identification typically requires evaluation of adult specimens under high microscope magnification by an entomologist. Identification is restricted to adult specimens because there are no adequate keys for the identification of eggs, larvae or nymphs.

Adult onion thrips are pale yellow to dark brown, approximately 1 to 2 mm long, with brownish blotches on the body (thorax) and center of the abdomen. Females have two pairs of yellow wings fringed with long hairs; males are wingless. They have three simple eyes (ocelli) on their heads and gray antennae with seven segments. Coloration around the eyes is usually gray, never red.

Males are rare, sometime nonexistent. Females typically lay eggs throughout the season without mating (parthenogenesis).

Eggs are very small, approximately 0.2 mm long, and kidney shaped. Initially white, they change to an orange tint as development continues. Each egg is deposited individually just under the outer layer of surface tissue (epidermis) of leaves, flowers, stems or bulbs. One end of the egg will be near the surface of the tissue to allow the larvae to emerge. They hatch in 4 to 5 days. A single female will lay between 20 and 80 eggs in her lifetime.

Larvae are 0.3 mm to 1.2 mm long with elongated, elliptical, slender bodies that resemble adults, but without wings. Initially white, they become yellowish as they age. Their antennae are short, eyes are dark in color, and they crawl quickly when disturbed. Larvae are usually found between the young leaf blades at the top of the plant. Development is complete in 10 to 14 days.

Nymphs are pale yellow to brown and look like larvae except they have small, nonfunctional wing buds. They are relatively inactive and are non-feeding. They are primarily found in the soil around the infested plants but may also be found at the base of the onion plant neck or underneath bulb scales. Nymph development is complete in 4 to 7 days.

Adults and larvae overwinter in the soil or under plant litter on the ground. Nymphs overwinter in the soil. In most cases thrips are not a problem in the rainy season because the rain washes the tiny insects from the plant. At the end of the hot dry season, thrips populations are at their maximum. Onion thrips have been found in most countries throughout the world but do not do well in the wet tropics. They can infest crops from sea level up to 2,000 meters above sea level.

Impact: Thrips populations increase rapidly under hot, arid conditions and can lead to economic crop loss. Most economic injury is to onions, garlic and chives. However, onion thrips can infest a wide range of different plant species (polyphagous) that includes asparagus, beans, broccoli, cabbage, carnations, carrots, cauliflower, Chinese broccoli, chrysanthemums, cotton, cucumbers, head cabbage, leeks, melons, orchids, papayas, peas, pineapples, potatoes, pumpkins, roses, squash, strawberries, sugar beets, tobacco, tomatoes, and turnips.

Onion thrips can also act as vectors for purple blotch fungus, iris yellow spot virus, tomato spotted wilt virus, and impatiens necrotic spot virus.

Migration/Secondary Transport: In the spring when temperatures warm, adults fly to new onion fields. They are not good flyers, but they move long distances with the aid of wind currents. Migration also occurs through shipment/transportation of eggs, larvae, nymphs or adults on infested plant material (e.g., flowers, fruits, leaves, seedlings, stems, shoots) and growing medium accompanying infested plants.

Signs of Infestation: In general, during the early part of the season there are more thrips in the plants on the borders of a crop field than in the center. Thrips are easily detected by visual inspection. Adults fly readily when disturbed. Host plants should be examined to observe the inner, youngest leaves, which the thrips prefers. Affected plants may display yellowing or dropping of leaves, buds, or flowers. Leaf damage may include silvery-white, mottled lesions; distortion; and curl upward. Plant growth may be stunted. Plant surfaces may be spotted with small black specks of fecal matter.

C21-A012

Oriental Fruit Fly
Bactrocera dorsalis
EPPO: DACUDO

Adults are larger than a housefly, approximately 6 mm to 8 mm long, with highly variable coloration. The middle part of the body (thorax) is usually dark with two prominent yellow stripes on top and yellow marks on each side. The abdomen is yellowish with two horizontal black stripes and a longitudinal median stripe. These markings often form a "T"-shaped pattern, but this is variable. The wings are approximately 7 mm long, hyaline, with a narrow brown band along the edge. Female have pointed, slender egg-laying organs (ovipositors).

Eggs are white to yellow-white, elongate ellipses, approximately 1 mm long by 0.2 mm wide. Eggs are laid in cluster of 3 to 50, 1 mm to 3 mm below the fruit surface. A single female can lay more than 3,000 eggs. Eggs hatch in approximately 2 days, although this can be delayed up to 20 days in cool conditions.

Larvae are creamy white, legless, and resemble an elongated cone with the mouth at the pointed end. They go through three stages of development (instars) and ultimately grow to a length of 7 mm to 11 mm with a width of 1 mm to 2 mm. Depending on temperature, after 6 to 35 days larvae exit the fruit, drop to the ground, and burrow 2 cm to 3 cm into the soil to form a pupa.

Pupa are barrel-shaped, tan to dark brown, and about 5 mm long. Adults emerge in approximately 10 to 12 days. This developmental period may be extended up to 90 day under colder conditions. Oriental fruit flies breed continually, producing several generations each year.

Originally from Southern Asia, they now inhabit much of tropical Asia and Sub-Saharan Africa. Other than Hawaii, they have been "detected but not established" in the United States as well as several countries in South America. They have been eradicated in most Oceania countries except for Palau, Tahiti, Papua New Guinea, and the Northern Mariana Islands.

Impact: Oriental fruit flies attack over 300 species of commercial and wild hosts (polyphagous). While avocado, mango and papaya are the most commonly infested fruit crops, they also attack ambarellas, apples, apricots, bananas, bitter melons, butter nuts, cashews, cherries, coffee, cumquats, figs, grapes, grapefruits, guavas, honeydews, loquats, lychees, macadamias, mangoes, marulas, oranges, passion fruits, peaches, pears, peppers, persimmons, pineapples, plums, rose apples, tangelos, tomatoes, tropical almonds, walnuts, wampees, yellow mombins, and yellow oleander.

Migration/Secondary Transport: Adult fly migration occurs through flying, reaching distances of up to 50 kilometers. Major migration occurs through shipment/transportation of eggs and larvae in infested fresh produce, fruit, and vegetables; and pupas in transposed soil (e.g., growing medium accompanying plants). Adults may also be present in cargo vessels such as trucks, railcars and aircraft. They may also be found on packing material and in vehicles (e.g., aircraft, railcars, trucks, etc.).

Signs of Infestation: Adult flies are large and easily seen. Holes left in fruit from egg laying are visible and there may be some necrosis around the puncture mark. Infested young fruit becomes distorted, callused, and usually drop; mature attacked fruits develop a water-soaked appearance. The larval tunnels provide entry points for bacteria and fungi that cause the fruit to rot.

C21-A013

Rice Armyworm
Mythimna unipuncta
EPPO: PSEDUN

Adult moths vary in color from pale-beige to a dark reddish-brown, with a distinctive single white spot on each forewing. Wing spans measure approximately 4 cm. The forewing is somewhat pointed with a transverse line of small black spots that become a black line at the wing tip. The hind wings are grayish.

Eggs are deposited in clusters consisting of two to five rows, in tight places, such as between the leaf sheath and stem, or in cracks on dry vegetation, making eggs difficult to locate under field conditions. Eggs are initially white to pale-yellow spheres approximately 0.5 mm in diameter. They progressively darken to a metallic gray just prior to hatching. Moths cover the egg clutch with an opaque secretion that becomes transparent when it dries. As it dries, the secretion draws the substrate foliage together, almost completely hiding the eggs. Eggs hatch in 3 to 24 days depending on temperature.

Larvae range from 4 mm to approximately 35 mm in length, depending on age. Young larvae are pale-green whereas the more mature larvae vary from gray-green to yellow-brown, depending on diet and climatic conditions. Three dorsal and two lateral longitudinal lines run the length of the body.

Mature larvae burrow 2 to 10 cm below the surface of the soil to form a pupa, 12 to 19 mm long and 5 to 6 mm wide. Initially pale-amber, pupa become a mahogany brown color as they mature. Duration of the pupal stage is 7 to 40 days, depending on the season.

Rice armyworms do no survive winter in cold climates. However, in intermediate climates they overwinter as larvae. In warm weather areas, all stages may be found during the winter. There are two to six generations per year, depending on latitude.

Adults are nocturnal, feeding early in the nighttime on nectar or ripe and decaying fruit. Larvae are the agriculturally significant phase due to their destructive feeding. After hatching, young larvae tend to move upward on the plant to find food and feed throughout the day. If disturbed, they immediately extrude silk and spin down to the soil. Older larvae are active only at night, seeking shelter during the day in leaf axils, or under dense vegetation and debris on the ground. Mature larvae are also gregarious and highly mobile. They often form large 'armies' that appear abruptly and cause a high level of defoliation.

Originally a neotropical species, rice armyworms are found in many areas of the world, including North, Central, and South America; southern Europe; central Africa; and western Asia.

Impact: The impact of larval feeding will vary with density of the larvae population, environmental conditions and the timing of infestation. Larvae prefer to feed on grasses (Poaceae) preferentially, only moving to other types of plants after this food supply is exhausted. As such, barley, millet, oats, rice, rye, corn, sorghum, sugarcane, wheat, and alfalfa are highly susceptible. Other susceptible species include apples, artichoke, beans, beets, cabbage, carrots, celery, cucumbers, lettuce, onions, parsley, parsnips, peas, peppers, potatoes, radishes, raspberries, strawberries, sweet potatoes, turnips, and watermelons.

Migration/Secondary Transport: Adults are strongly migratory. In locations where climatic conditions do not permit permanent occupation, temporary populations are established annually through the arrival of immigrants. In the autumn, the newly formed adults undertake a southward migration to avoid lethal winter conditions.

Signs of Infestation: Irregular defoliation of leaves. Depending on the age of the larvae, leaves may be skeletonized, exhibit chewed holes, or be missing entirely. In grains, seed heads may be cut off. There may be significant quantities of excrement (frass pellets) on the ground near plants or in leaf axis.

C21-A014

Spiny Bollworm
Earias insulana
EPPO: EARIIN

Adults moths are 8 mm to 12 mm long with a green head and thorax. The forewings can show seasonal differences (polymorphism) that depend on the temperature, humidity and foliage density. During the hotter summer months, wings may be bright green, while during the autumn they become a brownish-yellow. Throughout the year the hindwings are dull white with a brown line near the edge of the wing (subterminal). The wingspan is approximately 20 mm to 22 mm.

Eggs are 0.5 mm to 0.6 mm in diameter. They deposited at night singly or in small groups of four to eight on flower buds, fruit, in leaf axil or on the lower surface of leaves. They are sky blue, semi-spherical to square, with a wavy crown shaped top.

Larvae are initially only 1 mm long, but grow to a full length of up to 18 mm. The heads are dark and shiny while the bodies are initially gray or a glassy blue that is hard to see. As the larvae grows, they become greenish-white with black spots then finally yellow-green. The body of the larvae develop numerous small rounded projections (tubercles) bearing short thorny hairs or spines. There are no longitudinal bands on the body. The larval stage is typically 10 to 12 days.

Mature larvae weave a white, felt-like cocoon shaped like an inverted boat. The cocoon is attached to plant parts, which may or may not fall off the injured plant. The pupa itself is round at both ends and 12 mm to 14 mm long. Overall a dark brown, the abdomen area is reddish on the back and yellowish on the front. Typically, the pupal stage takes 9 to 15 days, but may extend to up to 2 months if development is delayed by low temperatures.

Feeding, mating and egg laying (oviposition) occur at night. Each female can deposit from 150 to 300 eggs on the young bolls, shoot tips and buds of host plants. When the larvae emerge, they bore into the host plant and feed. They prefer developed cotton bolls, especially the seeds. Towards the end of their development, larvae often move from boll to boll, thus increasing the amount of damage done during a single generation.

Moths will shift between crops with different growing seasons (e.g., okra, cotton), so there is no interruption to their food supply. There are up to 6 generations per year in the Middle East. Spiny bollworms are found in most of Africa, southern Europe, the Near and Middle East, in Japan, Taiwan, the Philippines, and in Australia.

Impact: The larvae feed on cotton and okra, but have also been recorded on rice, sugarcane and corn. Initially, larvae tunnel into the buds of the host plant. In cotton, the tunnel often enters the bolls from below, at a slight angle to the stalk (peduncle). The entrance holes are neatly rounded and approximately 1 mm in diameter. Larvae fill the tunnel opening with excrement (frass pellets) as they bore inward. Small bolls, up to 1 week old, turn brown, rot, and drop off the plant. Larger bolls 2 to 4 weeks old may not drop, but will open prematurely and may be so badly damaged they cannot be harvested. Secondary invasion by fungi and bacteria sometimes occurs. In the absence of flowers and bolls, larvae will penetrate the shoots of a young cotton plant and hollow them out. Impacted shoots wither and die.

Okra is attacked similarly. Initially, the larvae bore into the terminal shoots. With severe tunneling, the top leaves wilt and the apex of the plant may droop. When fruiting starts, larvae move to the flower buds, immature fruit and eventually the mature pods. When attacking the fruit, the larvae feed on the milky seeds and other contents of the pod leaving excrement-filled tunnels. Severely damaged pods may be completely hollowed out.

Migration/Secondary Transport: Adults will fly to a new territory. Migration also occurs through shipment/transportation of produce or commodities containing larvae or pupa. More locally, track out/transport of crop residual containing pests.

Signs of Infestation: Extensive tunneling results in wilting of the top leaves and the collapse of the apex of the main stem, which may turn blackish-brown and die. This can result in bunched growth in young plants and death of the growing point in a mature plant. As the buds and flowers appear, they wither and are shed. Stem borers only attack unripe bolls, which are vulnerable up to 6 weeks of age. They usually have a conspicuous hole where the larva has entered. Larvae tend to move from boll to boll and the damage caused may be disproportionate to the number of pests. Secondary invasion by fungi and bacteria may conceal the infestation.

C21-A015

Striped Rice Stem Borer
Chilo suppressalis
EPPO: CHILSU

Adult moths are 11 mm to 15 mm long with wing span of 20 mm to 30 mm. They vary in color; their forewings are dirty-white to yellow-brown, with brown to dark-brown specks scattered irregularly, and their hindwings are white to yellowish-brown. Although not always present, there is typically a row of black dots along the outer margin of the forewing. The front of the head is conical and strongly protruding forward beyond the eyes. Females are larger than males with paler forewing and fewer dark specks. Adults are nocturnal and become active early in the evening.

Eggs are approximately 0.9 mm by 0.5 mm and resemble translucent white to dark-yellow fish scales. They are laid in flat clusters of overlapping rows on the lower leaves or leaf sheaths of a plant. Each cluster contains up to 70 eggs. Each female can lay up to 550 eggs and there may be up to six generations per year.

Larvae go through as many as nine stages (instars) of development depending on environmental conditions. The first stage is approximately 1.5 mm long and grayish-white with a black head. Full-grown larvae are tapered slightly toward each end, have brown heads with 2 short horns, are approximately 20 mm to 26 mm long, and are yellowish to dirty-white with five longitudinal purplish-brown stripes running down the back (dorsal side) of the body.

After first preparing an exit hole so the adult can emerge from within the plant, larvae form a pupa within the plant stem. They do not make a cocoon. Pupa are reddish-brown, approximately 9 mm to 14 mm long, 3 mm wide, and have two short horns on the head. The terminal spine of the abdomen (cremaster) bears several small spines.

Initially, larvae remain in groups feeding on the leaf epidermis and live in a moist pulp of chewed plant debris and excrement (frass pellets). Later, they become solitary and tunnel into leaf sheaths to feed and continue development. When larvae reach the stem, they hollow it out one internode at a time, penetrating through successive nodes. One larva can destroy several plants. At harvest, larvae are typically found 10 cm to 15 cm aboveground. The full-grown larvae overwinter in a protective state of suspended development (diapause) in stubble and straw.

Under optimal conditions, the life cycle is completed in 35 to 60 days; in colder climate, development may take more than a year.

Striped rice stem borers are found in Australasia, the Pacific Islands, Asia, and Europe. They tolerate very low temperatures and this enables them to adapt to other regions. They prefer cultivated agricultural land but also inhabit undeveloped riverbanks and wetlands.

Impact: Striped Rice Stem Borer primarily infest rice, but also attacks maize, millet, sorghum as well as native plants such as cattails, elephant grass, various reeds, bulrushes, and wild rice.

Migration/Secondary Transport: Adults migrate by flying, and can travel up to 16 kilometers. They can cover longer distances if carried by the wind. Young larvae can migrate to adjoining plants to find food. They can also be dispersed to other plants by ballooning on extruded silk threads or by floating on leaf fragments. Water birds can also occasionally carry the eggs to remote locations. Migration also occurs through shipment/transportation of eggs, larvae, pupa, or adults on infested plant material.

Signs of Infestation: Symptoms vary, depending on the age of the plant. In young plants, growing points and the surrounding leaves die-off (dead hearts). In older plants, leaves dry out and turn yellow. They may curl and ultimately fall off. Heavy infestation can lead to completely hollowed-out stems, and to the formation of empty panicles (white heads). Borer entry and exit holes may be visible in stalks. Stems weakened by stem borers may also fall over (lodge). Stems should be opened to look for larvae and pupa.

C21-A016

Sunn Pest

Eurygaster integriceps
EPPO: EURYIN

Adults have wide, oval shaped bodies approximately 10 mm to 13 mm long. Heads are semi-triangular and convex. The plate at the base of the head (pronotum) is ornamented with black punctures. The back (scutum) is shield-like, rounded at the apex. They have varying coloration (polymorphism) from grayish-yellow to reddish or black, but are more often light brown.

Eggs are spherical and approximately 1 mm in diameter. They are deposited in clusters of 5 to 15, primarily on the undersides of smooth leaves but also on developing seeds or ears. Nearly translucent, they are initially green but become pink with age. Black dots appear on the eggs when they are approximately 1 week old. Red dots appear when eggs are close to hatching. Eggs develop in 6 to 28 days. A single female will lay an average of 200 eggs.

Nymphs go through five stages of development (instars). Initially 1.5 mm long and light-green, they become black within hours of hatching. These young nymphs are not very active and are often found clustered near the remains of their eggs. During the second stage, nymphs begin actively feeding on plant leaves. As they develop, nymphs switch from feeding on leaves and buds to feeding on the developing grain. The second and third stages of development have light abdomens with dark backs and heads. During the fourth stage nymphs begin to develop their wings. By the final stage of development, nymphs are 8 mm to 10 mm long and dark brown with black dots arranged in rows. Depending on temperature, nymphs develop into adults in 20 to 45 days after hatching.

Sunn pests overwinter (diapause) as adults in soil around grass roots, under plant litter, or in dense, compact low-growing vegetation. Adults migrate from overwintering sites to grains fields once the air temperature reaches 55°F. They may also hibernate during the summer (aestivation) in cooler areas at higher elevations. The Sunn pest can survive temperatures between −22°F and 113°F.

Sunn pests produce only one generation per year (univoltine). The developmental time from egg to adult ranges from 35 to 60 days depending on environmental conditions and availability of food. Both adults and nymphs need a food source with a high-water content. They also feed on dry grains, provided that sufficient moisture is available (e.g., weeds, dew, rain). Adults can survive up to 9 months without food.

Densities on crops can reach up to 120 individuals per square meter in outbreak conditions, and have been reported as high as 1,000 individuals per square meter in overwintering sites.

Sunn pests are found throughout the Mediterranean basin, in North Africa, and in the northern regions of the Middle East, which include Turkey, Syria, Lebanon, and Jordan.

Impact: Sunn pests are one of the most destructive pests of wheat in the Middle East. While they primarily feed on wheat, they also attack barley, corn, millet, oats, rye, sorghum, as well as other wild cereal grasses, damaging all parts of the plant.

Migration/Secondary Transport: Adults migrate by flying to new fields and can travel up to 250 kilometers. Migration also occurs through shipment/transportation of eggs, nymphs or adults on infested plant material (e.g., straw, hay), and plant soil. Adults have also been found on baggage, permit cargo, and stores.

Signs of Infestation: Small clusters of eggs are visible on the upper surface of the leaves and occasionally the stems of grasses. Feeding causes yellowing and white or brown discoloration on leaves, stems, and ears. More severe symptoms include dead or withered center leaves (dead heart); grain heads without kernels (white ears), or kernels with a dark pinhole surrounded by a contrasting pale halo. Wilting and death of new growth before the onset of flowering is also common. Kernels attacked during late maturity largely escape damage but can be slightly shriveled. Feeding may ultimately result in die back of the whole plant.

C21-A017

Tarnished Plant Bug
Lygus lineolaris
EPPO: LYGULI

Adults are 5 mm to 6 mm long and 2 mm to 3 mm wide, with a flattened oval body that is generally shiny or brassy in appearance. They are predominantly copper-brown, mottled with small, irregular patches of white, yellow, reddish-brown and black. There is a yellow "V-"shaped mark on the back (dorsal side) just behind the head, that is tipped with a black dot in the center. Wings are reddish brown.

Males are generally darker than females. Colors and markings also vary between overwintering and summer adults. Summer adults range from pale yellow with few black markings to reddish brown, or almost completely black, with few pale-yellow markings. Overwintered adults are much darker. Eggs are approximately 1 mm long and 0.25 mm wide, creamy white, flask shaped, and truncated, with a flat cap. They are inserted into the plant at the base of a leaf blade; or into leaf petioles, flowerlets or blossoms; with only the cap exposed. Eggs are usually deposited singly, but occasionally in clusters, and hatch in 7 to 10 days. Tarnished plant bugs do not reproduce on trees. Young nymphs are yellowish-green and approximately 1 mm long. They resemble adults but smaller and without wings. Nymphs go through five stages of development (instars). As they age, nymphs become green and develop yellow, green or black spots. After the third stage, nymphs have four distinct spots on the back (thorax) and one at the base of the abdomen. Full-grown nymphs have wing pads and are approximately 4 mm to 5 mm long.

In colder climates. tarnished plant bugs overwinter as adults and can be found in dead weeds, leaf litter, under tree bark, and in rock piles in fields, timber margins, stream and ditch banks, and road rights-of-ways. In warmer climates, adults will remain active throughout the year. The heaviest infestations are usually observed in mid-August. Development from egg to adult takes approximately 4 weeks and there are up to five generations per year.

Tarnished plant bugs are found in all Canadian provinces, the continental United States, and most of the states of Mexico.

Impact: The tarnished plant bug is among the most damaging of the true bugs and attacks at least 385 plants (polyphagous). Nymphs and adults feed on alfalfa, apples, artichoke, asparagus, beans, beets, blackberries, broccoli, cabbage, carrots, cauliflower, celery, chard, cherries, coriander, corn, cotton, cowpeas, cucumbers, eggplants, endive, escarole, fennel, grapes, horseradish, lettuce, mustard, onions, parsnips, parsley, peas, peaches, pears, pepper, potatoes, radish, raspberries soybeans, spinach, squash, strawberries, sweet potato, tomatoes, turnips, and watermelons. They also feed on ornamental flowers including asters, chrysanthemums, dahlias, impatiens, and marigolds; and on conifers including Douglas-firs, loblolly pines, lodgepole pines, pecans, and white spruce.

Tarnished plant bugs can also act as vectors for plant diseases.

Migration/Secondary Transport: Adults will fly to a new territory. Migration also occurs through shipment/transportation of eggs on cut flowers and plants for planting new fields.

Signs of Infestation: Tarnished plant bugs are relatively large insects and plant bug populations in crops tend to be aggregated or clumped. Symptoms of feeding include yellowed or distorted terminal growth, which reduces plant growth and causes plants to appear unhealthy. Leaves may be ragged, crinkled, and discolored. Buds, flowers, and fruit may drop prematurely. Fruit may be disfigured with rough, corky areas and sunken spots (cat-facing). There may be an increased number of vegetative branches with multiple crowns, elongation of plant stems between two of the nodes from which leaves emerge (internodes), split stem lesions, and swollen nodes. Characteristic plant damage has led to several colorful descriptive identifier terms including crazy cotton, stop-back, bush-head, and bushy-top.

C21-A018

Taro Caterpillar
Spodoptera litura
EPPO: PRODLI

Adults moths are 15 mm to 20 mm long with gray-brown bodies and a wingspan of 30 mm to 38 mm. Their forewings are gray to reddish-brown with pronounced irregular patches and streaks. The hindwings are grayish-white with gray margins, often with dark veins. Males may have bluish areas at the base of the wings and on their tips. Adult taro caterpillars are almost indistinguishable from cotton leafworms (see C21-A002).

Eggs are 0.6 mm in diameter. They are typically pale orange-brown or pink, with a flattened spherical shape. They are laid in clusters of 100 to 300 eggs. The egg clusters are approximately 4 mm to 7 mm in diameter and covered with hair scales from the female moth, which gives the mass a golden-brown color. Eggs hatch in 2 to 3 days in warm conditions, and 11 to 12 days when it is colder. Each female can lay up to 2,600 eggs, and there may be up to 12 generations per year.

Larvae are hairless with eight pairs of legs. Initially 1 mm, they can grow to a length of 45 mm. Young larvae are whitish or light green, but become dark green or brown with age. Their bellies remain a lighter color. There is a bright-yellow stripe along the length of the back. Other markings are variable, but there are typically dark and light longitudinal bands on the sides of the body and two dark semilunar spots on the back of each body segment. Coloration and color patterns can change dramatically during development and vary greatly between individual specimens.

Mature larvae form a pupa in the soil close to the plant they were feeding on. Pupa are 15 mm to 20 mm long and red-brown. There are two small spines on the tip of pupal abdomen; but these are easily broken off and may not be present. Adults emerge from the pupa after approximately 7 to 10 days.

Taro caterpillars are primarily found in Asia, Australia, and the Pacific Islands. The species is often confused with the cotton leafworm (see C21-A002). Principal habitat includes agricultural land, forests, plantations, orchards, and grasslands. They may also be found along riverbanks and in wetlands.

Impact: Taro caterpillars can infest over 120 different plant species (polyphagous) including alfalfa, beans, cabbage, corn, cotton, eggplants, flax, groundnuts, jute, peanut, peppers, potatoes, rice, soybean, sweet potato, taro, tea, tobacco, tomatoes, and many ornamental plants. Larvae are voracious defoliators.

Affected plants are also susceptible to damage from fungus and bacteria.

Migration/Secondary Transport: Adults migrate by flying, traveling up to 1.5 kilometers each night. Larvae can crawl short distances to infest an adjacent field in search of food. Migration also occurs through shipment/transportation of eggs, larvae, pupa or adults on infested plant material (e.g., flowers, vegetables, seedlings) and growing medium accompanying infested plants.

Signs of Infestation: Young larvae make scratch marks on the leaf surface as they feed. Older larvae chew large areas of a leaf and can completely strip a plant. The stems may be mined and young grains may be damaged. There may be yellowish-green to dark-green excrement around the holes eaten in the leaves. Older larvae feed only at night and are usually found in the soil around the base of plants during the day.

C21-A019

Two Spotted Spider Mite
Tetranychus urticae
EPPO: TETRUR

Two spotted spider mites are barely visible with the naked eye and appear as tiny spots on leaves and stems. Adult females are oval and 0.4 mm to 0.5 mm long. They are typically pale green or greenish-yellow, but may be darker due to feeding on some types of plants. They have two darker spots on their bodies and eight legs. Males are more active than females and have a smaller, narrower, more pointed body.

Only female spider mites overwinter and they become bright orange or red during these months of inactivity.

Eggs are spherical, pearl-like, approximately 0.13 mm to 0.15 mm in diameter, and laid in webbing on the underside of leaves. They are initially transparent but gradually turn yellowish over time. Eggs hatch after 3–15 days. Each female can lay up to 120 eggs and, depending on geographic region, there may be as many as 25 generations per year.

Larvae are initially colorless and have only six legs. They go through three stages of development (instars) within 5 days. At the beginning of each stage, larvae often lack the characteristic body spots associated with these insects. Larvae become green as they develop.

Nymphs are pale yellowish to green, with the two characteristic dark spots that give them their name, and, like the adults, they have eight legs. They go through two stages of development in as little as 5 days before becoming adults.

Spider mites live in colonies within the webs they construct to protect them from predators and adverse environmental conditions. During the winter, females leave the infested plant to hibernate in cracks and crevices in the soil or within glasshouse structures.

Spider mites live in temperate and subtropical regions throughout the world. They have been found in most countries in Europe; Asia; Africa; Australasia; the Pacific and Caribbean islands; North, Central, and South America. They are also a pest in glasshouses beyond their natural geographic range.

Impact: Two spotted spider mites can infest a wide variety of over 200 crops (polyphagous) including alfalfa, apples, Asian pears, beans, beets, blackberries, blueberries, corn, cotton, cucumbers, red and black currant, eggplants, gourds, grapes, groundnut, hops, jujubes, lettuce, melons, okra, papaws, passion fruit, peaches, peppers, plantains, sorghum, squash, star fruit, soybeans, strawberries, sunflower, tea, tomatoes, zucchini, as well as ornamental plants such as arborvitae, azalea, camellia, carnations, chrysanthemums, citrus, evergreens, hollies, orchids, pyracantha, rose, and viburnum. They also infest trees and may damage maple, elm, and redbud. They have also been reported on ash, black locust, and popular.

Migration/Secondary Transport: Two spotted spider mites migrate by ballooning on extruded silk threads, detaching from the infested host plant and being blown to new locations by winds. They are also spread via transport on farmers clothing, footwear and tools. Major migration occurs through shipment/transportation of eggs, larvae, nymphs, or adults on infested plants and plant material (e.g., growing medium, seedlings, flowers, stems, leaves), as well as infested fruit.

Signs of Infestation: Two spotted spider mites can be found on all areas of the plant, but usually colonizes the lower side of leaves. First visible symptoms are small, white to pale green spots, mainly around the midrib and larger veins of leaves. When the spots merge, the entire area may have a whitish, bronze or silvery transparent appearance. With ongoing infestation, leaves may collapse, turn yellow, wilt, and finally fall. At very high infestation levels, reddish-brown masses of mites can be seen hanging from the tips of leaves. Complete defoliation can occur; plants may be stunted or killed. Petals of open flower become brown and withered; flowers may develop gold flecks.

Spider mites spin fine strands of webbing that may cover the under surface of leaves. In cases of extreme infestation, their webbing can cover the entire plant.

C21-A020

Western Corn Rootworm
Diabrotica virgifera verifier
EPPO: DIABVI

Adults beetles are greenish-yellow, approximately 5 mm to 7 mm long, with three longitudinal black stripes (vittae) on their forewings. The stripes on the females are distinct, while the stripes on males tend to merge. Heads are shiny and black, with thread-like antennae resembling a string of beads that are slightly longer than the body. The plate at the base of the head (pronotum) is almost square (subquadrate), and sometimes has orange markings. Their legs can be yellow or dark. Males are generally smaller and darker than females. Adults are most active at dawn and dusk.

Eggs are white or beige flattened ovals, approximately 0.5 mm long. They are found in the soil near corn plants to a depth of 35 cm below the surface. The majority of eggs are generally concentrated in the top 15 cm of soil. A small percentage of eggs may not hatch until the second spring due to extended hibernation (diapause). A single female can lay up to 1,000 eggs.

Larvae are slender, elongated, white to pale yellow, with a yellowish-brown head capsule and a brownish plate on the last abdominal segment. Larvae go through three stages of development (instars). Newly-hatched larvae are approximately 3 mm long and feed primarily on fine foot hairs of roots. As they develop, they burrow into the main roots of the plant. They ultimately reach a length of 10 mm to18 mm. Larvae are usually found in the top 15 cm of soil.

Pupa are found in earthen cells in the soil near plant roots. They are 3 mm to 4 mm long, and look similar to adult beetles. They are initially white, but turn brownish before emerging as adults. The pupal stage lasts approximately 7 to 10 days.

Western corn rootworms have one generation per year (univoltine). They overwinter as eggs. In regions with warm, dry summers, the number of beetles declines rapidly in mid-August. In climates experiencing cooler summers, adult beetles may be found as long as green corn plants are available; remaining active in fields until frost.

Western corn rootworms likely originated from Central America but are now a major pest throughout North America. The have become a threat to all of Europe, and are likely to survive and develop wherever corn is grown.

Impact: Western corn rootworms are economically the most important corn pest worldwide. While they primarily affect corn, they also infest barley, millet, pumpkin, soybeans, sunflowers, wheat, and wild grasses including clover.

Western corn rootworms can transmit maize chlorotic mottle virus. They also quickly develop resistances to pesticides.

Migration/Secondary Transport: Adults are strong flyers and can migrate up to 25 kilometers in a single flight; they can be carried farther by storms. Other major migration pathways include transportation of infested soil containing eggs, larvae or pupa; or adults on parts of corn plants (e.g., fodder, green manure, cobs, consignments of green maize).

Signs of Infestation: Due to their size adults can be counted visually. Adult beetles feed principally on pollen, silk (silk clipping) and young kernels but do not cause any particularly characteristic symptoms. In instances where there is insufficient pollen or silk to support the beetle population, they may feed on young maize leaves leaving a window-pane appearance.

Plants infested with larvae can easily be uprooted; heavily infested plants will fall over (lodge) as they lose roots. Corn stalks with only partially damaged roots will continue growing after lodging, which produces a characteristic bent or twisted shape (goose necking).

Upon inspection, root tips appear brown and often contain tunnels. Roots may be chewed back to the base of the plant and the entire root system may be destroyed. Larvae may be found tunneling in larger roots and occasionally in the plant crown. Larvae may also burrow through plants near the base, causing stunting or death of the growing point and frequently causing production of side shoots (tillering).

C21-A021

Wheat Thrips
Haplothrips tritici
EPPO: HAPLTR

Because of their small size, the various species of thrips cannot readily be identified in the field, even with a hand lens. Identification typically requires evaluation of adult specimens under high microscope magnification by an entomologist. Identification is restricted to adult specimens because there are no adequate keys for the identification of eggs, larvae, or nymphs.

Adult wheat thrips are black-brown to black, approximately 1 mm to 2 mm long, with elongated, thin bodies. They have rounded, translucent wings without ribs that have a fringe of hair on the margins. Heads are longer than wide with large dark-brown eyes that are approximately one-third the size of the head. They have black antennae with eight segments.

Males are smaller than females and may be rare.

Eggs are very small, approximately 0.3 to 0.5 mm long, pale-orange, and elliptical shaped. They are usually laid in small clusters of four to eight, but sometimes individually. Eggs are deposited on the interior sides of an immature ear of grain. They hatch in 6 to 11 days. A single female will lay between 13 and 30 eggs in her lifetime.

Larvae resemble adults but without wings. Initially greenish-yellow, their bodies become bright red as they age. The antennae, head, legs and tip of abdomen are black. Development is complete in 23 to 27 days. Once mature, larvae fall to the ground and borrows into the soil 10 to 30 cm.

Nymphs look like larvae except they have small, nonfunctional wing buds. The antennae, head, legs, and tip of abdomen have changed from black to white.

Although wheat thrips flourish in dry and warm weather, they are vulnerable to desiccation in drought conditions.

In hotter climates, nymphs leave plants and dig into the soil where they enter a protective state of suspended development (diapause) to avoid the heat of summer. They emerge in the fall after sufficient rainfall and then overwinter in thatch and plant litter. In colder climates where the life cycle is delayed until later in the summer, larvae overwinter in the soil and emerge in the spring when the soil surface warms to 61°F to continue metamorphosis to adults. Adults begin to appear on plants when the air temperature reaches 65°F. There is only one generation per year.

Wheat thrips are found across eastern Europe and adjoining areas of Asia into Western Europe and north Africa.

Impact: Wheat thrips are a pest of small grains (oligophagous), including barley, cornflower, oats, rye, sunflower, wheat, and other cereals. Spring wheat is the most favorable for insect development. In the spring adults feed on the tenderest leaves, sucking up the cell contents. Once larvae emerge, they suck sap from ear scales, flower glumes, and then from the grain. Peak population density of larvae occurs during the milk stage of seeds (R3) and begins to decrease during the early dough stage (R4). Feeding can lead to partial or complete white ear effect, drying of flag leaves, partial ear fertilization, incomplete grain filling, and even abortion of the grain.

Affected ears are also susceptible to damage from fungus and bacteria.

Migration/Secondary Transport: Adults migrate by flying, traveling greater distances with the aid of wind currents. The most intensive flight coincides with the beginning of ear formation of spring wheat where the great bulk of adults concentrate.

Signs of Infestation: Infested plants show streaks on leaves, with lesions and brown spots on seeds. Grain ears show deepening and widening of grain furrows with brown coloration in their depths. Grain will have bright areas noticeable in the places punctured by the larvae. Grain may be deformed, underdeveloped, and undersized with deep wrinkles and folds.

BIBLIOGRAPHY

Biological and Toxin Weapons Convention (1998). – Working paper submitted by the Islamic Republic of Iran: Vectors and Pests. BWC/AD HOC GROUP/WP.322, 12th Session, 14 September – 9 October 1998, Geneva. United Nations, Geneva.

Bokan, Slavko. "Biological Warfare Agents, Toxins, Vectors and Pests as Biological Terrorism Agents." In *Technology for Combating WMD Terrorism*. Dordrecht, Netherlands: Springer, 2004.

Centre for Agriculture and Bioscience International. *Invasive Species Compendium*. 2020 [https://www.cabi.org/isc]. December 31, 2020.

Chaudhry, Fahad Nazir, Muhammad Faheem Malik, Mubashar Hussain and Nayab Asif. "Insects as Biological Weapons". *Journal of Bioterrorism & Biodefense* 9, no. 1 (2017): 10.4172.

European and Mediterranean Plant Protection Organization. *EPPO Activities on Plant Quarantine*. 2020 [https://www.eppo.int/ACTIVITIES/quarantine_activities]. December 31, 2020.

European and Mediterranean Plant Protection Organization. *EPPO Global Database*. 2020 [https://gd.eppo.int/]. December 31, 2020.

European and Mediterranean Plant Protection Organization. "Diagnostic Standard PM 7/3 (3) *Thrips palmi*." *OEPP/EPPO Bulletin* 48(2018): 446–460.

Food and Agriculture Organization of the United Nations (FAO) International Plant Protection Convention. "International Standard for Phytosanitary Measures 27, Diagnostic Protocols for Regulated Pests, DP 1: *Thrips palmi* Karny." March 2010.

Food and Agriculture Organization of the United Nations (FAO) International Plant Protection Convention. "International Standard for Phytosanitary Measures 27, Diagnostic Protocols for Regulated Pests, DP 3: *Trogoderma Granarium* Everts." March 2012.

Food and Agriculture Organization of the United Nations (FAO) International Plant Protection Convention. "International Standard for Phytosanitary Measures 27, Diagnostic Protocols for Regulated Pests, DP 29: *Bactrocera dorsalis*." February 2019.

Hodges, Amanda, Scott Ludwig, Lance Osborne and G.B. Edwards. *Pest Thrips of The United States: Field Identification Guide*. United States Department of Agriculture Cooperative State Research, Education, and Extension Service Integrated Pest Management Centers, July 2009.

Invasive Species of Idaho. *Insects*. 2020 [https://invasivespecies.idaho.gov/insects]. December 31, 2020.

Invasive Species Specialist Group of the International Union for Conservation of Nature Species Survival Commission. *Global Invasive Species Database*. 2005 [http://issg.org/database/species/search.asp?sts=sss&st=sss&fr=1&x=37&y=12&sn=&rn=&hci=-1&ei=-1&lang=EN]. December 31, 2020.

Invasive Species Specialist Group of the International Union for Conservation of Nature Species Survival Commission. *100 of the World's Worst Invasive Alien Species*. 2020 [http://www.iucngisd.org/gisd/100_worst.php]. December 31, 2020.

Lockwood, Jeffery A. *Six-Legged Soldiers: Using Insects as Weapons of War*. New York, New York: Oxford University Press, 2009.

Ministry of Environment and Food in Denmark, The Danish Agricultural Agency. Plants, *Harmful Pests: Insects and Mites*. 2020 [https://eng.lbst.dk/plants/harmful-pests/insects-and-mites/]. December 31, 2020.

Monthei, Derek, Scott Mueller, Jeffrey Lockwood and Mustapha Debboun. "Entomological Terrorism: A Tactic in Asymmetrical Warfare." *The Army Medical Department Journal* April-June (2010): 11–21.

Mayor, Adrienne. *Greek Fire, Poison Arrows, and Scorpion Bombs: Biological and Chemical Warfare in the Ancient World*. Woodstock, New York: The Overlook Press, 2003.

National Institute for Environmental Studies, Japan. *Invasive Species of Japan: Insect Alert List*. 2018 [https://www.nies.go.jp/biodiversity/invasive/DB/aetoc6_insects.html]. December 31, 2020.

North America Plant Protection Organization's Phytosanitary Alert System. 2020 [https://www.pestalerts.org/]. December 31, 2020.

Oregon Department of Agriculture. *Pest Alerts*. 2020 [https://www.oregon.gov/ODA/programs/IPPM/InsectsSpiders/Pages/PestAlerts.aspx]. December 31, 2020.

Royal Botanical Gardens, Kew. "State of the World's Plants 2017." January 18, 2019: 64–70.

Stibick, J. *New Pest Response Guidelines: False Codling Moth Thaumatotibia leucotreta*. USDA–APHIS–PPQ–Emergency and Domestic Programs, Riverdale, Maryland, 2006.

Sullivan, M. CPHST Pest Datasheet for *Thaumatotibia leucotreta*. USDA-APHIS-PPQ-CPHST, Revised January 2014.

United States Congress, Office of Technology Assessment. *Alternative Coca Reduction Strategies in the Andean Region, OTA-F-556*. Washington, DC: U.S. Government Printing Office, July 1993.

United States Department of Agriculture. *National Invasive Species Information Center.* 2019 https://www. invasivespeciesinfo.gov/terrestrial-invasives/terrestrial-invertebrates]. December 31, 2020.

United States Department of Agriculture Animal and Plant Health Inspection Service. *The Threat.* 2019 [https://www.aphis.usda.gov/aphis/resources/pests-diseases/hungry-pests/The-Threat?utm_keyword=/ the-threat/]. December 31, 2020.

University of California Agriculture & Natural Resources Statewide Integrated Pest Management Program. *Insects, Mites, Mollusks, Nematodes.* 2019 [http://ipm.ucanr.edu/PMG/menu.invertebrate.html]. December 31, 2020.

University of Florida's Entomology and Nematology Department, and the Florida Department of Agriculture and Consumer Services' Division of Plant Industry. *Featured Creatures.* 2019 [http://entnemdept.ufl. edu/creatures/]. December 31, 2020.

University of Georgia's Warnell School of Forestry and Natural Resources, College of Agricultural and Environmental Sciences – Dept. of Entomology, Center for Invasive Species and Ecosystem Health, United States Department of Agriculture Animal and Plant Health Inspection Service, United States Forest Service, United States Department of Agriculture Identification Technology Program, and United States Department of Agriculture National Institute of Food and Agriculture. *Invasive and Exotic Species Profiles & State, Regional and National Lists.* October 2018 [https://www.invasive.org/ species.cfm]. December 31, 2020.

University of Hawaii, Manoa. *EXTension ENTOmology & UH-CTAHR Integrated Pest Management Program.* August 30, 2011 [http://www.extento.hawaii.edu/kbase/crop/Type/Croppest.htm]. December 31, 2020.

Wright, Susan. "Bioweapons: Cuba Case Tests Treaty." *Bulletin of The Atomic Scientists* 53, no. 7 (1997): 18–19.

Glossary

Ablation: removal of material from an object by an erosion-type process.

Absorption: drawing of a gas or liquid into the pores of a permeable solid (not to be confused with adsorption).

Acetylcholine: primary neurotransmitter of the parasympathetic nervous systems.

Acetylcholinesterase: enzyme that normally hydrolyzes acetylcholine, thereby stopping its activity. This enzyme is inhibited by nerve agents.

ACGIH: see American Conference of Governmental Industrial Hygienists.

ACH/ACh: see Acetylcholine.

AChE: see Acetylcholinesterase.

Acropetal: bottom to top fashion.

Active Ingredient: substance in a pharmaceutical drug or pesticide that is biologically active.

Acute Exposure Guideline Levels: describe the risk from single, non-repetitive exposures to airborne chemicals in a once-in-a-lifetime event. They represent threshold exposure limits for the general public and are applicable to emergency exposure periods ranging from 10 minutes to 8 hours. See Explanatory Notes for additional details.

Acute Exposure: exposure to a hazardous substance over a relatively short period of time.

Adenopathy: swelling or enlargement of the lymph nodes.

Adsorption: adhesion of a substance onto the surface of a solid or liquid (not to be confused with absorption).

Aerobic Bacteria/Aerobe: microorganism that can live and grow in the presence of oxygen.

Aerosol: suspensions of solid or liquid particles in air. Although aerosols are not gases, they are not necessarily visible. Solid-particle aerosols include dusts (formed by mechanical disintegration of a parent material – size range from submicron to visible) and fumes (produced by condensation of vapors or gaseous combustion products – less than 1 μm). Liquid-droplet aerosols include mists (formed by condensation or atomization – size range from submicron to 20 μm) and fogs (visible mist with high particle concentration). Solid/liquid particle aerosols include smokes (visible aerosol resulting from incomplete combustion – less than 1 micron) and smog (photochemical reaction products, usually combined with water vapor – less than 2 μm). See also Particulate Matter.

Aestivation: state of dormancy for some insects that occurs during the summer or dry season.

Aging (chemical): biochemical process that occurs when a nerve agent loses a second leaving group after the initial attachment to cholinesterase. Because of structural differences, the rate of aging varies for each nerve agent. Aging is irreversible and the agent-enzyme complex becomes refractory to oxime treatment.

AI: see Active Ingredient.

AIIR: see Airborne Infection Isolation Room.

Airborne Infection Isolation Room: room that has special air handling and ventilation capacity including negative air pressure relative to the surrounding area, air exhausted directly to the outside or passed through high-efficiency particulate air filtration before being re-circulated to the room, and with a minimum of six air exchanges per hour.

Alkaline: corrosive material that is a base. Aqueous alkaline solutions have a pH greater than 7.

American Conference of Governmental Industrial Hygienists: charitable scientific organization that advances occupational and environmental health. They develop and publish recommendations for industrial exposure limits including threshold limit values.

Amoeba: unicellular microorganism that does not have a definite shape.

Anaerobic Bacteria/Anaerobe: microorganism that can live without oxygen. In some cases, oxygen is toxic to these organisms.

Analgesic: compound capable that relieves pain by altering perception of painful stimuli without producing anesthesia or loss of consciousness.

Anaphylaxis: hypersensitivity or abnormal reaction to a foreign substance. It is an extreme form of allergy that often causes swelling and can be fatal.

Anhydrous: substance that contains no water.

Anisocoria: eyes have unequally sized pupils (anisocoria).

Anorexia: loss of appetite.

Anosmic: unable to detect odors.

Anoxia: absence or reduced supply of oxygen in inspired gases, arterial blood, or tissues.

Anteroom: outer chamber or waiting room situated before the main work room. The anteroom typically contains facilities for changing into protective suits prior to entry into the containment work area and facilities for removing protective suits and showering after exiting the containment work area.

Antibody: protein complex found in the blood or other bodily fluids of humans and animals. Antibodies are used by the immune system to identify and neutralize foreign objects, such as bacteria and viruses.

Anticonvulsant: agent that prevents or arrests seizures.

Antidote: any substance or other agent that inhibits or counteracts the effects of a poison.

Antigen: any substance that induces a state of sensitivity or immune responsiveness and that reacts with antibodies or immune cells produced.

Antitoxin: antibody formed in response to exposure to a biological poison and capable of neutralizing it. A serum containing these antibodies that is prepared from animals vaccinated with the poison.

Anuria: unable to urinate.

Aphasia: inability to communicate verbally or use written words.

Aphonia: inability to speak.

Apnea: temporary cessation of breathing.

Aqueous: solution in which the solvent is water.

Arbovirus: shortened form of arthropod-borne virus. Group of viruses transmitted to humans or animals from insects (e.g., mosquitoes) and arachnids (e.g., ticks).

Areflexia: absence of reflexes.

Arenavirus: group of viruses carried by rodents that cause hemorrhagic fever or meningitis (e.g., Lassa fever, Argentine hemorrhagic fever, lymphocytic choriomeningitis).

Arsenical: chemical compound containing arsenic.

Arthralgia: joint pain.

Arthrogryposis: permanent fixation of a joint in an extended or flexed position.

As: chemical symbol for arsenic.

Ascites: fluid in the abdomen.

Asthenia: weakness or debility.

Asthma: condition in which airways narrow, swell and produce extra mucus, making it difficult to breathing.

Asymptomatic: without signs or symptoms of disease or illness.

Ataxia: muscle incoordination.

Atelectasis: complete or partial collapse of the entire lung or lobe of the lung; occurring when the alveoli become deflated or filled with fluid.

Atomization: process of converting a stream of liquid into a fine spray or aerosol.

Atropine: alkaloid obtained from *Atropa belladonna* used as an antidote for nerve agent poisoning. It inhibits the action of acetylcholine at the muscle junction by binding to acetylcholine receptors.

Audiogenic: sound induced.

Australia Group: dating to June 1985, it is an informal forum of countries that seeks to ensure that exports do not contribute to the development of chemical or biological weapons. Recommendations by the group are not legally binding to the participant countries.

Autonomic Nervous System: part of the nervous system that governs involuntary functions, such as heart rate, reflexes, and breathing. It consists of the sympathetic and parasympathetic nervous system.

Alveoli: tiny air sacs within the lung.

Bacillus: rod-shaped bacteria.

Bacteria: one-celled microorganism that has no chlorophyll and reproduces by dividing in one, two, or three directions of space.

Batch Operation: method of production that has distinct start and end points to the process. Following collection of the products, the process can be restarted.

Binary Chemical Agent: highly toxic agent produced when two or more chemical substances, which individually have relatively little toxicity, react.

Binary Chemical Munition: munitions designed to hold segregated containers of chemical precursors that subsequently rupture while inflight to the target. The precursor components then mix and form a standard chemical agent just prior to reaching the intended target.

Binary Fission: reproduction of a bacterium where the microorganism divides itself into two independent cells.

Biocontainment: containment of highly pathogenic microorganisms within a well-defined, strictly controlled area (i.e., cabinets or rooms), typically used during laboratory research or treatment of an infected individual.

Biodegradation: decomposition of material by microorganisms.

Biological Agent: microorganism or toxin that causes either disease in man, plants, or animals; or deterioration of materiel.

Biological Half-Life: time required for half the quantity of a drug or other substance deposited in a living organism to be metabolized or eliminated by normal biological processes.

Biological Weapon: delivery vehicle (e.g., shell, bomb, sprayer) that releases microorganisms or toxins to casus damage to people, animals, agriculture or material.

Bioregulators/Modulators: biological agents that are biochemical compounds, such as peptides, that occur naturally in organisms.

Bioremediation: any process using microorganisms, fungi, green plants, or their enzymes to remove contaminants from the environment.

Biosafety Levels: sets of biocontainment precautions required to isolate dangerous biological agents in an enclosed laboratory facility. The levels range from the least protective level 1 to level 4. See Appendix 3 for additional information.

Biovar: biological variant. Division of species and subspecies into different strains with identical genetic but different biochemical or physiological characters.

Biphasic Disease: disease that has two distinct phases. After experiencing an initial set of symptoms, the patient seems to have recovered. Shortly afterward, a second set of symptoms develops that may be more severe than the initial ones.

Blebbing: breakup of the cell's nuclear envelope.

Blepharospasm: involuntary blinking.

Blister Agent: see Vesicant.

Blood Agent: compounds that affect the ability of individual cells to use oxygen carried by the blood or damage individual blood cells so that they can no longer transport oxygen from the lungs to the rest of the body. See C07 – COX Inhibiting Blood Agents, C08 – Arsine Blood Agents, and C09 – Carbon Monoxide Blood Agents.

B-NICE: acronym for types of weapons of mass destruction: biological, nuclear, incendiary, chemical, explosives.

Br: chemical symbol for bromine.

Bradycardia: slow heart rate, usually fewer than 60 beats per minute in an adult human.

Bronchitis: inflammation of the mucous membrane of the bronchi.

Bronchopneumonia: pneumonia involving inflammation of the lungs that spreads from and after infection of the bronchi.

Bronchospasm: constriction or spasm of the airway making it difficult to breath.

Bronchus (pl. Bronchi): either of two main branches of the trachea, leading directly to the lungs.

BSL (1,2,3,4): see biosafety levels.

Bubo: inflamed, tender swelling of a lymph node, especially in the area of the armpit or groin, that is characteristic of certain infectious diseases.

Bulla (pl. Bullae): large blister (diameter greater than 0.5 cm) caused by the presence of serum or occasionally by an injected substance.

bv: see Biovar.

BWC: Biological Weapons Convention (1972, 1975).

C: chemical symbol for carbon.

c: see Ceiling.

Carbuncle: boil-like painful localized bacterial infection of the skin and subcutaneous tissue.

Carcinoma: malignant tumor.

Cardiac Arrhythmia: irregular heartbeat.

Cardiotoxic: poison that attacks the heart muscles.

Carrier: individual who harbors specific disease organisms, without showing clinical signs or symptoms, and serves as a means of conveying the infection to others (e.g., Typhoid Mary).

CAS: see Chemical Abstract Service Number.

Casualty: any person who is inhibited from performing their assigned duty due to being killed, wounded, injured, diseased, or captured.

Catalyst: substance that causes a chemical reaction to happen more quickly or under milder conditions than otherwise required.

Caustic: solution of a strong alkali; a base.

CBRNE: acronym for types of weapons of mass destruction: chemical, biological, radiological, nuclear, explosives.

CDC: see Centers for Disease Control and Prevention.

CEGL: see Continuous Exposure Guidance Level.

Ceiling: industrial chemical exposure level that should never be exceeded.

Cellulitis: localized subcutaneous skin infection.

Centers for Disease Control and Prevention: national public health institute in the United States under the Department of Health and Human Services. Its main goal is to protect public health and safety through the control and prevention of disease, injury, and disability; focusing national attention on infectious disease, food borne pathogens, environmental health, occupational safety and health, health promotion, injury prevention, and educational activities.

Central Nervous System: part of the human nervous system that consists of the brain and spinal cord. Supervises and coordinates the activity of the entire nervous system.

CFU: See Colony Forming Units.

ChE: see Cholinesterase.

Chemical Abstract Service Number: unique identification number assigned to each chemical by the chemical abstract service of the American Chemical Society. They have worldwide use for the quick retrieval of matching chemical and safety information.

Chemical Agent Symbol: military code designation of any chemical agent. It is usually a combination of one to three letters or letter-number combinations. It is not the same as, and should not be confused with, the chemical symbol or formula.

Chemical Agent: solid, liquid, or gas which, through its chemical properties, produces lethal or damaging effects on man, animals, plants, or material.

Chemical Asphyxiant: referred to as blood poisons, these are compounds that interrupt the flow of oxygen in the blood or the tissues by binding to the blood cells preventing the transfer of oxygen, destroying the blood cells directly, or interfering with the enzymes that transfer oxygen from the blood to the cells.

Chemical Neutralization: method of destroying chemical agents that involves mixing the agent with hot water, or hot water and sodium hydroxide, to convert the agents into less-harmful chemicals.

Chemical Warfare Agent: see Chemical Agent.

Chemical Weapon: chemical warfare agent loaded into a delivery system such as a munition or sprayer.

Chemoprophylaxis: prevention of disease by the use of chemicals or drugs.

Chemostat: bioreactor in which constant growth conditions for microorganisms are maintained over prolonged periods of time by supplying the reactor with a continuous input of nutrients and continuous removal of growth medium and waste products.

Chimera: organism which results when the genetic material of two or more organisms are combined to enhance the harmful capabilities or survivability of the resulting new organism.

Chloroplast: specialized sub-cellular compartment within plant or algae cells where photosynthesis takes place.

Chlorotic: yellowed.

Choking Agent: see Pulmonary Agent.

Cholecystitis: inflammation of the gallbladder.

Cholinergic: relating to nerve cells or fibers that employ acetylcholine as their neurotransmitter.

Cholinesterase: enzyme that catalyzes the hydrolysis of acetylcholine to choline and acetic acid.

Chromosome: mixture of nucleic acid (DNA or RNA) and protein located in the nuclei of cells.

Cirrhosis: growth of scar tissue in the liver.

Cl: chemical symbol for chlorine.

Clean Room: room with a controlled environment and a constant, highly filtered airflow with very low levels of pollutants, dust, airborne microbes, aerosol particles, or chemical vapors.

CNS: see Central Nervous System.

Coagulopathy: excessive bleeding and a lack of clotting.

Colony-Forming Units: used in microbiology to estimate the number of viable organisms that can clone themselves in a given population of cells. They are not a measure for individual organisms since a single organism may be able to form a colony or it may require a mass of cells or spores.

Colorimetric: technique in which the identity or concentration of a chemical compound in a mixture is determined by measuring a color change or the absorption of light in a particular color range.

Commensalism: association between two organisms in which one benefits and the other derives neither benefit nor harm.

Communicable/Contagious Disease: infectious disease capable of being transmitted from one person to another by close contact, exposure to contaminated fomites, or by an infected vector.

Condensation: change of the physical state of matter from a gas into a liquid.

Condenser: device used to remove heat from a liquid or vapor, causing it to cool and transform to a solid or liquid.

Conjunctiva (pl. Conjunctivae): mucous membrane investing the anterior surface of the eyeball and the posterior surface of the lids.

Conjunctival Injection: bloodshot eyes.

Conjunctivitis: inflammation of the membrane on the inner part of the eyelids and the membrane covering the white of the eye (cornea).

Contagion: spread of disease from one person to another.

Continuous Exposure Guidance Level: ceiling concentrations designed to avoid adverse health effects, either immediate or delayed, of continuous exposure for up to 90 days. Developed by the U.S. Navy for use in submarines.

Convulsants: compounds that induce brief, muscular shock-like jerks (myoclonic seizures) that progress to generalized tonic-clonic (grand mal) convulsions and death.

Convulsion: abnormal violent and involuntary contraction or series of contractions of the voluntary muscles.

Corrosive: chemical that will destroy or damage other substances it comes in contact with; typically an acid or a base.

Cremaster: little hook-like processes on the posterior extremity of some pupa, by which they suspend themselves during pupation.

Crispr-CAS: system of molecular tools and a laboratory technique that facilitates cutting of an organism's DNA or RNA in a robust and sequence specific manner. Subsequent cellular repair of this break in the nucleic acids can result in the incorporation of a change or an edit to the nucleic acid sequence.

Crystallization: process of forming solid crystals from a gaseous or liquid solution.

Ct Value: measure of vapor or gas exposure by inhalation. It is the product of the concentration of the gas – usually expressed in mg/m^3 – and the duration of exposure in minutes. The resulting units are $mg\text{-}min/m^3$. These estimates are not valid over either very short or very long timeframes.

Cultivar: race or variety of plant that has been created or intentionally chosen and maintained in a population through cultivation. For example, farmers may preferentially grow a particular cultivar of rice that produces a higher crop yield or is more resistant to environmental stressors (e.g., temperature).

Cumulative Dose: total dose resulting from repeated exposures to small amounts of a chemical that build up in the body over an extended period of time reaching a point where the eventual effect is similar to a single large acute dose.

Cutaneous Exposure: pertaining to the skin.

Cyanobacteria: bacteria that obtain nutrients through photosynthesis. These bacteria are also known as blue-green algae.

Cyanosis: dark bluish or purplish coloration of the skin, mucous membranes, lips, or fingernails due to deficient oxygenation of the blood.

Cytotoxin: toxin that directly damages and kills the cell with which is makes contact.

Dead-End Host: any host from which a pathogen cannot escape to continue its life cycle. Once in a dead-end host, the pathogen is not transmissible from that host to another organism.

Defervescence: end of the fever.

Dehydration: in the context of freeze- or spray-drying, dehydration refers to the removal of water or other fluids from a solid by evaporation.

Deliquescent: solid that absorbs moisture from the atmosphere and becomes liquified. See also Hygroscopic.

Density: mass per unit volume.

Deoxyribonucleic Acid: self-replicating material present in nearly all living organisms that carries a species' genetic information.

Dermal Exposure: pertaining to the skin surface.

Dermatitis: inflammation of the skin.

Desiccant: substance that has an affinity for water.

Desquamation: skin sheds, peels, or comes off in scales; exfoliation.

Dewatering: process of removing water from a solid by centrifugation or filtration.

Diapause: temporary pause in the growth and development of an insect due to adverse environmental conditions.

Diaphoresis: excessive sweating.

Diffusion: process of spontaneous intermixing of different substances due to molecular motion, which tends to produce uniformity of concentration.

Dimorphic: existing or occurring in two distinct forms.

Diplopia: double vision.

Discal Cell: large cell in the basal half of the wing of a butterfly or moth.

Disease: deviation from the normal state or function of a cell, an organ, or an individual.

Disinfection: destruction of pathogenic and other types of microorganisms. Disinfection is less rigorous than sterilization because it destroys most recognized pathogenic microorganisms but not necessarily all microbial forms (e.g., bacterial spores).

Distillation: separation technique based on differences in boiling points between different components of a chemical mixture.

Diuresis: excessive urination.

DNA: see Deoxyribonucleic Acid.

Dopant: trace impurity element that is inserted into a material in order to alter the electrical or optical properties of that material. Also known as a doping agent.

Dose: amount of agent or energy that is taken into or absorbed by the body; the amount of substance, radiation, or energy absorbed in a unit volume, an organ, or an individual.

Dose Response: characteristics of exposure to a substance and the spectrum of effects.

Drift Spraying: spraying application where the particles generated are intended to "drift" for an extended period of time, covering a wide area. Drift spraying typically uses smaller particles.

Dusts: see Aerosol.

Dusty Agents: liquid chemical agent impregnated onto a finely ground carrier material (e.g., silica, talc, diatomaceous earth). This facilitates aerosol formation and enhances the effectiveness of the base chemical warfare agent by increasing the coverage area, the risk of inhalation injury from nonvolatile agents, and penetration of protective garments.

Dysarthria: unable to speak normally.

Dysgeusia: altered sense of taste.

Dysphagia: difficulty swallowing.

Dysphonia: altered voice production.

Dyspnea: difficult or labored breathing; shortness of breath.

Dystonia: neurological movement disorder characterized by sustained muscle contractions, which result in repetitive movements and uncontrollable twisting.

Dysuria: difficult or painful urination.

Ecchymosis: purplish patch occurring on the skin or a mucous membrane as a result of hemorrhage, differing from petechiae only in size (larger than 3 cm diameter).

ED: see Effective Dose.

Edema: accumulation of an excessive amount of watery fluid in cells, tissues, or serous cavities.

Effective Dose: dose that produces the desired effect. When followed by a subscript, it denotes the dose having the effect on that specified percentage of the test animals. For example, an ED_{50} indicates that the dose produces the desired effect in 50% of the test animals.

Effluent: liquid or gas that flows out from a larger body of liquid or gas.

Electrochemical Detection: sensor that indicates a change in the electric potential of a solution or thin film when the target chemical is absorbed. These types of sensors are influenced by environmental factors such as temperature and humidity.

Electrolysis: method of using a direct electric current to drive a chemical reaction.

Elytron (pl. Elytra): modified, hardened forewing of certain insects (e.g., beetles).

Empyema: pus in the chest cavity.

Emulsion: suspension of small globules of one liquid that is insoluble in a second liquid (e.g., oil and vinegar shaken together).

Enanthema: eruption or rash occurring on a mucous membrane, usually found in conjunction with a skin eruption (i.e., exanthema).

Encephalitis: inflammation and swelling of the brain.

Encephalomyelitis: inflammation and swelling of the brain and spinal cord.

Endemic: native to, or prevalent in, a particular district or region. An endemic disease has a low incidence but is constantly present in a given community.

Endocarditis: infection of the heart.

Endocardium: heart lining.

Endotoxin: toxin produced in the cell walls of certain Gram-negative bacteria and liberated only when the organism dies and disintegrates (e.g., shigella toxin).

Enteritis: inflammation of the intestinal tract.

Enterotoxin: bacterial toxin that affect the intestines, causing diarrhea (e.g., cholera toxin).

Enzootic: affecting or peculiar to animals of a specific geographic area.

Enzyme: biological molecules capable of causing chemical changes to take place quickly at body temperature by catabolic action. A biological catalyst.

Epidemic: outbreak of a contagious, infectious disease affecting a region in a country or a group of countries. The disease can be transmitted from an infected individual to a non-infected individual directly or by a vector.

Epiphytotic: outbreak of disease among plants. It is analogous to an epidemic in man.

Epistaxis: profuse bleeding from the nose.

Epizootic: outbreak of disease among animals. It is analogous to an epidemic in man.

Erythema: reddening of the skin.

Erythema Multiforme: lesions with a pink-red center, pale border ring and an outer pink-red ring.

Erythema Nodosum: reddish, painful, tender lumps.

Erythrocyte: mature red blood cell.

Erythroderma: flushing of the skin.

Erythropoiesis: formation of red blood cells.

Eschar: dry scab or slough formed on the skin as a result of a burn (thermal or chemical) or from a bacterial skin infection.

Evaporation: change of a liquid into a gas at any temperature below its boiling point.

Exanthema: skin eruption occurring as a symptom of an acute viral or coccal disease.

Excipient: inert substance mixed with the active ingredient in a drug formula to improve preservation, bulk up formulations that contain active ingredients in small amounts, or to enhance the therapeutic activity of the final dosage form.

Exotoxin: toxin excreted by a microorganism into the surrounding environment.

Expansion Ratio: ratio of the volume of a given mass of liquified material compared to the theoretical volume of the same mass of material in the gaseous state at the same temperature and pressure. For example, opening a 1 liter cylinder of liquified gas with an expansion ratio of 1,000 will displace all the air in an area equal to 1,000 liters or 1 cubic meter.

Extraction: process of removing a select chemical from a mixture or compound. In the context of purification, this term often refers to use of immiscible solvents.

Exudate: oozing.

F: chemical symbol for fluorine.

Facultative Anaerobe: bacteria that can survive in either aerobic or anerobic conditions.

Fasciculation: involuntary contractions or twitching of groups of muscle fibers; a coarser form of muscular contraction than fibrillation.

Febrile: fever or feverish.

Fermenter: bioreactor that enables optimal fermentation conditions to be maintained under sterile conditions.

F-Gas: synonym for V-series nerve agents.

Fibrillation: fine, rapid twitching of individual muscle fibers with little or no movement of the muscle as a whole.

Filtrate: following a filtration process, the part of a solution that is left after unwanted material is removed.

Flame Photometric Detector: detector used to detect compounds containing sulfur or phosphorus utilizing chemiluminescent reactions. It can also be used to detect certain metals.

Flash Point: lowest temperature at which the application of an ignition source causes the vapors of a liquid to ignite. Below this temperature, materials do not present a significant fire hazard. The lower the flash point, the greater the fire risk.

Flocculant: substance added to a suspension to clarify it by enhancing the aggregation of the suspended particles, causing them to drop out of the solution.

Fomite: inanimate objects contaminated with infectious organisms and serves as a means of disease transmission.

Frass: insect excrement; fecal matter.

Freeze Drying: see Lyophilization.

Fume: see Aerosol.

Fungicide: chemical substance that kills or inhibits the growth of fungi.

Fungus (pl. Fungi): any one of a group of thallophytic plants, including molds, mildews, rusts, smuts, and mushrooms. These plants do not contain chlorophyll, and reproduce mainly by sporulation.

Gamma Globulins: see Antibody.

Gas Chromatography: analytical technique used to separate and analyze samples that can be vaporized without thermal decomposition. The gaseous mixture is forced through a treated column that causes the individual components to separate.

Gastroenteritis: inflammation of the lining of the intestines.

Genetically Modified: organism whose genetic material (DNA or RNA) has been modified in a laboratory.

Genome: entirety of an organism's hereditary information found in a cell and encoded by DNA or RNA.

Germicidal: capability to kill or reduce germs, especially pathogenic microorganisms.

Germs: disease-producing microorganisms including bacteria, viruses, and fungi.

Gnat: see Midge.

Gram Staining: procedure used in classifying bacteria and is related to the composition of the bacterial cell wall. Gram-negative bacteria generally possess a thin layer of peptidoglycan between two membranes. Gram-positive bacteria generally have a single membrane surrounded by a thick peptidoglycan.

G-Series Nerve Agent: class of organophosphorus nerve agents so named because German scientists were the first to synthesize them. Classic examples include tabun, sarin, and soman.

H: chemical symbol for hydrogen.

Halogens: family of elements consisting of fluorine, chlorine, bromine, and iodine.

Hantavirus: group of viruses carried by rodents that cause hemorrhagic fever and severe respiratory infections in humans; includes Sin Nombre, Andes, Hantaan, and Dobrava-Belgrade viruses.

Hemagglutinin: cause red blood cells to clump together.

Hematemesis: vomiting blood.

Hematoma: bruise.

Hematopoietic: pertaining to or effecting the formation of blood cells.

Hematuria: blood in the urine.

Hemiparesis: weakness of one entire side of the body.

Hemolytic: destructive to red blood cells.

Hemolysis: alteration, dissolution, or destruction of red blood cells in such a manner that hemoglobin is liberated into the medium in which the cells are suspended.

Hemolytic Anemia: low level of red blood cells due to destruction of these cells.

Hemolytic Uremic Syndrome: hemolytic anemia and thrombocytopenia occurring with acute renal failure.

Hemoptysis: coughing up blood.

Hemorrhage: to bleed.

Hemorrhagic: profuse bleeding.

Hemorrhagic Gastroenteritis: inflammation and hemorrhage in the stomach and intestines.

HEPA: see High-Efficiency Particulate Air.

Hepatitis: inflammation of the liver.

Hepatomegaly: enlargement of the liver.

Hepatosplenomegaly: enlargement of both the liver and the spleen.

High Vacuum: vacuum with a pressure range between 10^{-11} psi and 10^{-5} psi.

High-Efficiency Particulate Air: air filter or air filtration system rated to remove at least 99.7% of airborne particles with a diameter of 0.3 μm or greater.

Host Cell: cells that viruses use to reproduce since they lack the ability to reproduce or replicate on their own.

Host Range: geographical and/or species distribution that is susceptible to an infectious agent under natural conditions.

Host: organism that is susceptible to or harbors a pathogen under natural conditions.

Hydrolysis: interaction of a chemical with water to produce a different material. When used as part of a decontamination effort the intent is to yield a less toxic product or products.

Hydropericardium: fluid in the sac surrounding the heart.

Hydrothorax: fluid in the chest cavity.

Hygromas: lymph-filled cystic cavities.

Hygroscopic: material that easily absorbs moisture from the atmosphere or surrounding environment. See also Deliquescent.

Hyperemia: presence of an increased amount of blood in a part or organ.

Hyperesthesia: abnormal acuteness of sensitivity to touch, pain, or other sensory stimuli.

Hyperestrogenism: excessive amounts of estrogen.

Hyperpigmentation: excess pigment in the skin.

Hypertension: high blood pressure.

Hypertonia: extreme tension of the muscles or arteries.

Hypotension: low blood pressure.

Hypothermia: low body temperature.

Hypovolemia: low blood volume.

Hypoxemia: low blood oxygen.

Hypoxia: low oxygen levels in the blood or tissue.

IC_{50}: see Incapacitating Concentration.

ID_{50}: see Incapacitating Dose.

Idiopathic: denoting a disease of unknown cause.

Idiosyncratic Reaction: individuals genetically determined abnormal reactivity to a chemical.

IDLH: see Immediately Dangerous to Life or Health.

Ig: Immunoglobulin.

Immediately Dangerous to Life or Health: exposure to airborne contaminants as vapors or aerosols that poses an immediate or delayed threat to life, that could cause irreversible adverse health effects, or that would interfere with an individual's ability to escape unaided from the hazardous atmosphere, in a short time frame. Originally developed to ensure that workers would have time to escape from a contaminated environment in the event of a failure of their respiratory protection equipment. In these calculations the time frame is set at 30 minutes or less.

Immiscible: does not dissolve in the specified solvent.

Immunity: protection against an infectious disease either by natural immune system response or an induced response following a previous infection.

Immunoglobulins: see Antibody.

Impervious: describes a characteristic of a material that does not allow penetration of a substance through the material.

Impregnated: describes a material in which another substance has been embedded throughout to reduce porosity (i.e., blocks pores so that fluids do not penetrate the material), or incorporated to neutralize hazardous materials, kill microorganisms, or repel insects.

In situ: Latin phrase meaning "in place." In biology, this term refers to examining an event exactly in the place it occurs.

In vitro: Latin phrase meaning "an artificial environment," referring to a process or reaction occurring therein, as in a test tube or culture media.

In vivo: Latin phrase meaning "in the living body," referring to a process or reaction occurring therein.

Incapacitating Agent: compounds that produces temporary physiological or mental effects, or both, which will render an individual incapable of concerted effort. Effects continue after individuals are removed from exposure to the agent. Compare with Riot Control Agents. See C12 – Mind-Altering Agents.

Incapacitating Concentration: concentration of a vapor or aerosol necessary to incapacitate or disable exposed and unprotected individuals through inhalation of the agent. A subscript number indicates the theoretical percentage of the exposed population that would be expected to experience the proscribed effects.

Incapacitating Dose: amount of liquid or solid agent necessary to incapacitate or disable exposed individuals through some pathway other than inhalation (e.g., ingestion, injection, percutaneous). A subscript number indicates the theoretical percentage of the exposed population that would be expected to experience the proscribed effects.

Incendiary Agent: compound that generates sufficient heat to cause destructive thermal degradation or destructive combustion.

Incineration: destruction of chemicals or other materials by burning at very high temperatures.

Incubation Period: interval of time between infection by a microorganism and the first signs or symptoms of the illness.

Infarction: tissue death.

Infection Control Zone: area of infection control, usually a room in a hospital, most frequently used to isolate a patient with an illness easily spread by physical contact or through the air. An infection control zone can also protect a patient from infection by setting filtered air to flow out of a room (positive pressure).

Infectious Dose: amount of a pathogen that it takes to cause infection. The dose varies in different animal species and by route of exposure. Variables likely to alter the infectious dose include sex, age, nutritional status, health status, immune competence, previous exposure to the agent, use of medications, and immunizations.

Infectivity: ability of a pathogen or a biological agent to establish an infection.

Inhalable Particulate Mass: see Threshold Limit Values for Particulates.

Inhibition: process of preventing, stopping, or retarding a chemical reaction.

Inoculation: process of introducing a pathogen into a living organism. When referring to production of large quantities of a pathogen (e.g., growing bacteria in a fermenter), it can also refer to the process of introducing a pathogen to suitable media to facilitate growth.

Inoculum: biological sample used as starting material for growth of microorganisms or viruses. Inocula are used for legitimate production of pharmaceutical or bacterial products, but also can be used for illegitimate production of a biological weapon agent or toxin.

Inspiratory Dyspnea: difficulty inhaling.

Instars: stages of development in the life of a juvenile insect between molts.

Interferons: are proteins used to communicate between cells to trigger protective actions of the immune system to destroy pathogens or cancerous cells.

International Task Force 25: was formed in March 1994 as a joint work group of the United States, Canada, and Great Britain to determine whether there were hazards from the release of industrial chemicals in a military situation, and, if so, to develop criteria for assessing the hazards and develop a list of chemicals of concern. The final report was delivered on March 18, 1996. See Explanatory Notes for additional details.

International Task Force 40: was formed in 2001 as a joint work group of the United States, Canada, and Great Britain to develop appropriate tools for commanders to identify, assess and control the risks from exposures caused by releases of industrial chemicals in a military situation. It was tasked with reviewing, revising and validating the report of the International Task Force 25 and then creating a prioritized threat list for toxic industrial chemicals including agricultural and commercial chemicals. See Explanatory Notes for additional details.

Ion Mobility Spectroscopy: analytical method used to separate and identify ionized molecules in the gas phase based on differences in their mobility in a carrier gas.

Iritis: inflammation of the iris.

ITF-25: see International Task Force 25.

ITF-40: see International Task Force 40.

Jaundice: yellow discoloration of the skin and eyes.

K Agents: another name for incapacitating agents.

Kelvin: absolute temperature scale and designated by °K. Temperatures in Fahrenheit (°F) can be converted to Kelvin by the following formula: $°K = 273 + (°F - 32)/1.8$.

Keratitis: inflammation of the cornea.

Key Component of Binary or Multicomponent Chemical Systems: Chemical Weapons Convention defines them as the precursor which plays the most important role in determining the toxic properties of the final product and reacts rapidly with other chemicals in the binary or multicomponent system.

Lachrymation: tearing.

Lamina: upper side of the leaf.

Laryngitis: swelling, irritation, and/or inflammation of the larynx (voice box). This is usually associated with hoarseness and/or loss of voice.

Lassitude: feeling of weariness or diminished energy.

Latent period: time of seeming inactivity between the incidence of exposure and the onset of symptoms. For example, effects from most blister agents do not become apparent until after a latent period of 6 to 24 hours after exposure.

LC_{50}: median lethal concentration of vapor or aerosol in air for a hypothetical population. See Lethal Concentration.

LD_{50}: median lethal dose for a hypothetical population. See Lethal Dose.

Lethal Concentration: concentration of a vapor or aerosol necessary to kill exposed and unprotected individuals through inhalation of the agent. Subscripts identify the theoretical percentage of an exposed, unprotected, population that would be killed by inhaling the specified concentration of agent.

Lethal Dose: amount of liquid or solid necessary to kill exposed individuals. The pathway of exposure is also often indicated (e.g., injection, ingestion). Subscripts identify the theoretical percentage of an exposed, unprotected, population that would be killed by being exposed through the expressed pathway to the specified amount of agent.

Lethargy: lack of energy.

Leukocytosis: elevated white blood count.

Leukopenia: low white blood cell count.

LOAEL: see Lowest-Observed Adverse Effect Level.

Local Effect: occurs at the point of direct contact on the body such as the skin, eyes, nose, throat, and airways.

Low Vacuum: vacuum with a pressure range between 0.44 psi and 1.45 psi.

Lower Explosive Limit: lowest concentration, or percentage, in air of a vapor or gas, that will sustain or support combustion when an ignition source is introduced. Below this limit the vapor is too lean to burn.

Lower Flammability Limit: see Lower Explosive Limit.

Lowest-Observed Adverse Effect Level: lowest exposure level at which there are statistically or biologically significant increases in frequency or severity of adverse effects between the exposed population and an appropriate control group.

Lung-Damaging Agent: see Pulmonary Agent.

Lymphadenitis: infection of the lymph system.

Lymphadenopathy: painful lymph nodes.

Lymphangitis: red streaking with swollen lymph nodes.

Lymphocytosis: increased white blood cells.

Lymphopenia: reduction, relative, or absolute, in the number of lymphocytes in the circulating blood.

Lyophilization: dehydration process used to stabilize nearly any perishable material in order to increase its shelf life and reduce its sensitivity to environmental stresses. Also known as freeze drying.

Macular (pl. Macules): flat spots.

Macular Rash: rash comprised of small blemishes or discolorations that do not rise above the skin surface. Individual blemishes may resemble freckles.

Maculopapular: rash of mixed flat and raised spots.

Malaise: vague feeling of bodily discomfort.

Malodorant: compounds with unpleasant odors causing strong, repulsive responses including nausea, gagging, and/or vomiting. See C15 – Malodorants.

Meander: sideways spread of a cloud caused by shifting air currents and horizontal turbulence.

Mechanical Vector: organism, such as a fly or cockroach, that transmits a disease by inadvertently carrying the pathogen on its body to a new location. See also Vector.

Medium Vacuum: vacuum with a pressure range between 10^{-5} psi and 0.44 psi.

Melena: black tar-like stools.

Melting Point: temperature at which a solid changes to a liquid. The melting point is the same as the freezing point.

Meninges: any membrane; specifically, one of the membranous coverings of the brain and spinal cord.

Meningitis: inflammation of the membranes that surround the brain and spinal cord.

Meningoencephalitis: inflammation of the brain, spinal cord, and the membranes covering them (i.e., meninges).

Methemoglobinemia: oxidation of hemoglobin so it can no longer transport oxygen. At concentrations of 3%–5%, cyanosis occurs. At concentrations above 25%, the central nervous system is affected. Concentrations above 70% can rapidly lead to death.

Metrorrhagia: non-menstrual bleeding from the uterus.

mg/m³: see Milligrams per Cubic Meter.

Micron: abbreviated as μm. One micron is equal to 10^{-6} meters.

Microorganism: also known as a "microbe." An organism that is too small to be seen with the human eye. Microorganisms are very diverse and include bacteria, fungi, protists, microscopic plants (green algae), and microscopic animals (e.g., plankton). Most microorganisms are single-celled organisms. Viruses meet the size definitions of microorganisms, but they are considered nonliving when separated from the host they require for replication.

Midge: small flying insect that closely resembles a mosquito.

Mid-Spectrum Agents: toxic chemicals of biological origin (i.e., toxins or bioregulators). They fall under the general clauses of both the Chemical Weapons Convention, and the Biological and Toxin Weapons Convention. Some references also include man-made synthetic viruses and genocidal agents (i.e., agents that target a specific ethnic group) in this category.

Milligrams per Cubic Meter: expression of concentration (mass) of an agent in air. It is the appropriate measure for the airborne concentration of an aerosol, mist, fume or dust. See Conversion Factor in Explanatory Notes for additional details.

Miosis: excessive contraction of the pupil.

Miscible: soluble in all proportions in the specified solvent.

Mist: see Aerosol.

Mole: unit of measurement in chemistry defined as 6.022×10^{23} atoms or molecules. One mole of a chemical is equal to its molecular/formula weight in grams.

Monoclonal Antibody: globular proteins used by the immune system of humans and animals to identify and neutralize foreign objects, such as bacteria and viruses.

Morbidity Rate: describes the average percentage of individuals that become ill when exposed to a pathogen.

Moribund: dying; at the point of death.

Mortality Rate: describes the average percentage of individuals that die after becoming ill when exposed to a pathogen.

Mucopurulent: discharge of mucus and pus.

Myalgia: muscle pain.

Mycotoxin: toxin from fungi (e.g., T2 toxin).

Mydriasis: excessive dilation of the pupil.

Myelitis: inflammation of the spinal cord.

Myocarditis: inflammation of the heart muscle.

Myoclonic Seizure: brief, muscular shock-like jerks.

N: chemical symbol for nitrogen.

NAAK: see Nerve Agent Antidote Kit.

Narcosis: condition of deep stupor or unconsciousness produced by a drug or other chemical substance.

National Institute for Occupational Safety and Health: U.S. federal agency responsible for conducting research and making recommendations for the prevention of work-related injury and illness. It is part of the Centers for Disease Control and Prevention within the Department of Health and Human Services.

Nausea: stomach sickness; tendency to vomit.

Neat Agent Equivalent: actual volume of chemical agent that will be formed when containers of a binary agent's precursors are mixed.

Neat Chemical Agent: non-diluted, full-strength (as manufactured) chemical agent. A chemical agent manufactured by a binary synthesis route is also considered a neat agent regardless of purity.

Necrosis: death of a cell or group of cells.

Nephritis: kidney inflammation.

Nerve Agent: chemical warfare agent that affects bodily functions by reacting with the enzyme acetylcholinesterase, resulting in accumulation of acetylcholine. This produces continual stimulation of the parasympathetic nervous system, as well as affecting other parts of the autonomic nervous system. See C01 – Organophosphorus Nerve Agents and C02 – Carbamate Nerve Agents.

Nerve Agent Antidote Kit: consists of two prefilled autoinjectors for the rapid administration of (1) atropine and (2) pralidoxime chloride.

Neurotoxin: a poison affecting nerve tissue.

Neurotransmitter: chemical produced in nerve cells that transmits signals between nerve cells.

Neutralent: residue or material remaining after the chemical neutralization of chemical warfare agents.

Neutralization: altering the chemical, physical, or toxicological properties of a chemical agent to render it ineffective as a weapon.

New World: geographic term referring to North and South America collectively.

NIOSH: see National Institute for Occupational Safety and Health.

NOAEL: see No-Observable Adverse Effects Level.

Node: small mass of swollen tissue.

Nodule: small mass of rounded or irregular shape.

Nonpersistent Agent: chemical agent that dissipates or loses its capability to cause casualties after 10 to 15 minutes.

Nontraditional Agents: generic U.S. military term for a chemical or biochemical materiel that is not U.S. type-classified as a chemical agent but assessed to be in development by foreign threat elements and that could emerge as a threat agent.

No-Observed Adverse Effects Level: highest exposure level for a material at which there are no statistically or biologically significant increases in the frequency or severity of adverse effects.

Nosocomial: denoting a new disorder (not the patient's original condition) associated with being treated in a hospital; a hospital-acquired infection.

NTAs: see Nontraditional Agents.

Nuchal Rigidity: neck stiffness.

Nucleic Acid: large macromolecule that encodes genetic information in the form of DNA or RNA.

Nystagmus: rapid involuntary movements of the eyes.

O: chemical symbol for oxygen.

Obligate Intracellular: can only reproduce within a host's cells.

Occupational Safety and Health Administration: agency of the U.S. Department of Labor created to assure safe and healthful working conditions by setting and enforcing standards, including exposure limits to industrial airborne chemicals.

Ocellus: simple eye of an insect consisting of a number of sensory cells and often a single lens.

Odontalgia: toothache or pain.

Old Chemical Weapons: Chemical Weapons Convention defines them as chemical weapons which were produced before 1925; or chemical weapons produced in the period between 1925 and 1946 that have deteriorated to such extent that they can no longer be used as chemical weapons.

Old World: geographic term referring to Europe, Asia, Africa, and Australia collectively.

Oligophagous: feeding on only a limited number or range of foods.

Oliguria: scant urine production.

Ootheca: hardened egg case of an insect.

Opacity: cloudiness.

OPCW: see Organization for the Prohibition of Chemical Weapons.

Ophthalmoplegia: paralysis of the motor nerves of the eye.

Opisthotonos: severe spasm with the head and feet drawn backward and the spine arching forward.

Orchitis: inflammation of the testicles.

Organization for the Prohibition of Chemical Weapons: implementing body for the Chemical Weapons Convention and oversees the global endeavor to permanently and verifiably eliminate chemical weapons.

Organophosphorus: describing organic compounds containing carbon-phosphorus bonds.

Oropharynx: portion of the throat that lies posterior to the mouth behind the tongue.

Orthopnea: only able to breathe while erect.

OSHA: see Occupational Safety and Health Administration.

Osteomyelitis: inflammation of the bone marrow and adjacent bone.

Outbreak: occurrence of disease greater than expected for a particular time and place. Outbreaks may be an epidemic (i.e., affecting a region in a country or a group of countries), or a pandemic (i.e., affecting populations globally).

Overwinter: to live through or pass the winter.

Ovipositor: tubular structure that many female insects have to deposit eggs.

Oxide: compound that contains oxygen and one other element.

Oxidizer: compound that accepts an electron from another compound; also known as an oxidizing agent. Oxidizers support and accelerate combustion.

P: chemical symbol for phosphorous.

Pancaking: term to describe the effect observed when the vapor created by a chemical munition settles, then spreads downward and outward to cover a target.

Pandemic: disease outbreak affecting or attacking the population of an extensive region, a country, a continent, or the world. An extensive epidemic.

PAPR: powered air-purifying respirator.

Papule: small, circumscribed, solid elevations (pimples) up to 1 cm in diameter on the skin.

Papulovesicular: pimple and blister rash.

Parasitemia: presence of parasites in the circulating blood; used especially with reference to malarial and other protozoan forms, and microfilariae.

Parasympathetic Nervous System: part of the autonomic nervous system that decreases pupil size, heart rate, and blood pressure, and increases functions such as secretion of saliva, tears, and perspiration.

Paresis: partial paralysis.

Paresthesia: burning, prickling, itching, or tingling skin sensation.

Parotitis: inflammation of the salivary glands.

Parthenogenesis: form of reproduction in which an unfertilized egg develops into a new individual.

Particulate: extremely small amount of solid or liquid matter. Particles of biological interest have aerodynamic diameters of less than or equal to 100 μm.

Particulate Matter: generic term applied to chemically heterogeneous discrete liquid droplets or solid particles suspended in the air. The particulate matter in an aerosol can range in size from 0.001 to greater than 100 μm in diameter. See also Aerosol.

Parts per Million: relative ratio of volume of agent to volume of air. One percent is equal to 10,000 parts per million. See Conversion Factor in Explanatory Notes for additional details.

Pathogen: microorganism that causes disease in humans, plants, or animals. Manifestations of the disease may be due to a toxin produced by the organism.

Pathogenicity: potential of a particular microorganism to cause disease.

Pathognomonic: symptom or finding that is specially or decisively characteristic of a particular disease.

Pathovar: bacterial strain or set of strains with similar characteristics but distinctive pathogenicity to one or more plant hosts. Named as a ternary or quaternary addition to the species binomial name.

PCR: see Polymerase Chain Reaction.

Peduncle: stalk bearing a solitary flower in a cluster of flowers arranged on a stem.

PEL: see Permissible Exposure Limit.

Pemphigus Vulgaris: autoimmune disorder affecting the outer layer of the skin and mucous membranes, producing large fluid-filled, rupture-prone blisters. It has a high mortality rate.

Peptide: short polymer of amino acids. A polypeptide is a small protein.

Percutaneous: denoting the passage of substances through unbroken skin. Often occurs without visible damage to the skin.

Pericardium: sac surrounding the heart.

Peritoneum: inflammation of the lungs, air sacs, liver, heart, spleen, kidneys, and the lining of the abdomen.

Perivascular: surrounding a blood or lymph vessel.

Permeation: movement of chemicals at the molecular level through intact suit material. It is usually expressed as breakthrough times for a given chemical or chemical class.

Permissible Exposure Limit: in the United States, a time-weighted average concentration that must not be exceeded during any 8-hour work shift of a 40-hour workweek. They are established by the Occupational, Safety and Health Administration and are designed to protect workers exposed to industrial chemicals on a daily basis. See also Time-Weighted Average.

Persistency: expression of the duration of effectiveness of a chemical agent. It is dependent on the physical and chemical properties of the agent, weather, methods of dissemination, and conditions of the terrain. Under battlefield conditions, non-persistent agents generally lose their effectiveness approximately 10–15 minutes after deployment. However, if the same agent is released indoors, the persistency is greatly increased. See also Relative Persistency.

Persistent Agent: chemical agents that do not hydrolyze or volatilize readily and retain their effectiveness as a toxic threat for hours or days.

Petechia (pl. Petechiae): minute crimson, purple, or livid spots, of pinpoint to pinhead size, occurring on the skin or a mucous membrane as a result of hemorrhage.

Petiole: leaf stem.

Pharyngeal: relating to the throat.

Pharyngitis: sore throat.

Phlebitis: vein inflammation.

Photoionization Detection: type of gas detector that measures volatile compounds in very small concentrations by ionizing them and detecting the resulting electric current.

Photophobia: sensitivity to light.

Photosensitization: an abnormally heightened reactivity to light.

Photosynthesis: process used by plants and other organisms to convert light energy into chemical energy.

Phycotoxin: toxin from algae (e.g., anatoxin A).

Phytotoxin: toxin from a plant (e.g., ricin).

Plasmid: a small circular DNA molecule.

Pleural Effusion: fluid in the chest cavity.

Pleurisy: inflammation of the two membranous sacs (pleura), each of which lines one side of the thoracic cavity and envelops the adjacent lung, they reduce the friction of respiratory movements to a minimum.

Pleuritic: sharp, stabbing, or burning pain.

Pneumonitis: inflammation of lung tissue.

Poaceae: family of plants known as grasses and includes the plants that produce cereal grains.

Pollakiuria: frequent urination.

Polyarthritis: painful arthritis appearing in multiple joints simultaneously.

Polymerase Chain Reaction: laboratory technique used to make multiple copies of a segment of DNA. It is very precise and can be used to amplify, or copy, a specific DNA target from a mixture of DNA molecules.

Polymorphism: occurring in more than one form.

Polyphagous: feeding on many different kinds of food.

Polysaccharide: polymer of carbohydrates used to store energy or provide structure. Examples include starch or cellulose.

Polyuria: excessive production of urine.

Posterior Paresis: partial paralysis of the hind legs.

Potentiometric: describing a type of chemical analysis used to determine the concentration of a particular component of a solution by measuring the voltage of the solution.

ppm: see Parts Per Million.

PR: see Protective Ratio.

Precursor: any chemical reactant that takes part in the production of a toxic chemical. This includes the key component of a binary or multicomponent chemical system.

Probiotic: microorganism that is believed to provide health benefits when consumed.

Prodrome: early symptom indicating the onset of a disease, usually followed by a period of apparent recovery before the full onset of illness.

Pronotum: dorsal plate of the prothorax in insects.

Protection Factor: the level of protection against agent vapors afforded by a piece of protective equipment. The protection factor is essentially a measure of the reduction in cumulative exposure to an aerosol afforded by the equipment. The protection factor for an ensemble is affected by the fit, the design of its seals and closures, and some physical aspects of the individual wearing the suit.

Protective Ratio: factor expressing the effectiveness of an antidote at reducing the toxic effects produced by a poison. For example, a protective ratio of 1 would mean that the antidote had no effect, whereas a protection factor of 2 would mean that it takes twice as much agent to harm an individual.

Protein: any of numerous naturally occurring, extremely complex substances that consist of amino-acid residues joined by peptide bonds; they include many essential biological compounds or immunoglobulins.

Proteinuria: protein in the urine.

Protists: eukaryotic organisms that are not fungi, plants, or animals and are chiefly unicellular or colonial. Includes protozoans, certain algae, oomycetes, and slime molds.

Pruritus: severe itching.

Ptosis (pl. Ptoses): drooping of the eyelids.

Ptyalism: drooling.

Pulmonary: pertaining to the lungs.

Pulmonary Agent: compounds that cause irritation and inflammation of bronchial tubes and lungs. Their primary physiological action is limited to the respiratory tract with injury extending to the deepest part of the lungs. See C10 – Pulmonary Agents.

Pulmonary Edema: swelling or excessive accumulation of fluid in the lungs.

Puparium: pupa included within the last larval skin of an insect.

Purpura: dermal hemorrhages.

Pustule: small, pus-filled blister or an elevation of a hair cuticle with an inflamed base.

pv: see Pathovar.

Pyrexia: high fever.

Pyrogenic: causing fever.

Pyrophoric: substances that, even in small quantities, are liable to ignite within five minutes after coming into contact with the oxygen in air.

Pyuria: pus in the urine.

R: generic chemical symbol for an unspecified organic group. The group is generally a hydrocarbon but may contain other functional groups within the chain. Different groups in a series are designated by a sequence of primes (e.g., "or").

Race: in the context of plant biology, a group of plants having similar characteristics that distinguish them from other plants within the same species. The species is divided into race based on the host range and further subdivided into biovar to distinguish between subtle differences in metabolic pathways that contribute to the pathogenicity of the agent.

Rales: noncontinuous clicking or rattling sounds, typically in the lungs.

Reactant: also known as a precursor, reactants are combined and undergo a chemical reaction to generate a new chemical known as the product.

Reagent: chemical substances used in a chemical reaction to produce a new material.

Recombination: process by which pieces of an organism's genome (DNA or RNA) are broken and rejoined.

Releases Other Than Attacks: threat of exposure of a military unit to nuclear, biological or chemical substances coming from an accidental or deliberate release of hazardous industrial materials; a hazmat incident.

Relative Persistency: mathematical comparison of the evaporation rate of water at 68°F (20°C) to the evaporation rate of a chemical at ambient temperature. Values greater than 1 indicate that the material evaporates slower than a similar spill of water. See Explanatory Notes for additional details.

Reservoir (pathogens): refers to any animal, plant, or nonliving material in which pathogens normally live and multiply – usually without causing harm to the host – until it is passed to an organism that is susceptible to the disease. Reservoirs serve as a continual source of infection for spreading a disease.

Residual Contamination: that amount of hazardous material that remains after decontamination.

Respirable Particulate Mass: see Threshold Limit Values for Particulates

Retinitis: inflammation of the retina.

Retrosternal: behind the sternum.

Rhinitis: inflammation of the mucous membrane of the nose.

Rhinorrhea: excessive and persistent watery mucus discharge from the nose.

Rhonchi: continuous low pitched, rattling lung sounds that often resemble snoring.

Ribonucleic Acid: genetic material present in all living cells that acts as an intermediary carrying instructions from DNA to synthesize proteins in the cell. Some viruses have RNA as the primary carrier of genetic information in place of DNA.

Rickettsia: Gram-negative, nonmotile, intracellular, parasitic bacteria that were once considered a separate type of organism. They are intermediate in size between other bacteria and viruses.

Riot Control Agent: agent that produces only a temporary irritating or incapacitating effect. This class of agents includes both tear and vomiting agents. Effects last minimally longer than contact with the agent. Compare with Incapacitating Agents. See C13 – Irritating and Lachrymatory Agents and C14 – Vomiting/Sternatory Agents

Risk Group (pathogens): classification of infective microorganisms used only in a laboratory setting. It is not a general overall ranking of the threat posed by a pathogen in the environment. See Appendix 3 for additional information.

RNA: see Ribonucleic Acid.

ROTA: see Releases Other Than Attacks.

Ruminants: various hoofed, even-toed mammals that chew a cud (e.g., cattle, sheep and goats).

S: chemical symbol for sulfur.

Saprophyte: feeds only on dead or decaying organic matter.

Schedule 1 of the Chemical Weapons Convention: include chemicals that pose a high risk and have been used as warfare agents or weaponized for war, as well as some key precursors for the agents in this schedule. Two toxins are specifically listed: ricin and saxitoxin.

Schedule 2 of the Chemical Weapons Convention: includes chemicals that pose a significant risk for use as a chemical weapon, key chemicals for the synthesis of nerve and blister agents, and three specifically listed chemicals: the pesticide amiton, perfluoroisobutylene (PFIB), and BZ.

Schedule 3 of the Chemical Weapons Convention: includes chemicals that pose a risk for use as a chemical weapon, and includes commercial materials that have been used as warfare agents or their precursors.

Scutum: shield-shaped dorsal plate of certain insects.

Secondary Contamination: contamination that occurs due to contact with a contaminated person or object rather than to direct contact with bulk agent; cross-contamination.

Sepsis: presence of pathogenic microorganisms or their toxins in the blood or other tissues.

Septic Shock: shock associated with pathogenic organisms or their toxins in the bloodstream.

Septicemia: systemic disease caused by pathogenic organisms or their toxins in the bloodstream.

Sequela (pl. Sequelae): condition following as a consequence of a disease or chemical exposure.

ser: see Serovar.

Serovar: subcategory into which a bacterium is placed based on its serological activity. Also called serotype.

Serum: clear, watery fluid from tissue, or the clear yellowish fluid obtained upon separating whole blood.

Seta: stiff hair-like bristle on an insect.

Shock: upset in the body caused by inadequate amounts of blood circulating in the bloodstream. It can be caused by marked blood loss, overwhelming infection, severe injury to tissues, emotional factors, etc.

Short-Term Exposure Limits: industrial chemical exposure level that represents the maximum average concentration a person may be exposed to over a short period of time, usually 15 minutes, as long as there are only four such exposures in an 8-hour day and each exposure is separated by a minimum 1-hour interval.

Simple Asphyxiant: inert gas that displaces the oxygen necessary for breathing, or dilutes the oxygen concentration below the level that is useful for the human body.

Simulant: material that appears and acts like a chemical or biological agent.

Sinusitis: sinus infection.

Solubility: ability of a substance to dissolve in a given solvent – usually a liquid – and form a homogeneous solution.

Solute: substance dissolved in another substance, a component of a solution. Can be a solid, liquid or a gas.

Solvent: liquid or other substance in which a solute is able to be dissolved to form a solution.

Somnolence: semiconsciousness.

Spastic Paralysis: condition in which muscles undergo persistent spasms and exaggerated reflexes because nervous system control of the muscles has been disrupted or altered.

Species: in biology, a group of living organisms with similar characteristics that is capable of exchanging genetic material and interbreeding. It is a taxonomic unit that ranks below a genus.

Spiracle: breathing hole of an insect.

Splenomegaly: enlargement of the spleen.

Spore: inactive, dormant form of a bacteria or reproductive form of a fungus. Bacterial spores can survive for long periods in harsh environmental conditions and reactivate upon exposure to an environment that supports growth (e.g., an environment containing water and nutrients). Fungal spores are not resistant to disinfection like bacterial spores.

Sporulation: the process by which bacteria or fungi form spores.

ssp: subspecies.

Staphylococcus Scalded Skin Syndrome: staphylococcus skin infection characterized by widespread erythema with subsequent peeling and necrosis of the skin.

STB: see Super Tropical Bleach.

STEL: see Short-Term Exposure Limits.

Sterilization: process used to render a product free of all forms of viable microorganisms. Sterilization is more powerful than disinfection because it destroys all microbial forms of microorganisms (e.g., bacteria, spores, viruses).

Sternutator: agent that produces sneezing. See also Vomiting Agents.

Stomata: pore found on the surface of a plant that is used to control gas exchange.

Stomatitis: inflammation of the mouth.

Strabismus: inability of the eyes to point in the same direction.

Strain: in biology, a highly specific taxonomic rank used to describe a group of organisms of the same species with similar characteristics that are distinct from other members of the same species.

Stridor: high-pitched, noisy respiration. A sign of respiratory obstruction, especially in the trachea or larynx.

Sublimation: process of changing from a solid to a gas without passing through an intermediate liquid phase.

subsp: subspecies.

Super Tropical Bleach: mixture of calcium oxide and bleaching powder used as a decontaminating agent.

Superantigen Toxin: causes systemic effects by activating the immune system in a nonspecific way (e.g., staphylococcal enterotoxin B).

Surface Acoustic Wave Spectroscopy: in the context of chemical monitors, the detection of gases by measuring acoustic vibrational changes that occur when the material is absorbed on the sensor.

Sympathetic Nervous System: network of nerves that trigger certain involuntary and automatic bodily functions, such as constricting blood vessels, widening the pupils, and speeding up the heartbeat.

Symptoms: functional evidence of disease; a change in condition indicative of some mental or bodily state.

Synapse: site at which neurons make functional contacts with other neurons or cells.

Syncope: loss of consciousness.

Syndrome: set of symptoms that occur together.

Synergistic: working together, having combined cooperative action that increases the effectiveness of one or more of the components' properties.

Synthetic Biology: field of study that applies engineering principles to the fundamental components of biology. Goals include the design and construction of new biological parts, devices, and systems and the redesign of existing natural biological systems for useful purposes.

Systemic Effect: occurs when a chemical gets into the blood system and is distributed throughout the body to affect critical organs and tissues at a distance from the point of entry.

Tachycardia: rapid heart rate, conventionally applied to rates over 100 per minute in an adult.

Tachypnea: rapid breathing.

Taxonomy: branch of science concerned with classification, especially of organisms. Biological classification is a method used to group organisms with other similar organisms. The groups most typically used to classify organisms are (from largest to smallest): domain, kingdom, phylum, class, order, family, genus, and species.

Tear Agents: compounds that cause a copious flow of tears and intense (although temporary) eye pain. In high concentrations, they are irritating to the skin and cause a temporary burning and itching sensation. Very high concentration can cause dermal burns and pulmonary edema.

Tenesmus: feeling of the need to pass stool even when the bowels are empty.

Theoretical Man: individual assumed to be a healthy adult male between the ages of 18 and 35 years old and weighing 154 pounds (70 kg). Used by the military when calculating response to exposure to chemical agents.

Therapeutic Index: ratio of the median lethal dose (LD_{50}) compared to the median effective dose (ED_{50}) of a drug. A small value indicates that there is a narrow margin between a dose that produces the desired response and a dose that causes life-threatening effects. Therefore, the greater the index value the lesser the risk of an accidental overdose.

Thoracic Particulate Mass: see Threshold Limit Values for Particulates.

Thorax: second or middle region of the body of an insect lying between the head and the abdomen.

Threshold Limit Value: industrial exposure recommendation developed by the American Conference of Governmental Industrial Hygienists for the maximum average airborne concentration of a hazardous material to which healthy adult workers can be exposed during an 8-hour workday and 40-hour workweek without experiencing significant adverse health effects.

Threshold Limit Values for Particulates: are expressed in three forms: a). Inhalable Particulate Mass – for those materials that are hazardous when deposited anywhere in the respiratory tract (particles with aerodynamic diameters up to 100 μm); b). Thoracic Particulate Mass – for those materials that are hazardous when deposited anywhere within the lung airways and the gas-exchange regions (particles with aerodynamic diameters up to 25 μm); and c). Respirable Particulate Mass – for those materials that are hazardous when deposited in the gas-exchange region (particles with aerodynamic diameters up to 10 μm).

Thrombocytopenia: low blood platelet count.

Thrombosis: blood clots.

TIC: see Toxic Industrial Chemical.

TIM: see Toxic Industrial Material.

Time-Weighted Average: expression of industrial exposure describing the average concentration of a chemical that a normal, healthy worker can be continuously exposed to during a workday (typically defined as either 8 hours or 10 hours long) and a 40-hour workweek without showing any adverse health effects.

Tissue Culture: growth of cells or tissues outside of an organism, often in a Petri dish, flask, or bioreactor. Cultured cells may be used to grow viruses.

TLV: see Threshold Limit Value.

Tobacco Reaction: flat, metallic taste when smoking tobacco after exposure to small amounts of toxic agents such as phosgene, hydrogen cyanide, or sulfur dioxide. Used as an indicator of exposure by troops in the field during World War I.

Torticollis: sideways twitching of the neck.

Toxemia: condition caused by the circulation of toxins in the blood.

Toxic Chemical: as defined by the Chemical Weapons Convention it is any chemical which through its chemical action on life processes can cause death, temporary incapacitation or permanent harm to humans or animals. This includes all such chemicals, regardless of their origin or of their method of production, and regardless of whether they are produced in facilities, in munitions or elsewhere.

Toxic Epidermal Necrolysis: potentially deadly skin disease that usually caused by a reaction to a medication. Symptoms can be non-specific such as cough, aching, headaches, and feverishness, which can progress to a red rash across the face and the trunk of the body that spreads to other parts of the body. The rash can form into blisters, and these blisters can form in areas such as the eyes, mouth and vaginal area. The mucous membranes can become inflamed, and layers of skin, hair, and nails can be easily pulled off. The affected skin may look burned.

Toxic: containing poison or being poisonous in a way that is capable of causing serious harm or death.

Toxic Industrial Chemical: industrial chemicals that are manufactured and used in large quantities that also possess dangerous chemical or toxicological properties making them potential weapons.

Toxic Industrial Material: generic term for toxic or radioactive substances in solid, liquid, aerosolized, or gaseous form that may be used or stored for industrial, commercial, medical, military, or domestic purposes.

Toxicity: the potency or degree to which a toxic substance or compound can damage an organism.

Toxicological Effects: can be a). Additive: Situation in which the combined effect of two chemicals is equal to the sum of the effect of each agent given alone (e.g., 3 + 3 = 6); b). Synergistic: Situation in which the combined effect of two chemicals is much greater than the sum of the effect of each agent given alone (e.g., 3 + 3 = 20); c). Potentiated: Situation in which one substance does not have a toxic effect, but when it is added to another chemical, it makes the latter much more toxic (e.g., 0 + 3 = 10); or d). Antagonistic: Situation in which two chemicals given together interfere with each other's actions (e.g., 3 + 3 = 0, or 0 + 3 = 1). Antidotes to poisons fall into this last class.

Toxicology: branch of biology and medicine concerned with the study of the adverse effects of chemicals on living organisms.

Toxin: any poisonous substance from microorganisms, plants, or animals. See C16 – Toxins.

Toxoid: modified bacterial toxin that has been rendered nontoxic (commonly with formaldehyde) but retains the ability to stimulate the formation of antitoxins (antibodies) and thus producing an active immunity. Examples include botulinum, tetanus, and diphtheria toxoids.

Tracheitis: inflammation of the lining membrane of the trachea.

Training Agents and Compounds: agents or simulants authorized for use by the military in training to enhance proficiency for operating in a chemical environment.

Triage: system of sorting casualties when the number of victims overwhelms available medical resources. It is designed to produce the greatest benefit by giving treatment to those who may survive with proper treatment and not to those who have no chance of survival or to those who will survive without assistance. A Priority 1 (also known as Class 1, Immediate, or Red) casualty is one who needs treatment very quickly to save their life. A Priority 2 (also known as Class 2, Delayed, or Yellow) casualty is one who may have a major injury, but who can wait for care. A Priority 3 (also known as Class 3, Minimal, or Green) casualty is one with a minor injury who can wait for treatment, and then can be treated quickly. A Priority 4 (also Class 4, Expectant, or Black) casualty is one who cannot be saved with the resources on hand or who probably could not be saved even if transported to a major hospital.

Tubercle: in botany, a tuber-like swelling or nodule.

TWA: see Time-Weighted Average.

Ultra-high Vacuum: definitions can vary, but typically considered vacuum with a pressure range between 10^{-14} psi and 10^{-11} psi.

Ultrasonic Nozzle: device used, primarily in pharmaceutical spray dryers, to atomize a liquid feedstock by subjecting the liquid feed to high-frequency vibrations.

Ultraviolet Light: light waves shorter (and therefore with higher energy) than the visible blue-violet waves, but longer (and therefore with less energy) than X-rays. Ultraviolet light is very effective in killing microorganisms.

UN Number: four-digit codes used to identify hazardous substances and articles in transport. The master list of UN numbers is the dangerous goods list of the united nations recommendations on the transport of dangerous goods.

Uncertainty Factor: factor used to operationally derive a standard or a reference dose from experimental data. Usually expressed as a multiple of 10, these factors are intended to account for a). variations in sensitivity among members of the human population; b). uncertainty in extrapolating animal data to humans; c). uncertainty in extrapolating from data obtained in a study involving less-than-lifetime exposure; d). uncertainty in using "lowest-observed adverse effect level" data rather than "no-observed adverse effects level" data; e). inability of any single study to address adequately all possible adverse outcomes in man.

Unitary Chemical Munition: munitions designed to contain a single-component chemical agent for release on a target.

Univoltine: having only one brood or generation per year.

Upper Explosive Limit: highest concentration, or percentage in air, of a gas or vapor that will sustain or support combustion, when an ignition source is introduced. Above this limit the vapor is too rich to burn.

Upper Flammability Limit: see Upper Explosive Limit.

Urticant: compounds that produce instant, almost intolerable pain and cause immediate local destruction of skin and mucous membranes. Sensations caused by exposure to these agents range from mild prickling to almost intolerable pain resembling a severe bee sting. See C05 – Urticants.

Urticaria: eruption of itching wheals, usually of systemic origin. May be due to a state of hypersensitivity to foods or drugs, foci of infection, physical agents (heat, cold, light, friction), or psychic stimuli.

Vaccine: biological preparation that enhances the vaccinated subject's immunity to a particular disease. A vaccine stimulates the subject's immune system with a weakened pathogen so the immune system becomes prepared to recognize and destroy the virulent disease-causing pathogen.

Vacuum Distillation: method of distillation where the pressure is significantly reduced. This either allows separation of compounds that normally have very similar boiling points or allows for distillation when the compound would normally thermally decompose before it begins to boil.

Vapor Density: comparison of the weight of a gas to an equal volume of air. Air is given a relative vapor density of 1. Substances with relative vapor densities less than 1 are lighter and will rise in air. Substances with relative vapor densities greater than 1 are heavier and will sink in air, collecting in low-lying areas or close to the ground.

Vapor Pressure: pressure exerted by vapor against the atmosphere and is dependent on temperature. The greater the vapor pressure, the faster a material will evaporate. When the vapor pressure reaches atmospheric pressure, the material boils.

Vapor: gaseous form of substances in equilibrium with its solid or liquid state.

Vaporization: change of a substance from a liquid into a gas.

Vasculitis: inflamed blood vessels.

Vector Amplification: situation that occurs when an infected vector (e.g., mosquitoes, ticks) transmits a disease to a new host, who in turn passes the disease on to multiple new, uninfected vectors. These new vectors then continue the cycle, amplifying the rate and spread of the infection.

Vector: organism – typically an insect such as a mosquito, flea, or tick – that carries and transmits a disease from one host to another, often via a bite.

Vegetative Cell: bacteria in their growing state.

Velogenic: describes the virulence of a virus that generally causes lethal infection in its host.

Venom: poisonous mixture of toxins and other natural chemical produced by animals (e.g., snakes, spiders). See also Toxin.

Vesicant: also known as blister agents, they affect the eyes, lungs and skin by destroying cell tissue and causing inflammation and blisters. In addition, vapors also attack the respiratory system with the most sever impacts on the upper tract. Many vesicants are insidious in action and there is little or no pain at the time of exposure. The development of casualties is somewhat delayed (6–24 hours). See C03 – Sulfur and Nitrogen Vesicants and C04 – Arsenic Vesicants.

Vesication: process of forming of blisters.

Vesicle: blister with a diameter of less than 0.5 cm.

Viability: ability of a microorganism to survive, grow, and reproduce.

Viremia: presence of virus in the bloodstream.

Virion: individual virus particle.

Virulence: ability of an organism to invade the tissue of a host. It is the capacity of a microorganism to produce disease or a measure of the severity of the disease that is produce.

Virus: infectious agent generally composed of genetic material packaged within a protein coat. Viruses must infect host cells in order to replicate and spread the infection.

Viscosity: property of a liquid that describes its resistance to flow. Liquids with high viscosity tend to be thick do not flow easily. Viscosity is inversely proportionate to temperature.

Vittae: streak or band of color.

Volatility: tendency of a chemical to vaporize or give off fumes. The volatility of an agent varies with temperature and is expressed as the weight of vapor present in a given volume of air. Volatility is often confused with relative persistency. See Explanatory Notes for additional details.

Vomiting Agent: compounds that cause vomiting and may produce coughing, sneezing, nasal discharge, tears, and pain in the nose and throat. See C14 – Vomiting/Sternatory Agents.

V-Series Nerve Agent: type of nerve agent first discovered in the 1950s that are more toxic and persistent than the G-series agents.

Weaponization (for pathogens): involves processing pathogens into a material that can be easily dispersed at the target. Agents are either wet (liquids with high concentrations of the pathogens) or solids (either spores or freeze-dried vegetative pathogens). Solids must be milled to the appropriate size (usually 1–10 microns) and may also be coated with anti-static materials.

Weapons of Mass Destruction: United States defines this term to mean any destructive device that includes explosive, incendiary, or poison gases; bombs; grenades; rockets having a propellant charge of more than four ounces; missiles having an explosive or incendiary charge of more than one-quarter ounce; mines; solid projectiles with a diameter of more than one-half inch (50 cal.) in diameter; any weapon that releases toxic chemicals or their precursors, a disease or organism, or radiation at levels dangerous to human life.

WMD: see Weapons of Mass Destruction.

X: generic chemical symbol for a halogen (i.e., fluorine, chlorine, bromine, or iodine).

Xylem: vascular tissue plants use to transport water and minerals.

Zoonosis: disease of animals that may be transmitted to man.

Zoonotic: describes a pathogen that is capable of being transmitted from animals to humans or from humans to animals. The latter is sometimes referred to as reverse zoonosis.

Zootoxin: toxin or poison of animal origin such as the venom of snakes, spiders, and scorpions.

Appendix A: Agent Delivery

This appendix provides an overview of the characteristics of the various means of releasing chemical and biological agents onto a target, and the relationship of the design of the munition to the potential class of agent it could disperse. Examples in each category provide some historical perspective of the evolution of these weapons over the decades since World War I and also give a perspective of the size of the weapons that were developed; ranging from millimeter diameter pellets used in assassinations up through ballistic missiles. The broad design categories discussed include the major successful methods developed over the years. This list is by no means all-inclusive. It does not, for example, include designs that were thoroughly researched but then never made it into the field, whether for practical, economic, or political reasons.

Since simply delivering an agent to a target does not fully explain the difference between a successful attack and a failure, this appendix begins with a discussion of the characteristics of the agent cloud immediately after a release. This is followed by a discussion of the impacts of the local environment, particularly weather and terrain, on the movement the cloud, and persistence of the agent.

THE AGENT CLOUD

The primary agent cloud generated by a chemical or biological weapon will be a gas, an aerosol of either solid particles or liquid droplets, or a mixture of gas and aerosol. Immediately after release, clouds of gases behave like fluids. The buoyancy of this early cloud is based on its relative density as compared to air. Initially, gases with greater relative densities tend to stay close to the ground and collect in low places such as trenches or basements. If sufficient gas is available, it is even possible to displace the oxygen to the point where there is a suffocation hazard. These heavier gases also follow the contours of the land and tend to flow downhill just as a liquid would. Clouds with relative densities less than air will more rapidly rise off the ground and dissipate into the surrounding air.

If left undisturbed by the wind, gas clouds will still spread over time because of the diffusion of the individual molecules through the surrounding air. The rate of this diffusion is related to molecular weight of the chemical that makes up the gas; heavier gases will move more slowly than lighter ones. If the gas is confined to a specific area, such as a sealed room, the gas will ultimately diffuse and spread to uniformly fill the entire area, regardless of its relative vapor density.

Although particulate aerosols behave in a manner similar to gases, they are heavier and these types of clouds tend to retain their forms longer. Particulate matter, solids or liquids, in an aerosol can range in size from an aerodynamic diameter of greater than 100 microns down to 0.001 microns. The smallest particle visible to the unaided human eye is approximately 50 microns in diameter. The disposition of individual particles within the cloud is affected by their aerodynamic diameter. Particles 100 microns or larger quickly precipitate out of the cloud with a ballistic-like trajectory. As the aerodynamic diameter of a particle decreases, the period it can remain suspended increases. For example, a spherical particle 4 microns in diameter will take approximately 30 minutes to settle 1 meter in still air, whereas a particle with a diameter of 1 micron can take up to 8 hours.

Size also affects how far particles can penetrate into the human respiratory system. Particles with a diameter of less than 100 microns can be inhaled, but only particles with a diameter of less than 25 microns can be drawn into the lower airway. To move into the gas-exchange region of the lungs containing the alveoli, particles must have a diameter of no more than 10 microns. Particles smaller than 0.5 microns are easily respired and do not remain in the lungs.

Agents that condense out of the cloud can evaporate over time to create a secondary vapor cloud. The concentration of vapors will change with the ambient temperature; as the temperature increases, the concentration of vapor above the liquid or solid increases. Because of their high vapor densities, these secondary clouds tend to remain near the ground surface. The risk of re-aerosolization of nonvolatile particulates is generally minimal; however, it is still possible and depends on the size of the particle, its electrostatic characteristics, and the nature of the surface it has settled on.

ENVIRONMENTAL IMPACTS

For the first 30 seconds, the size, shape, and movement of the agent cloud are determined by the design and characteristics of the munition that created it. Once the cloud temperature reaches equilibrium with the surrounding air, the primary factors influencing the shape and dynamics of the cloud become wind and terrain. The greater the density of the initial cloud, the longer it resists these influences.

If the wind is calm, the cloud will remain near the point of release. If not, the wind pushes the cloud along the ground in a rolling motion; lower wind speeds produce higher dosages but smaller area coverage. From an offensive point of view, optimal downwind coverage occurs when the wind speed is between 5 and 12 kilometers per hour. If the wind is greater than 20 kilometers per hour, it tends to rapidly tear apart and disperse the cloud. The leading edge of the primary agent cloud can travel at up to 1.5 times the speed of the wind while the trailing edge moves at approximately one-half the wind speed. Hence, the cloud lengthens as it travels, dispersing and breaking up over time and distance. Horizontal air currents cause the cloud to spread sideways or meander, while air turbulence tends to stretch, tear, and dilute the cloud. Convection currents, created when solar radiation heats the earth, can cause the cloud to rise and break up. Cities, due to the construction materials used, tend to absorb a great deal of solar radiation, resulting in strong upward thermals during the day. Cities also retain this heat longer into the evening than agricultural or hinterlands.

As the cloud moves, the base suffers drag as it interacts with the ground. This slows the base with respect to the upper levels and causes additional elongation along the axis of travel. The amount of drag varies depending on the nature of the terrain: flat lands with only grasses induce 10% drag while gently rolling terrain with crops, bushes, or scattered trees can cause approximately 20%. Broken, undulating or highly vegetated terrain will not only produce significant drag, but will also increase wind turbulence and create cross currents that help break up the cloud and substantially reduce its downwind travel.

Predicting the path and dissipation of the cloud in an urban area is very complicated as it is a function of building spacing; relative heights of the buildings to each other; individual building height, width, and shape; as well as the angle the wind is blowing in relation to a particular cluster of buildings. Buildings tend to increase the velocity of the wind near them and cause gusting that can create significant wind anomalies due to the complicated flows that develop. Wind direction and velocity at ground level may not reflect the prevailing wind near the top of the building. Single or even multiple turbulent patterns can develop intermittently between adjoining buildings and redirect the cloud. Cross currents can cause fragments of the cloud to move up side streets that are perpendicular to the apparent axis of the prevailing wind pattern. In some instances, counterflows can cause portions of the cloud to move short distances upwind in recirculation zones created along the sides or top of a building. If the current is slow moving, the cloud can become trapped in alleys or streets. Recessed areas in buildings such as courtyards and alcoves can also trap and hold a segment of the cloud for a short time after the main plume has passed.

Buildings themselves can also become temporary cloud repositories. Gases and particulates will enter buildings based on how fast air on the outside of a building exchanges with air on the inside. Excluding forced air ventilation systems, this air exchange is a function of the quality of buildings construction, particularly around openings such as doors and windows. In buildings with a low level of air exchange, it takes longer for the cloud to penetrate into the structure and build up a noteworthy concentration. However, once inside, it conversely takes longer for the building to clear itself of the residual agent without the aid of mechanical ventilation.

Other weather factors also affect the concentration and diffusion of the primary cloud as it travels. Cold air temperatures, especially less than 32 degrees Fahrenheit, can significantly attenuate a cloud because the agent will condense into droplets or crystals that fall to the ground. While this reduces that amount of agent in the cloud, it subsequently increases the amount and persistence of local ground contamination. Even if the air is warm, contact with cold surfaces will also condense agent out of the air.

Rain can flush the agent from the air and wash contamination from the soil. Depending on the properties of the agent, this runoff can either spread contamination farther over and into the soil, or the agent may react with the water and decompose. Even high humidity can hydrolyze some agent vapors if they readily react with water. For example, vapors of either phosgene (C10-A003) or lewisite (C04-A002) will rapidly break down if the humidity is greater than 70%.

Agents that have precipitated or condensed out of the cloud can later evaporate to produce a secondary cloud. Depending on agent volatility, these secondary clouds can have a relatively low concentration or can reach levels nearly as high as the original cloud. In either case, this new cloud of agent vapors tends to remain near the ground surface due to the high vapor density of the agent. As the temperature of the contaminated surface increases, the evaporation rate will increase proportionately. Wind blowing over the contaminated surface will also increase the rate of agent evaporation. Evaporation due to the wind is proportional to both the air speed and the surface area of the contaminated object.

Agent vapors, and any liquid spread by the munition itself, will also be absorbed into porous materials. These captured agents will begin desorbing once all the surface contamination has evaporated. This migration of agent back to the surface will prolong the life of the secondary cloud, albeit at a significantly diminished concentration.

In contrast to the outdoor environment, an indoor release poses a greater hazard because the building circumvents most of the environmental issues discussed above. Also, the integral heating and ventilation systems will rapidly spread the airborne contamination throughout adjacent section of the building. There is a significant amount of porous material, such as carpets, wood, and wallboard, available indoors to absorb liquids and vapors and increase the persistence of the contamination. The confined nature of the building interior also focuses the release, trapping the agent and maximizing its effects. Individuals occupying the buildings may be trapped in, or have to pass through, contaminated areas in order to exit the structure.

WEAPON DESIGN

Whether chemical or pathogen, an agent dispersed by a weapon must follow the basic laws of chemistry and physics that dictate the properties for its physical state of matter. If, for example, an agent must be inhaled to produce the desired physiological response and it is not a gas or vapor, then some form of energy must be introduced to cause it to become airborne so it can be inspired. This is usually accomplished by supplying either an outside physical or mechanical force, such as pressure from an attached cylinder of inert gas or using explosives to propel the material into the air, or through thermal aerosolization by either direct heating or using hot gases to distill the agent. While chemical agents can be solids, liquids, or gases, biological agents are living entities and, despite their small size, never exist as gases or vapors. To become airborne, they must be transported on small solid or liquid particles in the form of an aerosol. Despite these restrictions, it is possible to manipulate or modify the physical characteristics of an agent and change its properties. Examples include such things as increasing the persistence of a liquid by adding thickeners to reduce its vapor pressure or coating solid particulates with a material that resists static buildup enabling them to remain airborne for longer periods.

Chemical and biological agents are essentially poisons, and an appropriate dose must be delivered by the weapon to achieve the desired effect, which is typically either death or incapacitation. The implications of delivering an insufficient amount of agent are obvious. If excess agent is used and the desire is to produce fatalities, then it is simply inefficient and may result in the need for extensive decontamination if the area is to be occupied by friendly forces subsequent to the attack. If the desired result is limited to incapacitation of the target population, however, then delivery of excessive agent can result in unintended consequences including physical trauma, permanently altered mental capacity, or even death. While the primary routes of exposure on the battlefield are through inhalation and dermal exposure, some weapons designers have attempted to introduce an injection pathway by incorporating contaminated shrapnel into a traditional fragmentary weapon. For example, during World War II the Japanese developed the HA bomb, which contained 1,500 metal pellets immersed in 0.5 liters of an emulsion of either *Bacillus anthracis* (C17-A001) or *Clostridium tetani* (C17-A016), with the goal of increasing the lethality of otherwise superficial wounds.

Chemical and biological weapons are intended to cause casualties by releasing their contents in a controlled manner. They are designed, in general, to either produce immediate casualties by releasing an inhalable cloud or to create a percutaneous and potentially long-term exposure hazard by contaminating the target area. However, they are rarely successful in selectively achieving either of these goals and actually produce a combination of results that varies according to the characteristics of the agent and the specifics of the weapon. No matter the design, there is typically some contamination directly around the point of release and some aerosolization of at least a fraction of the agent.

A chemical or biological weapon can be designed to release its contents in all directions simultaneously or in a single direction like a syringe. Because aerosols and vapors can penetrate and spread throughout an area, the device does not need to have line-of-sight access to the intended target. The basic components of these weapons are the agent, a container, a means of dispersing the agent, and an initiator. The agent itself must be stable during storage, in a physical form that allows for effective dissemination, able to survive the mechanism of release, and minimally affected by the environmental conditions at the target. The container can be sacrificial and destroyed during the release, or it can be recoverable, such as a gas cylinder. In addition to the physical properties of the agent itself, other mechanisms used for dispersal include explosives, thermal distillation, and high-pressure or mechanical aerosolization. In the case of some pathogens, vectors such as mosquitoes, fleas, and ticks are also a potential means of delivering the agent. The initiator is comparable to any other remote controlled or time delayed device.

Efficiency of munitions is typically expressed in either of two ways. The first is the ratio of mass of agent to mass of the weapon. This method was particularly useful for militaries when evaluating the logistics of a mission, as when comparing the payload a bomber could carry and the actual amount of agent that payload would deliver to the target. For example, the AN-M79 1,000-pound bomb was in the U.S. arsenal after World War II. It was filled with either phosgene (C10-A003), cyanogen chloride (C07-A003), or hydrogen cyanide (C07-A001); all nonpersistent agents. Due to the difference in density of the various liquified agents, the weight of the munition varied based on the contents. Using phosgene, each bomb was filled with 190 kilograms of agent for a total weight of 430 kilograms. The efficiency of this configuration would therefore be 44%. When it was filled with cyanogen chloride, a bomb contained 160 kilograms of agent and the total weight was 400 kilograms, for an efficiency of 40%. The hydrogen cyanide filled bomb was the least efficient. Each bomb contained only 90 kilograms of liquified agent for a total munition weight of 330 kilograms, resulting in an efficiency of only 27%.

The second method is an expression of the ratio of the mass of agent that will be delivered in the desired form (i.e., aerosolization versus contamination) compared to the mass of agent that will be left in the munition, destroyed during dispersal, driven into the ground upon detonation, etc. Most military chemical munitions designed to create an aerosol are estimated to be approximately 90%

efficient. Improvised chemical munitions can be expected to achieve less than 40% efficiency. Military biological munitions, other than sprayers, are typically less than 4% efficient in aerosolizing viable, living pathogens. Sprayers, operating under optimal conditions, can achieve approximately 25% efficiency when dispersing biological agents suspended in a liquid media, and approximately 40% efficiency with dry, powdered agents.

Munitions can range in size from containing thousands of kilograms of agent down to containing mere grams, and include missiles, bombs, rockets, sprayers, mines, grenades, submunitions, and bullets. Weapon systems and the size of the munition employed are selected based on such considerations as the intent of the attack (i.e., producing immediate casualties or causing area contamination), target characteristics (e.g., open terrain, urban environment, a bunker), and meteorological conditions including temperature, wind, humidity, cloud cover, and precipitation.

Initially, chemical weapons relied on the weather to actually deliver the agent to the target. During World War I, cylinder attacks were delayed for days or even weeks waiting for the wind to blow in the right direction. Even using artillery shells, early belligerents often added a smoke-producing component to the toxic payload to allow observers to correct the bombardment based on the way the wind was moving the gas cloud over the target. Some modern systems still rely on wind to assist in disbursing the agent payload over the target. For example, both the bigeye bomb and cruise missiles were designed to disperse their agent payloads while in flight over the edge of a target. This would create a line of contamination that traveled with the wind before it settled down and blanketed the area. Most modern systems, however, deliver the payload directly to the target.

In many cases, to evenly distribute an agent over a large area it is more effective to use multiple smaller ordnance, or a weapon that distributes a payload of submunitions, than it is to use a single large shell. This also minimizes the immediate impact of meteorology on the agent cloud. Consider the results of throwing a single gallon bag of paint from the top of a building versus simultaneously throwing 16 cup-sized bags that spread out over a larger area before impact. While the density at the point of impact of the gallon bag is greater, the area impacted, as well as the uniformity of the coverage, is significantly less than with the multiple smaller bags.

The primary weapon systems that have been developed to accomplish this are multiple launch systems and cluster bombs. Multiple launch systems are weapons platforms with a number of barrels that discharge their munitions, either simultaneously or in a rapid controlled sequence, to saturate the target area. The advantage of modern military versions of these systems over traditional artillery batteries is the ability to rapidly inundate a target area using a minimal number of weapons, which can then be rapidly relocated to prevent counterbattery fire. Examples include:

- The Livens projector, employed by the Allies during World War I, was a simple smoothbore single-shot mortar, consisting of a tube, a base plate, a shell, and a smokeless powder propelling charge. The projectors were dug in behind the front lines and fired simultaneously in batteries containing thousands of individual tubes. They were a saturation weapon that was not designed to be accurate. It had a wide dispersion of shots in both range and deflection with an unpredictability of where individual drums would land due to the fact that they tumbled in the air and had an erratic flight path. They delivered the highest concentration of gas on a target of any weapon system used during the war. At the point of impact, the concentration of gas could rapidly overwhelm the filtering capability of enemy gas masks. Gas densities even reached levels that displaced the oxygen in dugouts or low areas to the point where soldiers were asphyxiated instead of poisoned. Their major disadvantages were lack of mobility and the time required to dig in the tubes. At the close of the war, the Allies were developing metal sledges, each containing 16 loaded projector tubes and towed into place behind Mark IV tanks, that allowed soldiers to prepare and launch an attack within 4 hours.

- During World War II, the Germans fielded the Nebelwerfer 41, a rocket system consisting of six 15-centimeters barrels arranged in a circular pattern and mounted on a wheeled cart. The rockets had a range of approximately 6 kilometers. All the rockets were discharged in series over a 5-second interval. The system could be reloaded and prepared to fire again in 90 seconds. Although a nerve agent warhead was available for the weapon, it was never used during the war.
- After World War II, the United States adapted the T66, a 4.5-inch multiple rocket launcher with 24 tubes and a maximum range of approximately 5 kilometers, to fire rockets with chemical warheads containing either 1.5 kilograms of GB (C01-A002) or 1.8 kilograms of HD (C03-A001). The rockets were equipped with a point fuze that exploded on impact.
- Some nonmilitary multiple launch systems that disperse lachrymatory agents for crowd control include the Thunderstorm MBL12 38-millimeter launcher system (which can disperse 36 tear gas capsules in one single burst), the CTS Venom (a modular launching system that accepts three cassettes, each loaded with ten 37-millimeter cartridges that are fired in immediate succession), and the Cougar 12 (a 12-barreled launcher that fires 56-millimeter munitions, singularly or in salvos of either 4 or 12).

Cluster bombs are essentially large containers of bomblets that are detonated above the target, discharging the bomblets over a wide area. Just prior to or upon impact, the bomblets release their contents to generate a large, uniform area of contamination. Examples include:

- The M34A1 was a United States cold war era cluster bomb that contained 76 M125A1 ten-pound bomblet submunitions; each consisting of a body, payload of 1.2 kilograms of nerve agent GB (C01-A002), parachute, 250 grams tetryl burster, and fuze. At the appropriate altitude over the target, the M34A1 opened and ejected the submunitions. After a brief free fall, the parachute contained on each submunition would open, orienting them and slowing their decent to prevent cratering into the ground. The widely dispersed bomblets exploded on impact releasing the agent payload in a horizontal aerosol over the target.
- The CBR-16A/A dispenser system consisted of the SUU-13/A cluster bomb and 1,280 BLU 50/8 bomblets. Each bomblet was the size of a D-cell flashlight battery. The cluster bomb discharged the bomblets at an altitude that allowed them to rain down widely dispersed over the target. As they were discharged from the cluster bomb, a 5-second pyrotechnic delay fuze in each bomblet was ignited and burned down until it in turn ignited the blend of BZ (C12-A001) and pyrotechnic mixture. The hot gases from the burning payload were expelled through a small orifice causing the bomblet to skitter over the ground, increasing the overall dispersion of the aerosolized agent. The bomblets would continue to burn and disseminate agent for approximately 17 seconds.

FRANGIBLE CONTAINERS

The simplest type of device is a frangible container that disperses its contents when it impacts a hard surface and breaks open. This type of device is not pressurized, other than the vapor pressure from the chemical agent itself, nor does it have an explosive burster to assist in dispersal. Containers are typically glass, thin metal or, in some cases, even paper. The concentration of the primary cloud created when the container ruptures depends on the velocity of the container and the physical/chemical properties of the agent filler (e.g., liquefied gases, liquids with a high vapor pressure, finely powdered solids). The agent dispersal pattern is dependent on the trajectory of the container. Agents released from containers traveling with a trajectory perpendicular to the target have a circular dispersal pattern with the greatest concentration of agent at the point of impact. Otherwise, agents are released with an elliptical or tear-drop shaped dispersal pattern flowing along the direction of flight. The greatest concentration of agent is again at the initial point of impact, while the length and density of the elliptical contamination is proportional to the angle and velocity of the container prior to impact. These types of devices can be used with all classes of agents. Examples of frangible weapons include:

- Drums filled with 30 to 90 kilograms of sulfur mustard (C03-A001) were dropped from Italian aircraft during the war between Italy and Ethiopia in 1935. The drums ruptured on impact, creating puddles of contamination.
- Just before World War II, the United States experimented with the use of commercial 1-gallon rectangular metal cans that were filled with sulfur mustard (C03-A001) and dropped from low flying aircraft at high speed. The cans ruptured on impact and were found to give excellent distribution of the agent.
- Japanese World War II hand grenades resembling a glass-baseball filled with liquified hydrogen cyanide (C07-A001).
- During World War II, the Germans stockpiled glass bulbs filled with chemical agents, called bodenkugeln, that were intended to be used in minefields or as booby traps.
- Light bulbs filled with *Bacillus subtilis* (C17-A003), a biological warfare simulant, were broken in the New York city subway system as part of a 1966 test by the U.S. military. The wind created by the movement of the trains carried the spores throughout the subway system.
- The XM28 bagged agent dispenser used by the United States during the Vietnam War consisted of a container carried under a UH-1 helicopter that was filled with hundreds of paper bags filled with micro pulverized CS2 (C13-A009) powder. When activated, the device discharged the bags through a trap door and they fell to the ground where they burst on impact. This created a significant dust cloud along with residual contamination over an area of up to 30 meters wide by 210 meters long.
- In 2000, the U.S. Navy patented a frangible projectile, similar to paint balls, that would break open on impact delivering a variety of riot control agents. The intent was to allow soldiers or police to target specific individuals during the melee of a riot. As designed, the projectile was effective out to a range of approximately 50 meters.

PRESSURIZED CYLINDERS

The first effective means of delivering chemical agents employed during World War I was pressurized cylinders. The most effective cylinder gas of World War I was a mixture of 50% chlorine (C10-A001) and 50% phosgene (C10-A003). Agents effectively dispersed by this method are limited to compressed gases, liquified gasses, or fine particulates of agents, either liquids or solids, suspended in an inert carrier gas. A shortcoming of cylinders is that as the gas is ejected, the tank cools causing condensation or even ice to form at the outlet. This can significantly decrease or possibly stop the discharge of agent from the cylinder.

Cylinders used during World War I varied in size from standard commercial bottles, approximately one meter tall and weighing nearly 90 kilograms when full, to smaller more maneuverable ones that were just over 500 centimeters tall and weighed just over 23 kilograms when full. The Germans employed cylinders that contained approximately 20 kilograms of gas. They found it took 10 cylinders per meter of front to create an effective gas cloud. As a general rule, the clouds were effective to a depth equal to the width of the front of the total discharge.

The M-1A1 portable chemical cylinder was developed by the United States shortly after World War I. It was 20 centimeters in diameter, approximately 50 centimeters long, and weighted 8 kilograms empty. It was capable of carrying up to 14 kilograms of liquefied gas depending on the agent selected. It was fitted with a carrier system consisting of two band-iron hooks formed to fit a man's shoulders. The bands were pointed so the cylinders could be suspended on a parapet wall. It was equipped with a nozzle that not only eliminated the need for discharge piping, but was also designed to limit the potential for condensation or freezing as the gas left the cylinder.

CHEMICAL REACTION DEVICES

Another rudimentary type of dispersal device relies on mixing two or more chemicals to create a toxic cloud that then flows out of the weapon into the surrounding area. This is basically the same concept used when executions were conducted employing a gas chamber. A container of each of the reactive components is placed within an outer shell. A burster, either a mechanical device or a small explosive charge, ruptures the contains and allows the contents to mix and react. Examples of this type of weapon include:

- A British World War I experimental aerial bomb that produced arsine (C08-A001) by mixing a powdered magnesium arsenic alloy with sulfuric acid when it hit the ground. The bomb was an attempt to overcome the tendency for the highly flammable gas to ignite if an explosive burster was used.
- In 1995, shortly after Aum Shinrikyo cult operatives shut down the Tokyo subway by releasing sarin (C01-A002), other members of the cult left binary devices containing a cyanide salt and dilute sulfuric acid in subway restrooms. A timer system was use to mix the ingredients to produce hydrogen cyanide (C07-A001).
- The mubtakkar was an improvised device that was reportedly developed by al-Qaeda for use in subway systems. It consisted of two or more containers sealed inside a perforated outer container. In one version, one of these inner containers was filled with a cyanided salt and the other with an acid. In another, the internal containers were filled with the acid and the cyanide salt was simply poured around them filling the outer shell. A radio or timed fuse was used to break the inner containers, allowing the two ingredients to mix and produce hydrogen cyanide (C07-A001) that then flowed out of the perforations into the surrounding environment.

EXPLOSIVE DISSEMINATION

Devices employing explosive dissemination can be directional or nondirectional depending on the design of the munition and the placement of the explosive charge. The agents themselves can be located in an isolated container with an explosive charge attached, or they can be intermixed with the explosive or with the shrapnel. Variation in the properties of the explosive chosen for the burster, such things as high or low order, brisance, and heat of detonation, can significantly affect the device efficiency. The amount, type, and placement of the explosive charge also determines the ratio of area contamination to aerosolization of the agent, as well as the size of the particles within the aerosol cloud. Regardless of the design of the device, it is highly inefficient to use explosives to disseminate biological agents as the heat, pressure, toxic gases, and sheer forces generated by the blast reduces the recovery of viable pathogens to between one-tenth and 4% of the payload.

In order to achieve an effective aerosol cloud that poses a respiratory hazard, the explosives must be powerful enough to disperse the agent into particles of the appropriate size, typically 1 to 10 microns, but not so large or powerful as to destroy the agent through heat or shock. Clouds created by an explosive initially expand due to the force and heat of the detonation, then grow cooler and heavier before settling back toward the ground. This is referred to as pancaking. If the vapor density of the agent is less than that of air, these clouds will quickly rise back up off the ground and dissipate. If the vapor is heavier than air, then the cloud tends to flatten and flow over the ground in a downwind, downhill direction.

Conversely, if the intent of the device is to produce area contamination instead of an aerosol, the explosive charge will be small in comparison to the mass of the agent. The actual ratio will determine if the resulting coating is thick and confined to the immediate area around the detonation or thinner with a greater area of impact. In either case, the contamination will not be evenly spread over the target area; there may even be some sections that are not contaminated. In general, the heaviest concentration of agent will be found at the epicenter of the explosion with the contamination reducing exponentially as the distance increases.

Examples of explosive dispersal weapons include:

- The Livens projector had a shell that was a light, thin-walled cylinder that was rounded at both ends and had a burster that ran through the longitudinal axis of the shell. Since the shell flew end-over-end in flight, it was equipped with either a time delay fuse or an all-ways detonator. It contained approximately 14 kilograms of liquified phosgene (C10-A003). Fired electronically, the range was determined by the amount of smokeless powder used and extended out to approximately 1,300 meters.

- After World War I, the U.S. Army standardized the one-gallon chemical land mine. It consisted of a commercial one-gallon rectangular metal can filled with 4.5 kilograms of HD (C03-A001). Originally, a small paper cartridge containing pressed nitrostarch was used as the burster. This was modified to several wraps of detonating cord around the can. Depending on the amount of detonating cord employed, the mine effectively contaminated an area between 14 and 23 meters in diameter.

- During World War II, the United States tested a modified M74 base-ejection incendiary bomb filled with powdered ricin (C16-A031). On impact, an explosive charge in the nose of the bomb drove the agent backward through the bomb and expelled it out the tail section. It was eventually abandoned because of uneven dispersion of the agent due to the formation of aggregate clumps in the particulate cloud.

- The M-25 hand grenade was developed by the United States toward the end of World War II for use at prisoner-of-war camps. It was plastic (to minimize the fragmentation hazard), spherical and approximately 7 centimeters in diameter. It weighed approximately 225 grams and was filled with 80 grams of micropulverized CN (C13-A008). The bursting radius was approximately 4.5 meters and it was designed for point contamination instead of area coverage. Two modifications, the A1 and A2, were developed that updated the fusing, plastic body and payload capacity. The agent payload was also expanded to include either CS (C13-A009) or DM (C14-A003).

- After World War II, the United States standardized the M47 chemical bomb. It was 1.3 meters long and had a diameter of 20 centimeters. It had a central burster that ran the length of the body and employed a super-quick fuse to ensure minimal agent was lost in the bomb crater. It contained 33 kilograms (26 liters) of HD (C03-A001) that contaminated an area in a circular pattern with a radius of approximately 30 meters.

- The U.S. M23 chemical mine was a modified high-explosive M15 anti-tank mine. Most of the high-explosive filling was removed and replaced with 4.8 kilograms (7.6 liters) of nerve agent VX (C01-A017). It was circular, with a diameter of approximately 33 centimeters, and was 13 centimeters tall. It used approximately 0.4 kilograms of explosive to disperse the agent in a circular pattern approximately 19 meters in diameter. An antipersonnel modification was available that caused the mine to bound approximately 20 meters into the air before detonating. This modification increased the area coverage and created a greater downwind hazard.

- The Gshch-304 was a submunition for a cluster bomb or rocket developed and stockpiled by the former Soviet Union for dispersal of biological agents. These aluminum bomblets were round and approximately 12 centimeters in diameter. They weighted approximately 450 grams and used three small TNT burster charges to release the agent after deployment.

- The U.S. E134 submunition was very similar to the Soviet Gshch-304 except it was designed to carry nerve agent GB (C01-A002). Made of thin steel and approximately 11 centimeters in diameter, it weighed 1.5 kilograms and was filled with 500 grams of agent. It employed an RDX burster that activated on impact.

- The XM96 was used by the United States during the Vietnam War. It was a shoulder-fired, fin-stabilized 66-millimeter rocket, 53 centimeters long with a gross weight of approximately 1.4 kilograms. It contained approximately 200 grams of CS2 (C13-A009) that was dispersed as a particulate cloud when the rocket detonated.
- The U.S. MK 116 Mod 0, Weteye 500-pound chemical bomb had an aluminum body approximately 220 centimeters long weighing approximately 34 kilograms. The interior of the bomb was divided into three compartments separated by perforated baffles. There was a central burster that ran the length of the body through all of the compartments. The bomb could be filled with either 129 kilograms (129 liters) of agent VX (C01-A017) or 157 kilograms (142 liters) of agent GB (C01-A002). When filled with VX, the bomb was intended to cause persistent ground contamination and was configured with a low-explosive burster that limited the spread of the agent to a circular area with a radius of approximately 18 meters. When filled with GB, the bomb was intended to create an aerosol cloud that would travel with the wind across the target and was configured with a high-explosive burster that created a cloud with an initial radius of approximately 50 meters.

BINARY MUNITIONS

Binary explosive munitions are a unique subgroup of this category of weapons. The payload consists of two or more non-toxic compounds – either liquids or solids – that react to form a standard chemical agent when mixed. These components are stored in separate canisters or compartments within the munition. The barriers between these containers are broken and the components allowed to mix as the munition travels to the target. If the munition should detonate while the components are still segregated, the blast will not mix the precursors prior to dispersing them into the atmosphere and conversion to the final chemical agent does not occur.

These types of munitions are considered safer than those containing unitary chemical agents because of the non-toxic nature of the individual components. However, non-toxic is a relative term and many of these components are extremely hazardous in their own right, just significantly less so than the traditional agent they combine to produce. Another safety advantage comes from the ability to store the individual agent containers away from the balance of the munition. Thus, if there is a leak, a canister is far easier to deal with than an assembled munition.

These types of munition are more complex than unitary munitions in that they must be designed to rapidly facilitate mixing of the components at the appropriate time and also be able to withstand the heat and pressure generated by the exothermic reaction of the components. An example of how this latter concern can cause problems occurred when the United States was attempting to develop the BLU-80/B Bigeye bomb toward the end of the Cold War. Liquid precursor QL (C01-C064) was the first binary component and was stored in compartments within the main body of the munition. Just prior to loading the bomb onto an aircraft, a ballonet filled with powdered sulfur (C01-C061), the second binary component, was inserted into a holding well that extended the length of the bomb's longitudinal axis. As originally designed, the pilot would remotely rupture the two containers and initiate the chemical reaction that would begin to generated the crude VX (C01-A017) as he approached the target area. At the target, he would release the bomb and it would begin to discharge the VX through exhaust ports on either side of the main body.

Initial testing was conducted only with simulants to prevent release of actual nerve agent. During these tests, when the plane was flying at high altitudes, the low atmospheric temperatures caused the reaction to run smoothly and generated the simulated nerve agent in short order. However, when the bomb was tested at the lower altitudes where combat missions would actually occur, the higher ambient temperature caused a much more vigorous reaction and the internal pressure rapidly increased to in excess of 650 psi. The ballonet was forcefully ejected along with a mixture of the two staring materials and crude simulated VX, contaminating the aircraft and theoretically threatening the pilot. Development of the Bigeye bomb was cancelled before all of the technical issues could be corrected.

An example of a successful binary munition design was the U.S. M687 155-millimeter artillery projectile. It was a modified M485A1 base-ejecting illumination round that was loaded with two canisters of agent precursor – one filled with 6.6 kilograms (9 liters) of OPA (C01-C063) and one filled with 4.6 kilograms (3.3 liters) of DF (C01-C053). When the howitzer fired the projectile, inertial forces ruptured the containers and allowed the components to interact. The in-flight spin of the shell mixed the DF and OPA, creating a crude form of the nerve agent GB (C01-A002). Approximately 70% conversion was achieved after a 10-second flight time. Once on target, the detonator expelled the crude agent out the base of the projectile. For safety reasons, the M687 projectiles were stored and shipped with only the OPA canister in place.

SPRAYERS

Sprayers, also called dispersers, are a simple and highly effective means of delivering either liquids or finely powdered solids of all categories of chemical and biological agents. Delivery with these types of systems is, in fact, significantly less stressful on pathogens than most other methods. To disperse their contents, sprayers can utilize a simple hand-pumped mechanism, incorporate a mechanical system such as a spinning disk generator, employ a turbine jet venturi system, or make use of auxiliary high-pressure gas cylinders to expel their contents. These systems can be a small handheld unit, a backpack man-portable unit, or so large that they must be mounted on vehicles. Examples include:

- Handheld pump sprayers were used by the Rajneeshee religious cult in September 1984 to clandestinely spray *Salmonella typhi* (C17-A033) bacteria over salad bars at multiple restaurants in the Oregon town of The Dalles. The purpose of the attack was to influence the results of a local election by making residence too sick to go to the polls to vote. Approximately 750 people became sick as a result of the attack.
- The M3 backpack disperser was employed by the United States during the Vietnam War. It resembled a World War II flamethrower, consisting of twin agent tanks and a pressure tank mounted on a tubular frame equipped with shoulder straps. The operator used a disperser gun attached to the system via a hose. Each agent tank could hold up to 4.5 kilograms of powdered agent – typically powdered CS (C13-A009) – that could be delivered in bursts or in a single continuous stream lasting up to 30 seconds. It was usually discharged a minimum of 15 meters away from the target to prevent the rapid buildup of hazardous concentrations or cause excessive dermal injury.
- The vortex ring gun fires a short pulse of chemical-ladened high-pressure air similar to a smoke ring. The vortex, which occurs when the outer layer of a blast of air moving down the gun tube is dragged backward relative to itself until a large part of the flow swirls into a spinning doughnut, entrains agent from a reservoir near the end of the barrel before exiting the weapon. The intent is to deliver agent to a specific target rather than to a general area. None of these weapons have been produced.
- The AB2K uses high-pressure injection of a smoke solution containing Capsaicin (C13-A006) or CS (C13-A009) into a heat exchanger operating at 1,250°F to create an aerosol of agent that can flood an area out a range of approximately 30 meters. Aerosol particle are typically 0.2 to 0.4 microns.
- A trailer mounted unit was employed by the Japanese during World War II. The type 94-A gas sprayer was towed by a miniature tank and consisted of an elliptically shaped 260-liter tank with a cylinder of compressed air. The compressed air forced the agent out through a spray nozzle on the outside top-rear of the trailer. The sprayer was controlled from a panel board inside the rear hatch of the tankette, providing comparative safety for

the operator. It was used to contaminate an area with persistent agent HD (C03-A001) and could contaminate a path 6 meters wide and up to 100 meters long.

- During the 1950s, the United States conducted tests to determine the susceptibility of coastal cities to a potential attack with biological agents. Employing agent simulants, the military mounted systems on ships that either remained stationary off the coast to simulate a point-source attack or simulated a line attack by releasing the agent as they traveled in a line parallel to the coast. The results of the tests indicated that nearly 100% of the population in a targeted area would have inhaled an infectious dose of a pathogen using either pattern of attack.

Despite their relative simplicity, sprayer systems do have a number of inherent problems that commonly occur including pressure variations, leaks, and clogging of the discharge orifice. There are also operational requirements related to the agent itself that can severely affect their performance including the consistency, viscosity, and turbidity of liquids; and the particle size, clumping, and caking of powdered solids. Any of these issues can render a system unusable. An example of how critical these issues can become occurred in 1993 when the Aum Shinrikyo cult attempted a terrorist attack by spraying a broth containing *Bacillus anthracis* (C17-A001) from the roof of a building in Tokyo. On numerous occasions over several days, the nozzles of their sprayer system rapidly clogged with fine solid particles suspended in the broth forcing them to stop the attack, depressurize, disassemble, clean, and then reassemble their equipment. They eventually abandoned their efforts.

For the military, aerial sprayers have the widest application as these systems can cover the greatest amount of territory in the least amount of time without overly endangering friendly forces. Once the attack is complete, the aircraft can even jettison the sprayer system to minimize the amount of decontamination the aircraft requires and avoid potential contamination of its home runway. The military began experimenting with aerial spraying soon after World War I and quickly came up with operational systems. The civilian potential was also rapidly recognized and in 1921 the first agricultural crop-dusting application occurred when U.S. Army pilot Lieutenant John Macready used a modified Curtiss JN-6 biplane to spray a grove of catalpa trees with lead arsenate to kill sphinx moth larva infesting the trees.

For liquid agents released from an aircraft, the characteristics of the initial cloud is dependent on size of the droplets created by the sprayer as well as both the altitude and speed of the aircraft at the time of release. At higher altitudes or faster speeds, the agent has to be forced from the sprayer in larger drops to prevent excessive atomization. Also, the higher the altitude the more the discharge is susceptible to the velocity and direction of the regional wind currents as it precipitates to the ground. The greater the speed of the air current and the greater the angle that it is oblique to the line of flight of the aircraft, the wider the dispersal pattern.

From low-flying aircraft discharging liquid agents, the sprayer does not need to be pressurized and the agent can simply run out of the holding tank when the discharge vent is opened. The aircraft slip stream will break up the liquid, which then falls the short distance to the ground like rain or mist. In general, optimal conditions include deployment at altitudes of less than 150 meters, air speeds not exceeding 300 knots, and cross winds between 2 and 7 meters per second. However, these flying conditions can also expose the aircraft and crew to intense ground fire. For example, aircraft flying Operation Ranch Hand herbicidal spraying missions during the Vietnam War often returned to base with multiple bullet holes from small arms fire. On April 30, 1964, one C-123 earned the name "Patches" after landing with forty 0.50-caliber bullet holes stitched along the wings and body of the aircraft. Over the 3-year service life of this one aircraft as it flew defoliant operations, it was hit more than 600 times by enemy ground fire.

Examples of aerial liquid sprayer systems include:

- The AERO 14B airborne chemical spray tank was an externally mounted, pneumatically controlled spray system that was filled with approximately 320 liters of chemical agent, primarily VX (C01-A017) or thickened GB (C01-A002). It consisted of a nose assembly, a center section containing the agent reservoir, and a tail section with a discharge nozzle assembly. The nose assembly contained an 1,800-psi compressed air tank with a regulator that reduced the pressure to 100 psi for operation. The pilot activated a valve that released the agent through the discharge tube. Although the tank was reusable, it could be jettisoned in flight.

- The TMU-28/B spray tank was 4.5 meters long with a diameter of 57 centimeters and held approximately 600 kilograms of agent VX (C01-A017). A dissemination nozzle in the tail section of the tank was held in a retracted position during flight, then extended just prior to discharging the tank contents. When the system was activated, a small explosive charge opened a ram air scoop on the forward half of the spray tank that forced air into the tank and thereby forced the contents out the rear dissemination nozzle at a rate of approximately 75 liters per second. The slip steam created by the aircraft broke the liquid into a fine mist. Once empty, the spray tank was released from the aircraft.

Solid agents can also be dispersed by aerial sprayer system. For example, during the Vietnam War the United States used an M4 Disperser, which was often mounted in a helicopter. It was 0.8 meters long, 0.6 meters wide, 1.2 meters tall, weighed approximately 73 kilograms, and consisted of a pressure tank, regulator, and cone-shaped hopper. The hopper could be filled with up to 45 kilograms of CS1 (C13-A009), 54 kilograms of CN1 (C13-A008), or 54 kilograms of DM1 (C14-A003). The agent was released through a hose assembly attached to the landing struts and spread by the rotor wash. It took approximately 20 seconds to discharge the entire contents of the hopper. In addition to two pilots to fly the aircraft, it required a two-man crew to operate the system. At least one of the pilots had to wear a protective mask whenever they were spraying because air currents could carry the agent back into the aircraft.

To avoid the hazards to the flight crews of either accidental contamination acquired during the spraying operation or due to the threat from enemy ground fire directed at the slow, low-flying aircraft, spray systems were developed for rockets, cruise missiles, and gliding bombs. The aforementioned BLU-80/B Bigeye was an example of a gliding bomb. As designed, it had spring loaded, folding wings that deployed after being released from the aircraft giving it a wingspan of approximately 44 centimeters. The body was 2.3 meters long with a diameter of approximately 34 centimeters and, after the conversion to VX (C01-A017) was complete, contained approximately 87 liters of crude agent. When released from the aircraft, the bomb would glide without internal guidance or propulsion systems for several kilometers before passing over its intended target. At that point, explosive charges would open dissemination ports in the body of the bomb to allow droplets of the nerve agent to stream from the weapon creating an elevated line source that rained down on the enemy position.

THERMAL GENERATORS

Many chemical agents do not evaporate readily and heating them to the point where they will begin to vaporize causes them to decompose. Pyrotechnic devices, historically referred to as candles, burn an internal fuel source to produce hot gases, such as water and carbon dioxide, that are then directed over an otherwise nonvolatile agent so that they distill it into the air and create an aerosol cloud. The particulate matter generated by this distillation process have an extremely small aerodynamic diameter caused by the rapid condensation of the agent vapor. Heat from the distillation process initially causes the cloud to rise up, but it quickly cools and stabilizes, then settle back to earth. This technique has been incorporated into grenades, rockets, bombs, and cluster munitions. While these types of devices are very effective, there are some obvious limitations – they do not work with easily ignited chemical agents, they cannot be used with pathogens, and collateral fires are a potential secondary hazard.

The first pyrotechnic munition was the M-device, developed by the British toward the end of World War I but first employed in 1919 during the Russian civil war. The British Air Force used them to bomb the Red Army near Archangelsk, making the M-device the first chemical weapon delivered by an airplane. It consisted of two metal cans bolted together and separated by an asbestos gasket. A small circular flue was the only opening between the bottom and top containers. The bottom compartment was filled with a cake of smokeless powder and fitted with a wire ignition device, while the top was filled with approximately 4 kilograms of DM (C14-A003). The combustion gases from the burning smokeless powder passed over the surface of the cake of DM, distilling the agent into the smoke. The M-device would burn for up to 4 minutes and produced a heavy cloud of extremely irritating smoke.

Other examples of pyrotechnic devices include:

- The M7 hand grenade was first issued around 1935 and was the standard tear gas grenade for the United States until after World War II. It was approximately 6 centimeters in diameter and 11 centimeters tall without the fuze. It weighed approximately 480 grams and contained approximately 290 grams of a pyrotechnic mixture consisting of a blend of CN (C13-A008) and smokeless powder. It had a 3-second delay fuse and burned for up to 60 seconds. When functioning, flames would emit from the gas ports and tended to ignite nearby combustible materials. On occasion, it was prone to excessive flaming and even to explode. After various modifications, including updating the tear agent to CS (C13-A009), changing the composition of the pyrotechnic fuel mixture, and segregating the agent from the fuel, the M7A3 was standardized.
- The XM651E1 was used by the United States during the Vietnam War. It was a 40-millimeter tactical round for the M79 grenade launcher and contained 20 grams of CS (C13-A009) blended with approximately 30 grams of incendiary fuel. On impact, the pyrotechnic mixture was ignited and a cloud of CS was emitted for approximately 25 seconds. Under ideal conditions the initial gas cloud would be 14 meters in diameter. It was accurate up to a range of 180 meters, but area targets could be engaged out to 400 meters. It would penetrate 2 centimeters of plywood at 200 meters and release its CS payload after penetration.
- The United States used the E8 launcher during the Vietnam War. It was a single use multiple-rocket launcher designed for use against small area targets. Consisting of a box-like launcher module with an attached firing platform, it was approximately the size of a backpack and weighed 15 kilograms. It had padded shoulder straps so that a soldier could carry the unit into combat on his back. The launcher consisted of 16 tubes, each containing four 35-millimeter rockets, with a total CS (C13-A009) pyrotechnic-mixture payload of approximately 1 kilogram. When the weapon was fired, all 64 rockets were discharge within 1 minute in a series of four 5-second bursts (16 rockets per burst). If the target was in an open area, it took 30 seconds to build up an incapacitating concentration of tear gas. The size of the impact area varied with range. Because of the large number of rockets impacting in the same area, the size of the initial cloud was relatively independent of meteorological conditions.
- XM629 was a 105-millimeter howitzer round used by the United States during the Vietnam War. The entire cartridge was 76 centimeters long and weighed 19 kilograms; the projectile itself weighed 15 kilograms. It was equipped with a variable time fuze that could be set for impact or a time delay. A small charge ejected four canisters, each filled with a pyrotechnic mixture of 150 grams of CS (C13-A009) intermixed with 225 grams of incendiary fuel, out of the base of the projectile. The pyrotechnic mix burned for 60 seconds and under ideal conditions produced an initial gas cloud that was 20 meters in diameter. It had a maximum range of 14 kilometers with an optimum air burst height of between 90 and 150 meters.

- BLU-39 bomblet was the standard U.S. submunition for a number of air force cluster bombs during the Vietnam War. These bomblets were approximately the size of a D-cell battery (6 centimeters long and 3 centimeters in diameter) and weighed approximately 55 grams. Each contained approximately 37 grams of a pyrotechnic mixture of CS (C13-A009) intermixed with an incendiary fuel. Each bomblet had a 5- to 6-second delay fuse that was ignited by the expelling charge that dispersed them from the cluster bomb. The pyrotechnic mix in each cartridge burned for 20 seconds after ignition. The discharging exhaust gas was forced out of the bomblet through an off-center port that caused the canister to move randomly across the ground. This action further dispersed the bomblets throughout the target area and made it difficult for anyone to pick them up or kick them away.

VECTORS

A delivery system that is unique to pathogen agents is the use of vectors to carry and transmit the disease throughout the target area. Examples of vectors that transmit diseases among people include flies, mosquitoes, ticks, fleas, and lice. (See the Vector Index for a list of vector-borne diseases discussed in this handbook.) Vectors can be either the reservoir or an intermediate host for the pathogen. Vertical transmission – where the mother passes the infection onto her eggs so that the young are born able to transmit the disease – can occur with some pathogens.

Most vectors transmit the pathogen when they bite a new host; but there are other instances, such as with lice carrying *Rickettsia prowazekii*, which causes typhus (C17-A030), where the pathogen is excreted in the vector's feces as it feeds and forced into the wound by the victim when scratching the bite. A single vector can cause multiple infections in the target population. This can, in turn, lead to vector amplification, a situation that occurs when an infected vector transmits a disease to a new host, who in turn passes the disease on to multiple new, uninfected vectors. These newly infected vectors then continue the cycle, amplifying the rate and spread of the infection.

The vector not only serves as the delivery system, its body also protects the pathogen from elements of the environment, such as desiccation, heat, or sunlight, that would otherwise kill the microbe. Vectors often remain infected for life – up to 2 months for mosquitoes, 7 months for fleas, and several years for ticks – thus increasing the persistency of the outbreak.

Vectors are still living creatures and the method of releasing them into the target area must not kill or incapacitate them. They would not survive an explosion that generated inordinate heat, pressure or shock; nor would they survive the discharge from a high-speed aircraft where the wind turbulence and sheer would kill them. They are also subject to seasonal weather limitations that would affect their viability or mobility.

An example of a vector weapon was the Japanese use of fleas infected with plague during World War II. In addition to sprayers filled with the arthropods that they discharged from low-flying airplanes; they also developed the 25 kilograms Uji bomb. It was made of porcelain and used a very low-order explosive to rupture the casing without killing a significant portion of the payload. Each bomb had a 10-liter capacity, and was filled with approximately 30,000 infected fleas. They were designed to detonate at a height of 200 meters to 300 meters above the target, thus dispersing the fleas over a wide area. Up to 80% of the fleas were reported to survive dispersal from the bombs.

The United States also researched the use of vectors, particularly mosquitoes, during test programs such as Operation Big Itch, Operation Magic Sword, and Operation Bellwether II. Similar programs were also investigated by the former Soviet Union.

ASSASSINATIONS AND ATTEMPTS

Although this final category could employ weapons with any of the characteristics described above, it differs for a number of reasons. Among these is a desire to focus the effect on a targeted individual instead of producing mass casualties. That is not to say that indiscriminate devices could not be employed if broader collateral damage was acceptable, just as explosive devices have been employed to kill hard to acquire targets along with their entourage and accompanying protective detail. There is also typically a desire for some aspect of deniability, or at least uncertainty, as to the ultimate perpetrator of the assault. Finally, the motive is political, and not related to acquiring a tactical advantage on the battlefield.

In addition to classical poisoning of food and drink, which has been the fear of politicians and other powerful individuals for centuries, there have been a number of innovative techniques employed with the advent of modern chemical weapons:

- Bulgarian dissident Georgi Markov was assassinated on Waterloo Bridge in London in the late morning of Thursday, September 7, 1978. The assassin used a specially engineered weapon disguised as an umbrella to injected a platinum-iridium alloy pellet, approximately 1.7 millimeters in diameter, into the muscle of his thigh. The pellet had two perpendicular holes drilled through its center that were filled with ricin (C16-A031). Thinking the poke to his thigh was an accident, he proceeded on with his day. Early the following morning, he became sick with a high fever and was seen by his family physician. He was admitted to a hospital shortly before midnight on September 8, and then died Sunday night, September 11. Despite the puncture mark on his thigh, his general poor health led authorities to discount foul play and no post-mortem examination was conducted.

    Shortly after his death and the circumstances surrounding it were made public, Vladimir Kostov, another Bulgarian dissident who lived in Paris, came forward with the claim that he had been attacked in a similar manner. He had been shot in the back with a pellet from an air gun 3 weeks earlier on August 27. Unlike Markov, the pellet lodged in a fat deposit in the overweight Kostov and the ricin never entered his bloodstream. On September 28, physicians working with Scotland Yard removed the pellet from the back of Kostov. An examination of the wound in Markov's thigh produced an identical pellet and the cause of death was changed to murder. The assassin was never caught.

- Kim Jong-nam, brother to North Korean dictator Kim Jong-un, was assassinated on February 13, 2017. He was accosted by two unwitting assassins in rapid succession at Kuala Lumpur International Airport in Malaysian. The first assailant, whose hands were coated with the first component for a binary version of the nerve agent VX (C01-A017), walked up behind Kim, reached over his shoulders and rubbed his face. In rapid succession, the second assailant, carrying a rag soaked with the second component, rubbed his face with the rag. Kim reported the incident to airport police. He began feeling dizzy and was taken to a secure room while an ambulance was called. He died while en route to a local hospital. The two female assailants, one Indonesian and the other Vietnamese, were quickly captured by the police. They told authorities they were unaware their actions had hurt Mr. Kim and thought it was another in a series of pranks they had been hired to perform where they rubbed baby oil on the faces of strangers at airports and in shopping malls.

    During their lengthy trial, the women maintained that they had been duped by North Korean operatives into unknowingly committing the killing. Nearly 2 years into the trial, the charges against the Indonesian woman were dropped after Indonesian President Joko Widodo personally discussed the issue with Malaysian Prime Minister Mahathir Mohamad. Two months later, the Vietnamese woman was freed after pleading guilty to lesser charges and allowed to return to Vietnam. The North Korean government maintained it had no involvement with the incident.

- Sergei Skripal, a former colonel in Russia's military intelligence service who had historically passed intelligence to Britain's MI6, and his daughter Yulia were found unresponsive on a bench at the Maltings shopping precinct in Salisbury, England, on March 4, 2018. They were transported to a hospital as victims of an unidentified poison. The subsequent investigation over the next week determined that they had been exposed to a Novichok series nerve agent earlier on the 4th. The agent had been sprayed onto the handle of the front door to their house by Russian operatives and they were exposed when they touched it. Although rapid acting if inhaled, it can take several hours to work percutaneously.

After 3 weeks in critical condition, Yulia regained consciousness. She was discharged on April 9th. It took longer for Sergei to regain consciousness but he was ultimately discharged on May 18th. During the initial investigation, a police officer was also exposed to the agent when he touched the door handle and required treatment in the intensive care unit. He was discharged on March 22nd.

As the authorities continued their investigation into who the Russian operatives were and how they had applied the nerve agent to the door handle, Mr. Charlie Rowley was treasure hunting in dumpsters around Salisbury where he found a discarded perfume dispenser labeled Nina Ricci Premier Jour. He brought the bottle home to the nearby town of Amesbury and later, on June 30, 2018, presented it as a gift to his girlfriend, Dawn Sturgess. In the process of attaching the dispenser to the bottle, he spilled some of the contents on his hands. After he got it affixed properly, she sprayed the contents directly onto her wrists. Shortly thereafter, they both collapsed with symptoms of pinpoint pupils and frothing at the mouth. Both were taken to a local hospital where they were treated for nerve-agent poisoning. The contents of the bottle were analyzed and identified as including an agent in the Novichok series. On July 8th, Ms. Sturgess died as a result of her exposure. Mr. Rowley emerged from a coma on July 10th and slowly recovered.

Largely as a result of this incident, the Organisation for the Prohibition of Chemical Weapons added the Novichok class of agents to Schedule 1 of the Chemical Weapons Convention on November 27, 2019.

- On August 20, 2020, Alexei Navalny, a prominent critic of Russian President Vladimir Putin, lost consciousness while on a flight from Tomsk to Moscow. The jet made an emergency landing in the city of Omsk, where he spent more than a day in a local hospital. He was eventually transported to Berlin, Germany, on a medical evacuation flight on August 22nd. Originally classified as poisoning by an unknow cholinesterase inhibitor, toxicology tests carried out at a German military laboratory ultimately identified the toxin as a member of the Novichok class of nerve agents. Although some agent residue was found on a discarded bottle of water that he had consumed prior to departing for the airport to board the flight, it was later learned that operatives had actually coated the lining of his underwear with the agent, which was then absorbed through his skin. When he was transferred from the flight to the hospital in Omsk, more operatives collected his clothes, decontaminated them, and then returned the items to his hospital room in an attempt to prevent identification of the toxin. Eventually, after intensive treatment in Germany, he recovered and was discharged on September 22nd. The Russian government maintained throughout the incident that it had no knowledge of the events, questioning if he had actually been intentionally poisoned. President Putin even quipped in response to one reporter's question that had the Russian Federal Security Service actually wanted to poison Mr. Navalny, then they would have been successful.

BIBLIOGRAPHY

Brown, F. J. *Chemical Warfare: A Study in Restraints*. Princeton, New Jersey: Princeton University Press, 1968.

Cole, L. A. *Clouds of Secrecy: The Army's Germ Warfare Tests Over Populated Areas*. Totowa, New Jersey: Rowman & Littlefield, 1988.

Cook, Tim. *No Place to Run: The Canadian Corps and Gas Warfare in The First World War*. Vancouver, Canada: University of British Columbia, 1999.

Crowley, Michael. *Tear Gassing by Remote Control*. London, United Kingdom: Remote Control Project, December 2015.

Haber, Ludwig Fritz. *The Poisonous Cloud: Chemical Warfare in the First World War.* Oxford: Clarendon Press, 1986.

Hedden, E. M. and D. C. Hawkins. *CBU-30/A Incapacitating Munitions System, AFATL-TR-67-178.* Florida: Air Force Armament Laboratory, Eglin Air Force Base, October 1967.

Leitenberg, Milton and Raymond A. Zilinskas. *The Soviet Biological Weapons Program: A History.* Cambridge, Massachusetts: Harvard University Press, 2012.

MSI Delivery Systems Inc. *AB2K MMADS User and Safety Manual.* 3rd Edition. Rocky Mount, North Carolina: MSI Delivery Systems Inc., February 1, 2016.

Parsch, Andreas. *Designations of U.S. Aeronautical and Support Equipment.* June 23, 2006. http://www.designation-systems.net/usmilav/aerosupport.html#_ASETDS_Component_Listings_Alpha. December 31, 2020.

Patrick, W.C. "Biological Terrorism and Aerosol Dissemination." *Politics and The Life Sciences,* 15 (1996): 208.

Plunkett, Geoff. *Chemical Warfare in Australia.* Loftus, Australia: Australian Military History Publications, 2007, 420-422, 450-516.

Prentiss, Augustin M. *Chemicals in War: A Treatise on Chemical Warfare.* New York, New York: McGraw-Hill Book Company, Inc., 1937.

Richter, Donald. *Chemical Soldiers: British Gas Warfare in World War I.* Lawrence, Kansas: University Press of Kansas, 1992.

Robinson, Julian Perry. *The Rise of CB Weapons.* Stockholm, Sweden: Almqvist & Wiksell, 1971.

Spassky, Nikolai, ed. "Chemical Weapons" Part 17 of *Russia's Arms Catalog Volume VII: Precision Guided Weapons and Munitions.* Moscow, Russia: Military Parade, Ltd., 1997.

Spiers, E. M. *Chemical Weaponry: A Continuing Challenge.* London, United Kingdom: The Macmillan Press, LTD, 1989.

Swearengen, Thomas F. *Tear Gas Munitions: An Analysis of Commercial Riot Gas Guns, Tear Gas Projectiles, Grenades, Small Arms Ammunition, and Related Tear Gas Devices.* Springfield, Illinois: Charles C Thomas, Publisher, 1966.

United Nations Monitoring, Verification and Inspection Commission. "Compendium of Iraq's Proscribed Weapons Programmes in the Chemical, Biological and Missile Areas." June 2007.

United States Army Chemical Materiel Destruction Agency. *Old Chemical Weapons: Munitions Specification Report.* Washington, D.C.: Government Printing Office, September 1994.

United States Army Headquarters. *Unexploded Ordnance (UXO) Procedures, Field Manual No. 21–16.* Washington, D.C.: Government Printing Office, August 30, 1994.

United States Army Headquarters. *Army Ammunition Data Sheets for Grenades, Technical Manual No. 43–0001-29.* With Change 1. Washington, D.C.: Government Printing Office, June 30, 1994.

United States Army Headquarters. *Chemical Bombs and Clusters, Technical Manual No. 3–400.* Washington, D.C.: Government Printing Office, May 8, 1957.

United States Army Headquarters. *Field Behavior of NBC Agents, Field Manual No. 3–6.* Washington, D.C.: Government Printing Office, November 1986.

United States Army Headquarters. *NBC Defense, Chemical Warfare, Smoke and Flame Operations, Field Manual No. 3–100.* Washington, D.C.: Government Printing Office, May 23, 1991.

United States Army Headquarters. "Obsolete Hand Grenades," Appendix E in *Grenades and Pyrotechnic Signals, Field Manual No. 23–30.* Washington, D.C.: Government Printing Office, September 1, 2000.

United States Army Headquarters. "Riot Control Agent Munitions and Delivery Systems," Appendix B in *Flame, Riot Control Agents and Herbicide Operations, Field Manual No. 3–11.* Washington, D.C.: Government Printing Office, August 19, 1996.

United States Army Headquarters. "Riot Control Agents, Equipment, and Munitions," Chapter 2 in *Employment of Riot Control Agents, Flame, Smoke, and Herbicides in Counterguerrilla Operations, Training Circular No. 3-16.* Washington, D.C.: Government Printing Office, July 11, 1966.

United States Army Headquarters. "Riot Control Agents, Equipment, and Munitions," Chapter 2 in *Employment of Riot Control Agents, Flame, Smoke, Antiplant Agents and Personnel Detectors in Counterguerrilla Operations, Training Circular No. 3–16.* Washington, D.C.: Government Printing Office, April 09, 1969; Reprint, Stockholm, Sweden: Swedish Medical Aid Committee for Vietnam, 1970.

United States Department of the Army. *Employment of Chemical Agents, Field Manual No. 3–10.* Washington, D.C.: Government Printing Office, March 31, 1966.

United States Army Chemical Materials Activity, Recovered Chemical Materiel Directorate, Recovered Chemical Warfare Materiel Integrating Office. "Old Chemical Weapons and Related Materiel Reference Guide." May 2018.

United States Army Program Manager for Chemical Demilitarization, Office of the Project Manager for Non-Stockpile Chemical Materiel. "Old Chemical Weapons Reference Guide." May 1998.

Waitt, Alden H. *Gas Warfare: The Chemical Weapon, Its Use, and Protection Against It.* Revised Edition. New York, New York: Duell, Sloan and Pearce, 1944.

Woodall, Robert and Felipe Garcia. "Frangible Payload-Dispensing Projectile" United States Patent 6145441, November 14, 2000.

Zajtchuk, R. and Bellamy, R. F., Eds. *Medical Aspects of Chemical and Biological Warfare*, Textbook of Military Medicine Part 1. Washington, D.C.: Government Printing Office, 1997.

Appendix B: Summary of Military Chemical Munition Markings

Countries typically label their chemical munitions with special markings, colors, and/or bands in order to facilitate their use on the battlefield. With the exception of Cold War–era NATO and Warsaw Pact countries, or countries that purchased weapons instead of producing them themselves, no two countries employed identical systems. Further, over the course of history, countries have introduced changes to their respective system. For this reason, some knowledge of the date of manufacture is required to accurately identify the appropriate markins. In this appendix, the time periods are divided into World War I, World War II, and the Cold War. The countries identified are: Germany, Great Britain, France, Italy, Japan, the United States, and the Soviet Union. Due to a lack of publicly available information or to the absence of a consistent marking system, other countries know or suspected to have had chemical and biological warfare programs are not detailed. These include: China, Egypt, India, Iraq, Libya, North Korea, South Korea, Spain, and Syria.

WORLD WAR I

FRANCE

Toxic, Irritant, or Harassing Agents
- Shells were dark green or blue-gray with a black ogive.
- Chemical fills were indicated by numbers, usually on the ogive.
  - No. 4 for a mixture of hydrogen cyanide, arsenic trichloride, stannic chloride, and chloroform (C07-A).
  - No. 4B for a mixture of cyanogen chloride and arsenic trichloride or a mixture consisting of hydrogen cyanide and arsenic trichloride (C07-A).
  - No. 5 for a mixture of phosgene and arsenic trichloride; or a mixture consisting of phosgene and stannic chloride (C10-A).
  - No. 6 for a mixture of chloromethyl chloroformate and stannic chloride (C13-A010).
  - No. 7 for a mixture of chloropicrin and stannic chloride (C10-A006).
  - No. 8 for a mixture of acrolein and stannic chloride or titanium tetrachloride (C13-A).
  - No. 9 for a mixture of bromoacetone, chloroacetone, and possibly stannic chloride (C13-A).
  - No. 9B for a mixture of bromomethylethyl ketone, chloromethylethyl ketone, and stannic chloride (C13-A).
  - No. 10 for a mixture of iodoacetone and stannic chloride (C13-A013).
  - No. 11 for a mixture of benzyl bromide and stannic chloride (C13-A002).
  - No. 12 for a mixture of benzyl iodide, benzyl chloride, and possibly stannic chloride (C13-A).
  - No. 13 for a mixture of ethyl chlorosulfonate and stannic chloride (C10-A010).
  - No. 16 for a mixture of dimethyl sulfate and chlorosulfonic acid or methyl chlorosulfonate (C03-A).
  - No. 14 for a mixture of benzyl bromide and titanium tetrachloride (C13-A).
  - No. 15 for a mixture of thiophosgene and stannic chloride (C13-A017).
  - No. 20 for mustard gas. In addition to the numerical 20, additional letters that might be present include "Yt", which indicated the mustard is dissolved in carbon tetrachloride; or "Yc", which indicated it is dissolved in chlorobenzene (C03-A001).

- No. 21 for a mixture of bromobenzyl cyanide and chloropicrin (C13-A004).
- Colored bands encircling the shell body indicated persistence.
  - White bands indicated a nonpersistent fill.
  - After January 1, 1918, shells filled with a persistent filling were marked with an orange-yellow band.

Incendiary and Smoke Munitions
- Shells were dark green.
- The upper half was sometimes red.

GERMANY

Early shells were gray, then latter in the war they were blue with a yellow ogive. A series of Geneva crosses indicated the physiological effects produced and not the specific agent fill of the munition. Trench mortar shells could be marked with colored bands instead of crosses.

Toxic Agents
- Green cross for pulmonary agents.
- Yellow cross for vesicant agents.

Irritant or Harassing Agents
- Blue cross for vomiting/sternutatory agents.
- White cross for lachrymatory agents.

Additional Markings and Bands
- Double crosses of the same color indicated the agent was packed with a high explosive charge to increase aerosolization of the agent and enhance its lethality.
- A number could be added beside the cross to further identify agent fill.
- A large letter on the shell was used to denote "Stoff" agent fills.
  - B for bromoacetone (C13-A003).
  - Bn for bromomethylethyl ketone (C13-A005).
  - C for chloromethyl chloroformate (C13-A010) and mixtures.
  - D for phosgene (C10-A003).
  - K for chloromethyl chloroformate (C13-A010) and mixtures.
  - N for sulfur trioxide (C11-A156).
  - T for benzyl bromide (C13-A002) and mixtures.

GREAT BRITAIN

Toxic, Irritant, or Harassing Agents
- Shells were gray.
- Series of one to four white and/or red bands indicated the agent fill.
  - One white band for chloropicrin.
  - One white band and one red band for Vinciennite, a mixture of hydrogen cyanide, stannic chloride, and arsenic trichloride.
  - One white band, one red band, and one white band for phosgene.
  - One white band and two red bands for AK, a mixture of hydrogen cyanide and ethyl iodoacetate.
  - Two white bands for PG, a mixture of phosgene and chloropicrin.
  - One red band for CBR, a mixture of phosgene and arsenic trichloride.

- Three red bands for JBR, hydrogen cyanide in chloroform.
- Four red bands for mustard gas.
- Military agent symbol was on the side of the munition.

Incendiary Munitions
- Shells were red.
- No other key markings were on the shell.

Smoke Munitions
- Shells were light-green.
- White phosphorous was indicated by the word "PHOS" on the body in black.

Additional Markings and Bands
- A red band encircling the nose of the shell indicated it was loaded with explosive burster.
- A brown band near the nose of the shell indicated the body was made of cast iron.

UNITED STATES

The marking system employed at this time did not differentiate between toxic and lachrymatory/sternatory agents. No chemical symbols or other markings were used to identify the specific agent fill of the munition. Under this system, newer, more effective gases were substituted for older, discontinued ones without altering the sequence of bands.

Shells were gray.

Toxic or Irritant Agents
- White bands denoted a nonpersistent gas. Increasing the number of bands indicated greater relative persistency of the agent compared to others in that class.
- Red bands denoted a persistent gas. Increasing the number of bands indicated greater relative persistency of the agent compared to others in that class.
- A red band and a white band combination denoted a semipersistent gas.
- A white, red, white combination denoted a nonpersistent gas mixed with a semipersistent gas.
- A yellow band indicated the mixture also contained a smoke producing agent.
- Munition were marked with descriptive words "SPECIAL GAS" or "GAS" in black.

Incendiary Munitions
- Munitions had one purple band.
- Munitions were marked with descriptive words "SPECIAL INCENDIARY", "INCEND" "OIL", or "THERMITE" in black.

Smoke Munitions
- Munitions had yellow bands to indicate persistence.
  - One band for nonpersistent.
  - Two bands for persistent.
- Munitions were marked with descriptive words "SPECIAL SMOKE" or "SMOKE".

WORLD WAR II

GERMANY

Initially used the same system of Geneva crosses employed during World War I but switched to a system of colored bands early in the war. In general, the crosses/bands indicated the physiological effects produced and not the specific agent fill of the munition. Ground systems and aerial bombs employed different marking systems.

Artillery, Rockets, and Mortars

Shells were blue with a yellow ogive.

Toxic Agents

- Green bands for lethal agents (including vesicant agents) with a medium burster.
- Yellow bands for sulfur mustard with a small burster.
- A green band and a yellow band for sulfur mustard with a large burster.
- A gas identification code was in black on the side of the munition.
  - B for a mixture of sulfur mustard and lewisite [winterlost] (C03-A010).
  - C for a mixture of sulfur mustard and lewisite [winterlost] (C03-A010).
  - D for thickened sulfur mustard (C03-A001).
  - E for thickened sulfur mustard (C03-A001).
  - F for phosgene (C10-A003).
  - G for tabun (C01-A001).
  - GA for a solution of tabun in chlorobenzene (C01-A001).
  - H for diphosgene (C10-A004).
  - K for nitrogen mustard 3 (C03-A018).
  - L for a mixture of sulfur mustard and anthracene (winterlost) (C03-A001).
  - for sulfur mustard (C03-A001).
  - P for hydrogen cyanide (C07-A001).

Irritant or Harassing Agents

- Blue bands for vomiting/sternutatory agents.
- White bands for lachrymatory agents.
- A gas identification code was in black on the side of the munition.

Additional Markings and Bands

- A number could be added below the band to further identify agents.
- Shells filled with sulfur mustard and equipped with a small burster had a Gb or G/B in yellow on the ogive.
- Shells filled with sulfur mustard and equipped with medium burster had a Gb or L/O in green on the ogive.

Aerial Bombs

Bombs had either a gray or tan body with colored band(s) encircling the nose and also the center of the body indicating the effects produced by the agent fill.

Toxic Agents

- Green bands for lethal agents (pulmonary, blood, nerve).
- One green band and one yellow band for sulfur mustard with a large burster.
- One yellow band for winterized sulfur mustard.
- Two yellow bands for thickened sulfur mustard.

Irritant or Harassing Agents

- Blue bands for vomiting agents.
- White bands for lachrymatory agents.

Additional Markings and Bands

- The bomb designation, including a code for the bands on the bomb, was in black on the side of the bomb (i.e., type of bomb followed by weight class followed by band code).
  - Gb for one yellow band.
  - IIGb for two yellow bands.
  - Gr for one green band and one yellow band.
  - IGr, IIGr or IIIGr for one, two or three green bands, respectively.
  - IBu or IIBu for one or two blue bands, respectively.
  - W for white bands.
- Within a broken circle on the side of the bomb was a code indicating the type of fuze used.

Mines

Had one or two concentric yellow bands on top to indicate sulfur mustard or thickened sulfur mustard (C03-A001), respectively.

GREAT BRITAIN

Utilized a series of colored bands encircling the lower-middle body of the shell to indicate the physiological effects produced and persistency. The military agent symbol was painted in black on the band indicating the exact filling.

Toxic Agents (vesicant, blood, pulmonary)

- Shells were gray.
- Green bands for nonpersistent agents.
- Yellow bands for persistent agents.

Irritant or Harassing Agents (lachrymation, vomiting/sternatory)

- Shells were gray.
- One black band.

Incendiary Munitions

- Shells were red.
- No other key markings were on the shell.

Smoke Munitions

- Shells were green
- No other key markings were on the shell unless it contained white phosphorous, then either "SMOKE W PHOS" or "PHOS" was printed on the body in black.

Additional Markings and Bands

- Addition of a narrow white band indicated the fill had an arsenic component.
- Addition of a red band indicated the agent was thickened.

ITALY

Munitions had a yellow body. A series of Geneva crosses indicated the physiological effects produced and not the specific agent fill of the munition.

Toxic Agents
- White cross for pulmonary agents.
- Green cross for vesicant agents.

Irritant or Harassing Agents
- Black cross for vomiting/sternutatory agents.
- Red cross for lachrymatory agents.

Additional Markings and Bands
- Munitions producing multiple effects (e.g., pulmonary and lachrymation) had a colored cross for each effect produced.

JAPAN

The system employed by the Japanese was complex. It varied by branch of service, ground-based weapon systems versus aerial bombs, and even the time period when the munition was manufactured during the war. The color banding system indicated the physiological effects produced and not the specific agent fill of the munition.

Artillery, Mortars, Mines, and Candles
Artillery shells were gray (the exception being chloropicrin shells, which were red). A red tip on the ogive meant the explosive components were installed. A blue band just below the red tip would indicate that the chemical components were installed. A yellow band just below the blue band would indicate a high explosive bursting charge was used in the munition. Early in the war, a band at the base of the shell just forward of the driving band indicated the material used to make the shell body: white for high grade steel, green for low grade steel, and no band for cast iron. Wider colored bands toward the center of the shell and just aft of the center of gravity indicated the chemical fill of the munition.

Information on naval munitions is limited. Many had a maroon body. A green band just blow the fuse indicated the explosive components were installed. A colored band at the middle of the shell indicted the center of gravity. Wider colored bands toward the center of the shell and aft of the center of gravity indicated the chemical fill of the munition.

Mortar rounds had a black body. A red band just blow the fuse indicated the explosive components were installed. A blue band just below the red tip indicated the chemical components were installed. Early in the war, a band at the junction of the body and tail assembly indicated the material used to make the shell body: white for high grade steel, green for low grade steel, and no band for cast iron. Colored bands halfway between the nose and tail assembly indicated the chemical fill of the munition.

Candles (pyrotechnic aerosol grenades or pots) had a green, brown, or gray body and only contained irritant or harassing agents with the appropriate colored band around the body of the device.

Mines were gray with a yellow band encircling the body at the midway point.

Toxic Agents
- Blue band for pulmonary agents.
- Brown band for cyanide blood agents.
- Yellow bands for vesicant agents.
- The general effects classification of a specific agent could be changed by adjusting the burster charge in the shell (e.g., vesicants with large burster charges would be primarily intended as pulmonary agents and have the respective banding).

- Shells could have additional markings to further identify the agent.
  - Mk III for vesicant agents.
  - Mk IIIA for sulfur mustard (C03-A001).
  - Mk IIIB for lewisite (C04-A002).

Irritant or Harassing Agents
- Green band for lachrymators.
- Red band for vomiting agents.
- Shells could have additional markings to further identify the agent.
  - Mk I for lachrymatory agents.
  - Mk II for vomiting agents.
  - Mk IIA for diphenylcyanoarsine (C14-A002).
  - Mk IIB for diphenylchloroarsine (C14-A001).

Aerial Bombs

Army bombs were gray with a red tip. A yellow band encircling the body just forward of the center of gravity would indicate a high explosive burster. Early in the war, a white or green band encircled the body at the center of gravity and indicated the quality of steel used to construct the bomb: white for high grade steel, green for low grade steel, and no band for cast iron. Later the use of this marking band was discontinued. Colored bands encircling the bomb aft of the suspension lug indicated the chemical fill and used the same color scheme as artillery shells described above.

Naval bombs were gray. Early in the war, bombs had a red tip with a broad colored band on both the nose and tail struts indicating agent fill. Later in the war the nose of the bomb was green, indicating the explosive components were installed. The navy was also reported to have a Special Mark 7 bacillus (bacterial warfare) bomb that had a green-purple band with gray-purple markings.

Soviet Union

Munitions were gray.

Toxic Agents
- Blue or green band for pulmonary agents.
  - Two bands for more persistent agents.
- Yellow band for hydrogen cyanide (C07-A001).
- Red bands for vesicant agents (number of bands indicated specific agent not persistency).
  - One band for sulfur mustard (C03-A001).
  - Two bands for lewisite (C04-A002).
  - Three bands for sulfur mustard/lewisite mixtures (C03-A010).

Irritant or Harassing Agents
- One white band for lachrymatory agents.
- One white band with a blue or green band for vomiting agents.

Additional Markings
- P-5 for sulfur mustard (C03-A001).
- P-10 for phosgene (C10-A003).
- P-1-0 or 1-0-0 for diphosgene (C10-A004).
- P-15 for adamsite (C14-A003).
- PC for lewisite (C04-A002).

- TO for nitrogen mustard 3 (C03-A018).
- ВИР indicated the agent was thickened.

UNITED STATES

Munitions were gray. The color banding system indicated the relative persistency and physiological effects produced, not the specific agent fill of the munition.

Toxic Agents (vesicant, blood, pulmonary)
- Green bands indicated persistence.
  - One band for nonpersistent.
  - Two bands for persistent.
- Agent name or military agent symbol in green indicated exact filling.
- Also marked with descriptive word "GAS" in green.

Irritant or Harassing Agents (lachrymation, vomiting)
- Red bands indicated persistence.
  - One band for nonpersistent.
  - Two bands for persistent.
- Agent name or military agent symbol in red indicated exact filling.
- Also marked with descriptive word "GAS" in red.

Incendiary Munitions
- One purple band.
- Agent name or military agent symbol in purple indicated exact filling.
- Also marked with descriptive word "INCENDIARY" or "INCEND" in purple.

Smoke Munitions
- One yellow band.
- Agent name or military agent symbol in yellow indicated exact filling.
- Also marked with descriptive word "SMOKE" in yellow.

COLD WAR

SOVIET UNION

Munitions were gray. Submunitions in cluster bombs or rocket packs were usually unmarked. Ground systems and aerial bombs employed different marking systems. The color banding system indicated the relative persistency, physiological effects produced, and additional weapon characteristics (e.g., fragmentation, cluster submunitions). Banding did not indicate the specific agent fill of the munition.

General Ordnance Markings
- 3 for incendiary.
- Д for smoke.
- OX for fragmentation and chemical fill.
- X for chemical fill.

Artillery, Mortars, and Rockets
Toxic, Irritant, or Harassing Agents
- Number of bands indicated persistence.
  - One band for nonpersistent.
  - Two bands for persistent.

- Military agent symbol indicated exact filling.
  - P-35 for sarin (C01-A002).
  - P-55 for soman (C01-A003).
  - P-5 for sulfur mustard (C03-A001).
  - P-74 for sulfur mustard (C03-A001).
  - PC for lewisite (C04-A002).
  - P-43 for lewisite(C04-A002).
  - PK-7 for sulfur mustard/lewisite mixture (C03-A010).
  - PЮ for phosgene (C10-A003).
  - P-2 for hydrogen cyanide (C07-A001).
  - P-15 for adamsite (C14-A003).

Incendiary Munitions
- One red band.
- Military agent symbol indicated exact filling.
  - P-4 for white phosphorous.
  - TP for thermite.

Smoke Munitions
- One black band.

Aerial Bombs
Toxic, Irritant, or Harassing Agents
- Green nose band with green body band for nonpersistent chemical.
- Green nose band with blue and green body bands for nonpersistent chemical with fragmentation.
- Red nose band with green body band for persistent chemical.
- Military agent symbol indicated exact filling.
  - P-35 for sarin (C01-A002).
  - P-55 for soman (C01-A003).
  - P-5 for sulfur mustard (C03-A001).
  - P-74 for sulfur mustard (C03-A001).
  - PC for lewisite (C04-A002).
  - P-43 for lewisite(C04-A002).
  - PK-7 for sulfur mustard/lewisite mixture (C03-A010).
  - PЮ for phosgene (C10-A003).
  - P-2 for hydrogen cyanide (C07-A001).

Incendiary Munitions
- Red nose band with blue body band for incendiary.
- Red nose band with red body band for incendiary dispenser with submunitions.

Additional Markings and Bands
- ЗАБ for incendiary.
- ХАБ for chemical fill.
- КРАБ for toxic smoke.
- ДАБ for incendiary smoke.
- AOX for fragmentation and chemical fill.

UNITED STATES/NATO

Munitions gray unless otherwise indicated. Submunitions in cluster bombs or rocket packs were usually unmarked. The color banding system indicated the relative persistency and physiological effects produced, not the specific agent fill of the munition.

Toxic Agents (nerve, vesicant, blood, pulmonary)
- Green lettering and green bands.
- Number of bands indicated persistence.
  - Munitions manufactured prior to 1960.
    - One band for nonpersistent.
    - Two bands for persistent.
  - Munitions manufactured between 1960 and 1976.
    - One band for nonpersistent.
    - Two bands for persistent.
    - Three bands for nerve agents.
  - Munitions manufactured after 1976.
    - One band for all categories (nerve, vesicant, blood, pulmonary); both nonpersistent and persistent.
    - Dashed band for binary agents.
- Agent name or military agent symbol indicated exact filling.
- Munitions manufactured prior to 1976 were also marked with descriptive word "GAS".

Incapacitating Agents (mind-altering)
- Prior to 1960, there was no incapacitating agent category.
- Colored lettering and bands.
  - Munitions manufactured before 1976 used red.
  - Munitions manufactured after 1976 used purple.
- All munitions marked with two bands indicating persistent agents.
- Agent name or military agent symbol indicated exact filling.
- No descriptive wording on munitions from any manufacturing era.

Irritant or Harassing Agents (lachrymation, vomiting, riot control)
- Red lettering and red band.
- Munitions marked with one or two bands to indicate persistence of agent.
- Agent name or military agent symbol indicated exact filling.
- Also marked with descriptive word.
  - Munitions manufactured prior to 1960 used "GAS".
  - Munitions manufactured after 1960 used "RIOT".

Incendiary Munitions
- Only munitions manufactured prior to 1960 were marked with a purple band.
- Colored lettering on contrasting colored background.
  - Munitions manufactured prior to 1960 were gray with purple lettering.
  - Munitions manufactured after 1960 were red with black lettering.
- Agent name or military agent symbol indicated exact filling.
- Also marked with descriptive word "INCENDIARY" or "INCEND".

Smoke Munitions
- Only munitions manufactured prior to 1960 were marked with a yellow band.
- Colored lettering on contrasting colored background.

- Munitions manufactured prior to 1960 were gray with yellow lettering.
- Munitions manufactured after 1960 were light-green with black lettering.
- After 1960 white phosphorus was considered to be dual used (smoke/incendiary). These munitions were light-green with light-red lettering.
- M18 smoke grenades were olive-drab green with a light green band and lettering.
- Agent name or military agent symbol indicated exact filling.
- Also marked with descriptive word "SMOKE".

Additional Markings and Bands

- Munitions manufactured after 1960 had an additional band indicating type of burster or expelling charge.
  - Yellow band indicated high explosive.
  - Dark red band indicated low explosive.
- Munitions manufactured prior to 1960 did not indicate explosive content.

BIBLIOGRAPHY

Headquarters, American Expeditionary Forces. "Chemical Warfare Service Intelligence Bulletin: Notes on Toxic Shells (On the Charging, Marking and Storing of Special Shell)." July 22, 1918.

Hirsch, Walter. Soviet BW and CW Preparations and Capabilities, *Section 1*. Translated by Zaven Nalbandian. United States Army Chemical Warfare Service, Edgewood, Maryland, 1951, 67–68.

Naval Explosive Ordnance Disposal Technology Division . Iraq Purple Book – Operational Support Guide for Joint Service EOD for Iraq Area of Operations, Version 1.0. November, 2001: iv–viii.

Prentiss, Augustin M. *Chemicals in War: A Treatise on Chemical Warfare*. New York, New York: McGraw-Hill Book Company, Inc., 1937.

United States Army Headquarters. "Ammunition Color Codes," Appendix A in *Unexploded Ordnance (UXO) Procedures, Field Manual No. 21–16*. Washington, D.C.: Government Printing Office, August 30, 1994.

United States Army Chemical Materiel Destruction Agency. "Old Chemical Weapons: Munitions Specification Report." September 1994.

United States Army Chemical Materials Activity, Recovered Chemical Materiel Directorate, Recovered Chemical Warfare Materiel Integrating Office. "Old Chemical Weapons and Related Materiel Reference Guide." May 2018.

United States Army Program Manager for Chemical Demilitarization, Office of the Project Manager for Non-Stockpile Chemical Materiel. "Old Chemical Weapons Reference Guide." May 1998.

Appendix C: Biological Risk Groups and Biosafety Levels

Risk groups define the perceived risk posed by a given microorganism as a combination of the probability of the occurrence of harm coupled with the severity of the harm (e.g., infection, allergy, toxicity) it could cause in a given area, regardless of the source (i.e., unintentional exposure, accidental release or loss, theft, misuse, diversion, unauthorized access or intentional unauthorized release). Risk groups are assigned based on the pathogenicity of the organism, modes of transmission, and susceptible host range (i.e., number and kinds of hosts the microorganism might target). The weight assigned to each of these factors may be influenced by existing levels of immunity, density and movement of each host population, presence of appropriate vectors, standards of environmental hygiene, availability of effective preventive measures, and availability of effective treatment (e.g., passive immunization, postexposure vaccination, antibiotics, chemotherapeutic agents). Environmental and population factors that are considered include virulence of the pathogen, biological stability and ability to survive in the environment, local standards of environmental hygiene, availability of effective preventive measures, and treatments. Further considerations arise if a vector plays a role in transmission and include such factors as mobility of the vector (e.g., crawling vs. flying,), migration range, means of transmission, and resistance to pesticides.

There is general worldwide agreement on the four-tier classification system of microorganism risk groups; however, there is some disagreement over how to assign individual pathogens to a given risk group. This results from reginal geographic and climatic considerations, pathogen reservoirs and vectors, and sometimes economic or even political considerations. The most widely accepted classification of risk groups is:

> **Risk Group 1** (no or low individual, no or low community risk): A microorganism unlikely to cause human or animal disease. Nonetheless, due care should be exercised along with safe work practices whenever handling these organisms.
> **Risk Group 2** (moderate individual risk, low community risk): A pathogen that can cause human or animal disease but is unlikely to be a serious hazard to laboratory workers, the community, livestock or the environment. Laboratory exposures may cause serious infection, but effective treatment and preventive measures are available and the risk of spread of infection is limited.
> **Risk Group 3** (high individual risk, low community risk): A pathogen that usually causes serious human or animal disease but does not ordinarily spread from one infected individual to another and therefore the risk to the public is low. Effective treatment and preventive measures are available.
> **Risk Group 4** (high individual, high community risk): A pathogen that usually causes serious human or animal disease and can be readily transmitted from one individual to another, directly or indirectly. Effective treatment and preventive measures are not usually available.

As an addition, a fifth group is used by some agencies to further define risks. *Environmental-risk microorganisms* are pathogens that offer a more severe threat to the environment than to man, and may be responsible for heavy economic losses in an area. This risk group is most applicable to plant pathogens that could devastate agricultural crops, but could also be used for some animal pathogens.

Biological risk groups correlate with, but are not the same as, biosafety levels. Risk groups are used to help assign microorganisms to the appropriate safety level, which in turn are used to ensure an appropriate amount of protection for laboratory workers and the environment. The assignment of an organism to a biosafety level is based on a risk assessment that evaluates the pathogenicity, mode of transmission, and host range of the organism, as well as the availability of effective preventive measures and treatment. Increasing containment and operational requirements come with each successive safety level and are met through a composite of design features, construction, containment facilities, equipment, practices, and work procedures.

Animal biosafety levels are comparable to, but not interchangeable with, standard biosafety levels. This is because the animals themselves, in addition to being infected with the disease under study, pose operational risks or hazards. Requirements and recommendations for facilities working with animals are identified for each biosafety level discussed below.

Arthropod containment level guidelines provide recommendations for facilities working with blood-sucking (hematophagous) arthropods and arthropod vectors of pathogenic agents to ensure the safety of researchers and the surrounding community. First published in 2003 by the American Committee of Medical Entomology and American Society of Tropical Medicine and Hygiene, these guidelines emphasize the use of site-specific risk assessments to complement the regulatory framework as opposed to a checklist approach. Guidelines, coupled with the risk assessment, address the questions of: 1) How to prevent the arthropods from escaping, and 2) what to do if they escape. Containment requirements are minimally set at those of the biological safety level required for the pathogen under study (in the United States, arthropods deliberately infected by Select Agents are themselves considered Select Agents) and adjusted to include additional physical barriers, protective equipment, procedures, and/or operational practices needed to ensure location and destruction of escaped arthropods.

Biosafety Level 1 is suitable for work involving well-characterized agents not known to consistently cause disease in healthy adult humans, and that present minimal potential hazard to laboratory personnel and the environment. Special containment equipment or facility design is not required and work is typically conducted on open bench tops using standard microbiological practices.

Animal Biosafety Level 1 utilizes standard animal facilities, although they should be separate from the general traffic patterns of the building. External facility doors are self-closing and self-locking. Doors to areas where infectious materials and/or animals are housed open inward, are self-closing, and are kept closed when experimental animals are present. Floors, walls, and ceilings are water resistant. Penetrations in floors, walls, and ceiling surfaces (e.g., openings around ducts, plumbing, light fixtures, doorframes) should be sealed to facilitate pest control and cleaning. Recirculation of exhaust air is not permitted and inward directional air flow is recommended. Standard animal care and management practices are implemented. Cages can be either washed mechanically or by hand, as long as they are properly disinfected.

Arthropod Containment Level 1 is suitable for work with uninfected arthropod vectors or those infected with a virus, bacterium, or other agent that does not cause disease. This includes arthropods that are already present in the local geographic region.

The insectary is separate from the general traffic patterns of the building. Door and windows in the area are designed to minimize escape and prevent the entrance of surreptitious arthropods or pests. An effective trapping program is established to monitor for escaping arthropods. Personnel working with the arthropods should wear white laboratory coats or gowns and personal protective equipment (e.g., gloves, masks, head covers) as needed.

Between uses or before disposal of primary containers, they are cleaned to prevent arthropod survival and escape (e.g., heated or chilled to a lethal temperature). Unneeded arthropods should be killed (e.g., heated or chilled to a lethal temperature) prior to disposal.

The primary container used to hold the arthropods should effectively prevent escape, even during feeding. When handling/removing vertebrate animals (i.e., hosts or blood sources) after exposure to arthropods, precautions must be taken to prevent the insects from escaping during the transfer process. Host animals must also be inspected closely for residual clandestine arthropods. All precautions must be taken to prevent accidental transfer of the arthropods to the host cages.

Biosafety Level 2 is suitable for work involving agents associated with human disease transmitted via percutaneous, mucous membrane or oral exposure, and that pose moderate hazards to personnel and the environment, but can be managed by good laboratory techniques. Biosafety level 2 practices, containment equipment, and facilities are recommended for activities using clinical materials and diagnostic quantities of infectious cultures of most biological warfare agents.

Access to the laboratories is through self-closing doors and restricted when work is being conducted. All personnel working in the laboratories must have specific training in handling pathogenic agents. They must be under medical surveillance and offered immunization against the pathogens they will be working with. Appropriate personal protective equipment (typically laboratory coats and gloves with respiratory protection as needed) must be utilized to reduce exposure to infectious agents and contaminated equipment. A means of decontaminating infected waste (e.g., autoclave) must be available.

All procedures involving high concentrations or large volumes of infectious agents, or that could generate infectious aerosols or splashes, must be conducted in biosafety cabinets or other physical containment equipment. High-efficiency particulate air (HEPA) filtered exhaust air from the biosafety cabinets can be recirculated back into the laboratory, discharged into the exhaust ventilation system, or discharged directly outside the laboratory.

While there are no specific requirements for ventilation systems, facilities should consider mechanical ventilation systems that provide an inward flow of air without recirculation to spaces outside of the laboratory.

Animal Biosafety Level 2 facilities have restricted access and personnel must have specific training in handling infected animals and the manipulation of pathogenic agents. Implementation of employee occupational health programs should be considered.

Doors to areas where infectious materials and/or animals are housed open inward, are self-closing, and are kept closed when experimental animals are present. Floors, walls, and ceilings are water resistant. Penetrations in floors, walls, and ceiling surfaces (e.g., openings around ducts, plumbing, light fixtures, doorframes) should be sealed to facilitate pest control and cleaning. Cages should be autoclaved or otherwise decontaminated prior to washing. Mechanical cage washers are used and have a final rinse temperature of at least 180°F.

There is a negative (inward) directional airflow, with a ducted exhaust air ventilation system. Exhaust air is discharged to the outside without being recirculated to other rooms. HEPA-filtered exhaust air from the biosafety cabinets can be recirculated back into the laboratory.

Arthropod Containment Level 2 is suitable for work with exotic and indigenous arthropods infected with biosafety level 2 agents that are associated with animal and/or human disease, or are suspected of being infected with such agents.

The insectary is separate from the general traffic patterns of the building and it is designed and maintained to enhance detection of escaped arthropods. It is recommended that the entrance to the insectary be through a double-door vestibule that prevents either flying or crawling arthropods from escaping. Self-closing doors are highly recommended. All penetrations (e.g., drains, pipe ducts, light fixtures) are sealed or have traps to prevent accidental release of arthropods or agents. Potential harborage and breeding areas are eliminated. All interior surfaces are light colored and smooth to enhance detection of loose arthropods. Ceilings are as low as possible. Door and windows in the area are designed to minimize escape and prevent the entrance of surreptitious arthropods or pests. Autoclave or appropriate decontamination system is available.

All personnel working with the arthropods must have specific training on safe maintenance and operation of the insectary and in handling pathogenic agents. They should be under medical surveillance and offered immunization against the pathogens they will be working with. Personnel working with the arthropods should wear white laboratory coats or gowns that minimize exposed skin. Gloves are worn whenever practicable and personal protective equipment (e.g., masks, head covers) is worn as needed. Laboratory garments should be checked for infestation before exiting the insectary.

The primary container used to hold the arthropods should effectively prevent escape, even during feeding. When handling/removing vertebrate animals (i.e., hosts or blood sources) after exposure to arthropods, precautions must be taken to prevent the insects from escaping during the transfer process. Host animals must also be inspected closely for residual clandestine arthropods. All precautions must be taken to prevent accidental transfer of the arthropods to the host cages.

An effective trapping program is established to monitor for escaping arthropods. Loose arthropods must be killed and disposed of or recaptured and returned to the container from which they escaped. Infected arthropods must not be killed with bare hands. They must be manipulated using filtered mechanical or vacuum aspirators, or other appropriate means.

Between uses or before disposal of primary containers, they are cleaned to prevent arthropod survival and escape (e.g., heated or chilled to a lethal temperature) and disinfected (e.g., autoclaved, incinerated, chemically treated). Unneeded arthropods should be killed (e.g., chilled to a lethal temperature) and then subjected to appropriate disinfection (e.g., autoclave, incineration, chemicals) prior to disposal.

Biosafety Level 3 is applicable to work performed with indigenous or exotic agents that may cause serious or potentially lethal disease through inhalation of the pathogen. Biosafety level 3 practices, containment equipment, and facilities are recommended for working with cultures, production volumes, or higher concentration cultures of most biological warfare agents.

Access to the laboratories is through two sets of self-closing and locking doors, and is restricted when work is being conducted. All personnel working in the laboratories must receive specific training in handling high-risk agents. They must be under medical surveillance and offered immunization against the pathogens they will be working with. Appropriate personal protective equipment (e.g., solid-front laboratory wear, shoe covers, gloves, eye and face protection, respiratory protection as needed) must be utilized to reduce exposure to infectious agents and contaminated equipment.

The laboratory is equipped with a ducted air ventilation system that draws air from clean areas toward contaminated areas. Exhaust air must be discharged from the laboratory away from occupied areas and building air intakes, or else passed through HEPA filters prior to being discharged. All vacuum lines are protected with HEPA filters.

All procedures involving the manipulation of infectious material are conducted within biosafety cabinets or by use of other physical containment equipment. Air from the biosafety cabinets is directed through HEPA filters before being recirculated into the laboratory, discharged into the exhaust ventilation system, or discharged directly outside the laboratory.

A means of decontaminating infected waste (e.g., autoclave) must be available. A comprehensive, integrated pest management program is required.

Enhanced Biosafety Level 3 has additional protective features such as all surfaces should be sealed and must be smooth to support wipe-down decontamination. The room must be capable of being sealed in case gaseous decontamination is required. Additional safety features may include having personnel enter and exit through a clothing-change/shower room, mandatory laboratory air ventilation system processed through HEPA filters, and more a extensive decontamination system for research material and waste.

Animal Biosafety Level 3 facilities have restricted access. Personnel working in the laboratory must have specific training in handling infected animals and the manipulation of pathogenic agents. Implementation of employee occupational health programs is required.

Entry is through two sets of self-closing and self-locking doors that open inward. Floors, walls, and ceilings are water resistant. Penetrations in floors, walls, and ceiling surfaces (e.g., openings around ducts, plumbing, light fixtures, doorframes) must be sealed to facilitate pest control, cleaning, and decontamination. Animal cages are designed to minimize infectious aerosols from the animals or their bedding. Cages should be autoclaved or otherwise decontaminated prior to washing in a mechanical cage washer. Final rinse temperature of the washer should be at least

180°F. An autoclave or other appropriate decontamination equipment must be convenient to the animal rooms.

Appropriate eye, face, and respiratory protection must be used in rooms containing infected animals. Restraint devices and practices are used to reduce the risk of exposure during animal manipulations. All procedures involving the manipulation of infectious material are conducted within biosafety cabinets or by use of other physical containment equipment.

The laboratory is equipped with a ducted air ventilation system that draws air from clean areas toward contaminated areas. Exhaust air must be discharged from the laboratory away from occupied areas and building air intakes, or else passed through HEPA filters prior to being discharged.

Biosafety Level 3-Agriculture was developed by the U.S. Department of Agriculture to address activities using agents designated as High Consequence Pathogens and specifically designed to protect the environment. Facilities are utilized for loose-housed animals that cannot be enclosed in primary containment caging (e.g., large animals). Each facility is designed so that the room itself acts as a primary containment barrier to prevent release of infectious agents into the environment. As a matter of course these facilities include almost all of the features ordinarily found in biosafety level 4 facilities. While such facilities can be a separate building, they are more often found as an isolated zone within a facility operating at a lower biosafety level, usually at biosafety level 3. This isolated zone has strictly controlled access with special physical security measures and functions on the box-within-a-box principle.

Arthropod Containment Level 3 is suitable for work with potential or known vectors that are or are likely to be infected with biosafety level 3 agents associated with human disease. If possible, these facilities should not be located in an area where the species is indigenous or if an alternative suitable vector is present since an escaped arthropod could introduce the pathogen into the local population. If this is not possible, then work with infected arthropods should only be conducted during a time of year when arthropods would not survive in the outside environment.

Access to the insectary is through two sets of self-closing and locking doors that open inward and cannot be opened simultaneously. Access is restricted when work is being conducted. The area between the two sets of doors contains a changing room and shower. Showers are plumbed to prevent arthropod escape. Doors within the insectary are self-closing. The spaces around these inner doors are sealed to facilitate decontamination. All penetrations (e.g., drains, pipe ducts, light fixtures) are sealed or have traps to prevent accidental release of arthropods or agents. Appropriate filters or barriers are installed within the ventilation system to prevent escape of arthropods. The insectary has an inward direction of airflow with a progressively negative pressure gradient maintained as distance from the main entrance increases.

The arthropod work area should be small to minimize searches for escaped arthropods. Ceilings are as low as possible. All interior surfaces are light colored and smooth to enhance detection of loose arthropods. Harborage and breeding areas are eliminated. All work is done within a primary barrier (e.g., biological safety cabinet, glove box) or within a designated, enclosed, and segregated area of a biosafety level 3 laboratory serving as a primary barrier. An autoclave is available in the rooms containing arthropods.

All personnel working in the insectary must receive specific training in handling high-risk agents as well as arthropod containment practices and procedures. They must be under medical surveillance and offered immunization against the pathogens they will be working with. When entering the laboratory, all personal clothing is removed in the changing room and exchanged for a complete set of laboratory clothing. Outer laboratory wear is white and has a wrap-around or solid front. Other personal protective equipment (e.g., eye and face protection, respirators) are utilized as needed to reduce exposure to infectious agents and contaminated equipment. Gloves are worn whenever working with arthropods, host animals or contaminated equipment. When exiting the laboratory, outer laboratory clothing is discarded in the main insectary before proceeding through the first set of doors and into the changing room. Used inner laboratory clothing remains in the changing room and must be decontaminated before laundering. A personal shower may be required prior to exiting the insectary.

The primary container used to hold the arthropods should effectively prevent escape, even during feeding. When handling/removing vertebrate animals (i.e., hosts or blood sources) after exposure to arthropods, precautions must be taken to prevent the insects from escaping during the transfer process. Host animals must also be inspected closely for residual clandestine arthropods. All precautions must be taken to prevent accidental transfer of the arthropods to the host cages.

An effective trapping program is established to monitor for escaping arthropods. Loose arthropods must be killed and disposed of or recaptured and returned to the container from which they escaped. Infected arthropods must not be killed with bare hands. They must be manipulated using filtered mechanical or vacuum aspirators, or other appropriate means. Pesticide for emergency use is available in areas in which escape of arthropods is likely.

Unneeded infected arthropods should be killed by freezing or other appropriate methods, followed by autoclaving or incineration. In the United States, a written inventory of each arthropod infected with a Select Agent must be maintained.

Between uses or before disposal of primary containers, they are cleaned to prevent arthropod survival and escape (e.g., heated or chilled to a lethal temperature) and disinfected (e.g., autoclaved, incinerated, chemically treated). All arthropod waste materials are autoclaved or incinerated.

Biosafety Level 4 is required for work with dangerous and exotic agents that pose a high risk of aerosol-transmitted laboratory infections and cause life-threatening disease that is frequently fatal or for which there are no vaccines or treatments. They are also used for agents with an unknown risk of transmission.

These laboratories are located in individual buildings or in an isolated and restricted section of existing facilities. Access is restricted and a logbook documenting the date and time of all persons entering and leaving the laboratory must be maintained. All personnel working in the laboratory must demonstrate high proficiency in standard and special microbiological practices and techniques necessary for working with extremely hazardous infectious agents. They must be under medical surveillance and be immunized against the pathogens they will be working with, if vaccines are available. An isolation and medical care facility capable of caring for personnel with potential or known laboratory-acquired infections must be available.

When entering the laboratory, all personal clothing must be removed in an outer clothing change room and exchanged for a complete set of laboratory clothing. When exiting the laboratory, all persons must discard their laboratory clothing in an inner clothing change room before proceeding through a decontamination shower to the outer clothing change room. Used laboratory clothing remains in the inner change room and must be decontaminated before laundering.

Laboratories are equipped with a double-door autoclave, fumigation chamber, or airlock that is used to pass materials, supplies, or equipment into or out of the lab. The double doors are interlocked in a manner that prevents opening of both doors at once and ensures the outer door cannot be opened unless the autoclave or fumigation chamber has processed through a decontamination cycle. Personnel within the laboratory can only retrieve incoming material by securing the outer doors prior to opening the inner doors.

Penetrations in floors, walls, and ceiling surfaces (e.g., openings around ducts, plumbing, light fixtures, doorframes) must be sealed to prevent agent migration and to facilitate pest control, cleaning and decontamination. Services and plumbing entering the laboratory must be fitted with two backflow prevention devices in series. A comprehensive, integrated pest management program is required.

The laboratory has a dedicated non-recirculating ventilation system, protected by two HEPA filters in series, that is designed to maintain the laboratory at negative pressure to the surrounding area. The air exhaust discharge must be located away from occupied spaces and building air intakes.

There are two types of biosafety level 4 laboratories. Cabinet laboratories where agents are only handled within a Class III biosafety cabinet (i.e., a gas-tight sealed container that is designed to allow for manipulation of objects using integrated gloves.) These cabinets are equipped with a HEPA-filtered air supply and have an in-series double HEPA-filtered exhaust.

Potentially contaminated material leaving the cabinet must pass through a double-doored autoclave, disinfectant dunk tank, fumigation chamber, or decontamination shower. The doors must be interlocked such that only one set of doors can be opened at a time and the decontamination system must have cycled prior to being able to open the outer doors.

Appropriate personal protective equipment (e.g., solid-front laboratory wear, eye, and face protection) must be utilized to reduce exposure to infectious agents and contaminated equipment. Respiratory protection may be required if deemed appropriate for the work being performed. Disposable gloves must be worn under the cabinet gloves.

Suit laboratories require personnel to wear a one-piece positive-pressure supplied air protective suit (i.e., totally encapsulating protection garments that force air to flow out if the suit is damaged) and perform work in a laminar flow biosafety cabinet. Air from the biosafety cabinets is directed through HEPA filters before being recirculated into the laboratory or discharged into the laboratory exhaust ventilation system. Workers wear appropriate laboratory clothing (e.g., surgical scrubs) and inner gloves under the protective suit. Entry into the laboratory is only through an airlock fitted with airtight doors.

In addition to the personal shower used to exit the laboratory, these laboratories also have a chemical shower used to decontaminate the surface of the protective suit prior to personnel leaving the work area and entering the exit decontamination zone.

Animal Biosafety Level 4 requires personnel working in the laboratory to be competent handling animals and to demonstrate high proficiency in standard and special microbiological practices and techniques necessary for working with extremely hazardous pathogens and infected animals, and in implementing procedures requiring animal biosafety level 4 containment.

Cabinet laboratories: agents and animals are only handled within a Class III biosafety cabinet. Restraint devices and practices that reduce the risk of exposure during animal manipulations must be used if practicable (e.g., physical restraint devices, chemical restraint medications, mesh gloves, etc.).

Suit laboratories: personnel must wear a one-piece positive-pressure supplied air protective suit. Infected animals should be manipulated within a primary containment/barrier system, such as a Class II biosafety cabinet or equivalent.

Arthropod Containment Level 4 safety guidelines are for the most dangerous pathogen-infected arthropods. Vectors infected with certain other pathogens (e.g., restricted animal pathogens) may also require this level. All of the standard practices of arthropod containment level 3 are in place, with the additional recommendations described here.

This containment level can be employed with either cabinet laboratories or suit laboratories. In either case, the area designated for arthropod research is small, light colored, and contains only items required for the study. Vectors must be safely contained at all times. Infected arthropods should never be handled outside of a primary containment barrier (e.g., cages are opened only in an arthropod-secure glove box). In a suit laboratory working with ticks or mites, a well-lit, stand alone handling table incorporating a moat to prevent arthropod escape may be sufficient. Infected arthropods are kept in primary and secondary secure containers. Unneeded arthropods must be killed, denominated, and discarded.

All personnel working in the laboratory must demonstrate high proficiency in standard and special microbiological practices and techniques necessary for working with extremely hazardous, infectious agents and infected arthropods. They must be competent in handling arthropods, agents and implementing procedures requiring arthropod biosafety level 4 containment. Since working with arthropods often requires the use of small instruments and hence requires considerable dexterity, personnel must be trained and practice extensively using either a Class III biosafety cabinet or one-piece positive-pressure supplied air protective suit before actually working with infected arthropods.

Every arthropod is counted and accounted for throughout every experiment. Personnel do not enter or leave the room until all arthropods are accounted for and all living arthropods have been secured in double-taped cages or vials and placed in secondary sealed holding trays. If even one arthropod is missing and cannot be found, the facility is shut down and treated with a pesticide.

BIBLIOGRAPHY

American Committee of Medical Entomology; American Society of Tropical Medicine and Hygiene. *Arthropod Containment Guidelines, Version 3.2.* Vector Borne and Zoonotic Diseases 19, #3, (2019): 152–173.

Belgian Biosafety Server. *Contained Use – International Classifications Schemes for Micro-Organisms Based on Their Biological Risks.* 2020 [https://www.biosafety.be/content/contained-use-international-classifications-schemes-micro-organisms-based-their-biological] December 31, 2020.

Chosewood, L. Casey and Deborah E. Wilson, ed. *Biosafety in Microbiological and Biomedical Laboratories.* 5th Edition. Washington, D.C.: US Government Printing Office, 2009.

World Health Organization. *Laboratory Biosafety Manual, 3rd Edition.* Geneva, Switzerland: World Health Organization, 2004.

Appendix D: International and Domestic Lists of Agents

This appendix contains a listing of the agents detailed in this handbook that are identified in the Chemical Weapon Convention or in lists of reportable or controlled materials. Within each table (i.e., chemicals, toxins, pathogens), agents are listed alphabetically by a common or scientific name followed by the agent index number as a means to cross reference these materials with detailed information in the appropriate chapter.

There are ten sources referenced in this appendix. The Chemical Weapons Convention (*CWC*) divides chemical agents and critical precursors into three schedules. Schedule 1 (**1**) lists chemicals that are deemed to pose a high risk to the purposes of the CWC, but have very limited, if any, commercial applications. Schedule 2 (**2**) lists chemicals that are deemed to pose a significant risk to the purposes of the CWC. It also includes toxic chemicals that could be used as chemical warfare agents as well as key precursors to the chemicals in Schedule 1. Schedule 3 (**3**) lists other chemicals that are considered to pose a risk to the purposes of the CWC including dual-use chemicals such as phosgene and all other regulated precursors for chemical warfare agents. Chemicals can be added or deleted from the various schedules as deemed appropriate. For more information see Organisation for the Prohibition of Chemical Weapons (OPCW). *Declarations Handbook for the Convention on the Prohibition of the Development, Production, Stockpiling and Use of Chemical Weapons and on their Destruction, Appendix 2*. 2019 [https://www.opcw.org/resources/declarations/handbook-chemicals] 2020.

The Australia Group (*AG*) is an informal forum of states established in 1985 with the goal of discouraging and impeding chemical/biological weapons proliferation by harmonizing national export controls on pathogen cultures, toxins, and precursor chemicals. The group recommends regulations for international transfers of agents and equipment utilizing three designated lists: An export control/core list (**C**), a warning list (**W**) of dual-use chemical weapons precursors and bulk chemicals, as well as microorganisms and toxins that could be used in a biological warfare program, and an awareness raising list (**A**) of pathogens. For more information see The Australia Group. *Common Control Lists*. February 28, 2020 [https://www.dfat.gov.au/publications/minisite/theaustraliagroupnet/site/en/controllists.html] 2020.

The World Health Organization for Animals *(OIE)* has the mission of ensuring transparency in and enhanced knowledge of the worldwide animal health situation. In 2005, it established a single list of notifiable terrestrial and aquatic animal diseases to replace the former A List (i.e., transmissible diseases that have the potential for very serious and rapid spread, irrespective of national borders, that are of serious socio-economic or public health consequence and that are of major importance in the international trade of animals and animal products) and B List (i.e., transmissible diseases that are considered to be of socio-economic and/or public health importance within countries and that are significant in the international trade of animals and animal products). The list is reviewed on a regular basis and any modifications come into force on January first of the following year. The list of diseases for 2020 is reported in this appendix. For more information see World Organization for Animal Health (OIE). *Listed Diseases, Infections and Infestations in Force in 2020* [https://www.oie.int/en/animal-health-in-the-world/oie-listed-diseases-2020/] 2020.

The U.S. Select Agents and Toxins list *(US SAT)* identifies biological agents and toxins that have been determined by the U.S. Departments of Health and Human Services and Agriculture to have the potential to pose a severe threat to both human and animal health, to plant health, or to animal and plant products. Tier 1 select agents and toxins (**1**) is a subset of these agents and toxins deemed to present the greatest risk of deliberate misuse with significant potential for mass casualties or devastating effect to the economy, critical infrastructure, or public confidence, and pose a severe threat to public health and safety. For more information see U.S. Federal Select Agent Program [https://www.selectagents.gov/SelectAgentsandToxinsList.html] 2020.

The U.S. Department of Homeland Security chemicals of concern *(DHS)* is a list of industrial chemicals that are deemed to have the potential to create significant human life or health consequences if released, stolen or diverted, or sabotaged/contaminated. Development of this list was a requirement of the Department of Homeland Security Appropriations Act of 2007. For more information see U.S. Department of Homeland Security. "6 CFR Appendix a to Part 27 – DHS Chemicals of Interest, Final Rule," *Federal Register* 72, No. 223 (November 20, 2007): 65421–65434.

The Australia Chemicals of Security Concern *(AU CSC)* list was generated as a result of a 2008 report recommending the Council of Australian Governments (now the National Cabinet) set up a process to assess, and where necessary, take action to reduce the risk of chemicals being used for terrorist purposes. The list encompasses explosive precursors or components and toxic agents. For more information see Commonwealth of Australia. "National Code of Practice for Chemicals of Security Concern, Appendix A." 2016: 14.

The U.S. Federal Bureau of Investigation (**FBI**) list of highly toxic chemicals and pesticides judged likely to be used by terrorists. This list includes toxic agricultural and industrial chemicals identified by the Centers for Disease Control, the World Health Organization and the U.S. Environmental Protection Agency. The main criteria for this list were high dermal or inhalation toxicity, common malicious use reported, and prior use by terrorists. For more information see United States Federal Bureau of Investigation. "Note to Poison Control Centers for Suspicious Pesticide/OP Nerve Gas Incidents", 2001.

The International Taskforce-25 (**ITF-25**) was formed in 1996 under a 1994 Memorandum of Understanding between the United States, United Kingdom, and Canada. This group of experts developed a list of agricultural and commercial industrial chemicals that could be significant in a military operation. Factors that were considered include global production, physical state of the material, history of previous use by the military or significant industrial accidents, potential impact on troops should a release occur, and a hazard ranking based on the toxicity, corrosivity, ignitability and reactivity of the material. Although the original study considered over 1,000 toxic chemicals, the focus was limited to chemicals that posed a risk through inhalation. For more information see Memorandum of Understanding Cooperative Program on Research, Development, Production and Procurement of Chemical and Biological Defense Material. *Final Report of International Task Force-25 Hazard from Toxic Industrial Chemicals.* March 18, 1996.

In 2008, the U.S. Naval Research Laboratory (**NRL**) was tasked by the Joint Project Manager for Individual Protection (now the Joint Project Manager for Protection) to develop a scientifically based prioritization of chemicals that selects a representative list of industrial chemicals for evaluation of chemical/biological defense systems. Chemicals were scored and prioritized according to their toxicity, stability, physical state, and production and distribution characteristics. For more information see Thomas Sutto "Prioritization and Sensitivity Analysis of the Inhalation/ Ocular Hazard of Industrial Chemicals." Naval Research Laboratory, Materials and Sensors Branch, Materials Science and Technology Division October 28, 2011; Thomas Sutto "Prioritization of the Percutaneous Hazard of Industrial Chemicals." Naval Research Laboratory, Materials and Sensors Branch, Materials Science and Technology Division October 28, 2011; and Thomas Sutto "Prioritization of the Oral (Ingestive) Hazard of Industrial Chemicals." Naval Research Laboratory, Materials and Sensors Branch, Materials Science and Technology Division October 28, 2011.

As part of regulations governing hazardous biological agents, the U.S. Centers for Disease Control (**CDC**) has developed a list of infectious agents and toxins that it believes pose a significant risk to public health. Agents may be added or deleted as deemed necessary to protect the public. Category **A** agents are high-priority agents that pose a risk to national security because they can be easily disseminated or transmitted from person to person, result in high mortality rates and have the potential for major public health impact, might cause public panic and social disruption or require special action for public health preparedness. Category **B** agents are moderately easy to disseminate, result in moderate morbidity rates and low mortality rates, and require specific enhancements of CDC's diagnostic capacity and enhanced disease surveillance. Category **C** agents are emerging pathogens that could be engineered for mass dissemination in the future because of availability, ease of production and dissemination, and potential for high morbidity and mortality rates with major health impact. For more information see Centers for Disease Control and Prevention. *Bioterrorism Agents/Diseases*. April 4, 2018 [https://emergency.cdc.gov/agent/agentlist-category.asp#catdef] 2020.

CHEMICAL WARFARE AGENTS, PRECURSORS, AND POTENTIAL INDUSTRIAL AGENTS

Chemical	CWC	AG	DHS	AU CSC	FBI	ITF-25	NRL
1-(3-Dimethylaminophenoxy)-3-(3-dimethylamino-5-dimethylcarbamoxyphenoxy)propane Dimethiodide C02-A012	1						
1-(3-Hydroxy)quinuclidinio-10-N-(3-dimethylcarbamoxy-α-picolinyl)-N,N-dimethylammoniodecane Dibromide C02-A042	1						
1-(4-Aldoximino)pyridinio-10-N-(3-dimethylcarbamoxy-α-picolinyl)-N,N-dimethylammoniodecene Dibromide C02-A039	1						
1-(4-Dimethylaminophenoxy)-2-(3-dimethylamino-5-dimethylcarbamoxyphenoxy)ethane Dimethiodide C02-A011	1						
1-(4-Dimethylcarbamoxy-2-dimethylaminophenoxy)-3-(4-dimethylaminophenoxy)propane Dimethiodide C02-A027	1						
1-(N,N,N-Tributylammonio)-10-N-(3-dimethylcarbamoxy-α-picolinyl)-N,N-dimethylammoniodecane Dibromide C02-A047	1						
1-(N,N,N-Trimethylammonio)-10-N-(2-dimethylcarbamoxybenzyl)-N,N-dimethylammoniodecane Dibromide C02-A033	1						
1-(N,N,N-Trimethylammonio)-10-N-(5-dimethylcarbamoxy)isoquinoliniodecane Dibromide C02-A029	1						
1-(N,N,N-Trimethylammonio)-8-N-(2-dimethylcarbamoxybenzyl)-N,N-dimethylammoniooctane Dibromide C02-A028	1						
1-(N,N-Dimethylamino)-10-N-(3-dimethylcarbamoxy-2-pyridylmethyl)-N-methylaminodecane Dimethobromide C02-A003	1						

(Continued)

Chemical	CWC	AG	DHS	AU CSC	FBI	ITF-25	NRL
1-(N,N-Dimethyl-N-cyanomethylammonio)-10-N-(3-dimethylcarbamoxy-α-picolinyl)-N,N-dimethylammoniodecane Dibromide C02-A031	1						
1-(N,N-Dimethyl-N-cyclohexylammonio)-10-N-(3-dimethylcarbamoxy-α-picolinyl)-N,N-dimethylammoniodecane Dibromide C02-A045	1						
1-(N-Methyl)pyrrolidinio-10-N-(3-dimethylcarbamoxy-α-picolinyl)-N,N-dimethylammoniodecane Dibromide C02-A036	1						
1,1,3,3-Tetraethyl-2-fluoro(methoxy) phosphorylguanidine (A-262) C01-A040	1						
1,1,3,3-Tetraethyl-2-fluoro(methyl) phosphorylguanidine (A-242) C01-A039	1						
1,10-Bis(2-dimethylcarbamoxybenzyl) propylaminodecane Dimethobromide C02-A025	1						
1,10-Bis(3-dimethylcarbamoxy-α-picolinyl) ethylaminodecane Dimethobromide C02-A022	1						
1,10-Bis{10-(3-dimethylcarbamoxy-α-picolinyl) methylaminodecylmethylamino}decane Tetramethobromide C02-A049	1						
1,10-Bis{N-1-(2-dimethylcarbamoxyphenyl)ethyl-N,N-dimethylammonio}decane-2,9-dione Tetraphenylboronate C02-A023	1						
1,10-Bismethyl-2-(3-dimethyl-carbamoxypyridyl) methylaminodecane Dimethobromide C02-A001	1						
1,10-BisN-(2-dimethylcarbamoxybenzyl)-N,N-dimethylammoniodecane-2,9-dione Dibromide C02-A019	1						
1,10-BisN-(3-dimethylcarbamoxy-α-picolyl)-N,N-dimethylammoniodecane-2,9-dione Dibromide C02-A017	1						
1,10-BisN-(3-dimethylcarbamoxy-α-picolyl)-N-ethyl-N-methylammoniodecane-2,9-dione Dibromide C02-A020	1						
1,11-Bismethyl-2(3-dimethylcarbamoxypyridyl) methylaminoundecane Dimethobromide C02-A009	1						
1,2-Dimethylhydrazine C11-A059						X	
1,3-Bis(2-chloroethylthio)-n-propane C03-A014	1		X				
1,3-Bismethyl-2(3-dimethylcarbamoxypyridyl) methylaminopropane Dimethobromide C02-A007	1						
1,4-Bis(2-chloroethylthio)-n-butane C03-A012	1		X				
1,4-Bismethyl-2(3-dimethylcarbamoxypyridyl) methylaminobutane Dimethobromide C02-A006	1						
1,5-Bis(2-chloroethylthio)-n-pentane C03-A013	1		X				
1,5-Bismethyl-2(3-dimethylcarbamoxypyridyl) methylaminopentane Dimethobromide C02-A005	1						
1,6-Bismethyl-2-(3-dimethylcarbamoxypyridyl) methylaminohexane Dimethobromide C02-A002	1						

Chemical	CWC	AG	DHS	AU CSC	FBI	ITF-25	NRL
1,7-Bismethyl-2(3-dimethylcarbamoxypyridyl) methylaminoheptane Dimethobromide C02-A010	1						
1,8-Bis(2-Dimethylcarbamoxybenzyl) ethylaminooctane Dimethobromide C02-A021	1						
1,8-Bis(3-dimethylcarbamoxy-α-picolinyl) ethylaminooctane Dimethobromide C02-A018	1						
1,8-Bismethyl-2(3-dimethylcarbamoxypyridyl) methylaminooctane Dimethobromide C02-A004	1						
1,8-BisN-(2-Dimethylcarbamoxybenzyl)-N,N-dimethylammoniooctane-2,7-dione Dibromide C02-A016	1						
1,8-BisN-(3-dimethylcarbamoxy-α-picolyl)-N,N-dimethylammoniooctane-2,7-dione Dibromide C02-A014	1						
1,9-Bismethyl-2(3-dimethylcarbamoxypyridyl) methylaminononane Dimethobromide C02-A008	1						
1-N-(2-Dimethylcarbamoxybenzyl)pyrrolinio-10-(N,N,N-trimethylammonio)decane Dibromide C02-A040	1						
1-N-(2-Dimethylcarbamoxybenzyl)pyrrolinio-8-(N,N,N-trimethylammonio)octane Dibromide C02-A030	1						
1-N-(3-Dimethylcarbamoxy-α-picolyl)-N,N-dimethylammonio-10-(N-carbamoxymethyl-N,N-dimethylammonio)decane Dibromide C02-A032	1						
1-N-(3-Hydroxy-α-picolyl)-N,N-dimethylammonio-10-N-(3-dimethylcarbamoxy-α-picolyl)-N,N-dimethylammoniodecane Dibromide C02-A044	1						
1-N-(4-Dimethylcarbamoxymethyl)benzyl-N,N-dimethylammonio-10-N-(3-dimethylcarbamoxy-α-picolinyl)-N,N-dimethylammoniodecane Dibromide C02-A048	1						
1-N,N-Di(2-hydroxy)ethyl-N-methylammonio-10-N-(3-dimethylcarbamoxy-α-picolinyl)-N,N-dimethylammoniodecane Dibromide C02-A038	1						
1-N,N-Dimethyl-N-(2-acetoxy-2-methylethyl) ammonio-10-N-(3-dimethylcarbamoxy-α-picolinyl)-N,N-dimethylammoniodecane Dibromide C02-A043	1						
1-N,N-Dimethyl-N-(2-butyroxyethyl)ammonio-10-N-(3-dimethylcarbamoxy-α-picolinyl)-N,N-dimethylammoniodecane Dibromide C02-A046	1						
1-N,N-Dimethyl-N-(2-hydroxy)ethylammonio-10-N-(3-dimethylcarbamoxy-α-picolinyl)-N,N-dimethylammoniodecane Dibromide C02-A034	1						
1-N,N-Dimethyl-N-(3-cyanopropyl)ammonio-10-N-(3-dimethylcarbamoxy-α-picolinyl)-N,N-dimethylammoniodecane Dibromide C02-A041	1						

(*Continued*)

Chemical	CWC	AG	DHS	AU CSC	FBI	ITF-25	NRL
1-N,N-Dimethyl-N-(3-hydroxy)propylammonio-10-N-(3-dimethylcarbamoxy-α-picolinyl)-N,N-dimethylammoniodecane Dibromide C02-A037	1						
1-Pyridinio-10-N-(3-dimethylcarbamoxy-α-picolinyl)-N,N-dimethylammoniodecane Dibromide C02-A035	1						
2-Chloroethanol C03-C041	2	C	X				
2-(Diisopropylamino)ethanethiol C01-C105	2	C	X				
2-(Diisopropylamino)ethanol C01-C104	2	C					
2-Chloroethylchloromethylsulfide C03-A008	1		X				
3-(Dimethylcarbamoyl)oxy-N,N-diethyl-N-methylanilinium Iodide C02-A050	1						
3-Hydroxy-1-methylpiperidine C12-C035		C					
3-Quinuclidinol C12-C032	2	C					
3-Quinuclidone C12-C033		C					
Acetone Cyanohydrin C11-A001			X			X	
Acrolein C13-A001			X			X	
Acrylonitrile C11-A002			X			X	
Aldicarb C11-A003				X	X		X
Allyl Alcohol C11-A004			X			X	
Allyl Chloroformate C11-A006						X	
Allyl Isothiocyanate C11-A007						X	
Allylamine C11-A005			X			X	
Aluminum Phosphide C11-A139			X	X			
Amiton C01-A014	2		X				
Ammonia C11-A009			X	X		X	X
Ammonium Bifluoride C01-C080		C					
Arsenic Pentoxide C11-A011				X			
Arsenic Trichloride C04-C006	2	C	X			X	
Arsenic Trioxide C11-A012				X	X		X
Arsine C08-A001			X	X		X	X
Azinphosmethyl C11-A013				X	X		X
Bendiocarb C11-A014				X			
Benzilic Acid C12-C034	2	C					
Beryllium Sulfate C11-A015				X			
Bis(2-chloroethyl) Ether C11-A017							X
Bis(2-chloroethylthio)methane C03-A011	1		X				
Bis(2-chloroethylthiomethyl)ether C03-A015	1		X				
Bis{α-(3-dimethylcarbamoxyphenyl)methylamino}-4,4'-biacetophenone Dimethobromide C02-A024	1						
Bis{α-(3-dimethylcarbamoxy-α-picolinyl)pyrrolidinio}4,4'-biacetophenone Dibromide C02-A026	1						
Boron Tribromide C11-A019			X			X	X
Boron Trichloride C11-A020			X			X	X
Boron Trifluoride C11-A022			X			X	X
Bromine C10-A002			X	X		X	X

Chemical	CWC	AG	DHS	AU CSC	FBI	ITF-25	NRL
Bromine Chloride C11-A021			X			X	
Bromine Pentafluoride C11-A023			X			X	
Bromine Trifluoride C11-A024			X			X	
Butyl Chloroformate C11-A026						X	
Butyl Isocyanate C11-A028						X	
BZ C12-A001	2						
Cadmium Fume C11-A030							X
Cadusafos C11-A031				X			
Calcium Cyanide C07-C007				X			
Carbofuran C11-A032				X	X		
Carbon Disulfide C11-A033			X	X		X	
Carbon Monoxide C09-A001				X		X	
Carbonyl Fluoride C11-A034			X			X	
Carbonyl Sulfide C11-A035			X			X	
Chlorfenvinphos C11-A037				X			
Chlorine C10-A001			X	X		X	X
Chlorine Dioxide C11-A038			X				X
Chlorine Pentafluoride C11-A039			X			X	
Chlorine Trifluoride C10-A015			X			X	X
Chloroacetaldehyde C11-A040						X	
Chloroacetone C13-A007						X	
Chloroacetonitrile C11-A041						X	
Chloroacetyl Chloride C11-A042			X			X	
Chloroform C11-A043			X				
Chloropicrin C10-A006	3			X			
Chlorosarin C01-C054	1		X				
Chlorosoman C01-C055	1		X				
Chlorosulfonic Acid C11-A045			X			X	X
Copper Acetoarsenite C11-A046					X		
Crotonaldehyde C11-A047			X			X	
Cyanogen Bromide C07-A002				X			
Cyanogen Chloride C07-A003	3		X	X		X	
Cyclosarin (GF) C01-A005	1				X		
Deltamethrin C11-A049							X
Diazinon C11-A051				X			
Diborane C11-A052			X			X	
Dichlorvos C11-A055				X			
Diethyl Chlorophosphite C01-C159		C					
Diethyl Ethylphosphonate C01-C090	2	C					
Diethyl Methylphosphonate C01-C089	2	C					
Diethyl Methylphosphonite C01-C	2	C	X				
Diethyl N,N-Dimethylphosphoramidate C01-C	2	C					
Diethyl Phosphite C01-C087	3	C		X			
Diethylamine C01-C097		C					
Diethylaminoethanol C01-C103		C					

(*Continued*)

Chemical	CWC	AG	DHS	AU CSC	FBI	ITF-25	NRL
Diisopropylamine C01-D174		C					
Diketene C11-A057						X	
Dimethyl Ethylphosphonate C01-C	2	C					
Dimethyl Methylphosphonate C01-C059	2	C					
Dimethyl Phosphite C01-C085	3	C		X			
Dimethylamine C01-C098		C	X				
Dimethylamine Hydrochloride C01-C098		C	X				
Dimethylmercury C11-A060				X			
Dimethylsulfate C03-A009				X		X	
Diphenylmethane-4,4'-diisocyanate C11-A113						X	
Disulfoton C11-A062				X	X		
Endosulfan C11-A063				X			
Ethion C11-A065				X			
Ethyl Chlorofluorophosphate C01-C		C					
Ethyl Chloroformate C11-A066						X	
Ethyl Chlorothioformate C11-A067						X	
Ethyl Dichlorophosphate C01-C160		C					
Ethyl Difluorophosphate C01-C161		C					
Ethyl Parathion C11-A133					X		X
Ethyl Phosphinyl Dichloride C01-C093	2	C					
Ethyl Phosphinyl Difluoride C01-C	2	C					
Ethyl Phosphonic Dichloride C01-C094	2	C					
Ethyl Phosphonothioic Dichloride C01-C164	2		X			X	
Ethyl Phosphonyl Dichloride C01-C094	2	C				X	
Ethylamine C11-A068							X
Ethyldiethanolamine C03-C044	3	C	X	X			
Ethylene Dibromide C11-A069						X	X
Ethylene Oxide C11-A071			X			X	
Ethyleneimine C11-A072			X			X	
Ethylmercury Chloride C11-A073				X			
Ethylphosphonyl Difluoride C01-C	1	C	X				
Fenamiphos C11-A074				X			
Fluorine C11-A076			X	X		X	X
Fluoroacetic Acid C11-A078				X			
Fluoroethyl Alcohol C11-A079				X			
Fluoroethyl Fluoroacetate C11-A080				X			
Fluorotrichloromethane C11-A081							X
Fonofos C11-A082					X		
Formaldehyde C11-A083			X			X	X
Germanium Tetrafluoride C11-A084							X
Hexachlorocyclopentadiene C11-A085						X	
Hexafluoroacetone C11-A086			X				X
HL C03-A010	1						
HQ C03-A004	1						
HT C03-A005	1						
Hydrogen Bromide C11-A087			X			X	X

Chemical	CWC	AG	DHS	AU CSC	FBI	ITF-25	NRL
Hydrogen Chloride C11-A088			X	X		X	X
Hydrogen Cyanide C07-A001	3		X	X	X	X	
Hydrogen Fluoride C11-A089		C	X			X	X
Hydrogen Iodide C11-A090			X			X	X
Hydrogen Selenide C11-A091			X			X	
Hydrogen Sulfide C07-A006			X	X		X	X
Iron Pentacarbonyl C09-A002			X			X	
Isobutyl Chloroformate C11-A092						X	
Isopropyl Chloroformate C11-A093			X			X	
Isopropyl Isocyanate C11-A094						X	
Lewisite 2 C14-A004	1		X				
Lewisite 3 C04-C007	1		X				
Lewisite C04-A002	1		X				
Magnesium Phosphide C11-A139			X	X			
Mercuric Chloride C11-A096				X	X		
Mercuric Nitrate C11-A097				X			
Mercuric Oxide C11-A098				X			
Mercurous Nitrate C11-A099				X			
Mercury Cyanide C11-A100				X			
Methamidophos C11-A103				X	X		
Methanesulfonyl Chloride C11-A104						X	
Methidathion C11-A105				X	X		
Methiocarb C11-A106				X			
Methomyl C11-A107				X	X		
Methoxyethyl Mercury C11-A108					X		
Methyl Benzilate C12-C036		C					
Methyl Bromide C11-A110					X	X	X
Methyl Chlorofluorophosphate C01-C		C					
Methyl Chloroformate C11-A111			X			X	
Methyl Chlorosilane C11-A112			X			X	
Methyl Dichlorophosphate C01-C162		C					
Methyl Difluorophosphate C01-C		C					
Methyl Fluoroacetate C11-A114				X			
Methyl Hydrazine C11-A116			X			X	
Methyl Isocyanate C11-A117			X			X	
Methyl Mercaptan C11-A118			X			X	
Methyl Parathion C11-A120				X	X		
Methylamine C11-A109							X
Methyldiethanolamine C03-C043	3		X	X			
Methylmercury Dicyandiamide C11-A119					X		
Methylphenyldichlorosilane C11-A121							X
Methylphosphinyl Dichloride C01-C065	2	C					
Methylphosphinyl Difluoride C01-C	2	C					
Methylphosphonic Acid C01-C060	2	C					
Methylphosphonic Dichloride C01-C052	2	C					
Methylphosphonic Difluoride C01-C053	1	C	X				

(Continued)

Chemical	CWC	AG	DHS	AU CSC	FBI	ITF-25	NRL
Methylphosphonothioic Dichloride C01-C066	2	C	X				
Mevinphos C11-A123				X	X		
Molybdophosphoric Acid C11-A124							X
Monocrotophos C11-A125					X		
Mustard Gas (H) C03-A001	1		X				
N,N-Diethylacetamidine C01-C107		C					
N,N-Diethylbutanamidine C01-C		C					
N,N-Diethylformamidine C01-C		C					
N,N-Diethylisobutanamidine C01-C		C					
N,N-diethyl-N'-fluoro(methoxy) phosphorylacetamidine (A-232) C01-A037	1						
N,N-Diethyl-N'-fluoro(methyl) phosphorylacetamidine (A-230) C01-A036	1						
N,N-Diethylpropanamidine C01-C		C					
N,N-Diisopropylaminoethanethiol Hydrochloride C01-C105	2	C	X				
N,N-Diisopropylbutanamidine C01-C		C					
N,N-Diisopropylformamidine C01-C		C					
N,N-Diisopropyl-β-aminoethane Thiol C01-C105	2	C	X				
N,N-Diisopropyl-β-aminoethanol C01-C104	2	C					
N,N-Diisopropyl-β-aminoethyl chloride C01-C	2	C					
N,N-Diisopropyl-β-aminoethylchloride Hydrochloride C01-C	2	C					
N,N-Dimethylacetamidine C01-C		C					
N,N-Dimethylaminophosphoryl Dichloride C01-C163	2	C	X				
N,N-Dimethylbutanamidine C01-C		C					
N,N-Dimethylformamidine C01-C		C					
N,N-Dimethylisobutanamidine C01-C		C					
N,N-Dimethylpropanamidine C01-C		C					
N,N-Dipropylacetamidine C01-C		C					
N,N-Dipropylbutanamidine C01-C		C					
N,N-Dipropylformamidine C01-C		C					
N,N-Dipropylisobutanamidine C01-C		C					
N,N-Dipropylpropanamidine C01-C		C					
N'-Ethoxy(fluoro)phosphoryl-N,N-diethyl-acetamidine (A-234) C01-A038	1						
Nitric Acid, Fuming C11-A126			X	X		X	X
Nitric Oxide C11-A127			X	X		X	X
Nitrogen Dioxide C11-A128			X			X	X
Nitrogen Mustard 1 C03-A016	1		X				
Nitrogen Mustard 2 C03-A017	1		X				
Nitrogen Mustard 3 C03-A018	1		X				
Nitrogen Trifluoride C11-A129							X
o-Anisidine C11-A010							X
Octyl Mercaptan C11-A130						X	

Chemical	CWC	AG	DHS	AU CSC	FBI	ITF-25	NRL
O,O-Diethyl Phosphorodithioate C01-C073		C					
O,O-Diethyl Phosphorothioate C01-C074		C					
Octamethylene-bis(5-dimethylcarbamoxyisoquinolinium Bromide) C02-A015	1						
O-Ethyl 2-Diisopropylaminoethyl Methylphosphonite C01-C064	1	C					
Omethoate C11-A131				X			
O-Mustard C03-A003	1		X				
Osmium Tetroxide C11-A146				X			
Oxamyl C11-A147				X			
Paraquat C11-A132				X	X		
Perchloric Acid C11-A135				X			
Perchloroethylene C11-A136							X
Perchloromethyl Mercaptan C10-A014			X				
Perfluoroisobutylene C10-A008	2						
Phenylmercuric Acetate C11-A137					X		
Phorate C11-A138				X	X		
Phosgene C10-A003	3		X	X		X	X
Phosphine C11-A139			X	X	X	X	X
Phosphorus Oxychloride C01-C083	3	C	X	X		X	X
Phosphorus Pentachloride C01-C082	3	C	X	X			
Phosphorus Pentafluoride C11-A140						X	
Phosphorus Pentasulphide C01-C084		C					
Phosphorus Trichloride C01-C081	3	C	X	X		X	X
Pinacolone C01-C102		C					
Pinacolyl Alcohol C01-C102	2	C					
Potassium Bifluoride C01-C078		C					
Potassium Cyanide C07-C008		C	X	X			X
Potassium Fluoride C01-C076		C					
Propoxur C11-A143				X			
Propyl Chloroformate C11-A144			X			X	
Sarin (GB) C01-A002	1		X		X		
s-Butyl Chloroformate C11-A027						X	
Selenium Hexafluoride C11-A149			X			X	
Sesquimustard (Q) C03-A002	1		X				
Schradan C11-A148							X
Silicon Tetrafluoride C11-A150			X			X	X
Sodium Arsenite C11-A151					X		
Sodium Bifluoride C01-C079		C					
Sodium Cyanide C07-C009		C	X	X	X		
Sodium Fluoride C01-C077		C					
Sodium Fluoroacetate C11-A152				X	X		
Sodium Hexafluorosilicate C01-C106		C					
Sodium Sulfide C03-C042		C					
Soman (GB) C01-A003	1		X		X		

(*Continued*)

Chemical	CWC	AG	DHS	AU CSC	FBI	ITF-25	NRL
Stibine C11-A153			X			X	
Strychnine C11-A159				X	X		
Sulfotepp C11-A161					X		
Sulfur Dichloride C03-C039	3	C		X			
Sulfur Dioxide C11-A154			X			X	X
Sulfur Monochloride C03-C038	3	C		X			
Sulfur Tetrafluoride C11-A155			X				X
Sulfur Trioxide C11-A156			X			X	X
Sulfuric Acid C11-A157			X	X		X	X
Sulfuryl Chloride C11-A158			X			X	
Sulfuryl Fluoride C11-A160					X	X	
Tabun (GA) C01-A001	1		X		X		
t-Butyl Isocyanate C11-A029						X	
Tellurium Hexafluoride C11-A162			X			X	
Terbufos C11-A163				X	X		
Tetrabromoethane C11-A164							X
Tetrachloroethane C11-A165							X
Tetraethyl Lead C11-A168						X	
Tetraethyl Pyrophosphate C11-A167					X	X	
Tetrafluoroboric Acid C11-A169							X
Tetramethyl Lead C11-A170			X			X	
Thallium Sulfate C11-A171				X	X		
Thiodiglycol C03-C037	2	C	X				
Thionyl Chloride C03-C040	3	C	X	X			X
Thiophosphoryl Chloride C01-C075				X			
Titanium Tetrachloride C11-A172			X			X	
Toluene 2,4-Diisocyanate C11-A173			X			X	X
Toluene 2,6-Diisocyanate C11-A174			X			X	
Trichloroacetyl Chloride C11-A175						X	
Triethanolamine C03-C045	3	C	X	X			
Triethanolamine Hydrochloride C03-C045	3	C	X	X			
Triethyl Phosphite C01-C088	3	C	X	X			
Trifluoroacetyl Chloride C11-A176			X			X	
Triisopropylphosphite C01-C165		C					
Trimethyl Phosphite C01-C086	3	C	X	X			
Trimethylamine C11-A178							X
Tungsten Hexafluoride C11-A179			X			X	X
Vinyl Chloride C11-A180			X				
V-sub-x (Vx) C01-A018	1						
VX C01-A017	1		X		X		
Zinc Cyanide C07-C010				X			
Zinc Phosphide C11-A139			X	X			

TOXINS

	CWC	AG	US SAT	CDC
Abrin C16-A021		C	X	
Aflatoxins C16-A022		C		
Conotoxins C16-A010		C	X	
Botulinum Toxins C16-A005		C	1	A
Cholera Toxin C16-A024		C		
Clostridium Perfringens Toxins C16-A025		C		
Diacetoxyscirpenol C16-A039		C	X	
HT-2 Toxin C16-A042		C		
Microcystin C16-A028		C		
Modeccin C16-A029		C		
Ricin C16-A031	1	C	X	B
Saxitoxin C16-A015	1	C	X	
Shiga Toxins C16-A032		C		
Staphylococcus Enterotoxins C16-A033		C	X	B
T-2 Toxin C16-A045		C	X	
Tetrodotoxin C16-A018		C	X	
Viscumin C16-A036		C		
Volkensin C16-A037		C		

PATHOGENS

Pathogen	AG	OIE	US SAT	CDC
1918 Influenza Virus C18-A006	C		X	
African Horse Sickness Virus C18-A001	C	X	X	
African Swine Fever Virus C18-A002	C	X	X	
Andean Potato Latent Virus C18-A003	C			
Andes Virus C18-A004	C			C
Aujeszky's Disease C18-A005	C	X		
Avian Influenza Virus C18-A006	C	X	X	
Bacillus anthracis (Anthrax) C17-A001	C	X	1	A
Bacillus cereus biovar *anthracis* (Anthrax) C17-A002	W		1	
Banana Bunchy Top Virus C18-A007	A			
Bluetongue Virus C18-A008	C	X		
Brucella abortus (Brucellosis) C17-A006	C	X	X	B
Brucella melitensis (Brucellosis) C17-A006	C	X	X	B
Brucella suis (Brucellosis) C17-A006	C	X	X	B
Burkholderia mallei (Glanders) C17-A007	C	X	1	B
Burkholderia pseudomallei (Melioidosis) C17-A008	C		1	B
Chapare Virus C18-A009	C		X	
Chikungunya Virus C18-A010	C			

(*Continued*)

Pathogen	AG	OIE	US SAT	CDC
Chlamydia psittaci (Psittacosis) C17-A009	C	X		B
Choclo Virus C18-A011	C			
Classical Swine Fever Virus (Hog Cholera) C18-A012	C	X	X	
Clavibacter michiganensis subsp. *sepedonicus* (Ring Rot of Potato) C17-A010	C			
Clostridium argentinense (Botulism) C17-A011	C		1	
Clostridium baratii (Botulism) C17-A012	C		1	
Clostridium botulinum (Botulism) C17-A013	C		1	
Clostridium butyricum (Botulism) C17-A014	C		1	
Clostridium perfringens (Gas gangrene) C17-A015	C			B
Clostridium tetani (Tetanus) C17-A016	W			
Coccidioides immitis (Desert fever) C19-A003	C			
Coccidioides posadasii (Coccidioidomycosis, non-California variant) C19-A004	C			
Cochliobolus miyabeanus (Rice brown spot) C19-A005	C			
Colletotrichum coffeanum var. virulans (Coffee brown blight) C19-A006	C			
Coniothyrium glycines (Soybean red leaf blotch) C19-A007			X	
Coxiella burnetii (Q fever) C17-A017	C	X	X	B
Crimean-Congo Hemorrhagic Fever Virus C18-A013	C	X	X	
Dobrava-Belgrade Virus C18-A015	C			C
Eastern Equine Encephalitis Virus C18-A016	C	X	X	B
Ebolavirus C18-A017	C		1	A
Ehrlichia ruminantium (Heartwater) C17-A018		X		
Far Eastern Tick-Borne Encephalitis Virus C18-A018	C		X	
Foot and Mouth Disease Virus C18-A019	C	X	1	
Francisella tularensis (Tularemia) C17-A020	C	X	1	A
Fusarium langsethiae (Head blight) C19-A008	W			
Fusarium sporotrichioides (Head blight) C19-A009	W			
Goatpox Virus C18-A020	C	X	X	
Guanarito Virus (Venezuelan hemorrhagic fever) C18-A021	C		X	C
Hantaan Virus (Korean hemorrhagic fever) C18-A022	C			C
Hendra Virus (Equine morbillivirus) C18-A023	C		X	
Japanese Encephalitis Virus C18-A024	C	X		
Junin Virus (Argentinean hemorrhagic fever) C18-A025	C		X	C
Kyasanur Forest Virus C18-A026	C		X	
Laguna Negra Virus C18-A027	C			
Lassa Fever Virus C18-A028	C		X	A
Legionella pneumophila (Legionnaire's disease) C17-A021	W			
Louping Ill Virus C18-A030	C			

Pathogen	AG	OIE	US SAT	CDC
Lujo Virus C18-A029	C		X	
Lumpy Skin Disease Virus C18-A031	C	X	X	
Lymphocytic Choriomeningitis Virus C18-A032	C			
Machupo Virus (Bolivian hemorrhagic fever) C18-A033	C		X	C
Magnaporthe oryzae (Rice blast) C19-A011	C			
Marburgvirus C18-A034	C		1	A
MERS-CoV C18-A051	C			
Microcyclus ulei (South American rubber leaf blight) C19-A012	C			
Moniliophthora roreri (Cocoa frosty pod rot) C19-A013	A			
Monkeypox Virus C18-A035	C		X	
Murray Valley Encephalitis Virus C18-A036	C			
Mycoplasma capricolum subsp. *capripneumoniae* (Contagious caprine pleuropneumonia) C17-A024	C	X	X	
Mycoplasma mycoides subsp. *mycoides* SC (Contagious bovine pleuropneumonia) C17-A025	C	X	X	
Newcastle Disease Virus C18-A037	C	X	X	
Nipah Virus (Barking pig syndrome) C18-A038	C	X	X	C
Omsk Hemorrhagic Fever Virus C18-A039	C		X	
Oropouche Virus C18-A040	C			
Peronosclerospora philippinensis (Sugarcane downy mildew) C19-A014	C		X	
Peste Des Petits Ruminants Virus C18-A041	C	X	X	
Phoma tracheiphila (Citrus wilt) C19-A015	A			
Porcine Teschovirus (Infectious porcine encephalomyelitis) C18-A043	C			
Potato Spindle Tuber Viroid C18-A044	C			
Powassan Virus C18-A045	C			
Puccinia graminis subsp. *graminis* var. *graminis* (Cereal stem rust) C19-A017	C			
Puccinia striiformis (Cereal stripe rust) C19-A018	C			
Rabies Virus C18-A046	C	X		
Ralstonia solanacearum race 3, bv. 2 (Bacterial wilt) C17-A027	C		X	
Rathayibacter toxicus (Gumming disease) C17-A028			X	
Rickettsia prowazekii (Epidemic typhus) C17-A030	C		X	B
Rift Valley Fever Virus C18-A047	C	X	X	
Rinderpest Virus C18-A048	C	X	1	
Rocio Virus C18-A049	C			
Sabia Virus (Brazilian hemorrhagic fever) C18-A050	C		X	C
Salmonella enterica subsp. *enterica* ser. *Typhi* (Typhoid fever) C17-A033	C			
SARS-Related Coronavirus C18-A051	C		X	

(Continued)

Pathogen	AG	OIE	US SAT	CDC
Sclerophthora rayssiae var. zeae (Downy mildew of corn) C19-A019	C		X	
Seoul Hemorrhagic Fever Virus C18-A052	C			C
Sheeppox Virus C18-A053	C	X	X	
Shiga toxin producing *Escherichia coli* C17-A035	C			B
Shigella dysenteriae C17-A036	C			B
Siberian Tick-Borne Encephalitis Virus C18-A054			X	
Sin Nombre Virus C18-A055	C			C
St Louis Encephalitis Virus C18-A057	C			
Swine Vesicular Disease Virus C18-A058	C		X	
Synchytrium endobioticum (Potato wart disease) C19-A020	C		X	
Thecaphora solani (Potato smut) C19-A021	C			
Tilletia indica (Wheat cover smut) C19-A022	C			
Variola Major Virus (Smallpox) C18-A056	C		1	A
Variola Minor Virus C18-A060	C		1	
Venezuelan Equine Encephalitis Virus C18-A061	C	X	X	B
Vesicular Stomatitis Virus C18-A062	C			
Vibrio cholerae (Cholera) C17-A037	C			B
West Nile Virus C18-A063		X		
Western Equine Encephalitis Virus C18-A064	C	X		B
Xanthomonas albilineans (Sugarcane leaf scald) C17-A038	C			
Xanthomonas citri subsp. *citri* (Citrus canker) C17-A039	C			
Xanthomonas oryzae pv. *oryzae* (Rice leaf blight) C17-A040	C		X	
Xylella fastidiosa (Pierce's disease) C17-A041	A			
Yellow Fever Virus C18-A065	C			
Yersinia pestis (Plague) C17-A042	C		1	A
Yersinia pseudotuberculosis (Yersiniosis) C17-A043	C			

Alphanumeric Indices

There are ten indices in this volume to allow easy access to specific agents in this handbook. These indices are the Alphabetical Index of Chemical Names, Military Codes, and Synonyms; the Chemical Abstract Service (CAS) numbers index; the United Nations (UN) numbers index; the European inventory of existing commercial chemical substances (EC) numbers index; the U.S. Food and Drug Administration unique ingredient identifiers (UNII) index; the 11th Revision of the International Statistical Classification of Diseases and Related Health Problems (ICD-11) codes; the European and Mediterranean Plant Protection Organization (EPPO) codes index; an index of the viruses listed in the handbook by their taxonomic families; a cross index of vector and the diseases listed in the handbook that they can carry; and a cross index of agriculturally significant flora and fauna with the diseases listed in the handbook that can impact them.

Each of these indices is in alphanumeric order and entries are cross-referenced to information about the specific agent through the Handbook Number. The first two digits of the handbook number following the "C" indicate the class of agent. The classes appear in order from C01 through C21 as the chapters of this handbook. They provide general information about each subgroup of agents. The letter following the hyphen (e.g., C01-**A**) indicates that the materials is primarily used as an agent (A), component or precursor in the manufacture of that class of agents (C), or is a significant decomposition product or impurity of that class of agents (D). The three digits that follow the letter indicate the specific agent in the order that it appears in this handbook (e.g., C01-**A001**). These numbers proceed sequentially in each chapter. Materials that are not individually detailed in this handbook, typically due to the absence of published chemical or toxicological information, are only cross-referenced to the appropriate class. These materials do not have numbers after the letter following the hyphen (e.g., C01-A).

The list of synonyms included in the alphabetical index is by no means exhaustive. Although it contains a large number of formal chemical and biological names, it primarily contains historical names, military code names, and common names for the agents, precursors, components, and degradation products included in this handbook. In some instances, names have been included that have been popularized in the media but have no other historical connection to the agent.

Some synonyms in the alphabetical index are followed by bracketed notations. These notations provide additional clarifying information about the entry such as composition, modifications to the agents (e.g., thickened, dusty, binary), or a note for historical context. For example, "White Star" was a gas blend that was employed by the British in World War I consisting of 50% phosgene and 50% chlorine. The entry appears as:

White Star (British WWI Cylinder Gas) {Phosgene (50%) and Chlorine (50%) Mixture}

ALPHABETICAL INDEX OF CHEMICAL NAMES, MILITARY CODES, AND SYNONYMS

- A -

- C -

- D -

EA 5302 {1-Methyl-4-piperidyl Isopropylphenylglycolate (33%) and 1-Methoxy-1,3,5-cycloheptatriene Mixture}	C12-A
Ear Blight of Rice	C19-A005
Earias chlorion	C21-A014
Earias frondosana	C21-A014
Earias gossypii	C21-A014
Earias insulana	C21-A014
Earias siliquana	C21-A014
Earias simillema	C21-A014
Earias smaragdina	C21-A014
Earias tristrigosa	C21-A014
Earth	C12-A007
Eastern Encephalitis	C18-A016
Eastern Equine Encephalitis	C18-A016
Eastern Subtype of TBE	C18-A018
EBA	C13-A021
EBLV	C18-A046
EBO	C18-A017
Ebola Hemorrhagic Fever	C18-A017
Ebolavirus	C18-A017
EBOV	C18-A017
EC	C17-A019
E-Capsaicin	C13-A006
Ecgonine, Methyl Ester, Benzoate	C12-A010
Echaudure des Feuilles	C17-A038
ECO157:H7	C17-A035
Ecstasy	C12-A012
ECZ	C13-A022
EDCF	C20-A021
Edema Disease	C17-A035
EDN1	C20-A020
EDN2	C20-A021
EDN3	C20-A022
EE1530000	C20-A003
EEE	C18-A016
Effentora	C12-A013
Eggplant Mosaic Virus	C18-A003
Egyptian Bollworm	C21-A014
Egyptian Cotton Leaf Worm	C21-A002
Egyptian Cotton Worm	C21-A002
Egyptian Stem Borer	C21-A014
EHEC	C17-A035
EHF	C18-A017
Ehrlichia ruminantium	C17-A018
EI6201400	C16-A007
Eibischspinnmilbe	C21-A019
Eiernestrups	C21-A018
Einalon S	C12-A014
EJ6800000	C15-A016
EJ6800000	C15-A017
EK6300000	C15-A002

- F -

- H -

- I -

- L -

- M -

Methyl 2-Hydroxy-2,2-diphenylacetate	C12-C036
Methyl 3-Methyl-1-(2-phenylethyl)-4-(N-propanoylanilino)piperidine-4-carboxylate	C12-A017
Methyl 3-Methyl-1-(2-phenylethyl)-4-[phenyl(propionyl)amino]piperidine-4-carboxylate	C12-A017
Methyl 4-(N-(1-Oxopropyl)-N-phenylamino)-1-(2-phenylethyl)-4-piperidinecarboxylate	C12-A009
Methyl 4-(N-Propionyl-N-phenylamino)-1-(2-phenylethyl)-4-piperidine-carboxylate	C12-A009
Methyl alpha-Hydroxydiphenylacetate	C12-C036
Methyl alpha-Phenylmandelate	C12-C036
Methyl Benzilate	C12-C036
Methyl Benzillate	C12-C036
Methyl Benzoylecgonine	C12-A010
Methyl Chloromethyl Ketone	C13-A007
Methyl cis-(-)-3-Methyl-4-((1-oxopropyl)phenylamino)-1-(2-phenylethyl)-4-piperidinecarboxylate	C12-A017
Methyl Diphenylglycolate	C12-C036
Methyl Hydroxy(diphenyl)acetate	C12-C036
Methyl α-Hydroxydiphenylacetate	C12-C036
Methyl α-Phenylmandelate	C12-C036
Methyl[(2S)-1-phenylpropan-2-yl]amine	C12-A020
Methyl-2-hydroxy-2,2-diphenyl-acetate	C12-C036
Methylaminoethanolcatechol	C20-A025
Methylamphetamine	C12-A020
Methylarterenol	C20-A025
Methylbenzilate	C12-C036
Methylbenzoylepgonine	C12-A010
Methylbenzyl bromide	C13-A023
Methyl-beta-phenylisopropylamine	C12-A020
Methylbutyric Acid	C15-A020
Methylenedioxy Methamphetamine	C12-A012
Methylethylacetic Acid	C15-A020
Methyl-N-vanillyl-6-nonenamide, (E)-	C13-A006
Methyl-β-phenylisopropylamine	C12-A020
Metofane	C12-A021
Metossiflurano	C12-A021
Metoxfluran	C12-A021
Metoxifluran	C12-A021
Metoxiflurano	C12-A021
Meuse Fever	C17-A005
Mexfly	C21-A009
Mexican Fruit Fly	C21-A009
Mexikanische Fruchtfliege	C21-A009
Meyve Sinekleri	C21-A007
Mezcalin	C12-A019
Mezcalina	C12-A019
Mezcaline	C12-A019
Mezcline	C12-A019
MG 36851	C12-A011
Microcyclus ulei	C19-A012
Microcystin LR	C16-A028
Microcystis aeruginosa Toxin	C16-A028
Microdots	C12-A018
Middelhavsfrugtflue	C21-A007

- N -

- O -

- W -

CHEMICAL ABSTRACT SERVICE (CAS) NUMBERS

CAS	Handbook	CAS	Handbook
50-36-2	C12-A010	114-49-8	C12-A025
50-37-3	C12-A018	116-53-0	C15-A020
50–67-9	C20-A032	139-47-9	C12-A020
51-34-3	C12-A025	146-56-5	C12-A023
51-41-2	C20-A031	151-67-7	C12-A015
51-43-4	C20-A025	153-98-0	C20-A032
51–57-0	C12-A020	257-07-8	C13-A011
51–61-6	C20-A011	302-27-2	C16-A001
51–63-8	C12-A011	302-31-8	C12-A022
52-26-6	C12-A022	314-19-2	C14-A005
52–86-8	C12-A014	333-93-7	C15-A017
53-21-4	C12-A010	404-86-4	C13-A006
54-04-6	C12-A019	437-38-7	C12-A013
55-16-3	C12-A025	462-94-2	C15-A016
57-27-2	C12-A022	463-71-8	C13-A017
58-00-4	C14-A005	513-44-0	C15-A004
58–82-2	C20-A003	513-53-1	C15-A003
60-24-2	C15-A001	520-52-5	C12-A024
62-31-7	C20-A011	532-27-4	C13-A008
64-31-3	C12-A022	533-90-4	C15-A019
69-23-8	C12-A023	537-46-2	C12-A020
71-62-5	C16-A020	541-31-1	C15-A010
75-66-1	C15-A005	544-40-1	C15-A007
76-38-0	C12-A021	578-94-9	C14-A003
76–89-1	C12-C036	592-88-1	C15-A012
76–93-7	C12-C034	596-15-6	C12-A022
77-10-0	C12-A007	596-19-0	C12-A022
77-10-1	C12-A007	598-31-2	C13-A003
78–95-5	C13-A007	600-07-7	C15-A020
79-31-2	C15-A019	612-23-7	C13-A016
83-34-1	C15-A021	620-05-3	C13-A020
86-28-2	C13-A022	620-13-3	C13-A023
89–92-9	C13-A023	623-24-5	C13-A024
91-13-4	C13-A024	623-48-3	C13-A012
100-39-0	C13-A002	626-15-3	C13-A024
100-44-7	C13-A019	630-81-9	C12-A022
104-81-4	C13-A023	630-82-0	C12-A022
105-36-2	C13-A021	642-73-9	C12-A019
107-02-8	C13-A001	645-31-8	C20-A011
107-92-6	C15-A018	712-48-1	C14-A001
109-79-5	C15-A002	816-40-0	C13-A005
110-60-1	C15-A017	832-92-8	C12-A019
110-66-7	C15-A008	956-90-1	C12-A007
111-31-9	C15-A013	990-73-8	C12-A013
112-55-0	C15-A014	996-30-5	C15-A019
113-72-4	C12-A019	1077-43-6	C12-A011

CAS	Handbook	CAS	Handbook
1152-76-7	C12-A019	6211-15-0	C12-A022
1162-65-8	C16-A022	6509-18-8	C16-A001
1165-39-5	C16-A022	6581-06-2	C12-A001
1193-65-3	C12-C033	6740-88-1	C12-A016
1353-70-4	C16-A001	6795-23-9	C16-A022
1393-62-0	C16-A021	6846-03-3	C20-A003
1402-68-2	C16-A022	6885-57-0	C16-A022
1407-47-2	C20-A001	7220-81-7	C16-A022
1407-85-8	C12-A011	7241-98-7	C16-A022
1462-73-3	C12-A011	7488-97-3	C12-A018
1476-39-7	C15-A016	8001-85-2	C15-A026
1511-16-6	C12-A014	8023-77-6	C13-A006
1619-34-7	C12-C032	8053-16-5	C12-A022
1679-09-0	C15-A011	8056-95-9	C12-A021
1690-86-4	C12-A020	8066-00-0	C20-A004
1714-38-1	C13-A014	8072-83-1	C12-A
1867-66-9	C12-A016	8076-99-1	C12-A
1931-26-6	C13-A014	9001-71-2	C20-A004
2084-19-7	C15-A009	9009-86-3	C16-A031
2229-61-0	C16-A018	9011-97-6	C20-A004
2270-40-8	C16-A039	9012-22-0	C20-A004
2698-41-1	C13-A009	9012-63-9	C16-A024
2746-81-8	C12-A023	9014-39-5	C16-A018
2885-00-9	C15-A015	9061-58-9	C16-A045
2981-31-9	C12-A007	9067-26-9	C16-A031
3019-04-3	C13-A013	11000-17-2	C20-A033
3093-66-1	C12-A023	11003-08-0	C16-A022
3148-09-2	C16-A050	11005-69-9	C16-A018
3420-02-8	C15-A023	11017-04-2	C16-A015
3554-74-3	C12-C035	11026-09-8	C16-A018
3570-55-6	C15-A006	11032-79-4	C16-A007
3684-26-2	C12-C032	11034-16-5	C20-A033
3731-38-2	C12-C033	11039-36-4	C16-A050
4368-28-9	C16-A018	11050-21-8	C16-A008
4478-53-9	C12-A001	11051-21-1	C16-A045
4779-86-6	C15-A002	11061-96-4	C16-A023
4808-22-4	C14-A	11076-62-3	C16-A020
5002-47-1	C12-A023	11077-03-5	C16-A014
5221-20-5	C12-A018	11100-45-1	C16-A033
5798-79-8	C13-A004	11103-42-7	C16-A009
5913-62-2	C12-A010	11128-99-7	C20-A001
5967-42-0	C12-A019	12152-42-0	C13-A014
6009-81-0	C12-A022	12584-83-7	C16-A009
6034-57-7	C16-A001	12585-40-9	C16-A033
6055-06-7	C12-A022	12626-86-7	C16-A018
6055-69-2	C16-A001	12678-42-1	C20-A033
6191-56-6	C14-A005	12687-39-7	C16-A007

(*Continued*)

CAS	Handbook	CAS	Handbook
12769-46-9	C16-A009	26400-47-5	C16-A045
12778-32-4	C16-A007	26758-53-2	C12-A003
13004-56-3	C12-A001	26934-87-2	C16-A042
13214-11-4	C16-A022	27261-02-5	C16-A022
14287-82-2	C16-A040	27552-17-6	C16-A041
14611-50-8	C12-A020	27640-92-2	C16-A045
14729-29-4	C16-A046	27987-13-9	C13-A019
14797-94-5	C12-A025	28258-59-5	C13-A023
15232-63-0	C12-A018	28392-39-4	C16-A041
16096-32-5	C15-A022	28807-97-8	C13-A002
16891-85-3	C16-A047	28901-63-5	C12-A025
17289-88-2	C16-A018	29125-55-1	C12-A003
17878-54-5	C16-A022	29611-03-8	C16-A022
17878-56-7	C16-A022	29630-03-3	C13-A014
17924-92-4	C16-A051	29924-73-0	C12-C033
18046-08-7	C16-A050	29924-74-1	C12-C032
18525-25-2	C20-A032	30066-61-6	C16-A040
18660-81-6	C16-A018	31223-79-7	C16-A022
18695-28-8	C16-A051	31362-50-2	C20-A002
18949-73-0	C12-A007	32204-36-7	C16-A041
19455-20-0	C15-A019	32215-02-4	C16-A022
20309-96-0	C16-A022	33507-63-0	C20-A028
20421-08-3	C16-A022	33643-45-7	C12-A016
20421-10-7	C16-A022	35320-66-2	C16-A043
20978-13-6	C16-A022	35523-89-8	C16-A015
21259-20-1	C16-A045	35554-08-0	C16-A015
21462-53-3	C16-A050	35554-08-6	C16-A015
22128-62-7	C13-A010	36169-16-1	C15-A002
22128-63-8	C13-A018	36519-25-2	C16-A043
22135-69-9	C16-A040	36653-66-4	C16-A045
22228-82-6	C15-A019	36900-77-3	C12-A
22916-10-5	C16-A045	37209-28-2	C16-A007
22916-18-3	C16-A045	37338-78-6	C16-A004
23255-69-8	C16-A041	37830-21-0	C12-A004
23255-72-3	C16-A041	38738-57-7	C16-A043
23282-20-4	C16-A044	39313-28-5	C16-A014
23509-16-2	C16-A004	39379-15-2	C20-A030
23525-22-6	C14-A002	39465-37-7	C16-A019
23601-34-5	C15-A002	40334-69-8	C14-A004
24147-91-9	C16-A022	41372-20-7	C14-A005
24572-08-5	C12-A001	41598-07-6	C20-A016
25059-10-3	C13-A014	42045-86-3	C12-A030
25152-34-5	C16-A045	42542-10-9	C12-A012
25168-05-2	C13-A019	42881-99-2	C12-A002
25314-61-8	C13-A001	45952-89-4	C12-A020
25482-84-2	C13-A014	47106-99-0	C12-A022
25902-60-7	C16-A020	47646-09-3	C12-A023
26153-10-6	C16-A041	47861-26-7	C16-A001

CAS	Handbook	CAS	Handbook
50291-32-2	C12-A022	61380-41-4	C12-A017
50361-06-3	C14-A004	61512-76-3	C20-A017
50722-37-7	C16-A038	61512-77-4	C20-A019
51028-73-0	C20-A025	61788-97-4	C12-A014
51481-10-8	C16-A038	62869-67-4	C12-A028
51938-46-6	C16-A015	62869-68-5	C12-A001
52019-39-3	C16-A016	62869-69-6	C12-A001
52819-96-2	C16-A022	63038-80-2	C16-A015
53034-67-6	C12-A002	63038-81-3	C16-A015
53126-63-9	C16-A048	63490-86-8	C12-A004
53126-64-0	C16-A049	64057-70-1	C12-A012
53573-17-4	C14-A	64285-06-9	C16-A002
53800-72-9	C12-A001	64296-20-4	C16-A013
53950-09-7	C12-A001	64296-25-9	C16-A011
54018-05-2	C16-A050	64296-26-0	C16-A011
54098-95-2	C13-A014	64314-16-5	C16-A002
54385-59-0	C16-A049	64471-12-1	C12-A029
54385-60-3	C16-A048	64520-33-8	C12-A029
54390-94-2	C12-A005	65319-33-7	C12-A025
54946-52-0	C12-A012	65350-20-1	C12-A007
55508-42-4	C20-A030	65350-21-2	C12-A007
55514-87-9	C13-A002	65988-34-3	C16-A012
55803-44-6	C16-A015	65988-88-7	C16-A029
56030-54-7	C12-A026	67157-74-8	C12-A007
58569-55-4	C20-A024	69044-01-5	C13-A014
58670-87-4	C12-A025	69049-06-5	C12-A008
58822-25-6	C20-A023	69610-10-2	C12-A012
59141-40-1	C20-A023	70135-17-0	C12-A001
59392-53-9	C16-A027	70323-44-3	C16-A030
59708-52-0	C12-A009	70470-07-4	C16-A002
60119-99-5	C16-A045	70761-70-5	C12-A001
60124-79-0	C12-A007	70879-28-6	C12-A008
60124-86-9	C12-A007	70957-06-1	C12-A001
60508-89-6	C16-A011	70957-07-2	C12-A001
60537-65-7	C16-A011	71160-24-2	C20-A012
60538-73-0	C16-A049	71195-58-9	C12-A008
60558-95-4	C13-A024	71710-32-2	C20-A004
60561-17-3	C12-A026	72025-60-6	C20-A013
60617-12-1	C20-A018	72870-45-2	C16-A027
60645-00-3	C12-A017	72996-54-4	C12-A026
60645-15-0	C12-A009	73513-00-5	C16-A048
60748-39-2	C16-A011	73603-72-2	C16-A015
60748-42-7	C16-A011	73836-78-9	C20-A014
61086-44-0	C12-A009	74050-97-8	C12-A014
61356-56-7	C16-A042	75321-25-4	C12-A006
61380-27-6	C12-A009	75321-26-5	C12-A001
61380-40-3	C12-A017	75321-27-6	C12-A001

(*Continued*)

CAS	Handbook	CAS	Handbook
75321-28-7	C12-A006	93384-47-5	C16-A005
75478-79-4	C12-A014	94716-94-6	C16-A034
75715-89-8	C20-A015	96638-28-7	C16-A031
75757-64-1	C16-A032	97275-25-7	C12-C035
76862-65-2	C16-A010	97914-41-5	C12-A001
77734-91-9	C16-A014	97991-84-9	C12-A025
77734-92-0	C16-A014	98112-41-5	C16-A006
78126-11-1	C12-A023	98225-48-0	C16-A006
79499-51-7	C12-A016	99748-78-4	C15-A001
79580-28-2	C16-A006	100477-72-3	C12-A016
80226-62-6	C16-A011	101043-37-2	C16-A028
80248-94-8	C16-A011	101756-17-6	C12-A019
80573-42-8	C12-A001	103170-78-1	C16-A003
80655-66-9	C12-A001	106375-28-4	C16-A010
80832-90-2	C12-A013	106955-89-9	C16-A006
81133-24-6	C16-A010	107231-15-2	C16-A005
82115-62-6	C20-A010	107452-89-1	C16-A010
82248-93-9	C16-A030	108341-18-0	C20-A031
82326-63-4	C12-A003	110503-18-9	C15-A009
82362-17-2	C20-A024	110953-03-2	C12-A007
82698-46-2	C12-A007	115797-06-3	C16-A010
82737-04-0	C12-C033	115825-61-1	C16-A038
82785-45-3	C20-A029	115889-63-9	C16-A041
82810-43-3	C16-A011	116163-63-4	C16-A042
82810-44-4	C16-A011	116163-69-0	C16-A045
83481-45-2	C16-A010	116163-75-8	C16-A045
83590-17-4	C16-A036	117208-80-7	C12-A007
85079-48-7	C16-A006	117399-93-6	C20-A022
85087-27-0	C16-A006	117399-94-7	C20-A020
85201-37-2	C12-A022	122139-78-0	C16-A011
85514-42-7	C16-A002	122452-72-6	C12-A007
86933-74-6	C20-A026	122564-82-3	C16-A002
86933-75-7	C20-A027	123562-20-9	C20-A021
87915-42-2	C16-A026	123759-90-0	C16-A019
89675-50-3	C12-A001	126298-90-6	C12-A001
89930-68-7	C12-A018	127848-41-3	C12-A001
90817-17-7	C12-A001	128657-50-1	C16-A028
91840-99-2	C15-A003	129129-65-3	C16-A010
91853-95-1	C12-A001	132879-26-6	C15-A012
91933-11-8	C16-A037	133098-03-0	C12-A015
92142-32-0	C16-A002	137348-11-9	C20-A028
92216-03-0	C16-A002	137444-35-0	C12-A006
92844-80-9	C16-A002	139086-75-2	C12-C033
93101-02-1	C12-A027	139341-09-6	C16-A008
93101-83-8	C12-A031	140678-12-2	C16-A010
93384-43-1	C16-A005	142185-85-1	C16-A008
93384-44-2	C16-A005	143599-50-2	C12-A001
93384-46-4	C16-A005	145427-90-3	C16-A043

CAS	Handbook	CAS	Handbook
145427-92-5	C16-A042	440676-21-1	C12-A003
145427-93-6	C16-A045	440676-22-2	C12-A003
147794-23-8	C16-A010	440676-23-3	C12-A003
148157-34-0	C20-A007	440676-24-4	C12-A003
153834-62-9	C16-A032	496916-34-8	C16-A001
155666-11-8	C16-A011	524945-49-1	C16-A008
160902-51-2	C12-A001	588806-89-7	C16-A017
164790-60-7	C12-A026	607368-40-1	C12-A018
174593-75-0	C16-A023	620190-09-2	C16-A032
187108-44-7	C16-A017	675600-78-9	C13-A015
187108-45-8	C16-A017	675600-78-9	C13-A015
194020-07-0	C12-C033	676570-37-9	C16-A017
201295-24-1	C12-A026	679806-14-5	C13-A021
210357-85-0	C12-C033	859759-62-9	C15-A014
215812-86-5	C12-A026	863713-90-0	C12-A022
216299-21-7	C16-A010	866409-26-9	C16-A019
220355-66-8	C16-A015	912457-62-6	C13-A006
226562-00-1	C12-A025	1325730-46-8	C12-A022
308067-66-5	C20-A006	1401918-81-7	C13-A002
308079-78-9	C20-A009	2022216-71-1	C15-A013
375375-20-5	C16-A017	2022216-73-3	C15-A014

UNITED NATIONS (UN) NUMBERS

UN #	Handbook	UN #	Handbook
1092	C13-A001	1738	C13-A019
1111	C15-A008	1769	C14-A001
1569	C13-A003	2347	C15-A002
1603	C13-A021	2347	C15-A003
1694	C13-A004	2474	C13-A017
1695	C13-A007	2529	C15-A019
1697	C13-A008	2653	C13-A020
1698	C14-A003	2745	C13-A010
1701	C13-A023	2820	C15-A018
1737	C13-A002	2966	C15-A001

EUROPEAN INVENTORY OF EXISTING COMMERCIAL CHEMICAL SUBSTANCES (EC) NUMBERS

EC	Handbook	EC	Handbook
200-032-7	C12-A010	204-145-2	C15-A020
200-033-2	C12-A018	205-674-1	C12-A023
200-058-9	C20-A032	205-796-5	C12-A015
200-090-3	C12-A025	206-121-7	C16-A001
200-096-6	C20-A031	206-243-0	C14-A005
200-098-7	C20-A025	206-375-9	C15-A017
200-106-9	C12-A020	206-969-8	C13-A006
200-110-0	C20-A011	207-113-6	C12-A013
200-111-6	C12-A011	207-329-0	C15-A016
200-112-1	C12-A011	207-341-6	C13-A017
200-155-6	C12-A014	208-162-6	C15-A004
200-167-1	C12-A010	208-165-2	C15-A003
200-190-7	C12-A019	208-294-4	C12-A024
200-225-6	C12-A025	208-531-1	C13-A008
200-320-2	C12-A022	208-668-7	C12-A020
200-360-0	C14-A005	208-774-3	C15-A010
200-398-8	C20-A003	208-870-5	C15-A007
200-464-6	C15-A001	209-433-1	C14-A003
200-527-8	C20-A011	209-775-1	C15-A012
200-582-8	C12-A022	209-928-2	C13-A003
200-702-9	C12-A023	209-982-7	C15-A020
200-758-4	C16-A020	210-300-5	C13-A016
200-890-2	C15-A005	210-623-1	C13-A020
200-956-0	C12-A021	210-625-2	C13-A023
200-991-1	C12-C036	210-781-1	C13-A024
200-993-2	C12-C034	210-796-3	C13-A012
201-161-1	C13-A007	210-931-6	C13-A024
201-195-7	C15-A019	211-436-8	C20-A011
201-471-7	C15-A021	211-921-4	C14-A001
201-660-4	C13-A022	212-431-3	C13-A005
201-951-6	C13-A023	212-626-3	C12-A019
202-042-7	C13-A024	213-588-0	C12-A013
202-847-3	C13-A002	214-776-5	C12-C033
202-853-6	C13-A019	216-022-0	C15-A016
203-240-6	C13-A023	216-578-4	C12-C032
203-290-9	C13-A021	216-843-4	C15-A011
203-453-4	C13-A001	217-484-6	C12-A016
203-532-3	C15-A018	218-224-4	C15-A009
203-705-3	C15-A002	218-873-3	C16-A039
203-782-3	C15-A017	220-278-9	C13-A009
203-789-1	C15-A008	220-744-1	C15-A015
203-857-0	C15-A013	221-161-5	C13-A013
203-984-1	C15-A014	222-609-2	C12-C035
204-050-6	C12-A025	222-671-0	C15-A006

(Continued)

U.S. FOOD AND DRUG ADMINISTRATION UNIQUE INGREDIENT IDENTIFIERS (UNII)

UNII	Handbook	UNII	Handbook
0066WLJ02E	C20-A022	45I1K5482V	C16-A022
0296055VE0	C17-A013	489PW92WIV	C15-A005
04N5737KTQ	C18-A017	492LCM0TUL	C17-A006
04WOYJF7QH	C12-A027	4G08121T5U	C16-A028
067FQP576P	C13-A017	4G601DAM77	C18-A065
0C5T02KI0E	C16-A010	4M0784008H	C17-A037
0F7ZNZ0O50	C18-A017	4TVX2I407K	C20-A009
0M6MOQ3BOU	C17-A029	514B9K0L10	C17-A019
10S3JA1XRE	C16-A022	529XY6M4QX	C15-A009
11S92G0TIW	C12-A008	531JR2TJ2Z	C17-A008
14O3QZ9K1F	C14-A004	59H156XY46	C12-A013
14R9K67URN	C15-A001	5FNY4416UE	C20-A014
19A86ZW520	C16-A008	5G7820U646	C21-A001
1DB78J7PUD	C16-A022	5W827M159J	C16-A051
1EP6R5562J	C17-A036	5WOP02RM1U	C16-A044
1FFI43DL2K	C16-A011	60G7CF7CWZ	C15-A012
1H39V3559B	C14-A001	60ZTR74268	C13-A007
1HGW4DR56D	C20-A012	675VGV5J1D	C20-A028
1KH2129A0Q	C15-A013	690G0D6V8H	C12-A016
1N74HM2BS7	C12-A008	692D4TH3BM	C12-C036
2193YCG7X9	C16-A021	6AK165L0RO	C13-A022
26IU670827	C15-A011	6C6W9LJZ84	C12-A030
2CU6TT9V48	C20-A013	6EF1RL8Z5O	C19-A010
2K62B8Z6XF	C20-A020	6YCF89P8OX	C16-A021
2MS0D8WA29	C16-A022	6YRL8BWD9H	C16-A013
2RV7212BP0	C12-A024	74F2WGZ4EL	C16-A022
2UKQ3B7696	C16-A008	751E8J54VM	C17-A016
30905R8O7B	C12-A021	760T5R8B3O	C17-A033
3338387XEA	C15-A022	76I7G6D29C	C12-A022
333DO1RDJY	C20-A032	7706W8B346	C16-A021
3833H3TURF	C16-A011	77OY909F30	C15-A002
388II04NR0I	C16-A020	7864XYD3JJ	C13-A001
3E3H471GA3	C15-A007	7A7G0PQI12	C15-A026
3KUM2721U9	C16-A018	7H7YQ564XV	C12-A017
3NDQ4L309U	C13-A023	7JP1R2F6C6	C13-A004
3O8L0EWR5Q	C13-A003	7L3E358N9L	C20-A011
3R8G5WNW24	C18-A017	7SKR7S646P	C16-A022
3S51P4W3XQ	C20-A018	7WNP51KA7M	C12-A018
3T2U70084B	C16-A010	7Z2LPD8YGH	C16-A011
3TK3LQP1N7	C17-A004	80023A73NK	C16-A002
40UIR9Q29H	C15-A018	80474J4P87	C14-A004
4258P76E76	C16-A003	83H19HW7K6	C13-A019
44RAL3456C	C12-A020	84700R84PB	C15-A016
451IFR0GXB	C12-A025	8606X4NXMR	C19-A016

(Continued)

UNII	Handbook	UNII	Handbook
8731H7US2V	C16-A021	I5Y540LHVR	C12-A010
88B5039IQG	C13-A008	IDH134C231	C17-A007
8A00247K9O	C16-A022	IT7Z319PFY	C17-A002
8CF93KW41W	C17-A003	J067L6I0TY	C18-A063
8EYT8ATL7G	C20-A015	J1DOI7UV76	C12-A007
8F1YMD36YS	C14-A004	J6292F8L3D	C12-A014
8F6J993XXR	C12-C034	JNF66W9FFW	C19-A003
8LL210O1U0	C15-A019	JT37HYP23V	C16-A038
8NA5SWF92O	C12-A018	L28X8242OW	C18-A017
8WOA0KP06H	C16-A010	L90BEN6OLL	C15-A016
92QVL9080Y	C20-A006	LA9DTA2L8F	C12-A009
94168F9W1D	C20-A026	M089EFU921	C20-A001
974MVZ0WOK	C12-C032	M4BNP1KR7W	C16-A020
98826FBF79	C16-A047	M9RL1JF99O	C16-A009
997F43Z9CV	C12-A020	MMK4SUN45E	C15-A010
9A3YK965F3	C15-A008	MUN5LYG46H	C12-A013
9FEM1MON3P	C16-A008	N21FAR7B4S	C14-A005
9FRJ55E5UL	C13-A024	NNR1301B0H	C17-A020
9H070UFP2X	C15-A004	NY80X52FQ2	C16-A023
9JEZ9OD3AS	C20-A024	O0HJ02QBWN	C19-A017
9N2N2Y55MH	C16-A022	O18YUO0I83	C12-A016
9NLW29GJAX	C19-A018	OEU4AZC07S	C15-A006
9P59GES78D	C16-A027	OL62X66O4I	C16-A050
9R2271TA4J	C17-A005	OQ17NC0MOV	C16-A014
9W945B5H7R	C15-A021	OQF85710LZ	C17-A006
A16IX59JOH	C17-A009	OVP6XX033E	C17-A015
ADE45928XQ	C12-A019	P050J5FWC5	C20-A010
AFE2YW0IIZ	C12-A026	PLZ86LH7A6	C16-A043
B1AGI4NK4Z	C16-A023	PX9AZU7QPK	C20-A003
BY7U39XXK0	C20-A029	Q0638E899B	C16-A015
C1Q77A87V1	C13-A011	Q2P66EIP1F	C12-A025
D20KFB313W	C13-A021	QI287G628L	C14-A003
D279BDR31E	C16-A023	QTM6OJZ56S	C15-A023
D8317IAV7Q	C13-A009	QVU94XE61A	C12-A030
DFU5RIN74Y	C16-A001	R7D54R07N5	C18-A051
DL48G20X8X	C12-A025	R9HH0NDE2E	C18-A006
DV0VFN5F4T	C19-A004	RD1A447UC0	C12-A029
E69DLR7470	C12-A001	RHO99102VC	C12-A019
F39049Y068	C14-A005	RI01R707R6	C20-A023
FB3Y68CF2Y	C17-A034	RS7296A9LB	C16-A017
FMU62K1L3C	C12-A023	RXY07S6CZ2	C20-A016
GDV6F3DW7S	C17-A037	S07O44R1ZM	C13-A006
GG6P4Q1KNT	C12-A010	S6JJH3XV1D	C17-A042
GKN429M9VS	C20-A032	S79426A41Z	C12-A023
GRY5SDU86N	C17-A017	S8ZJB6X253	C15-A014
HB8U484NPM	C13-A016	S9ZFX8403R	C12-A026
I3020O28I3	C16-A022	SF5K94TPOF	C14-A002
I3FL5NM3MO	C16-A045	T1AB847C4E	C16-A021

11TH REVISION OF THE INTERNATIONAL STATISTICAL CLASSIFICATION OF DISEASES AND RELATED HEALTH PROBLEMS (ICD-11) CODES

EUROPEAN AND MEDITERRANEAN PLANT PROTECTION ORGANIZATION (EPPO) CODES

EPPO	Handbook	EPPO	Handbook
AJELCP	C19-A010	LYGULI	C21-A017
ANSTLU	C21-A009	MAYEDE	C21-A005
APLV00	C18-A003	MICCUL	C19-A012
ARGPLE	C21-A003	MONPRO	C19-A013
ASPEFU	C19-A001	NEOVIN	C19-A022
BACISU	C17-A003	PHYTIN	C19-A016
BACITH	C17-A004	PPV000	C18-A042
BARNQU	C17-A005	PRODLI	C21-A018
BBTV00	C18-A007	PRSCPH	C19-A014
BRULME	C17-A006	PSDMS3	C17-A027
BRVPOU	C21-A004	PSEDUN	C21-A013
BRVPPY	C21-A004	PSTVD0	C18-A044
CCDIIM	C19-A003	PUCCGR	C19-A017
CERTCA	C21-A007	PUCCST	C19-A018
CHILSU	C21-A015	PYRIOR	C19-A011
COCHMI	C19-A005	RATHTO	C17-A028
COLLCO	C19-A006	RICKPR	C17-A029
CORBSE	C17-A010	RICKRI	C17-A031
COWDRU	C17-A018	RICKTY	C17-A032
COXIBU	C17-A017	SALLTP	C17-A033
DACHGY	C19-A007	SCPHRZ	C19-A019
DACUDO	C21-A012	SERRMA	C17-A034
DACUFE	C21-A012	SHIGDY	C17-A036
DEUTTR	C19-A015	SPODLI	C21-A002
DIABVI	C21-A020	SYNCEN	C19-A020
DICIMA	C21-A010	TETRUR	C21-A019
EARIIN	C21-A014	THPHSO	C19-A021
ERYSGR	C19-A002	THRIPL	C21-A008
ESCHCO	C17-A019	THRITB	C21-A011
EURYIN	C21-A016	TROGGA	C21-A006
FMDV00	C18-A019	VIBRCH	C17-A037
FRNSTU	C17-A020	XANTAB	C17-A038
FUSALA	C19-A008	XANTCI	C17-A039
FUSASR	C19-A009	XANTOR	C17-A040
HAPLTR	C21-A021	XYLEFA	C17-A041
LEGIPN	C17-A021	YERSPE	C17-A042
LIBEAF	C17-A022	YERSPS	C17-A043
LIBEAS	C17-A023		
LPTNDE	C21-A001		

VIRAL TAXONOMIC FAMILIES

ARENAVIRIDAE

Chapare Hemorrhagic Fever C18-A009
Flexal Fever Virus C18-A068
Guanarito Hemorrhagic Fever C18-A021
Junin Hemorrhagic Fever C18-A025
Lassa Fever C18-A028
Lujo Virus C18-A029
Lymphocytic Choriomeningitis C18-A032
Machupo Hemorrhagic Fever C18-A033
Sabia Hemorrhagic Fever C18-A050

ASFARVIRIDAE

African Swine Fever C18-A002

BUNYAVIRIDAE

Akabane Virus C18-A066
Andes Virus C18-A004
Choclo Virus C18-A011
Crimean-Congo Hemorrhagic Fever C18-A013
Dobrava Hemorrhagic Fever C18-A015
Hantaan Hemorrhagic Fever C18-A022
Laguna Negra Virus C18-A027
Oropouche Virus Disease C18-A040
Puumala Hemorrhagic Fever C18-A070
Rift Valley Fever C18-A047
Seoul Hemorrhagic Fever C18-A052
Sin Nombre C18-A055

CORONAVIRIDAE

SARS Associated Coronavirus C18-A051
Middle East Respiratory Syndrome Coronavirus C18-A051

FILOVIRIDAE

Ebola Hemorrhagic Fever C18-A017
Marburg Hemorrhagic Fever C18-A034

FLAVIVIRIDAE

Classic Swine Fever C18-A012
Dengue Fever C18-A014
Far Eastern Tick-Borne Encephalitis Virus C18-A018
Japanese Encephalitis C18-A024
Kyasanur Forest Disease C18-A026
Louping Ill C18-A030

Poxviridae

Camel Pox C18-A067
Goat Pox C18-A020
Lumpy Skin Disease C18-A031
Monkey Pox C18-A035
Sheep Pox C18-A053
Smallpox C18-A056
Variola Minor C18-A060

Reoviridae

African Horse Sickness C18-A001
Bluetongue C18-A008

Rhabdoviridae

Lyssavirus genus C18-A046
Rabies C18-A046
Vesicular Stomatitis Fever C18-A062

Togaviridae

Chikungunya C18-A010
Eastern Equine Encephalitis C18-A016
Venezuelan Equine Encephalitis C18-A061
Western Equine Encephalitis C18-A064

Tymoviridae

Andean Potato Latent Virus C18-A003

INSECT VECTORS AND DISEASES

Aphids

Banana Bunchy Top Virus C18-A007
Plum Pox C18-A042
Potato Spindle Tuber Viroid C18-A044
Ring Rot of Potatoes C17-A010

Beetles

Andean Potato Latent Virus C18-A003
Ring Rot of Potatoes C17-A010

Flies (biting)

Anthrax C17-A001
Bacillus cereus biovar *anthracis* C17-A002
Lumpy Skin Disease C18-A031

Tularemia C17-A020
Vesicular Stomatitis Fever C18-A062

FLIES (MECHANICAL VECTORS)

Bacillus cereus biovar *anthracis* C17-A002
Brucellosis C17-A006
Cholera C17-A037
Enterohemorrhagic *Escherichia coli* C17-A035
Goatpox C18-A020
Highly Pathogenic Avian Influenza C18-A006
Rift Valley Fever C18-A047
Sheeppox C18-A053
Shigellosis C17-A036
Venezuelan Equine Encephalitis C18-A061

FLEAS

Endemic Typhus C17-A032
Plague C17-A042

LEAFHOPPERS

Ring Rot of Potatoes C17-A010

LICE

Trench Fever C17-A005
Typhus C17-A030

MIDGES

African Horse Sickness C18-A001
Akabane Virus C18-A066
Bluetongue C18-A008
Oropouche Virus Disease C18-A040
Vesicular Stomatitis Fever C18-A062

MITES

Rickettsialpox C17-A029
Scrub Typhus C17-A026

MITES (MECHANICAL VECTORS)

Venezuelan Equine Encephalitis C18-A061

MOSQUITOES

African Horse Sickness C18-A001
Chikungunya C18-A010

Omsk Hemorrhagic Fever C18-A039
Powassan Encephalitis C18-A045
Q Fever C17-A017
Rocky Mountain Spotted Fever C17-A031
Siberian Tick-Borne Encephalitis C18-A054
Tularemia C17-A020

TICKS (MECHANICAL VECTORS)

Rift Valley Fever C18-A047

AGRICULTURAL ANIMAL DISEASES

CATTLE

Akabane C18-A066
Anthrax C17-A001
Aujeszky's Disease C18-A005
Bacillus cereus bv. *anthracis* C17-A002
Brucellosis C17-A006
Contagious Bovine Pleuropneumonia C17-A025
Foot-and-Mouth Disease Virus C18-A019
Gas Gangrene C17-A015
Heartwater C17-A018
Louping Ill C18-A030
Lumpy Skin Disease C18-A031
Malignant Catarrhal Fever C18-A069
Melioidosis C17-A008
Rabies C18-A046
Rift Valley Fever C18-A047
Rinderpest C18-A048
Tetanus C17-A016
Vesicular Stomatitis Fever C18-A062

GOATS

Akabane C18-A066
Anthrax C17-A001
Aujeszky's Disease C18-A005
Bacillus cereus bv. *anthracis* C17-A002
Brucellosis C17-A006
Contagious Caprine Pleuropneumonia C17-A024
Foot-and-Mouth Disease Virus C18-A019
Far Eastern Tick-Borne Encephalitis Virus C18-A018
Gas Gangrene C17-A015
Goatpox C18-A020
Heartwater C17-A018
Louping Ill C18-A030
Melioidosis C17-A008
Peste des Petits Ruminants C18-A041
Rabies C18-A046

SHEEP

Akabane C18-A066
Anthrax C17-A001
Aujeszky's Disease C18-A005
Bacillus cereus bv. *anthracis* C17-A002
Bluetongue C18-A008
Brucellosis C17-A006
Contagious Caprine Pleuropneumonia C17-A024
Far Eastern Tick-Borne Encephalitis Virus C18-A018
Foot-and-Mouth Disease Virus C18-A019
Gas Gangrene C17-A015
Goatpox C18-A020
Heartwater C17-A018
Louping Ill C18-A030
Melioidosis C17-A008
Peste des Petits Ruminants C18-A041
Rabies C18-A046
Rift Valley Fever C18-A047
Sheeppox C18-A053
Siberian Tick-Borne Encephalitis Virus C18-A054
Tetanus C17-A016
Tularemia C17-A020

AGRICULTURAL CROP DISEASES

ALMONDS

Pierce's Disease C17-A041
Plum Pox C18-A042

BANANAS

Bacterial Wilt C17-A027
Banana Bunchy Top Virus C18-A007

CITRUS

African Greening C17-A022
Asian Greening C17-A023
Citrus Canker C17-A039
Mal Secco C19-A015
Pierce's Disease C17-A041

COCOA TREES

Frosty Pod Rot C19-A013

RICE

Brown Spot C19-A005
Rice Blast C19-A011
Rice Leaf Blight C17-A040

RUBBER PLANTS

South American Leaf Blight of Rubber C19-A012

SORGHUM

Philippine Downy Mildew C19-A014

SOYBEANS

Red Leaf Blotch C19-A007

SUGARCANE

Leaf Scald of Sugarcane C17-A038
Philippine Downy Mildew C19-A014

TOMATOES

Bacterial Wilt C17-A027
Phytophthora Blight C19-A016
Potato Spindle Tuber Viroid C18-A044

Index